SUPERSYMMETRIC FIELD THEORIES

Adopting an elegant geometrical approach, this advanced pedagogical text describes deep and intuitive methods for understanding the subtle logic of supersymmetry, while avoiding lengthy computations.

The book describes how complex results and formulae obtained using other approaches can be significantly simplified when translated to a geometric setting. Introductory chapters describe geometric structures in field theory in the general case, while detailed later chapters address specific structures such as parallel tensor fields, G-structures, and isometry groups. The relationship between structures in supergravity and period maps of algebraic manifolds, Kodaira–Spencer theory, modularity, and the arithmetic properties of supergravity, are also addressed.

Relevant geometric concepts are introduced and described in detail, providing a self-contained toolkit of useful techniques, formulae and constructions. Covering all the material necessary for the application of supersymmetric field theories to fundamental physical questions, this is an outstanding resource for graduate students and researchers in theoretical physics.

SERGIO CECOTTI is Associate Professor at Scuola Internazionale Superiore di Studi Avanzati (SISSA), Trieste, Italy.

SUPERSYMMETRIC FIELD THEORIES

Geometric Structures and Dualities

SERGIO CECOTTI
Scuola Internazionale Superiore di Studi Avanzati (SISSA), Trieste

Shaftesbury Road, Cambridge CB2 8EA, United Kingdom

One Liberty Plaza, 20th Floor, New York, NY 10006, USA

477 Williamstown Road, Port Melbourne, VIC 3207, Australia

314–321, 3rd Floor, Plot 3, Splendor Forum, Jasola District Centre, New Delhi – 110025, India

103 Penang Road, #05–06/07, Visioncrest Commercial, Singapore 238467

Cambridge University Press is part of Cambridge University Press & Assessment,
a department of the University of Cambridge.

We share the University's mission to contribute to society through the pursuit of
education, learning and research at the highest international levels of excellence.

www.cambridge.org
Information on this title: www.cambridge.org/9781107053816

© S. Cecotti 2015

This publication is in copyright. Subject to statutory exception and to the provisions
of relevant collective licensing agreements, no reproduction of any part may take
place without the written permission of Cambridge University Press & Assessment.

First published 2015

A catalogue record for this publication is available from the British Library

ISBN 978-1-107-05381-6 Hardback

Cambridge University Press & Assessment has no responsibility for the persistence
or accuracy of URLs for external or third-party internet websites referred to in this
publication and does not guarantee that any content on such websites is, or will
remain, accurate or appropriate.

A Magda

Contents

Preface		xi
Notations		xiii
Part I	**How geometric structures arise in supersymmetric field theories**	**1**
1	Geometrical structures in (Q)FT	3
	1.1 (Gauged) σ–models	3
	1.2 Adding fields of arbitrary spin	8
	1.3 How strings come about	15
	1.4 Gauge dualities	16
	1.5 The emergence of modularity	27
	1.6 More dualities	34
2	Extended supersymmetry in diverse dimensions	42
	2.1 SUSY in diverse dimensions	42
	2.2 A little warm–up: $D=2$	48
	2.3 SUSY and the topology of $\mathcal{M}$	51
	2.4 Extended supersymmetry in 3D	59
	2.5 The language of G–structures. Flat (G_1, G_2)–structures	67
	2.6 Local extended supersymmetry in 3D	69
	2.7 Connections with algebraic geometry	78
	2.8 Supersymmetry in $D=4$ and 6 dimensions	79
	2.9 4D SUSY gauge theories. Special Kähler geometry	82
	2.10 4D supergravity	92
Part II	**Geometry and extended SUSY: more than eight supercharges**	**101**
3	Parallel structures and holonomy	103
	3.1 The holonomy group	103
	3.2 Symmetric Riemannian spaces	108

3.3	Berger's theorem	115
3.4	Parallel forms on $\mathcal{M}$	118
3.5	Parallel spinors and holonomy	120
3.6	G–structures and Spencer cohomology	123
3.7	The holonomy groups of Lorentzian manifolds	125

4 SUSY/SUGRA Lagrangians and U–duality — 129

4.1	Determination of the scalar manifold $\mathcal{M}$	129
4.2	Dimensional reduction and totally geodesic submanifolds	138
4.3	Four–fermion couplings vs. holonomy	139
4.4	Vector couplings in 4D SUGRA	143
4.5	The gauge point of view	146
4.6	The complete 4D Lagrangian: U–duality	147
4.7	U–duality, Z–map, and Grassmannians	150
4.8	6D chiral forms	154
4.9	Arithmetics of U–duality. Global geometry of $\mathcal{M}$	155

5 σ–Models and symmetric spaces — 160

5.1	Cartan connections on G	160
5.2	Maurier–Cartan forms	164
5.3	Invariant metrics on a compact group	166
5.4	Chiral models	168
5.5	Geometry of coset spaces G/H	169
5.6	Symmetric spaces	171
5.7	Classification of symmetric manifolds	174
5.8	Totally geodesic submanifolds. Rank	177
5.9	Other techniques	178
5.10	An example: $E_{7(7)}/SU(8)$	178
5.11	Symmetric and Iwasawa gauges	182

6 Killing spinors and rigid SUSY in curved spaces — 185

6.1	Review: space–time charges in General Relativity	185
6.2	AdS space	187
6.3	Killing spinors	189
6.4	The geometry of Killing spinors	198
6.5	Nester form of the space–time charges	205
6.6	The AdS/Poincaré SUSY algebra	207
6.7	Positive mass and BPS bounds	211
6.8	SUGRA Ward identities	214

7	Parallel structures and isometries	220
	7.1 Rigid SUSY: momentum maps	220
	7.2 T–tensors I	227
	7.3 Target space isometries in supergravity	230
	7.4 Holonomy vs. isometries	231
	7.5 The rigid case revisited. Superconformal geometries	236
	7.6 The Cartan–Kostant isomorphism	240
	7.7 The covariant momentum map	243
	7.8 T–tensors II. Generalized $\mathcal{T}$ in SUGRA	246
8	Gauging and potential terms	248
	8.1 Gaugings in rigid SUSY	248
	8.2 $\mathcal{N}$–extended (rigid) CS gauge theories	250
	8.3 Example: $\mathcal{N} = 4$ and the Gaiotto–Witten theorem	258
	8.4 A *puzzle* and its resolution	261
	8.5 World–volume theory of M2 branes: the ABJM model	265
	8.6 Gauged supergravities	266
	8.7 Symmetric target spaces	274
	8.8 *Gauged* supergravity in $D \geq 4$	280
	8.9 An example: $\mathcal{N} = 3$ supergravity in 4D	286
	8.10 Gauging maximal supergravity in D dimensions	287

Part III Special geometries 291

9	Kähler and Hodge manifolds	293
	9.1 Complex manifolds	293
	9.2 Kähler metrics and manifolds	300
	9.3 $U(n)$ manifolds	304
	9.4 Hodge theory in Kähler spaces	305
	9.5 Hodge manifolds	312
	9.6 Symmetric and homogeneous Kähler manifolds	314
10	$\mathcal{N} = 1$ supergravity in 4D	321
	10.1 $\mathcal{N} = 2$ supergravity in 3D	321
	10.2 $\mathcal{N} = 1$ D $= 4$ ungauged supergravity	325
	10.3 SuperHiggs. Flat potentials	327
	10.4 Gauged $\mathcal{N} = 1$ 4D supergravity	331
11	Flag manifolds. Variations of Hodge structures	335
	11.1 Hodge structures and Griffiths domains	335
	11.2 Geometry of reductive homogeneous spaces G/V	341
	11.3 Quick review of Kodaira–Spencer theory	345

11.4	Variations of Hodge structures (VHS)	348
11.5	The case of a Calabi–Yau 3–fold	358

12 Four–dimensional $\mathcal{N}=2$ supergravity — 368
- 12.1 The four geometric structures of $\mathcal{N}=2$ supergravity — 368
- 12.2 **K** ∩ **V** ∩ **Z**, or *projective* special Kähler geometry — 369
- 12.3 Formulas in projective special coordinates — 371
- 12.4 Aspects of projective special Kähler manifolds — 376
- 12.5 Coupling hypermultiplets to $\mathcal{N}=2$ supergravity — 380
- 12.6 Compactifying type II supergravity on a CY manifold — 382
- 12.7 Hypermultiplets: the c–map — 387

Appendix G–structures on manifolds — 390
References — 394
Index — 408

Preface

The focus of the present book is on the *geometric structures* underlying *all* supersymmetric field theories (classical and quantum). The language of geometric structures on smooth manifolds allows us to describe in a uniform and highly unified way all possible situations: rigid supersymmetry as well as local supergravity, in all space–time dimensions D, for all SUSY extensions $\mathcal{N}$, and all kinds of supersymmetries: superPoincaré, superconformal, and even rigid SUSY on general curved space–times.

This book evolved out of the lecture notes of a course in supergravity and supersymmetry taught at SISSA. The lectures were aimed at graduate students who already had a knowledge of supersymmetry and supergravity in the *standard* approaches (superfields, the Noether method, etc.), and the course was meant as an advanced (and perhaps deeper) topic. This explains why this book does not contain many materials that are fundamental tools for a physicist working in the field of supersymmetry but are more than adequately covered by existing books and reviews (see, e.g., the recent book *Supergravity* by D.Z. Freedman and A. van Proeyen (Cambridge University Press, 2012); our book instead focuses on the geometric aspects, with particular emphasis on the geometric structures that are *universal*, that is, that are present *mutatis mutandis* in all possible situations.

The geometric tools introduced in this book allow recovery of all the results obtained from the more classical approaches to SUSY, and typically more quickly and with less pain (however, for specific problems other viewpoints may be more efficient).

In our tale there are four main characters: (i) the Atiyah–Bott–Shapiro classification of Clifford modules; (ii) Berger's theorem on the Riemannian holonomy groups and the allied results on parallel tensor and spinor fields; (iii) Kostant theorem on the interplay of the holonomy and isometry groups, which describes the gauging of all SUSY field theories; (iv) Griffiths' theory of variations of Hodge structures, which gives a unifying view on the geometry of electromagnetic dualities. We pay particular attention to their arithmetic aspects, which are crucial for the quantum theory (and have never been discussed previously, to the best of our knowledge).

In particular, we give a new (simpler and more intrinsic) interpretation of rigid special Kähler geometry as a *flat (G, H)–structure* on the scalars' manifold. We also discuss, in the appropriate geometric setting, some recent major breakthroughs, such as the world–volume theory on a stack of M2–branes, including the Bagger–Lambert and ABJM models.

Some "phenomenological" topics, and some applications to other areas of theoretical physics, would deserve a more detailed discussion: perhaps it would be worthwhile to return to them in an enlarged (and corrected) second edition.

General references on supergravity include:

Van Nieuwenhuizen, P. (1981). Supergravity. *Phys. Rept.*, **68**, 189–398.
Salam, A., and Sezgin, E. (1989). *Supergravity in Diverse Dimensions*, vols. 1, 2. World Scientific.
Castellani, L., D'Auria, R., and Fré, P. (1991). *Supergravity and Superstrings: A Geometrical Perspective*, vols. 1, 2, 3. World Scientific.
Freedman, D.Z., van Proeyen, A. (2012). *Supergravity*. Cambridge University Press.
Fré, P. (2013). *Gravity, A Geometrical Course*, vol. 2. Springer.

Organization of the book

The book is divided into three parts. The purpose of Part I is to motivate the geometric structure approach by showing how differential geometric structures naturally appear in field theory in non–supersymmetric theories (Chapter 1) as well as in the supersymmetric ones (Chapter 2). Parts II and III are the body of the book, where the theory is developed in detail. In particular, Part II is the technical core of the book, where the general results are deduced and then illustrated in detail for the class of field theories having more than eight supersymmetries. In Parts II and III the geometry is discussed in full detail. Chapters in which geometry is presented in a rather rigorous way (stating explicitly Definitions, Lemmas, Theorems, etc.) are followed by physical chapters in which the geometry of the previous chapter is used to construct and understand supergravity and supersymmetric theories. As a rule, *starting from Chapter 3, odd–numbered chapters are purely geometric, while even–numbered ones contain physical applications and constructions*. Part III applies the general result to the theories having fewer than nine supersymmetries. The even chapters of this third part are somewhat sketchy, since they have a substantial overlap with existing literature to which the reader is referred. In the Appendix we present a quick review of the language of G–structures on smooth manifolds.

Notations

The math symbols we use are defined in the text. Recurring symbols are:
- $\mathbb{N}$, $\mathbb{Z}$, $\mathbb{Q}$, $\mathbb{R}$, $\mathbb{C}$, $\mathbb{H}$, and $\mathbb{O}$ denote respectively the natural, integer, rational, real, complex numbers, the Hamilton quaternions, and the Cayley octaves; $\mathbb{R}^\times$, $\mathbb{C}^\times$, ... the multiplicative group of non–zero elements in $\mathbb{R}$, $\mathbb{C}$, ...
- $V^\vee$ stands for the dual of the vector space V, $\otimes$ for the tensor product of vector spaces, $\odot$ for the *symmetric* tensor product of vector spaces, and $\wedge$ for the *antisymmetric* one. The same notation applies to vector bundles.
- The algebra of $n \times n$ matrices with entries in the algebra $\mathbb{F}$ is denoted $\mathbb{F}(n)$, the vector space of $n \times m$ matrices as $\mathbb{F}(n, m)$.
- $\mathbb{C}l(n)$ stands for the universal Clifford algebra in dimension n, $\mathbb{C}l^0(n)$ for its even subalgebra.
- If $G, H, K, L, \ldots, SU(n), SO(n), \ldots$ are Lie groups, the corresponding algebras are denoted as $\mathfrak{g}, \mathfrak{h}, \mathfrak{k}, \mathfrak{l}, \ldots, \mathfrak{su}(n), \mathfrak{so}(n), \ldots$.
- $Sp(2n)$ denotes the symplectic group with fundamental representation of dimension $2n$ corresponding to Cartan's Lie algebra C_n.
- Given a smooth manifold $\mathcal{M}$, its universal cover is denoted $\widetilde{\mathcal{M}}$, its tangent bundle $T\mathcal{M}$ and its cotangent bundle $T^*\mathcal{M}$.
- The kth Betti number of the manifold $\mathcal{M}$ is written $B_k(\mathcal{M})$, its Euler characteristic $\chi(\mathcal{M})$.
- The space of sections of the bundle/sheaf $\mathcal{E}$ over U is written $\Gamma(U, \mathcal{E})$ or simply $\mathcal{E}(U)$.
- The space of smooth k–forms over U is denoted $\Lambda^k(U)$.
- The sheaf of (germs of) holomorphic p–forms is denoted Ω^p.

Part I

How geometric structures arise in supersymmetric field theories

1
Geometrical structures in (Q)FT

Part I of this book is introductory in nature. Its purpose is to motivate our geometric approach to supersymmetric field theory. We show how geometric structures arise in classical and quantum field theories on quite general grounds. In Chapter 1 we consider the basic geometric structures which hold independently of supersymmetry. In Chapter 2 we specialize to the supersymmetric case (rigid and local) where more elegant structures emerge. Not being part of the technical body of the book, these chapters are rather elementary and sketchy. However, we show how dualities, modularity, and other stringy patterns are universal features of field theory.

Throughout this book, by a *field theory* we shall mean a *Lagrangian* field theory, that is, a classical or quantum system whose dynamics is described by a Lagrangian $\mathcal{L}$ with no more than two derivatives of the fields.

1.1 (Gauged) σ-models

Most quantum field theories (QFTs) have scalar fields. Usually we can understand a lot about the dynamics of a field theory just by studying its scalar sector. This is *a fortiori* true if the theory has (enough) supersymmetries, since in this case all other sectors are related to the scalar one by a symmetry. The understanding of the scalars' geometry is relevant even for theories, like quantum chromodynamics (QCD), that do not have fundamental scalar fields in their microscopic formulation. At low energy, QCD is well described by an effective scalar model whose fields represent pions (the lightest particles in the hadronic spectrum). Historically, this effective theory was the original σ-model. It encodes all current algebra of QCD, and its phenomenological predictions are quite a success [303, 304, 305]. Our first goal is to generalize this model. We begin by considering a theory with only scalar fields. In the next section we will add fields in arbitrary (finite) representations of the Lorentz group.

1.1.1 The target space $\mathcal{M}$

We consider a general field theory in D space–time dimensions whose Lagrangian description contains only scalar fields which we denote as ϕ^i, with $i = 1, 2, \ldots, n$. Let us write down the most general local, Hermitian, Poincaré–invariant Lagrangian having (at most) two derivatives; for $D \neq 2$ it has the form

$$\mathcal{L} = -\frac{1}{2} g_{ij}(\phi) \, \partial_\mu \phi^i \, \partial^\mu \phi^j + \text{terms with no derivative} \tag{1.1}$$

for some (field–dependent) real symmetric matrix $g_{ij}(\phi)$. For $D = 2$ we may add the P and T odd term $b_{ij}(\phi) \, \epsilon^{\mu\nu} \, \partial_\mu \phi^i \, \partial_\nu \phi^j$ with $b_{ij}(\phi)$ antisymmetric. For the moment we limit ourselves to P–invariant models and set $b_{ij} = 0$.

Unitarity requires the kinetic terms to be positive, so $g_{ij}(\phi)$ is a positive–definite matrix. Physical quantities are independent of the fields we use to parametrize the configuration, that is, observables are invariant under field reparametrizations of the form

$$\phi^i \to \varphi^i = \varphi^i(\phi). \tag{1.2}$$

Written in terms of the new fields φ^i, the Lagrangian takes the form

$$\mathcal{L} = -\frac{1}{2} \widetilde{g}_{ij}(\varphi) \, \partial_\mu \varphi^i \, \partial^\mu \varphi^j + \cdots \tag{1.3}$$

where

$$\widetilde{g}_{ij}(\varphi) \equiv \frac{\partial \phi^k}{\partial \varphi^i} \, g_{kl} \, \frac{\partial \phi^l}{\partial \varphi^j}. \tag{1.4}$$

The above equations have a simple geometric interpretation: the fields ϕ^i are local coordinates on a (smooth) manifold $\mathcal{M}$ and g_{ij} is a Riemannian metric for $\mathcal{M}$, which correctly transforms under diffeomorphisms as a symmetric tensor, Eq. (1.4). This interpretation allows us to describe the situation in more geometric terms: we have two manifolds, the *target* one $\mathcal{M}$, which can have a non–trivial topology,[1] and the space–time manifold Σ (which, for the moment, we take to be just Minkowski space $\mathbb{R}^{D-1,1}$). A classical field configuration is a (smooth) map

$$\Phi: \Sigma \to \mathcal{M} \tag{1.5}$$

which in local coordinates is given by the functions $\phi^i(x^\mu)$. The Lagrangian $\mathcal{L}$ is simply the trace (with respect to the space–time metric $\eta_{\mu\nu}$) of the pull–back (i.e., induced) metric $\Phi^* g$.

---

[1] Hence the fields ϕ^i are, in general, only locally defined on $\mathcal{M}$.

1.1 (Gauged) σ–models

(Q)FTs defined by maps $\Sigma \to \mathcal{M}$ and the Lagrangian

$$\mathcal{L} = -\frac{1}{2} g_{ij}(\phi)\, \partial_\mu \phi^i\, \partial^\mu \phi^j \qquad (1.6)$$

are called *σ–models*. We stress again that in such models *all* physical quantities, being reparametrization-independent, should be differential–geometric invariants of the Riemannian manifold $(\mathcal{M}, g)$. This simple observation, which we call the *Geometric Principle,* is quite powerful.

Example: the renormalization group β–functions

To show the power of the Geometric Principle, we discuss the one–loop β–functions of the σ–model. We take $D = 2$, the space–time dimension in which the model is power–counting renormalizable. We may consider the most general Lagrangian of the form (1.1), since the interactions with no derivative (the potential) is a soft term that does not affect the β–functions of the derivative couplings. We introduce the Planck constant, $\hbar$, as a loop–counting device; recall that in perturbative QFT the k–loop contribution to the β–function scales like $\hbar^{k-1}$. The action is

$$S = -\frac{1}{2\hbar} \int_\Sigma g_{ij}\, \partial_\mu \phi^i \partial^\mu \phi^j\, d^2z. \qquad (1.7)$$

We see that a rescaling $\hbar \to \lambda \hbar$ is equivalent to $g_{ij} \to \lambda^{-1} g_{ij}$, so the weak-coupling limit $\hbar \to 0$ is just the large volume limit for $\mathcal{M}$,

$$\mathrm{vol}(\mathcal{M}) \propto \hbar^{-\dim \mathcal{M}/2}. \qquad (1.8)$$

A general σ–model has an *infinite* number of coupling constants, $g_{i_1 i_2 \cdots i_l}$,

$$S = -\frac{1}{2\hbar} \int d^2z \sum_{l=2}^{\infty} g_{i_1 i_2 \cdots i_l}\, \phi^{i_3} \phi^{i_4} \cdots \phi^{i_l}\, \partial_\mu \phi^{i_1} \partial^\mu \phi^{i_2},$$

namely, the Taylor coefficients[2] of the metric $g_{ij}(\phi)$. We can conveniently assemble the infinite set of β–functions into a symmetric tensor, $\beta_{ij}(\phi)$, whose Taylor coefficients are the β–functions of the couplings $g_{i_1 i_2 \cdots i_l}$. The renormalization group (RG) flow then takes the form

$$\mu \frac{\partial}{\partial \mu} \frac{g_{ij}(\phi)}{\hbar} = \beta_{ij}(\phi). \qquad (1.9)$$

---
[2] Assuming the metric is of class C^ω.

The Geometric Principle implies that $\beta_{ij}(\phi)$ is a covariant symmetric tensor on $\mathcal{M}$ made out of the metric g_{ij} and its derivatives. Moreover, $\beta_{ij}(\phi)$ should vanish for a flat metric, since in that case the QFT is free. Therefore, $\beta_{ij}(\phi)$ is a symmetric tensor which has an expansion as a sum of products of Riemann tensor covariant derivatives, $\nabla_{i_1} \cdots \nabla_{i_s} R^j{}_{klm}$, with the indices contracted in a suitable way using the effective inverse metric $\hbar\, g^{ij}$. Since $\nabla_{i_1} \cdots \nabla_{i_s} R^j{}_{klm}$ is invariant under $g_{ij} \mapsto \hbar^{-1} g_{ij}$, all $\hbar$ dependence arises from the inverse metric contractions. Counting indices to be contracted, we see that a term in this expansion scales with the volume as

$$\hbar^{r+s-1}, \tag{1.10}$$

where r is the number of Riemann tensors and $2s$ the total number of covariant derivatives. Since the one–loop contribution scales as $\hbar^0$, this leaves only one possibility: one Riemann tensor and no derivative. Thus

$$\beta_{ij}|_{\text{one loop}} = c_1 R_{ij} + c_2\, g_{ij} R, \tag{1.11}$$

for some constants c_1, c_2. We claim that $c_2 = 0$ while c_1 is a universal coefficient that does not depend on $\dim \mathcal{M}$. Indeed, take $\mathcal{M} = \mathbb{R}^n \times \mathcal{N}$. The fields of the flat factor are free, and then $\beta_{ij}|_{\mathbb{R}^m} = 0$, whereas Eq. (1.11) gives $\beta_{ij}|_{\mathbb{R}^m} = c_2 R\, \delta_{ij}$. So $c_2 = 0$. On the other hand, $\beta_{ij}|_{\mathcal{N}} = c_1 R_{ij}$ cannot depend on $m \equiv$ the number of flat directions, since they correspond to decoupled free fields. Hence c_1 is independent of $\dim \mathcal{M}$, and it may be computed using any convenient manifold.

The fixed points of the RG flow need not correspond to zeros of the β–function, i.e., to metrics such that $\beta_{ij} = 0$; more generally, they may correspond to a flow which acts on the metric as a diffeomorphism, so that the action is scale-independent up to field redefinitions. This requires

$$\beta_{ij} = \pounds_v\, g_{ij}, \tag{1.12}$$

where v is a vector field on $\mathcal{M}$ and $\pounds_v$ denotes the Lie derivative [325] along v. In the one–loop approximation, the LHS is proportional to the Ricci tensor, and the metrics which solve equation (1.12) are precisely the *Ricci solitons* [95].

As we shall see in Chapter 2, the σ–model admits a supersymmetry (SUSY) completion. The above discussion for the β–function extends to the SUSY case.

1.1.2 Symmetries, gaugings, and Killing vectors

The geometry says more. Assume our Lagrangian field theory is invariant under a continuous symmetry group G which acts on the scalar fields ϕ^i. G should be, in particular, a symmetry of the two–derivative terms in $\mathcal{L}$, Eq. (1.1), hence it should leave invariant the metric g_{ij}; that is, G should be a subgroup of the isometry group $\text{Iso}(\mathcal{M})$ of the Riemannian manifold $\mathcal{M}$.

The corresponding infinitesimal symmetries, $\phi^i \to \phi^i + \epsilon^A K_A{}^i(\phi)$, are generated by vector fields $K_A{}^i \partial_i$ ($A = 1, 2 \ldots, \dim G$) which satisfy the Killing condition

$$\pounds_{K_A} g_{ij} \equiv \nabla_i K_{Aj} + \nabla_j K_{Ai} = 0, \tag{1.13}$$

as well as the algebra

$$\pounds_{K_A} K_B = [K_A, K_B] = f_{AB}{}^C K_C, \tag{1.14}$$

where $f_{AB}{}^C$ are the structure constants of $\mathfrak{g}$ (the Lie algebra of G).

The existence of a non–trivial group of isometries – in particular a non–Abelian group – is a strong requirement on the geometry of $\mathcal{M}$. For instance, by the Bochner theorem, if $\mathcal{M}$ is compact and has negative Ricci curvature, it has *no* Killing vectors [163].

Gauging a subgroup of $\text{Iso}(\mathcal{M})$

One may wish to gauge a subgroup G of the isometry group $\text{Iso}(\mathcal{M})$. The minimal coupling of the gauge vector fields A^A_μ to the scalars is also dictated by the geometry of $\mathcal{M}$ through the corresponding Killing vectors $K_A{}^i \partial_i$. To gauge the symmetry, one replaces in $\mathcal{L}$ the ordinary derivative by the covariant one:

$$\partial_\mu \phi^i \to D_\mu \phi^i := \partial_\mu \phi^i - A^A_\mu K_A{}^i. \tag{1.15}$$

The infinitesimal gauge transformation then reads

$$\delta \phi^i = \Lambda^A K_A{}^i, \tag{1.16}$$

$$\delta A^A_\mu = \partial_\mu \Lambda^A + f^A{}_{BC} A^B_\mu \Lambda^C, \tag{1.17}$$

where the parameters Λ^A are arbitrary functions in space–time. Then

$$\begin{aligned}\delta D_\mu \phi^i &= \Lambda^A (\partial_j K_A{}^i) D_\mu \phi^j \\ &\quad + A^B_\mu \Lambda^C [K_B{}^j \partial_j K_C{}^i - K_C{}^j \partial_j K_B{}^i] - f_{BC}{}^A A^B_\mu \Lambda^C K_A{}^i \\ &= \Lambda^A (\partial_j K_A{}^i) D_\mu \phi^j.\end{aligned} \tag{1.18}$$

The covariance[3] of $D_\mu\phi^i$ follows from the closure of the gauge algebra, Eq. (1.14), while the invariance of the kinetic term $g_{ij}D_\mu\phi^i D^\mu\phi^j$ requires $\pounds_{(\Lambda^A K_A)} g_{ij} = 0$, i.e., the Killing condition (1.13). Indeed, from Eq. (1.18)

$$\delta(g_{ij}D_\mu\phi^i D^\mu\phi^j) = \Lambda^A(\nabla_i K_{Aj} + \nabla_j K_{Ai})D_\mu\phi^i D^\mu\phi^j. \tag{1.19}$$

We summarize what we have learned in the following statement:

General Lesson 1.1 *The physics of the (gauged) σ–model is controlled by the differential geometry of the target manifold $\mathcal{M}$.*

The physics is invariant under general reparametrizations of the target space in the same sense that General Relativity is invariant under reparametrizations of the space–time manifold. In General Relativity this invariance is often stated in the form of the *equivalence principle* [296]. The same principle holds for target space as well:

Corollary 1.2 (target space equivalence principle) *Any physical quantity which is local in $\mathcal{M}$ and depends only on the metric and its first derivative may be safely computed using a flat target space.*

In Section 1.3 we shall see the deep reason why the target space behaves as a physical space–time.

Exercise 1.1.1 Using Feymann graphs, compute the universal coefficient c_1 in Eq. (1.11) for the σ–model (1.7). Check its universality.

1.2 Adding fields of arbitrary spin

1.2.1 Couplings and geometric structures on $\mathcal{M}$

We have seen in Section 1.1.2 that the scalars' couplings to gauge vectors are specified by a set of vector fields K_A on the manifold $\mathcal{M}$ which satisfy certain differential–geometric constraints. This is a first example of a general pattern: *all* the couplings in a Lagrangian may be identified with suitable differential-geometric structures on the scalar manifold $\mathcal{M}$. To make our point, we work out the details of a specific example, in which only scalars and spin–1/2 fermions

---

[3] Note that $D_\mu\phi^i$ transforms under an infinitesimal gauge transformation of parameter Λ^A as the differentials $d\phi^i \in \Lambda^1(\mathcal{M})$ under the infinitesimal diffeomorphism generated by the vector field $\Lambda^A K_A{}^i \partial_i$. Indeed,

$$\pounds_{\Lambda^A K_A} d\phi^i \equiv (d\, i_{\Lambda^A K_A} + i_{\Lambda^A K_A} d)d\phi^i = d(\Lambda^A K_A{}^i) \equiv \Lambda^A(\partial_j K_A{}^i)d\phi^j.$$

1.2 Adding fields of arbitrary spin

are present. The reader can easily convince herself that the arguments are pretty general, and work, *mutatis mutandis,* for fields of arbitrary (finite) spin.

General 2D *model with fermions.*

Let us consider the general theory with scalars ϕ^i, and fermions ψ^a, where $i = 1, 2, \ldots, n$, and $a = 1, 2, \ldots, m$. We choose $D = 2$, which is the number of dimensions in which these models have the more interesting applications (as world-sheet theories of some superstring [200]) and in which they make sense quantum mechanically. The arguments, however, are manifestly dimension–independent (apart from questions of existence as quantum field theories).

Limiting ourselves to power–counting renormalizable theories, the most general Lagrangian is

$$\begin{aligned}
\mathcal{L} = &-\frac{1}{2} g_{ij}(\phi)\, \partial_\mu \phi^i \partial^\mu \phi^j + b_{ij}(\phi)\, \epsilon^{\mu\nu} \partial_\mu \phi^i \partial_\nu \phi^j + V(\phi) \\
&+ i h_{ab}(\phi)\, \bar{\psi}^a \gamma^\mu \partial_\mu \psi^b + i \tilde{h}_{ab}(\phi)\, \bar{\psi}^a \gamma_3 \gamma^\mu \partial_\mu \psi^b \\
&+ k_{abi}(\phi)\, \bar{\psi}^a \gamma^\mu \psi^b \partial_\mu \phi^i + \tilde{k}_{abi}(\phi)\, \bar{\psi}^a \gamma^\mu \gamma_3 \psi^b \partial_\mu \phi^i \\
&+ y_{ab}(\phi)\, \bar{\psi}^a \psi^b + \tilde{y}_{ab}(\phi)\, \bar{\psi}^a \gamma_3 \psi^b \\
&+ s_{abcd}(\phi)\, \bar{\psi}^a \psi^c \bar{\psi}^b \psi^d + \cdots
\end{aligned} \quad (1.20)$$

where the couplings

$$g_{ij}(\phi),\ b_{ij}(\phi),\ V(\phi),\ h_{ab}(\phi),\ \tilde{h}_{ab}(\phi),\ k_{abi}(\phi), \\ \tilde{k}_{abi}(\phi),\ y_{ab}(\phi),\ \tilde{y}_{ab}(\phi),\ s_{abcd}(\phi),\ \ldots \quad (1.21)$$

are arbitrary functions of the scalar fields ϕ^i. Each term in the Lagrangian (1.20) may be interpreted as a geometric structure on $\mathcal{M}$. We already know that $g_{ij}(\phi)$ is a Riemannian metric. The coupling $b_{ij}(\phi)$ is antisymmetric in the indices i,j and hence can be seen as a differential 2–form $b = \frac{1}{2} b_{ij}(\phi)\, d\phi^i \wedge d\phi^j$ on $\mathcal{M}$. The value of this contribution to the action S for a given field configuration $\Phi \colon \Sigma \to \mathcal{M}$ is given by the very geometrical (in fact functorial) formula

$$\int_\Sigma \Phi^* b. \quad (1.22)$$

This, in particular, implies that – up to space–time boundary phenomena – the physics should be invariant under the *target space* gauge invariance

$$b \to b + d\xi \quad (1.23)$$

for all 1-forms ξ on $\mathcal{M}$. Indeed, under this variation of the $b_{ij}(\phi)$ coupling, the action changes as

$$S \to S + \int_\Sigma \Phi^* d\xi = S + \int_\Sigma d\Phi^* \xi = S + \int_{\partial\Sigma} \Phi^* \xi, \qquad (1.24)$$

so $b \to b + d\xi$ is a symmetry whenever we are allowed to ignore boundary terms in the action, e.g. if the space–time Σ is closed. In such a situation, the physics should depend only on the gauge–invariant field-strength 3–form of b, $H = db$, as well as on the harmonic projection of b. In particular, when $H = 0$ – i.e., b is closed – the physics depends only on the cohomology class of b, and the coupling $\int \Phi^* b$ is purely topological: it measures the class $[\Phi^* b]$ as a multiple of the fundamental class of Σ; then $\int \Phi^* b$ does not change under continuous deformations of the map Φ.

The next coupling, $V(\phi)$, is easy. It is just a scalar field on $\mathcal{M}$. In many situations $V(\phi)$ is required to satisfy further geometric conditions. For instance, to gauge the isometry associated to a Killing vector K, we must have

$$\pounds_K V = 0, \qquad (1.25)$$

which is the geometric statement of gauge invariance. Other geometric structures related to the scalar potential will be presented in Chapters 6, 7, 8.

To discuss the other couplings, we change notation and use Majorana–Weyl fermions, $\psi^a_\pm = \pm \gamma_3 \psi^a_\pm$, which are the minimal spinors in 2D. Writing

$$h^\pm_{ab} = h_{ab} \pm \tilde{h}_{ab}, \qquad (1.26)$$

the second line of Eq. (1.20) reads

$$i h^+_{ab} \psi^a_+ \partial_- \psi^b_+ + (+ \leftrightarrow -), \qquad (1.27)$$

where, without loss of generality, we may assume the matrices $h^\pm_{ab}$ to be symmetric, since their antisymmetric part gives, up to a total derivative, terms of the form $\psi^a_+ \psi^b_+ \partial_- h^+_{ab}$ which contribute to the third line of Eq. (1.20). The Hermitian conjugate of (1.27) is

$$-i(h^+_{ab})^*(\partial_- \psi^b_+) \psi^a_+ + (+ \leftrightarrow -). \qquad (1.28)$$

Unitarity requires $h^\pm_{ab}$ to be positive–definite *real* symmetric matrices. The geometric interpretation of this last condition is obvious: the chiral fermions $\psi^a_\pm$ are *sections of vector bundles*[4] over Σ, which are the pull–backs $\Phi^* \mathcal{V}_\pm$ of real vector

---

[4] For the moment we take Σ to be flat Minkowski/Euclidean space. For the general case, see General Lesson 1.3 below.

bundles $\mathcal{V}_\pm \to \mathcal{M}$ of rank m (more generally, we may choose the two bundles $\mathcal{V}_+$ and $\mathcal{V}_-$ to have different ranks[5] m_+ and m_-) with *fiber metrics* $h_{ab}^\pm$. Being positive–definite, $h_{ab}^\pm$ are, in particular, invertible; we write $h_\pm^{ab}$ for the inverse metrics. The couplings in the third line of (1.20) then take the form

$$i k^+_{ab\,i}(\phi)\, \psi_+^a \psi_+^b\, \partial_-\phi^i + i k^-_{ab\,i}(\phi)\, \psi_-^a \psi_-^b\, \partial_+\phi^i, \qquad (1.29)$$

where only the the antisymmetric part $k^\pm_{[ab]\,i}$ of $k^\pm_{ab\,i}$ matters; that is, we are *free* to set the symmetric part to any convenient value.

Defining $\omega_i^{\pm\,a}{}_b := h_\pm^{ac} k^\pm_{cb\,i}$, and writing

$$D_\mu \psi_\pm^a := \partial_\mu \psi_\pm^a + \partial_\mu \phi^i\, \omega_i^{\pm\,a}{}_b\, \psi_\pm^b, \qquad (1.30)$$

we may combine the second and third lines of Eq. (1.20) into the very geometric expression

$$i h^+_{ab}\, \psi_+^a D_- \psi_+^b + (+ \leftrightarrow -), \qquad (1.31)$$

where now the covariant derivative D_μ contains a connection $\partial_\mu \phi^i\, \omega_i^{\pm\,a}{}_b$ which is just the pull–back, by the map Φ, of the connection on the vector bundles $\mathcal{V}_\pm \to \mathcal{M}$ given by $\omega^\pm_{i\,ab}$. In particular, the couplings in the third line of (1.20) *do* transform as a connection under arbitrary field redefinitions.[6] The arbitrary symmetric part of $k^\pm_{ab\,i}$ may be fixed so that

$$D_\mp(h^\pm_{ab}\, \psi_\pm^b) = h^\pm_{ab}\, D_\mp \psi_\pm^b, \qquad (1.32)$$

which makes Hermiticity of the Lagrangian manifest. Then $\omega^\pm_{i\,ab}$ becomes *a metric connection with respect to the fiber metric* $h_{ab}^\pm$. Thus the couplings in the second and third lines of Eq. (1.20), although *a priori* totally generic and structureless, may be reinterpreted as nice geometric structures on the scalar manifold $\mathcal{M}$, namely *real vector bundles with fiber metrics and compatible (orthogonal) connections.*

As for the remaining couplings, we rewrite the last two lines of Eq. (1.20) in the general form

$$y_{ab}(\phi)\, \psi_+^a \psi_-^b + s_{abcd}(\phi)\, \psi_+^a \psi_+^b \psi_-^c \psi_-^d \qquad (1.33)$$

($s_{abcd} = -s_{bacd} = -s_{abdc}$). It is evident that the Yukawa coupling y_{ab} is a section of the (pull–back of the) bundle $\mathcal{V}_+^\vee \otimes \mathcal{V}_-^\vee$, while the 4–Fermi coupling is a section of $\wedge^2 \mathcal{V}_+^\vee \otimes \wedge^2 \mathcal{V}_-^\vee$. Both these bundles have natural fiber metrics and connections

---

[5] Paying attention to cancel axial anomalies, if we wish to have a sensible quantum theory.
[6] Here by "arbitrary field redefinitions" we mean redefinitions that leave the Lagrangian local and power–counting renormalizable in 2D.

inherited from $\mathcal{V}_\pm$. In conclusion, all terms in the Lagrangian are naturally and invariantly identified with geometric structures on $\mathcal{M}$. In the supersymmetric case, which is the focus of the present book, the couplings (1.21) are not independent but are related by supersymmetry. In the present language the SUSY relations between different couplings become relations between different geometric structures on $\mathcal{M}$; SUSY itself is more intrinsically described in terms of *special structures* on the manifold $\mathcal{M}$.

The above analysis is adequate for a *flat* space–time Σ. For general Σ, the spinors themselves are sections of some spin–bundles $\mathcal{S}_\pm \to \Sigma$, endowed with a spin–connection related to the Levi–Civita connection by the usual formulae of Riemannian geometry. When both Σ and $\mathcal{M}$ are non–trivial, the spinor bundles get "twisted" by the (pull–back of) the above target–space bundles $\mathcal{V}_\pm \to \mathcal{M}$. In conclusion,

General Lesson 1.3 *Everything in (Q)FT is differential–geometric. Fields with non–zero spin are sections of vector bundles of the form*

$$\mathcal{S}_R \otimes \Phi^* \mathcal{V} \to \Sigma, \tag{1.34}$$

where $\mathcal{S}_R \to \Sigma$ is the vector bundle associated with the spin representation R appropriate for the given field,[7] and $\mathcal{V}$ is a target–space bundle describing the interactions of that field with the scalars. The derivative couplings are determined by covariant derivatives whose connection is the natural (functorial) one on the bundle (1.34). For higher spin fields there is more structure.

In the above discussion no supersymmetry is implied. Supersymmetry, when present, selects specific bundles $\mathcal{V}_\pm \to \mathcal{M}$. Unifying bosonic and fermionic interactions, SUSY implies that $\mathcal{V}_\pm \to \mathcal{M}$ should be God–given bundles whose properties are uniquely fixed by the Riemannian geometry of $\mathcal{M}$. A simple (and typical) SUSY relation is $\mathcal{V}_\pm \simeq T\mathcal{M}$, with the fiber metric and connection equal to the Riemannian ones (see Chapter 2). Thus, while *all* theories have couplings which are described by geometric objects, SUSY theories have couplings which are described by *canonical* geometric objects. This explains why the geometric approach is the most convenient one in the supersymmetric case.

---

[7] $\mathcal{S}_R$ is defined as follows: the Riemannian geometry defines a metric connection on the tangent bundle $T\Sigma$ taking values in $\mathfrak{so}(\dim \Sigma)$ and so defines an $SO(\dim \Sigma)$ principal bundle. If Σ is a *spin* manifold, this principal bundle can (by definition) be uplifted to a $Spin(\dim \Sigma)$ principal bundle $\mathcal{P}$. $\mathcal{S}_R$ is the vector bundle associated with $\mathcal{P}$ by the representation R of $Spin(\dim \Sigma)$.

1.2.2 Couplings as arrows

Sometimes we will find it convenient to rephrase the above ideas in fancier language. A covariant 2–tensor t_{ij} on a manifold may be seen as a map from the vector fields to the differential forms $v^i \partial_i \mapsto t_{ji} v^i dx^j$. In formal language, such a map is a *vector bundle morphism* between the tangent and the cotangent bundle, $T\mathcal{M} \xrightarrow{t} T^*\mathcal{M} := (T\mathcal{M})^{\vee}$. When t_{ij} is the metric g_{ij}, this is just lowering the index of v^i by contracting with g_{ij}.

This viewpoint may be extended to the couplings in $\mathcal{L}$. Consider, for instance, the Yukawa coupling $Y_{ab}\psi_+^a \psi_-^b$ and the 4–Fermi one, $S_{abcd}\psi_+^a \psi_+^b \psi_-^c \psi_-^d$. We know that they are, respectively, sections of $\mathcal{V}_+^{\vee} \otimes \mathcal{V}_-^{\vee}$ and $\wedge^2 \mathcal{V}_+^{\vee} \otimes \wedge^2 \mathcal{V}_-^{\vee}$. We may equivalently see them as *vector bundle morphisms*

$$\mathcal{V}_- \xrightarrow{Y} \mathcal{V}_+^{\vee} \xrightarrow{(h^+)^{-1}} \mathcal{V}_+, \tag{1.35}$$

$$\wedge^2 \mathcal{V}_- \xrightarrow{S} \wedge^2 \mathcal{V}_+^{\vee} \simeq \wedge^2 \mathcal{V}_+, \tag{1.36}$$

where each coupling Y_{ab}, S_{abcd}, and h_{ab}^+ is written as a morphism. It follows from General Lesson 1.3 that *all* couplings in a Lagrangian (Q)FT may be written in this "arrow" form. The "arrow" point of view is useful when the couplings, viewed as arrows, fit into a larger arrow diagram and we can work out the geometry of the couplings in a purely arrow–theoretical manner (*abstract nonsense*). This will save us work, especially in relation to dualities. For a nice application, see the example at the end of Section 1.4.

As a matter of language, we shall say that a bundle morphism is a *monomorphism* if it is fiberwise injective, an *epimorphism* if it is fiberwise surjective, and an *isomorphism* if it is both. A bundle morphism of the form $\mathcal{E} \to \mathcal{E}$ is called an *endomorphism*, and an *automorphism* if it is also an isomorphism. Bundle morphisms $\mathcal{A} \to \mathcal{B}$ can be (equivalently) seen as sections of $\mathcal{B} \otimes \mathcal{A}^{\vee}$. We shall write $\text{End}(\mathcal{E})$ for $\mathcal{E} \otimes \mathcal{E}^{\vee}$, and $\text{Aut}(\mathcal{E})$ for the corresponding group of automorphisms.

1.2.3 Global symmetries and parallel structures

This is the right place to make a general comment about symmetries. As an illustration, we consider a model of fermions coupled to scalars, having the general structure in Eq. (1.20). To keep formulae simple, we set to zero all parity–odd couplings, we suppress the $\pm$ superscript, and replace Majorana spinors by Dirac ones.

Geometrically speaking, the metric $h_{ab}(\phi)$ reduces the structure group of the fermionic bundle to a subgroup of $U(m)$ (or $SO(m)$ in the Majorana case).

Correspondingly, we have the group $\mathcal{G} \subset \mathrm{Aut}(\mathcal{V})$ of fiber isometries acting as

$$\psi^a \mapsto M^a_{\ b}(\phi)\,\psi^b \tag{1.37}$$

where $M^a_{\ b}(\phi)$ is a *unitary* section of the bundle $\mathcal{V} \otimes \mathcal{V}^\vee$,

$$h_{ab}(\phi)\,M^a_{\ c}(\phi)^*\,M^b_{\ d}(\phi) = h_{cd}(\phi). \tag{1.38}$$

Under the action (1.37),

$$D_\mu \psi^a \mapsto M^a_{\ b}(\phi)\big(D_\mu \psi^b + (M^{-1}D_i M)^b_{\ c}\,\psi^c\,\partial_\mu \phi^i\big), \tag{1.39}$$

where $D_i M$ is the covariant derivative computed using the connection on $\mathcal{V} \otimes \mathcal{V}^\vee$ induced by the one of $\mathcal{V}$. We see that (1.37) is a *global* symmetry of the Fermi kinetic terms iff the section $M^a_{\ b}(\phi)$ satisfies

$$D_i M^a_{\ b}(\phi) = 0. \tag{1.40}$$

A non–zero section M of $\mathcal{V} \otimes \mathcal{V}^\vee$ satisfying the last condition is called *parallel*. A vector bundle $\mathcal{E} \to \mathcal{M}$ together with a parallel section s is called *a parallel structure*. Then:

General Lesson 1.4 *A linear transformation $\psi^a \mapsto M^a_{\ b}\psi^b$ can be a global symmetry only if* $(\mathrm{End}(\mathcal{V}), M)$ *is a (unitary) parallel structure on $\mathcal{M}$.*

The relevance of this observation is that often we *know*, from physical considerations, that some global symmetry should be present. Then the above General Lesson 1.4 requires the existence of the corresponding unitary parallel structures on $\mathcal{M}$, which puts severe constraints on the possible scalars' geometries, and this fact helps to determine the Lagrangian of the model. Indeed, Eq. (1.40) leads to the integrability condition

$$0 = [D_i, D_j]M = [F_{ij}, M], \tag{1.41}$$

where F_{ij} is the curvature of the connection ω_i. Thus, the presence of a parallel section gives an algebraic constraint on the curvature of $\mathcal{V}$. In the SUSY case the curvature of F_{ij} will be related to the curvature of the scalar manifold $\mathcal{M}$, so the existence of a global symmetry leads to a restriction on the curvature tensor of $\mathcal{M}$; that is, a condition on the scalar metric $g_{ij}(\phi)$.

For a generic vector bundle $\mathcal{V}$ with metric h and connection ω, the only parallel sections of $\mathrm{End}(\mathcal{V})$ correspond to the field $\mathbb{C}$ ($\mathbb{R}$ in the Majorana case) whose associated symmetry is the Fermi number (Fermi number mod 2 in the Majorana case). Then, for generic $(\mathcal{V}, h, \omega)$, Eq. (1.39) implies that all Lagrangian symmetries associated with $\mathrm{Aut}(\mathcal{V})$ – different from the Fermi number – must be *gauge* symmetries. Concretely,[8] one takes field–dependent matrices $\tau^A(\phi)^a_b$ valued in the Lie algebra $\mathfrak{g} \subset \mathfrak{u}(m)$ of the symmetry group G and adds to the covariant derivative acting on fermions a further connection term

$$-A^A_\mu \, \tau^A(\phi)^a_b. \qquad (1.42)$$

Such that the gauge variation of A^a_μ compensates the extra term in the RHS of Eq. (1.39). Geometrically, one describes the situation as a further twisting of the Fermi bundle $\mathcal{S} \otimes \Phi^*\mathcal{V}$ by some space–time bundle $\mathcal{G}$ associated with the gauge principal bundle $P_G \to \Sigma$.

The situation is a bit more intricate when the symmetry acts non–trivially on *both* the fermions and the scalars (through an isometry of $\mathcal{M}$). We defer the discussion of this case to the analysis of SUSY models in Chapter 2.

1.3 How strings come about

This section is just a remark – albeit fundamental – about Section 1.2.

The fact that the couplings may be seen as tensor fields on the target space $\mathcal{M}$, with the correct transformation properties, is not only a useful mathematical nicety. It is a deep physical property. The couplings of any theory – in particular of any two–dimensional model – can be interpreted as local fields in some target space $\mathcal{M}$, in which we have general covariance (that is, General Relativity), gauge–transformations, local supersymmetry, etc., much as $\mathcal{M}$ was a physical space–time. In fact, the world–sheet approach to (super)string theory takes this interpretation seriously: it is the target space $\mathcal{M}$ which is the physical space, whereas the source space Σ is seen as a mathematical space of (perturbative) parameters [170, 258].

For this interpretation to hold, any theory should have this geometrical feature, without extra requirements on the couplings, like SUSY or conformal invariance. Such additional conditions give differential equations for the target–space fields, which are naturally reinterpreted as space–time equations of motion. For example, the 2D SUSY requirement becomes part of the on–shell condition for the target–space fermions. The equation of motion for the bosons can be seen as the vanishing of the β–function, which – at 1–loop and in the absence of other target–space fields – is just $R_{ij}|_{\text{target space}} = 0$ (cf. Section 1.1.1), i.e. the vacuum Einstein equations. This target–space interpretation is even more general than Lagrangian field theory. In 2D it follows from the Zamolodchikov analysis of the renormalization group [327].

---

[8] For simplicity we work in a unitary trivialization of the bundle $\mathcal{V}$.

1.4 Gauge dualities

The non–scalar fields live in some vector bundle $\mathcal{V} \to \mathcal{M}$. We wish to study in more detail the geometry of these bundles in some relevant case, related to a class of dualities which generalize the usual electromagnetic duality. Again we do not assume supersymmetry. However, historically, the following results were first obtained in the context of supergravity (SUGRA), as they are needed in order to construct the Lagrangian of the maximal extended $\mathcal{N} = 8$ supergravity in 4D (see Cremmer and Julia [98] and also [110, 150]).

We take space–time Σ to have even dimension $D = 2m$ and Minkowski signature $(-, +, +, \cdots, +)$. We consider field theory models whose fields are scalars ϕ^i, parametrizing some target manifold $\mathcal{M}$, and $(m-1)$–form gauge fields B^a, $a = 1, 2, \ldots, n$. The gauge symmetry acts on the fields B^a as

$$B^a(x) \to B^a(x) + d\Lambda^a(x), \tag{1.43}$$

where the $\Lambda^a(x)$ are arbitrary $(m-2)$–forms on Σ. For $m > 1$, the most general gauge–invariant two–derivative Lagrangian is

$$-\frac{1}{2} f_{ab}(\phi) F^a \wedge *F^b + \frac{1}{2} \tilde{f}_{ab}(\phi) F^a \wedge F^b - \frac{1}{2} * g_{ij}(\phi)\, \partial_\mu \phi^i \partial^\mu \phi^j + \cdots, \tag{1.44}$$

where $F^a = dB^a$ is the m–form field–strength of B^a, and $*$ is the Hodge dual with respect to the space–time metric. $f_{ab}(\phi)$ is some positive–definite symmetric matrix depending, in general, on the point $\phi \in \mathcal{M}$, while the coupling $\tilde{f}_{ab}$ is antisymmetric for m odd and symmetric for m even. The case $D = 2$ is special in the sense that zero–forms are "almost" the same thing as scalars. For the present purpose, a zero–form is a scalar σ endowed with a Peccei–Quinn invariance, $\sigma \to \sigma + \text{const}$. With this specification, our arguments apply also to two dimensions.

We are interested in the possible symmetries of the field theory (1.44). The tricky point is that not all the interesting symmetries are invariances of the Lagrangian $\mathcal{L}$. Some symmetries – called *dualities* – hold only at the level of equations of motion. They are also symmetries of the energy–momentum tensor $T_{\mu\nu}$ (Exercise 1.4.2), and hence are physical symmetries (even at the quantum level, when an appropriate quantization exists).

1.4.1 Duality transformations

Before discussing the *symmetries* of the Lagrangian (1.44), we have to discuss the *morphisms* of the formalism, namely the changes of field variables which produce Lagrangians of the same general form. These transformations are the analogue for this class of theories of the target diffeomorphisms for the σ–models, i.e.,

1.4 Gauge dualities

operations which leave invariant the structure of the Lagrangian $-g_{ij}\,\partial_\mu\phi^i\partial\phi^j$, but are not symmetries of the particular field theory unless the diffeomorphism is actually an *isometry* of the metric g_{ij}.

Since it will cost us no more work, we consider a more general Lagrangian,

$$\mathcal{L} = L(F^a, \phi^i, \partial_\mu\phi^j, \Psi^A, \partial_\mu\Psi^A), \tag{1.45}$$

allowing for actions which are not quadratic in the field–strengths F^a, but are still "algebraic" in the sense that they do not contain any derivative of F^a. Thus the arguments of this section apply as well, say, to 4D Dirac–Born–Infeld (DBI) Lagrangians [58]. We also allow generic couplings to other fields Ψ^A which are neither scalars nor m–forms (as spin-$\frac{1}{2}$ fermions, spin-$\frac{3}{2}$ gravitini, k–form fields, and the metric). For example, we allow Pauli–like couplings

$$H_{apq}(\phi)\,F^a_{\mu_1\mu_2\cdots\mu_m}\,\bar\chi^p\gamma^{\mu_1\mu_2\cdots\mu_m}\chi^q \tag{1.46}$$

or generalized Chern–Simons terms. The essential point is that the gauge fields B^a enter in the Lagrangian $\mathcal{L}$ only through their field–strengths F^a.

Since $d^2 = 0$, the m-form field–strengths satisfy the Bianchi identity

$$dF^a = 0. \tag{1.47}$$

We define the dual field–strengths G_a as

$$G_a = *\frac{\partial L}{\partial F^a}. \tag{1.48}$$

The G_a are also m–forms. It is convenient to work with field–strengths F^a, G_a which are eigenforms of the Hodge operator $*$. In a $D = 2m$ dimensional space of Minkowski signature, when acting on m–forms, $*^2 = (-1)^{m-1}$. Hence $*$ has eigenvalues ± 1 in $D = 4k + 2$, and $\pm i$ in $D = 4k$. We write $F^a_\pm$, $G_{a\pm}$ for the components of F^a, G_a on which $*$ acts as multiplication by ± 1 or, respectively, $\pm i$. In the $*$–diagonal basis Eq. (1.48) becomes

$$G_{a\pm} = \pm i\frac{\partial L}{\partial F^a_\pm} \qquad \text{for } D = 4k, \tag{1.49}$$

$$G_{a\pm} = \pm\frac{\partial L}{\partial F^a_\mp} \qquad \text{for } D = 4k + 2. \tag{1.50}$$

Chiral form–fields

From Eqs.(1.49) and (1.50) it follows that in $D = 4k$ one has $(F^a_\pm)^* = F^a_\mp$, while in $D = 4k + 2$, F^a_+ and F^a_- are real and independent. Hence in dimension $D = 4k + 2$ it makes sense to impose the (anti)self–dual condition $F^a_- = 0$ as a physical constraint. Form–fields whose field–strengths have only the self–dual or the anti–self–dual part are called *chiral*. They are tricky both classically (their equations of motion do not follow from an action) and quantum mechanically (they may lead to anomalies [11] and other subtleties [238]); nevertheless, chiral forms *do* enter into some of the most remarkable physical theories such as the world–sheet theory of the heterotic string [174, 175], the space–time theory of Type IIB superstrings [170, 258], and the exotic (2, 0) SCFT theories in six dimensions[9] [275, 314, 316].

The $Sp(2n, \mathbb{R})/O(n_+, n_-)$ structure

The equations of motion of the fields B^a are simply

$$dG_a = 0, \tag{1.51}$$

which have exactly the same form as the Bianchi identities (1.47). Therefore the combined system of equations (1.47) and (1.51) is invariant under any real linear transformation mixing the $2n$ m-forms (F^a, G_b).

We look for transformations which act on the scalars as a diffeomorphism $f \colon \mathcal{M} \to \mathcal{M}$ and on the field–strengths in the form

$$\begin{pmatrix} F^a \\ G_b \end{pmatrix}_\pm \to \begin{pmatrix} \hat{F}^a \\ \hat{G}_b \end{pmatrix}_\pm = \begin{pmatrix} A^a{}_c & B^{ad} \\ C_{bc} & D_b{}^d \end{pmatrix} \begin{pmatrix} F^c \\ G_d \end{pmatrix}_\pm, \tag{1.52}$$

where A, B, C, D are *real* constant $n \times n$ matrices. Such transformations rotate the equations of motion into the Bianchi identities and vice versa. However, not every transformation of the form (1.52) is consistent with the formalism. In fact, to recover the original formulation in the new basis, we also need a new Lagrangian $\hat{L}(\hat{F}^a, \cdots)$ such that

$$\hat{G}_a = *\frac{\partial \hat{L}}{\partial \hat{F}^a}. \tag{1.53}$$

In view of Eqs. (1.49) and (1.50), this condition may be rewritten as

$$\delta \hat{L} = \hat{\eta} := \begin{cases} -i\hat{G}_{a+}\,\delta\hat{F}^a_+ + i\hat{G}_{a-}\,\delta\hat{F}^a_- & \text{for } D = 4k, \\ \hat{G}_{a+}\,\delta\hat{F}^a_- - \hat{G}_{a-}\,\delta\hat{F}^a_+ & \text{for } D = 4k+2. \end{cases} \tag{1.54}$$

Interpreting the symbol δ as the de Rham differential in field–strength space, this equation states that the 1–form $\hat{\eta}$ is δ–*exact*; as an integrability condition, this requires $\hat{\eta}$ to be δ–closed, $\delta\hat{\eta} = 0$. Equivalently, one requires

$$\hat{\eta} - \eta = \delta\Phi \tag{1.55}$$

---

[9] In the conventions of the present book, these theories will be (4, 0) SCFTs (cf. Chapter 2).

where η is defined by the expression on the RHS of Eq. (1.54) with the hatted field–strengths replaced by the unhatted ones. In Eq. (1.55),

$$\Phi = \Phi(F, G, \Psi^A, \partial_\mu \Psi^A) \tag{1.56}$$

is some real function of the F^a, G_a and of the other fields of the theory Ψ^A (and their derivatives). Equation (1.55) is precisely the equation which defines the *symplectomorphisms* [76, 177] (a.k.a. canonical transformations in classical mechanics [18]) with respect to the symplectic pairing

$$\boldsymbol{\Omega} = \begin{cases} \Omega \otimes \sigma_3 \equiv (i\sigma_2 \otimes 1_{n\times n}) \otimes \sigma_3 & D = 4k, \\ \Sigma \otimes i\sigma_2 \equiv (\sigma_1 \otimes 1_{n\times n}) \otimes i\sigma_2 & D = 4k+2, \end{cases} \tag{1.57}$$

where the last factor in the tensor product refers to the $(+,-)$ index, while the tensor product in the parentheses corresponds to writing the $2n \times 2n$ matrices in terms of $n \times n$ blocks as in Eq. (1.52). In the context of classical mechanics, Φ is known as *the Hamilton–Jacobi function* [18].

We conclude that the transformation in Eq. (1.52) should be a real linear symplectomorphism, with respect to $\boldsymbol{\Omega}$, having the special form[10]

$$\mathbf{S} \otimes 1_{2\times 2} \equiv \begin{pmatrix} A & B \\ C & D \end{pmatrix} \otimes 1_{2\times 2}. \tag{1.58}$$

The real $2n \times 2n$ matrix $\mathbf{S}$ then satisfies

$$\mathbf{S}^t \Omega \mathbf{S} = \Omega \qquad \text{for } D = 4k, \tag{1.59}$$

$$\mathbf{S}^t \Sigma \mathbf{S} = \Sigma \qquad \text{for } D = 4k+2. \tag{1.60}$$

By definition, this means that

$$\mathbf{S} \in Sp(2n, \mathbb{R}) \qquad \text{for } D = 4k, \tag{1.61}$$

$$\mathbf{S} \in O(n, n) \qquad \text{for } D = 4k+2, \tag{1.62}$$

since they preserve, respectively, a non–degenerate antisymmetric pairing, Ω, and a symmetric pairing of signature (n, n), Σ.

Alternatively, Eqs. (1.61)(1.62) can be obtained by requiring the invariance of the physical energy–momentum tensor which, being an observable, should not change under field redefinition (see [150] §2.3 for details).

---

[10] The symplectomorphisms have a restricted form since, in order to commute with the Lorentz transformations, they are required not to affect the $\pm$ space–time indices; see Eq. (1.52).

Generalization to chiral field–strengths. As mentioned before, in dimension $D = 4k+2$ we may have chiral $2k+1$ field–strengths and hence, in general, the numbers of F_+^a and F_-^i need not be equal. Let n_+, n_- be, respectively, the number of self–dual and anti–self–dual field–strengths. Then the duality morphism group, which in the symmetric case $n_+ = n_- = n$ was $SO(n,n)$, gets replaced, quite naturally, by

$$SO(n_+, n_-). \tag{1.63}$$

To see this, start with the symmetric case with $n = \max\{n_+, n_-\}$ form fields and simply set to zero in all formulae the $|n_+ - n_-|$ field–strengths which are not present. In the particular case of the world–sheet theory of the heterotic string, the duality group $SO(n_+, n_-)$ arises from the Narain lattices [243, 244] (for a discussion along the present lines see ref. [90]). To simplify the notation, in the following we consider the symmetric case $n_+ = n_- = n$. One recovers the general chiral case by setting $|n_+ - n_-|$ field–strengths to zero.

Symplectic structures for $D = 4k$. An important consequence of the above discussion is that, for $m \equiv D/2$ even, the bundle over $\mathcal{M}$ associated with the $(m-1)$–form fields carries a natural symplectic structure. This is a very useful observation, since symplectic geometry is a rich and well–understood subject. When scalars and $(m-1)$–forms are related by extended supersymmetry, the symplectic nature of the form–fields induces a symplectic structure also on $\mathcal{M}$. In this case the scalar space $\mathcal{M}$ becomes a symplectic manifold (usually with additional structure), and *all* the couplings in the Lagrangian are defined by symplectic geometry (a.k.a. classical mechanics). This structure is powerful enough to lead to a non–perturbative solution of non–Abelian $\mathcal{N} = 2$ gauge theories in 4D; see Seiberg and Witten [273, 274].

Transformation of the Lagrangian

From Eq. (1.55), we see that under duality the Lagrangian is not invariant in value; instead the new and old Lagrangians, $\widehat{L}, L$, are related by

$$\widehat{L} = L + \Phi. \tag{1.64}$$

For a *generic*[11] linear symplectomorphism, the Hamilton–Jacobi function Φ can be chosen to be a *quadratic form* in the field–strengths $F_\pm^a$ and $\widehat{F}_\pm^a$. To save print, we write down explicitly the $D = 4k$ case, leaving $D = 4k + 2$ as an exercise

---

[11] Here *generic* means $\det B \neq 0$, where B is the upper right $n \times n$ block of the square matrix on the RHS of Eq. (1.52).

for the reader. For the same reason we write down only the + component of the field–strengths, and use matrix notation. Then

$$i\Phi = \frac{1}{2}F_+^t M F_+ + \hat{F}_+^t N F_+ + \frac{1}{2}\hat{F}_+^t P \hat{F}_+ + \text{(terms in } F_-, \hat{F}_-\text{)} \quad (1.65)$$

where M, N, P are real $n \times n$ matrices and $M^t = M$, $P^t = P$. From Eq. (1.54),

$$\hat{G}_+ = NF_+ + P\hat{F}_+, \quad (1.66)$$

$$G_+ = -MF_+ - N^t \hat{F}_+, \quad (1.67)$$

or, inverting,

$$\begin{pmatrix} \hat{F}_+ \\ \hat{G}_+ \end{pmatrix} = \begin{pmatrix} -(N^t)^{-1}M & -(N^t)^{-1} \\ N - P(N^t)^{-1}M & -P(N^t)^{-1} \end{pmatrix} \begin{pmatrix} F_+ \\ G_+ \end{pmatrix}. \quad (1.68)$$

It is easy to check that the real $2n \times 2n$ matrix on the RHS satisfies the condition (1.59). Comparing with Eq. (1.52), we get the identifications

$$A = -(N^t)^{-1}M \qquad B = -(N^t)^{-1}, \quad (1.69)$$

$$C = N - P(N^t)^{-1}M \qquad D = -P(N^t)^{-1}, \quad (1.70)$$

or

$$M = B^{-1}A, \qquad P = DB^{-1}, \qquad N = -(B^t)^{-1}. \quad (1.71)$$

Finally, we get

$$\hat{L} = L - i\left(\frac{1}{2}F_+^t B^{-1}A F_+ - \hat{F}_+^t (B^t)^{-1} F_+ + \frac{1}{2}\hat{F}_+^t DB^{-1} \hat{F}_+ - \text{H.c.}\right). \quad (1.72)$$

This equality holds in *value*; to get the effective functional form of the new Lagrangian $\hat{L}$ we need to re–express everything in terms of the $\hat{F}$s, by inverting the transformation.

1.4.2 Duality symmetries

We ask under which condition a duality transformation of the form discussed in Section 1.4.1, is actually a *symmetry* of the theory: that is, $\hat{L}$ and L are identical in their functional form, although they may differ in value. To make things easy, and in view of the applications to the physically relevant models, we assume L to be quadratic in the field–strengths F^a, but for the rest we remain totally general, and in particular we still allow the possibility of arbitrary couplings to other fields Ψ^A. To make the formulae more compact, in this subsection we adopt the following conventions:

(a) Ω stands for the symplectic matrix $(i\sigma_2) \otimes 1_{n\times n}$ if $D = 4k$, while it is the orthogonal matrix $\sigma_1 \otimes 1_{n\times n}$ if $D = 4k + 2$.
(b) Consequently $O_\Omega(2n)$ denotes the group $Sp(2n, \mathbb{R})$ or, respectively, $O(n, n)$.
(c) j is a complex number with value i for $D = 4k$ and 1 for $D = 4k + 2$. In all dimensions, $j\Omega$ is a Hermitian traceless involution, $(j\Omega)^2 = 1$. In particular, $P_\pm = \frac{1 \mp j\Omega}{2}$ are projectors on subspaces of half the dimension.
(d) The shorthand "other fields" stands for any composite m–form polynomial in the Ψ^A and their derivatives, *not* containing the F^a.
(e) We combine the field–strengths F^a and G_b into a column $2n$–vector

$$\mathcal{F} = \begin{pmatrix} F^a \\ G_b \end{pmatrix}. \tag{1.73}$$

The vielbein $\mathcal{E}$

For a Lagrangian quadratic in the F^a, Eqs.(1.49),(1.50) take the form

$$G_{a+} = -j\mathcal{N}_{ab}(\phi) F_+^b + \text{"other fields,"} \tag{1.74}$$

where $\mathcal{N}_{ab}(\phi)$ is some $n \times n$ matrix whose entries are functions of the scalars. Unitarity (positivity of the kinetic energy) implies that $j\mathcal{N}$ is

1. a symmetric matrix whose imaginary part is positive–definite for $D = 4k$;
2. a real matrix whose symmetric part is negative–definite for $D = 4k + 2$.

We rewrite equation (1.74) in a $O_\Omega(2n)$ covariant fashion as

$$(1 - j\Omega)\mathcal{E}(\phi)\mathcal{F}_+ = \text{"other fields."} \tag{1.75}$$

Here $\mathcal{E}(\phi)$ – called in the SUGRA literature the *vielbein*[12] [110] – is an element of the group $O_\Omega(2n)$ whose matrix elements depend on the scalars ϕ^i. The vielbein $\mathcal{E}(\phi)$ encodes all the couplings of the F^a with the scalars; indeed, writing

$$\mathcal{E} = \begin{pmatrix} \mathcal{A} & \mathcal{B} \\ \mathcal{C} & \mathcal{D} \end{pmatrix}, \tag{1.76}$$

where $\mathcal{A}, \mathcal{B}, \mathcal{C}, \mathcal{D}$ are $n \times n$ matrices, Eq. (1.75) becomes

$$G_+ = (j\mathcal{D} - \mathcal{B})^{-1}(\mathcal{A} - j\mathcal{C})F_+ + \text{"other fields,"} \tag{1.77}$$

---

[12] The name is appropriate. Geometrically speaking, $\mathcal{E}$ is a bundle isomorphism whose effect is to reduce the structure group from $O_\Omega(2n)$ to its maximal compact subgroup. It should be compared with Cartan's vielbein for a Riemannian manifold, whose purpose is to reduce the structure group of the tangent bundle from $GL(n, \mathbb{R})$ to its maximal compact subgroup $O(n)$.

so Eqs. (1.74) and (1.75) are equivalent provided

$$j\mathcal{N} = (\mathcal{B} - j\mathcal{D})^{-1}(\mathcal{A} - j\mathcal{C}). \tag{1.78}$$

We have still to show that the vielbein $\mathcal{E}$ is an element of $O_\Omega(2n)$. It follows from the following very well–known lemma.

Lemma 1.5 *(1) All complex symmetric matrices $i\mathcal{N}$ whose imaginary part is positive–definite can be written in the form* (1.78) *for some $\mathcal{E} \in Sp(2n, \mathbb{R})$; (2) all real matrices $\mathcal{N}$ whose symmetric part is negative–definite are of the form* (1.78) *for some $\mathcal{E} \in O(n, n)$.*

Proof Choose $\mathcal{E}$ in the special form

$$\mathcal{E} = \begin{pmatrix} Q^{-1}M & Q^{-1} \\ \mp Q^t & 0 \end{pmatrix} \quad \text{with} \quad \begin{cases} M^t = \pm M, \\ \det Q \neq 0. \end{cases} \tag{1.79}$$

It is easy to check that $\mathcal{E}$ is an element of $Sp(2n, \mathbb{R})$ (resp. $O(n, n)$) for the upper (resp. lower) sign. Then Eq. (1.78) gives

$$j\mathcal{N} = \begin{cases} M + iQQ^t & \text{for } Sp(2n, \mathbb{R}), \\ M - QQ^t & \text{for } O(n, n). \end{cases} \tag{1.80}$$
□

Next we ask when two distinct vielbeins $\mathcal{E}$ and $\widetilde{\mathcal{E}}$ correspond to the same coupling matrix $\mathcal{N}_{ab}$. Let $\mathbf{S} \in O_\Omega(2n)$ be an element commuting with $j\Omega$. From Eq. (1.75) it is obvious that $\mathcal{E}$ and $\mathbf{S}\mathcal{E}$ give equivalent constraints on the field–strengths $\mathcal{F}$ (provided we also multiply by $\mathbf{S}$ the "other fields" on the RHS). Hence multiplication on the left of the vielbein by an element of the centralizer of $j\Omega$ leaves $\mathcal{N}_{ab}(\phi)$ invariant. One has

Lemma 1.6 *The centralizer of $j\Omega$ in $O_\Omega(2n)$ is $U(n)$ in the $Sp(2n, \mathbb{R})$ case, and $O(n) \times O(n)$ in the $O(n, n)$ one. In both cases it is a maximal compact subgroup $K_\Omega(2n) \subset O_\Omega(2n)$.*

Proof A simple computation shows that an element of the normalizer should have the form

$$\begin{pmatrix} A & B \\ \mp B & A \end{pmatrix} \quad \text{with } A^t A + B^t B = 1, \ (B^t A)^t = \pm B^t A, \tag{1.81}$$

upper sign for $Sp(2n, \mathbb{R})$, lower one for $SO(n, n)$. So

$$(A \mp iB)^t (A \pm iB) = 1 \qquad Sp(2n, \mathbb{R}) \text{ case} \tag{1.82}$$

$$(A \pm B)^t (A \pm B) = 1 \qquad O(n, n) \text{ case}; \tag{1.83}$$

hence, in the first case $U = A + iB \in U(n)$, while in the second one $O_{\pm} = A \pm B$ belong to two distinct copies of $O(n)$. □

The crucial point is that the coupling matrices $\mathcal{N}_{ab}(\phi)$ allowed by unitarity are in one–to–one correspondence with the points of the coset spaces

$$\frac{Sp(2n, \mathbb{R})}{U(n)} \qquad D = 4k, \tag{1.84}$$

$$\frac{SO_0(n, n)}{SO(n) \times SO(n)} \qquad D = 4k + 2, \tag{1.85}$$

which are well–known rank n Riemannian *symmetric spaces*, whose differential geometry is well understood. The (non–linear) action of the group $O_\Omega(2n)$ on the coupling $\mathcal{N}_{ab}(\phi)$ is simply the canonical transitive action of $O_\Omega(2n)$ on the coset space $O_\Omega(2n)/K_\Omega(2n)$. Since the transitive symmetry $O_\Omega(2n)$ is the group of duality transformations (Section 1.4.1), which preserve the energy–momentum, hence the positivity of energy, the converse to Lemma 1.5 is also true: the matrices $\mathcal{N}$ defined by Eqs. (1.76) and (1.78) are automatically consistent with positivity of the kinetic terms for *all* $\mathcal{E} \in O_\Omega(2n)$.

Geometry of the spaces $O_\Omega(2n)/K_\Omega(2n)$

Let us recall the geometric meaning[13] of the two symmetric spaces (1.84), (1.85). Consider $\mathbb{R}^{2n}$ equipped with the symplectic structure given by our (constant) matrix Ω. A *Lagrangian subspace* $L \subset \mathbb{R}^{2n}$ is an n–dimensional linear subspace such that $\Omega|_L = 0$. $Sp(2n, \mathbb{R})/U(n)$ is the manifold parametrizing the family of all Lagrangian subspaces of $\mathbb{R}^{2n}$. Indeed, this identification is precisely what we proved in our two lemmas. By definition, a point in $Sp(2n, \mathbb{R})/U(n)$ is an equivalence class of vielbeins (two vielbeins $\mathcal{E}, \mathcal{E}'$ being equivalent iff $\mathcal{E}' \sim S\mathcal{E}$ with $S \in U(n)$); the lemmas say that specifying a class of vielbeins is equivalent to specifying which n–dimensional linear subspace $\mathfrak{F}$ of the $2n$–dimensional field–strength space we interpret as curvatures, $F^a = dA^a$. From Eq. (1.57) we see that $\Omega|_{\mathfrak{F}} = 0$, so $\mathfrak{F} \subset \mathbb{R}^{2n}$ is a Lagrangian subspace, and any Lagrangian subspace L may play the role of $\mathfrak{F}$ since the group of duality rotations, $Sp(2n, \mathbb{R})$, acts *transitively* on the set of Lagrangian subspaces. Equivalently, the vielbein class $[\mathcal{E}]$ specifies a choice of *mutually local* form–fields B^a; in the present context, a Lagrangian subspace $\mathfrak{F} \subset \mathbb{R}^{2n}$ encodes precisely a maximal set of mutually local form–fields.

---

[13] Further details on the geometry of these spaces are given in Section 1.5 and Chapters 5 and 11.

1.4 Gauge dualities

Seeing $Sp(2n, \mathbb{R})/U(n)$ as the family of Lagrangian subspaces, we have a *tautological* rank n vector bundle $\mathfrak{T} \xrightarrow{\pi} Sp(2n, \mathbb{R})/U(n)$

$$\mathfrak{T} := \{(L, x) \mid L \in Sp(2n, \mathbb{R})/U(n), \ x \text{ a point in } L\}, \tag{1.86}$$

$$\pi : \mathfrak{T} \to Sp(2n, \mathbb{R})/U(n) \quad \text{given by } (L, x) \mapsto L. \tag{1.87}$$

The tautological bundle $\mathfrak{T}$ is a first example of a *homogeneous* bundle on a symmetric space. These are the bundles we can construct using the isometry group of the coset space; they come equipped with canonical metrics and connections, and will play a crucial role in this book (see Chapters 5 and 11).

A similar description holds for $SO(n,n)/SO(n) \times SO(n)$. This space parametrizes the n–dimensional linear subspaces $N \subset \mathbb{R}^{2n}$ which are *null* (isotropic) with respect to the indefinite inner product Σ; that is, such that $\Sigma|_N = 0$. Again we have a *tautological* bundle $\mathfrak{T} \xrightarrow{\pi} SO(n,n)/SO(n) \times SO(n)$ defined as in Eqs. (1.86) and (1.87).

We summarize the results of this subsection in the following General Lesson:

General Lesson 1.7 Let $F^a = dA^a$ be the m–form field–strengths of a $D = 2m$ field theory whose Lagrangian $\mathcal{L}$ is quadratic in the F^a. The Lagrangian coupling $\mathcal{N}_{ab}(\phi) F^a F^b/2$ defines a map of Riemannian manifolds[14]

$$\mu : \mathcal{M} \to \begin{cases} Sp(2n, \mathbb{R})/U(n) & D = 4k, \\ SO(n_+, n_-)/SO(n_+) \times SO(n_-) & D = 4k+2, \end{cases} \tag{1.88}$$

given by[15]

$$\mu : \phi^i \mapsto \mathcal{E}^{-1} H, \qquad \text{where } H = \begin{cases} U(n) & D + 4k, \\ SO(n) \times SO(n) & D = 4k+2, \end{cases} \tag{1.89}$$

and $[\mathcal{E}]$ has a representative as in Eq. (1.79) with

$$M = \begin{cases} \text{Re}(i\mathcal{N}) \\ (\mathcal{N} - \mathcal{N}^t)/2 \end{cases} \quad Q = \begin{cases} (\text{Im}(i\mathcal{N}))^{1/2} & D = 4k, \\ (-(\mathcal{N} + \mathcal{N}^t)/2)^{1/2} & D = 4k+2. \end{cases} \tag{1.90}$$

CONVERSELY, *any such map μ specifies a unique Lagrangian coupling.*

All bundles on $\mathcal{M}$ which describe the couplings of the F^a to other fields $(\phi, \chi, \psi_\mu, \ldots)$ are the pull–back by μ of <u>homogeneous</u> bundles over the coset space in Eq. (1.88). In particular, the Abelian curvatures $F^a \equiv dA^a$, $G_a \equiv *\partial L/\partial F^a$ take value in the pull–back $\mu^*\mathfrak{T}$ of the <u>flat</u> bundle $\mathfrak{T}$.[16]

---

[14] We call it μ since in the Maxwell theory it corresponds to the magnetic susceptibility.
[15] We use $\mathcal{E}^{-1}$ instead of $\mathcal{E}$ to convert the right $O_\Omega(2n)$ action into the standard left action.
[16] In the language of [77] $\mathfrak{T}$ is the vector bundle associated by the fundamental representation with the *principal Higgs bundle* over $O_\Omega(2n)/K_\Omega(2n)$.

When a duality transformation is a symmetry

Armed with General Lesson 1.7, we can easily answer this basic question. A duality transformation $\mathbf{S} \in O_\Omega(n)$ acts on the field–strengths as

$$\mathcal{F}_\pm \to \mathbf{S}\mathcal{F}_\pm, \tag{1.91}$$

so, in view of Eq. (1.75), it can be compensated by a change in the vielbein $\mathcal{E}$ of the form

$$\mathcal{E} \to \mathcal{E}\mathbf{S}^{-1}. \tag{1.92}$$

Suppose there exists an isometry $\mathcal{M} \xrightarrow{s} \mathcal{M}$ of the scalar manifold such that

$$\mu \circ s = \mathbf{S} \circ \mu \tag{1.93}$$

that is, explicitly,

$$\mathcal{E}(s^i(\phi)) = h(\phi^i)\,\mathcal{E}(\phi^i)\,\mathbf{S}^{-1} \quad \text{where } h(\phi^i) \in K_\Omega(2n). \tag{1.94}$$

Then the transformation

$$\mathcal{F}_\pm \to \mathbf{S}\mathcal{F}_\pm, \qquad \phi^i \to s^i(\phi) \tag{1.95}$$

is a symmetry of both the scalars' and the forms' kinetic terms. In fact, it is an invariance of the *full* Lagrangian provided the expression on the RHS of Eq. (1.75), "*other fields,*" transforms covariantly under $K_\Omega(2n)$; that is, if the other fields live in the (pull–back of) right homogeneous bundles and the remaining couplings of the scalars preserve the isometry.

A prototypical example. In order to get a model with a duality *symmetry*, we have to construct Lagrangian couplings that satisfy Eq. (1.93). To find explicit solutions to equation (1.93) is a formidable task. Luckily there are instances where we get solutions for free. The obvious example, which is the general case for 4D supergravity with $\mathcal{N} \geq 3$, is when the scalar manifold $\mathcal{M}$ is itself a symmetric space G/H where G is a subgroup of $Sp(2n, \mathbb{R})$ or $O(n, n)$ and $H = G \cap U(n)$ or, respectively, $H = G \cap (SO(n) \times SO(n))$. The metric on $\mathcal{M}$ is taken to be the symmetric one, which has a group G of isometries acting by left multiplication. As the map μ one takes that induced by the subgroup embedding $G \xrightarrow{i} O_\Omega(2n)$. For, say, $D = 4k$ the map μ is defined by the commutative diagram

$$\begin{array}{ccc} G & \xrightarrow{i} & Sp(2n, \mathbb{R}) \\ \pi_{\text{can}} \downarrow & & \downarrow \pi_{\text{can}} \\ \mathcal{M} = G/(G \cap U(n)) & \xrightarrow{\mu} & Sp(2n, \mathbb{R})/U(n), \end{array} \tag{1.96}$$

where the vertical maps are canonical projections.

In this case the property (1.93) holds by construction, and we have a full group G of symmetries.[17] This duality structure is a first reason for the ubiquity of coset spaces G/H in supergravity, superstrings, etc. It is not the only one, as we shall see. Equation (1.96) is also an example of how Lagrangian couplings (here μ) are more easily understood as maps, since they fit in larger diagrams which make the properties of the coupling more transparent.

Exercise 1.4.1 Check that the matrix in Eq. (1.68) belongs to $Sp(2n, \mathbb{R})$.

Exercise 1.4.2 Work out the details of the transformation of the Lagrangian under duality for the case $D = 4k + 2$.

Exercise 1.4.3 Show that the energy–momentum tensor $T_{\mu\nu}$ is invariant *in value* under an arbitrary duality transformation.

1.5 The emergence of modularity

From the results of Section 1.4 it follows that many physical systems have a "modular" structure in a sense that we now make precise (more details in Chapters 4, 5, 11). For simplicity, here we limit ourselves to the case of $D = 4k$.

Without doubt the reader will have noticed that Eq. (1.78) resembles the modular transformation of the period matrix $\Omega_{\alpha\beta}$ of a Riemann surface. This is not a coincidence: it is a deep fact. We start by recalling the basics.

1.5.1 Period matrix of a Riemann surface

On a genus g Riemann surface σ we have g linearly independent holomorphic differentials ω_α, $\alpha = 1, 2, \ldots, g$. On $H_1(\sigma, \mathbb{Z}) \simeq \mathbb{Z}^{2g}$ we have a non–degenerate symplectic pairing $\langle \cdot, \cdot \rangle$, given by intersection, which is dual to the bilinear form $\int_\sigma \alpha \wedge \beta$ in cohomology. We choose a canonical (symplectic) basis of 1–cycles (A_α, B^β) $(\alpha, \beta = 1, 2, \ldots, g)$ satisfying

$$\langle A_\alpha, A_\beta \rangle = \langle B^\alpha, B^\beta \rangle = 0, \qquad \langle A_\alpha, B^\beta \rangle = -\langle B^\beta, A_\alpha \rangle = \delta_\alpha^\beta. \tag{1.97}$$

Note that two choices of canonical bases in $H_1(\sigma, \mathbb{Z})$ differ by an integral $Sp(2g, \mathbb{Z})$ transformation. The holomorphic differentials are normalized by fixing their

---

[17] This is true at the level of kinetic terms; other couplings may break part of G. In 4D SUGRA only the gauge couplings may spoil the G symmetry, as we shall explain later in the book.

integrals on the A–cycles,

$$\int_{A_\alpha} \omega_\beta = \delta_{\alpha\beta}. \tag{1.98}$$

The integrals over the B–cycles then define the period matrix $\Omega_{\alpha\beta}$,

$$\int_{B^\alpha} \omega_\beta = \Omega_{\alpha\beta}. \tag{1.99}$$

One has [136] the following theorem.

Theorem 1.8 (Riemann bilinear relations) *The period matrix is symmetric $\Omega_{\alpha\beta} = \Omega_{\beta\alpha}$ and the real quadratic form $\mathrm{Im}\,\Omega_{\alpha\beta}$ is positive–definite.*

Notice that these are exactly the properties of the coupling matrix $j\mathcal{N}$ for $D = 4k$ (see discussion after Eq. (1.74)).

The space of complex $n \times n$ symmetric matrices with positive–definite imaginary part is called *Siegel's upper half–space*, $\mathfrak{H}_n$ [69]. For $n = 1$ it is the upper half–plane $\mathfrak{h} = \{z \in \mathbb{C} \mid \mathrm{Im}\,z > 0\}$. Lemmas 1.5, and 1.6 then give

$$\mathfrak{H}_n \simeq \frac{Sp(2n, \mathbb{R})}{U(n)}. \tag{1.100}$$

The symplectic basis (1.97) is not unique. We get all of them by a linear transformation of the form

$$\begin{pmatrix} B^\alpha \\ A_\beta \end{pmatrix} \mapsto \begin{pmatrix} \mathcal{A}^\alpha{}_\gamma & \mathcal{B}^{\alpha\delta} \\ \mathcal{C}^\gamma_\beta & \mathcal{D}_\beta{}^\delta \end{pmatrix} \begin{pmatrix} B^\gamma \\ A_\delta \end{pmatrix}, \tag{1.101}$$

where the $2g \times 2g$ square matrix on the RHS should have integral entries (in order to transform *integral* cycles into integral cycles) and be symplectic (to preserve the pairing in Eq. (1.97)). Thus it is an element of the arithmetic subgroup $Sp(2g, \mathbb{Z}) \subset Sp(2g, \mathbb{R})$ called the *Siegel modular group*.

How does $Sp(2g, \mathbb{R})$ act on the period matrix Ω? In block–matrix notation,

$$\begin{pmatrix} \int_B \omega \\ \int_A \omega \end{pmatrix} \equiv \begin{pmatrix} \Omega \\ 1 \end{pmatrix} \int_A \omega \longrightarrow \begin{pmatrix} \mathcal{A} & \mathcal{B} \\ \mathcal{C} & \mathcal{D} \end{pmatrix} \begin{pmatrix} \int_B \omega \\ \int_A \omega \end{pmatrix} \equiv \begin{pmatrix} \Omega' \\ 1 \end{pmatrix} \int_{A'} \omega, \tag{1.102}$$

that is,

$$\Omega' = (\mathcal{A}\Omega + \mathcal{B})(\mathcal{C}\Omega + \mathcal{D})^{-1}. \tag{1.103}$$

Lemmas 1.5 and 1.6 state that all Ωs satisfying Riemann's bilinear relations can be obtained from any given one, say $\Omega^0_{\alpha\beta} \equiv i\delta_{\alpha\beta}$, by an $Sp(2g, \mathbb{R})$ transformation, while two transformations differing by multiplication (on the right) by an element

1.5 The emergence of modularity

of the $U(g)$ subgroup give the same periods. The periods Ω characterize the complex structure of the Riemann surface σ. However, we see that two distinct Ωs related by an *integral* $Sp(2g, \mathbb{Z})$ correspond to the *same* Riemann surface σ (and different choices of the canonical basis in $H_1(\sigma, \mathbb{Z})$). This $Sp(2g, \mathbb{Z})$ equivalence of the periods Ω is called *modular invariance*. It appears in physics in all cases in which a quantity f should depend only on the Riemann surface and not on the choice of a symplectic basis: in this case $f(\Omega) = f(\Omega')$ whenever Ω, Ω' are related by an $Sp(2g, \mathbb{Z})$ transformation.

Comparison with Eq. (1.78). The modular structure of the periods Ω is identical to the one we encountered in Eq. (1.78). To relate the two we have just to change our conventions. We interchange the role of F_+ and G_+, taking the dual field–strengths G_+ as fundamental. Then we write

$$\mathcal{F}_+ = \begin{pmatrix} \Omega G_+ \\ G_+ \end{pmatrix}, \tag{1.104}$$

where the "period matrix" $\Omega \equiv i\mathcal{N}^{-1}$. The coupling Ω transforms under $Sp(2n, \mathbb{R})$ exactly as in Eq. (1.103). To make contact with the original form of Eq. (1.78), notice that, since $\mathcal{N}$ is symmetric, Ω can be rewritten as

$$\Omega \equiv (-i\mathcal{N})^{-1} = (i\mathcal{D}^t - \mathcal{B}^t)(-i\mathcal{C}^t + \mathcal{A}^t)^{-1}, \tag{1.105}$$

which is exactly the right expression if we recall that a vielbein $\mathcal{E}$ should be interpreted as the duality transformation $\mathcal{E}^{-1}$ and

$$\begin{pmatrix} \mathcal{A} & \mathcal{B} \\ \mathcal{C} & \mathcal{D} \end{pmatrix}^{-1} = \begin{pmatrix} \mathcal{D}^t & -\mathcal{B}^t \\ -\mathcal{C}^t & \mathcal{A}^t \end{pmatrix}. \tag{1.106}$$

The Cayley transformation

The space $Sp(2n, \mathbb{R})/U(n)$ coincides with the *Siegel's upper half–space*[18]

$$\mathfrak{H}_n := \{Z \in \mathbb{C}(n) \mid Z^t = Z, \ \mathrm{Im}\, Z > 0\}. \tag{1.107}$$

For $n = 1$, this is just the upper half–plane $\mathfrak{H} = \{z \in \mathbb{C}, z = x + iy, y > 0\}$. As is well known, the upper half–plane $\mathfrak{H}$ is conformally equivalent to the open unit disk

$$\mathfrak{D} = \{w \in \mathbb{C} \mid ww^* < 1\}. \tag{1.108}$$

---

[18] For more details, see ref. [69] pages 221–230.

In this second representation, the compact $U(1)$ isometry of $\mathfrak{H}$ acts simply as a rotation $w \mapsto e^{i\alpha} w$. The analogue of the disk $\mathfrak{D}$ for $n > 1$ is the space

$$\mathfrak{D}_n := \{W \in \mathbb{C}(n) \mid W^t = W, \; 1 - WW^* > 0\}. \tag{1.109}$$

The convenience of $\mathfrak{D}_n$ with respect to $\mathfrak{H}_n$ is twofold. It block–diagonalizes the representation of the subgroup $U(V) \in Sp(2V, \mathbb{R})$, leading to more transparent formulae, and it replaces the *unbounded* domain $\mathfrak{H}_n \subset \mathbb{C}^{n(n+1)/2}$ with the *bounded* domain $\mathfrak{D}_n \subset \mathbb{C}^{n(n+1)/2}$, allowing us to use the powerful machinery of bounded domain technology to study it [71].

To map $\mathfrak{H}_n$ into $\mathfrak{D}_n$, we conjugate $\mathcal{E}$ in $Sp(2n, \mathbb{C})$. That is, we let $\mathcal{E} \mapsto C\mathcal{E}C^{-1}$, where C, the Cayley transform, is the $Sp(2n, \mathbb{C})$ matrix

$$C = \frac{1}{\sqrt{2i}} \begin{pmatrix} 1_n & -i1_n \\ 1_n & i1_n \end{pmatrix}. \tag{1.110}$$

The new $\mathcal{E}$ reads

$$\mathcal{E} = \begin{pmatrix} U & V \\ V^* & U^* \end{pmatrix}, \tag{1.111}$$

where $2U = (\mathcal{A} + \mathcal{D}) + i(\mathcal{B} - \mathcal{C})$ and $2V = (\mathcal{A} - \mathcal{D}) - i(\mathcal{B} + \mathcal{C})$. Now the action of $h \in H$ is simply matrix multiplication

$$V \mapsto hV, \qquad U \mapsto hU. \tag{1.112}$$

In particular, the projectors $P_\pm = \frac{1}{2}(1 \mp i\Omega)$ are now diagonal:

$$P_\pm = \frac{1}{2}(1 \pm \Sigma_3). \tag{1.113}$$

In the Cayley basis, $Sp(2n, \mathbb{R})$ is the set of complex matrices such that

$$\mathcal{E}^* = \Sigma_1 \mathcal{E} \Sigma_1, \qquad \mathcal{E}^t \Omega \mathcal{E} = \Omega, \tag{1.114}$$

where

$$\Sigma_1 = \begin{pmatrix} 0 & 1 \\ 1 & 0 \end{pmatrix}, \qquad \Omega = \begin{pmatrix} 0 & 1 \\ -1 & 0 \end{pmatrix}, \qquad \Sigma_3 = \begin{pmatrix} 1 & 0 \\ 0 & -1 \end{pmatrix}. \tag{1.115}$$

The symplectic property now reads

$$U^t V^* - V^\dagger U = 0, \tag{1.116}$$

$$U^t U^* - V^\dagger V = 1. \tag{1.117}$$

1.5.2 Hodge decomposition

We may give a different interpretation to the period matrix Ω. The complex vector space $H^1(\sigma, \mathbb{C}) \simeq \mathbb{C}^{2g}$ has a natural real structure given by the identification $H^1(\sigma, \mathbb{C}) \simeq H^1(\sigma, \mathbb{R}) \otimes \mathbb{C}$. $H^1(\sigma, \mathbb{C})$ decomposes into the direct sum of the vector spaces $H^{1,0}$ and $H^{0,1}$ spanned respectively by the holomorphic and anti–holomorphic 1–forms

$$H^1(\sigma, \mathbb{C}) = H^{1,0} \oplus H^{0,1} \quad \text{and} \quad H^{0,1} = \overline{H^{1,0}} \tag{1.118}$$

(the bar standing for complex conjugation with respect to the real structure). The Riemann relations imply that: (i) the real symplectic form $\langle \cdot, \cdot \rangle$ vanishes when restricted to $H^{1,0}$, while (ii) the Hermitian form $-i \langle \cdot, \overline{\cdot} \rangle$ is positive–definite on $H^{1,0}$. Giving a period matrix Ω is equivalent to specifying a decomposition of $H^1(\sigma, \mathbb{C}) \simeq \mathbb{C}^{2g}$, as in Eq. (1.118) satisfying conditions (i) and (ii). Such a decomposition is called a *Hodge decomposition* [173].

Therefore, the coupling matrix $i\mathcal{N}$ for the Abelian forms may be seen as the specification of a Hodge decomposition. The Hodge point of view is the most intrinsic and general, as we shall see later in the book.

Generalization to higher-dimensional varieties

The above discussion may be generalized from curves (i.e., Riemann surfaces) to algebraic varieties X of complex dimension n. The relevant definitions are given in Chapter 11.[19]

One considers the cohomology space in middle dimension $H \equiv H^n(X, \mathbb{C})$. For n odd the intersection pairing on H is a symplectic form, while for n even it gives a symmetric quadratic form of signature (m_+, m_-). $H \simeq H^n(X, \mathbb{R}) \otimes \mathbb{C}$ has a natural real structure and a Hodge decomposition

$$H = \bigoplus_{p+q=n} H^{p,q}, \quad \text{with} \quad \overline{H^{p,q}} = H^{q,p}, \tag{1.119}$$

satisfying the appropriate Riemann bilinear relations (Chapters 9, 11). Writing[20] $2m := B_n(X)$ for n odd (resp. $m_\pm := B_\pm(X)$ for n even) and setting $h^{p,q} := \dim H^{p,q}$, the space of Hodge decompositions with given dimensions $\{h^{p,q}\}$ is

---

[19] The reader may prefer to skip this subsection in a first reading.
[20] $B_k(X)$ denotes the kth Betti number of X, i.e., the dimension of the vectors space of harmonic k–forms on X. $B_+(X)$ (resp. $B_-(X)$) is the dimension of the space of self–dual (resp. anti–self–dual) harmonic n–forms.

parametrized by the cosets

$$Sp(2m, \mathbb{R}) \Big/ \prod_{p<n/2} U(h^{p,n-p}) \qquad n \text{ odd},$$

$$SO(m_+, m_-)_0 \Big/ SO(h^{n/2,n/2}) \prod_{p<n/2} U(h^{p,n-p}) \quad n \text{ even},$$
(1.120)

called the *Griffiths period domains* [173]. For n odd we have a forgetful map

$$Sp(2m, \mathbb{R}) \Big/ \prod_{p \leq (n-1)/2} U(h_{p,n-p}) \xrightarrow{\varpi} Sp(2m, \mathbb{R})/U(m), \qquad (1.121)$$

given by the *coarser* decomposition

$$H = F \oplus \overline{F}, \quad \text{where } F = \bigoplus_{q \text{ even}} H^{n-q,q}. \qquad (1.122)$$

This has the following consequence. Suppose we have a natural map ξ from the scalar manifold $\mathcal{M}$ to a Griffith period domain $Sp(2m, \mathbb{R})/H$. Composing with ϖ, we get a map $\mu := \varpi \circ \xi$ which, according to the previous section (in, say, 4D), defines a Lagrangian coupling between Abelian gauge vectors and scalars. The coupling μ is obtained by ignoring some extra structure encoded in the *finer* Hodge decomposition (1.119). Hence the resulting vector–scalar couplings will have extra properties besides those described in Section 1.4: we end up with *special* vector–scalar couplings. We shall show in Chapters 11 and 12 that the couplings for 4D $\mathcal{N} = 2$ SUGRA arise precisely in this way. The same remark applies to n even and $D = 4k + 2$.

1.5.3 Arithmetic structure and Dirac quantization

In the case of Riemann surfaces, the arithmetical subgroup $Sp(2n, \mathbb{Z}) \subset Sp(2n, \mathbb{R})$ played a distinguished role. Is there a similar arithmetic structure for gauge dualities? *Yes,* there is. For simplicity, we illustrate this in the special case of 4D, in which our theories are just Abelian gauge theories. In Section 1.4 we were somewhat naive: to fully specify an Abelian gauge theory; it is not enough to give a Lagrangian: in addition, one has to fix the *global* geometry of the gauge group G which may be compact, $U(1)^n$, or non–compact, say $G = \mathbb{R}^n$. The Lagrangian is the same in the two cases, but if the space–time Σ has non–trivial topology the two theories are physically inequivalent, since the path integral is defined as a sum over different topological sectors. In particular, for $G = U(1)^n$, the magnetic fluxes

1.5 The emergence of modularity

concatenated with all closed surfaces $\sigma_\ell \subset \Sigma$ are integrally quantized,

$$\frac{1}{2\pi}\int_{\sigma_\ell} F^a = n^a{}_\ell \in \mathbb{Z}, \qquad (1.123)$$

while for G non–compact the magnetic flux must vanish. Equation (1.123) just expresses the fact that the first Chern class of the line bundles associated with the $U(1)^n$ gauge symmetries is integral [60, 94, 186, 235].

In the compact case all symmetries must preserve the flux quantization (1.123). Then a necessary condition for non–trivial duality symmetries is that the dual fluxes $\int_{\sigma_\ell} *G_a$ are also integrally quantized: in General Lesson 1.11 it will be shown that this is automatically true. Then setting

$$\mathcal{F} = \begin{pmatrix} *G_a \\ F^a \end{pmatrix}, \qquad (1.124)$$

we have $\int_{\sigma_\ell} \mathcal{F}/2\pi \in \mathbb{Z}^{2n}$ for all closed surfaces σ_ℓ. Although the equations of motion are invariant under arbitrary linear $Sp(2n, \mathbb{R})$ transformations $\mathcal{F} \to \mathbf{S}\mathcal{F}$, only matrices $\mathbf{S}$ *with integral coefficients* also preserve Dirac's quantization condition. Thus, in the compact case, the actual group of duality symmetry must be a subgroup of Siegel's modular group $Sp(2n, \mathbb{Z})$.

Consider, in particular, the case in which the matrix $\mathcal{N}_{ab}$ is a constant (so that the theory is free). If the gauge group is non–compact, $\mathbb{R}^n$, all $\mathcal{N}_{ab}$ are physically equivalent since we may always reduce to the case $\mathcal{N}_{ab} = i\delta_{ab}$ by a $Sp(2n, \mathbb{R})$ field–redefinition; on the contrary, in the compact $U(1)^n$ case only $Sp(2n, \mathbb{Z})$ dualities are admissible field–redefinitions and the physically inequivalent $\mathcal{N}_{ab}$ are parametrized by the modular double coset

$$Sp(2n, \mathbb{Z})\backslash Sp(2n, \mathbb{R})/U(n). \qquad (1.125)$$

Physical observables are then Siegel's modular functions, and the partition function is a (non–holomorphic) Siegel's modular form [313].

The quantization of the dual fluxes may be alternatively understood as a consequence of Gauss's law: if the surfaces σ_ℓ are away from the sources, $\int_{\sigma_\ell} *G_a$ measures the electric charge inside σ_ℓ, which is integrally quantized. Indeed, this is precisely the argument of Dirac's quantization of electric charge in presence of magnetic monopoles [128, 238]. The typical situation in which the above discussion becomes relevant is when we have some microscopic theory that at energies $\ll \Lambda$ reduces to an effective Abelian gauge theory with gauge group $U(1)^n$. Then only the Abelian fields propagate at distances $\gg \Lambda^{-1}$, and the infrared physics is just Maxwell theory. Let us pick a large sphere S^2 of radius $R \gg \Lambda^{-1}$ and consider the integral

$$\frac{1}{2\pi}\int_{S^2} \mathcal{F} = \begin{pmatrix} e \\ m \end{pmatrix}, \qquad (1.126)$$

where e and m are, respectively, the electric and magnetic charges inside S^2. In this situation we have both particles carrying electric charges and particles carrying magnetic charges which should be quantized in integral Dirac units. Then the $2n$–vector (1.126) takes value in a lattice $\Gamma \simeq \mathbb{Z}^{2n}$. This lattice has a non–degenerate integral symplectic pairing

$$\langle \gamma, \gamma' \rangle = -\langle \gamma', \gamma \rangle \in \mathbb{Z} \quad \forall\, \gamma, \gamma' \in \Gamma, \qquad (1.127)$$

induced by the matrix Ω (cf. Eq.(1.57)) of the low–energy effective Abelian theory. The integral symplectic form (1.127) is called the *Dirac electromagnetic pairing*. Two particles are *mutually local*; that is, they may be simultaneously described by local (Lagrangian) fields, iff their charges γ, γ' have zero Dirac pairing, $\langle \gamma, \gamma' \rangle = 0$. From Eq. (1.126) it is obvious that an $Sp(2n, \mathbb{R})$ rotation $\mathcal{F} \to \mathbf{S}\mathcal{F}$ preserves the symplectic lattice Γ precisely if it belongs to the Siegel modular subgroup $Sp(2n, \mathbb{Z})$. In summary:

General Lesson 1.9 *The symmetry under $Sp(2n, \mathbb{R})$ or $SO(n_+, n_-)$ arises quite universally in Abelian gauge theories. Global topological restrictions or other interaction (e.g., gauge couplings) may spoil the invariance, leaving unbroken a discrete subgroup, congruent to $Sp(2n, \mathbb{Z})$ or $SO(n_+, n_-, \mathbb{Z})$. Thus the theory gets, on quite general grounds, modular invariance.*

An important application of this observation is to the analysis of the moduli space of compactifications of the superstring. It leads to the "target space modular invariance" [162].

Exercise 1.5.1 Show that the Cayley transformation C maps the region (1.107) of $\mathbb{C}^{n(n+1)/2}$ into the region (1.109).

1.6 More dualities

The dualities described in the previous sections are morphisms of the formalism which transform a Lagrangian into another one of the same "geometric" form. Under suitable circumstances, a (sub)group of dualities may be a physical symmetry. There are other dualities which map a Lagrangian with a given field content to another one with a *different* field content (and hence are *never* sysmmetries). The

1.6 More dualities

use of these dualities is to change the Lagrangian formulation into a more convenient one. For instance, symmetries which are hidden in one formulation may be explicit in a dual one, or structures that do not look "geometric" in one framework may appear so in dual variables. In this book we shall use systematically such dualities to put the Lagrangian in a canonical form, namely the one in which the physical/geometrical structures are easier to analyze.

1.6.1 Abelian dualities

Abelian dualities are a simple generalization of the one described in the previous section. The space–time, Σ, has dimension n and Minkowski signature. The field content consists of

- p–form fields A^a, with gauge invariance $A^a \sim A^a + d\Lambda^a$, $\Lambda^a \in \Lambda^{p-1}(\Sigma)$, which enter the Lagrangian only through their field–strengths $F^a \equiv dA^a$;
- scalars ϕ^i;
- non–scalar fields Ψ inert under the gauge transformation $A^a \to A^a + d\Lambda^a$.

The Bianchi identities and equations of motion of the A^a,

$$dF^a = 0 \quad \text{(Bianchi identity)} \tag{1.128}$$

$$d\left(*\frac{\partial L}{\partial F^a}\right) = 0 \quad \text{(equations of motion)}, \tag{1.129}$$

are symmetric under the interchange of the $(p+1)$–form F^a with the $(n-p-1)$–form $G_a \equiv *\partial L/\partial F^a$. Then the theory may be formulated replacing the "electric" p–forms A^a with the dual "magnetic" $(n-p-2)$–forms B_a, which solve the equations of motion (1.129)

$$G_a = dB_a. \tag{1.130}$$

One passes from a formulation to the dual one by performing a Legendre transform of the Lagrangian L. We recall the well–known procedure. In the action one introduces Lagrange multipliers $(n-p-2)$–forms B_a enforcing the Bianchi identity, Eq. (1.128). One gets

$$S = \frac{1}{2\pi}\int d^n x\, L(F,\phi,\Psi) + \frac{(-1)^{n-p-1}}{2\pi}\int B_a \wedge dF^a. \tag{1.131}$$

This allows us to replace the potentials A^a with their field–strengths F^a as the independent fields. Next one performs the path integral over the F^a; that is, one eliminates them using their equation of motion; note that the second term in Eq. (1.131) may be rewritten as $\frac{1}{2\pi}\int dB_a \wedge F^a$, so the F^a enter algebraically in the

Lagrangian, and hence are non–propagating fields. At the leading semi–classical order we get

$$S' = \frac{1}{2\pi} \int d^n x \, L\Big(F(G, \phi, \Psi), \phi, \Psi\Big) + \frac{1}{2\pi} \int_\Sigma G_a \wedge F^a(G, \phi, \Psi). \quad (1.132)$$

$$F^a(G, \phi, \Psi) \text{ is obtained by inverting } \frac{\partial L}{\partial F^a} = *G_a. \quad (1.133)$$

This result is *exact* quantum mechanically for L quadratic in the F^a, since in that case the path integral is Gaussian.

For instance, the above Legendre transform allows the replacement of 2–form gauge fields with Abelian gauge vectors in 5D, and Peccei–Quinn scalars[21] in 4D. The relevance of this duality for the (Q)FT geometric structures is better illustrated by an example.

Example: Tensor theories in 4D. Consider a model in 4D which has both scalars (living on some manifold $\mathcal{M}$) and gauge 2–forms, A^a ($a = 1, \ldots, m$). In this "mixed" formulation, the manifest geometric structure is essentially the differential geometry of the manifold $\mathcal{M}$. On performing a duality, the m 2–forms are replaced by m PQ scalars, and we get a larger scalar manifold $\widetilde{\mathcal{M}}$ of dimension $\dim \mathcal{M} + m$. Writing all degrees of freedom in terms of scalar fields, we put the Lagrangian in its *canonical* form, in which the *full* geometry $\widetilde{\mathcal{M}}$ is manifest. There are deep and useful interplays between the geometries of $\mathcal{M}$ and $\widetilde{\mathcal{M}}$ with its m PQ Killing vectors. The "dual" manifold $\widetilde{\mathcal{M}}$ is rather special, and often one uses the duality to construct Riemannian manifolds $\widetilde{\mathcal{M}}$ having some prescribed property (e.g., many isometries).

General Lesson 1.10 *To make the underlying geometric structure manifest, one puts the Lagrangian into canonical form by dualizing the higher–degree forms to lower–degree ones.*

Example: Vectors in 3D. 1–forms (= gauge vectors) in 3D behave much in the same way as 2–forms in four dimensions. Again we can replace the vectors by PQ scalars and study the geometry of the resulting bigger target manifold both locally and in the large (the global geometry being relevant for non–perturbative effects). However, there is an important difference: gauge vectors may be *non–Abelian*; the above procedure does not work in this case, so we cannot use it to replace non–Abelian vectors with scalars and we cannot make explicit all the underlying geometric structure. However, precisely in 3D, one may perform a non–Abelian

---

[21] A Peccei–Quinn scalar φ is one that enters into the Lagrangian only through its derivatives $\partial_\mu \varphi$. Hence it has a "gauge" symmetry $\varphi \to \varphi +$ const. which is called a PQ symmetry.

1.6 More dualities

duality, although less elementary than the one presented above. This will be the topic of Section 1.6.2.

Subtleties from Dirac quantization

In a theory with quantized fluxes the field–strengths F^a are d–closed $(p+1)$–forms whose cohomology class satisfies

$$\left[\frac{F^a}{2\pi}\right] \in H^{p+1}(X, \mathbb{Z}). \tag{1.134}$$

We rewrite the action (1.131) in terms of the gauge–invariant field–strength F^a and G_a as

$$S = \frac{1}{2\pi} \int d^n x \, L(F^a, \cdots) + \frac{1}{2\pi} \int G_a \wedge F^a. \tag{1.135}$$

The equation of motion of the dual gauge fields B_a should enforce the condition that $F^a/2\pi$ is a d–closed form whose cohomology class is quantized as in Eq. (1.134). Taking G_a to be a global d-closed $(n - p - 1)$–form will enforce the stronger condition that F^a is d–exact. The correct prescription in Eq. (1.135) is to take the dual field–strength G_a to be also a d–closed form with a quantized class $[G_a/2\pi] \in H^{n-p-1}(X, \mathbb{Z})$. Such a field–strength may be decomposed as

$$G_a = 2\pi \sum_s n_{a,s} \omega_s + dB_a, \tag{1.136}$$

where $\{\omega_s\}$ represent a basis of the lattice

$$H^{n-p-1}(X, \mathbb{Z})/H^{n-p-1}(X, \mathbb{Z})_{\text{torsion}}, \tag{1.137}$$

and $n_{a,s} \in \mathbb{Z}$, and B_a are globally defined $(n - p - 2)$–forms. Integrating away the dual degrees of freedom to recover the original theory amounts to integrating the field B_a (with the appropriate Fadeev–Popov measure for a gauge field) and *summing* over the integers $n_{a,s}$. Schematically, one gets

$$\sum_{n_{a,s} \in \mathbb{Z}} \exp\left(i \sum_s n_{a,s} \int \omega_s \wedge F^a\right) \int [dB_a] \exp\left(\frac{i}{2\pi} \int (*L + dB_a \wedge F^a)\right).$$

One has

$$\sum_{n_{a,s} \in \mathbb{Z}} \exp\left(i \sum_s n_{a,s} \int \omega_s \wedge F^a\right) = \prod_{a,s} \sum_{m_{a,s} \in \mathbb{Z}} \delta\left(\int \omega_s \wedge F^a - 2\pi \, m_{a,s}\right), \tag{1.138}$$

which is the right flux quantization for the original gauge fields. The equality (1.138) is equivalent to the *Poisson summation formula* [187], which states that the $\mathbb{Z}$–delta function

$$\delta_{\mathbb{Z}}(x) := \sum_{k\in\mathbb{Z}} \delta(x-k) \qquad (1.139)$$

is its own Fourier transform:

$$\delta_{\mathbb{Z}}(x) = \widehat{\delta_{\mathbb{Z}}}(x) := \int dp\, e^{2\pi i p x}\, \delta_{\mathbb{Z}}(p). \qquad (1.140)$$

General Lesson 1.11 *Quantization of fluxes reproduces itself under duality. At the level of path integrals this is realized by dual Poisson summations.*

Example: Abelian vectors in 3D. In three dimensions a gauge vector is dual to a scalar field with PQ symmetry $\phi \to \phi + \text{const}$. The periods of the dual field–strength $d\phi$ are of the form 2π times an integer. This means that the scalar ϕ is not globally defined, but rather its variation along a closed loop γ,

$$\Delta\phi = \int_\gamma d\phi = 2\pi\, n, \quad n \in \mathbb{Z}, \qquad (1.141)$$

is a multiple of 2π. In other words, ϕ is well–defined only mod 2π, and the scalar takes values in the compact space S^1, not in $\mathbb{R}$.

1.6.2 Non–Abelian duality in 3D

The reader may wonder why we focus on such a peculiar case: vector–scalar duality in 3D. The reason is twofold. There is a technical motivation and a physical one. The physical one is easier to explain: generalized (supersymmetric) 3D Chern–Simons–matter models are believed to correspond to the world–volume theory of a stack of coinciding M2 branes (the membranes of *M*–theory) [3, 30] so, in a sense, they are fundamental theories, and the duality we are going to describe is one of their deepest properties. On the technical side, the 3D non–Abelian duality is an essential ingredient of our geometric approach to SUSY. In this book we start by studying SUGRA in 3D, arguing that in that case *everything* is differential–geometric, and then we generalize to $D > 3$. However, as illustrated by the examples of Section 1.6.1, in 3D everything is explicitly geometric only in a formulation in which all the bosonic propagating degrees of freedom are represented by scalar fields. Otherwise we have access only to the geometry of the submanifold $\mathcal{M} \subset \widetilde{\mathcal{M}}$. Luckily, thanks to de Wit, Herger and Samtleben [120] (see also

[40, 250]), the duality can be done explicitly. The non–Abelian duality replaces vector fields with canonical kinetic terms of the form

$$-\frac{1}{4}M_{AB}(\phi)\,F^A_{\mu\nu}F^{B\,\mu\nu}$$

by scalars ϕ_A and vectors A^A_μ, $B_{A\,\mu}$ whose derivatives enter in the Lagrangian only in the form of Chern–Simons (CS) couplings. In suitable gauges, we can think of the scalars ϕ_A as the fields describing the physical local degrees of freedom.[22]

Following [120], we consider a general 3D Lagrangian, quadratic in the Yang–Mills field–strengths, having the form

$$\mathcal{L} = -\frac{1}{4}\sqrt{-g}(F^A_{\mu\nu} + O^A_{\mu\nu})M_{AB}(\Phi)(F^{B\,\mu\nu} + O^{B\,\mu\nu}) + \mathcal{L}'(A,\Phi), \qquad (1.142)$$

where A^A_μ are non–Abelian gauge fields,

$$F^A_{\mu\nu} = \partial_\mu A^A_\nu - \partial_\nu A^A_\mu - f_{BC}{}^A A^B_\mu A^C_\nu$$

are the corresponding field–strengths, and Φ stands for all other fields, possibly transforming in non–trivial representations of the gauge group G_{YM}. $O^A_{\mu\nu} \equiv O^A_{\mu\nu}(A,\Phi)$ are gauge covariant operators formed with the fields Φ, and A^A_μ. The field–dependent matrix M_{AB} also transforms covariantly under G_{YM}. $\mathcal{L}'$ is gauge–invariant (up to total derivatives), and contains A^A_μ only through the covariant derivatives of fields and, possibly, Chern–Simons terms.

The Bianchi identities and equations of motion read

$$D_\mu \tilde{F}^{A\,\mu} = 0, \qquad (1.143)$$

$$D_{[\mu}\left(M_{AB}(\tilde{F}_{\nu]} + \tilde{O}^A_{\nu]})\right) = J_{A\,\mu\nu}, \qquad (1.144)$$

where

$$\tilde{F}^A_\mu = \frac{1}{2}\sqrt{-g}\,\varepsilon_{\mu\nu\rho}\,F^{A\,\nu\rho}, \qquad (1.145)$$

$$\tilde{O}^A_\mu = \frac{1}{2}\sqrt{-g}\,\varepsilon_{\mu\nu\rho}\,O^{A\,\nu\rho}, \qquad (1.146)$$

$$J_{A\,\mu\nu} = \frac{1}{2}\sqrt{-g}\,\varepsilon_{\mu\nu\rho}\,\frac{\partial \mathcal{L}'}{\partial A^A_\rho}. \qquad (1.147)$$

---

[22] The pure CS model, $S = \int \mathrm{Tr}(AdA + \frac{2}{3}A^3)$, is a topological field theory which does not propagate any local degree of freedom [311] (for a review see [231]).

To perform the duality, we introduce new vector fields $B_{A\mu}$ and compensating scalars ϕ_A, both transforming in the (co)adjoint of G_{YM}. They are defined by the equation

$$B_{A\mu} - D_\mu \phi_A = M_{AB}(\tilde{F}^A_\mu + \tilde{O}^A_\mu), \tag{1.148}$$

which is invariant under the additional gauge transformation

$$\delta \phi_A = \Lambda_A, \qquad \delta B_{A\mu} = D_\mu \Lambda_A. \tag{1.149}$$

The generators of this new Abelian gauge group T transform in the (co)adjoint of G_{YM}, and hence the total gauge group is $G_{YM} \ltimes T$, with G_{YM}–covariant field–strengths $F^A_{\mu\nu}$ and $G_{A\mu\nu} = 2D_{[\mu}B_{A\nu]}$, which transform under T as

$$\delta F^A = 0 \quad \text{and} \quad \delta G_A = -\Lambda^C f_{AB}{}^C F^B. \tag{1.150}$$

The full $G_{YM} \ltimes T$–covariant derivatives of ϕ_A and dual field–strength are

$$\hat{D}_\mu \phi_A = D_\mu \phi_A - B_{A\mu}, \tag{1.151}$$

$$\hat{G}_{A\mu\nu} = 2 D_{[\mu} B_{A\nu]} + f_{AB}{}^C F^B_{\mu\nu} \phi_C. \tag{1.152}$$

We stress that the total gauge group $G_{YM} \ltimes T$ is *not* reductive (in particular, non–compact). A non–reductive gauge group is generally believed to be inconsistent with unitarity. This will be not an issue here, since the resulting theory is equivalent to the original one, Eq. (1.142), for which unitarity is manifest.

Consider the new Lagrangian

$$\mathcal{L}_{\text{new}} = -\frac{1}{2}\sqrt{-g}\,\hat{D}_\mu \phi_A\, M^{AB}\,\hat{D}^\mu \phi_B$$
$$+ \frac{1}{2}\varepsilon^{\mu\nu\rho}(F^A_{\mu\nu} B_{A\rho} - O^A_{\mu\nu}\hat{D}_\rho \phi_A) + \mathcal{L}', \tag{1.153}$$

where M^{AB} is the inverse of the matrix M_{AB}. The equations of motion obtained from this Lagrangian are

$$\frac{\delta \mathcal{L}}{\delta B^\mu_A} = 0 \rightarrow M^{AB}(D_\mu \phi_B - B^\mu_B) + \tilde{F}^A_\mu + \tilde{O}^A_\mu = 0 \equiv \text{Eq.}(1.148),$$

$$\frac{\delta \mathcal{L}}{\delta \phi_A} = 0 \rightarrow D^\mu(M^{AB}(D_\mu \phi_B - B^\mu_B) - \tilde{O}^A_\mu) = 0 \equiv \text{Eq.}(1.143),$$

$$*\frac{\delta \mathcal{L}}{\delta A^A_\rho} = 0 \rightarrow D_{[\mu}(B_{A\nu]} - D_{\nu]}\phi_A) = J_{A\mu\nu} \equiv \text{Eq.}(1.144),$$

together with the same equations as before for the other fields Φ. Therefore the two Lagrangians are physically *equivalent* at the classical level. The equivalence extends to the quantum theory: fixing in Eq. (1.153) the "unitary" gauge $\phi_A = 0$, and performing the Gaussian integral over the quadratic fields $B_{A\mu}$, we recover the original Lagrangian (1.142).

$\mathcal{L}_{\text{new}}$ is gauge–invariant under $G_{YM} \ltimes T$ up to a total derivative. The original Yang–Mills–like Lagrangian, Eq. (1.142), gets replaced by a Chern–Simons–matter Lagrangian with a larger gauge group and a new scalar manifold $\widetilde{\mathcal{M}}$ of dimension ($\dim \mathcal{M} + \dim G_{YM}$), equal to the total number of (bosonic) propagating degrees of freedom.

General Lesson 1.12 *In 3D we can always reduce to a scalar manifold of dimension equal to the effective (local) propagating degrees of freedom, at the price, in the presence of non–trivial gauge interactions, of introducing suitable Chern–Simons couplings and non–reductive gaugings.*

Exercise 1.6.1 We were sketchy in deriving Eq. (1.132). Fill in the details: (1) the functional measure; (2) proper treatment of the zero modes.

2
Extended supersymmetry in diverse dimensions

This chapter is also introductory/motivational in nature. After some warm–up in two space–time dimensions (and in one), we go to our preferred SUSY laboratory: $D = 3$. *Three* is a very nice number of dimensions. It is the first element in the magic sequence of real division algebras

$$\mathbb{R} \leftrightarrow \mathbb{C} \leftrightarrow \mathbb{H} \leftrightarrow \mathbb{O}$$

whose entries correspond, respectively, to SUSY in 3, 4, 6 and 10 dimensions. The relation between 3D supergravity and $D \geq 4$ SUGRA is like *diet Coke* versus the original drink: *no sugar, no caffeine, but all the flavor*. Our 3D *diet* SUGRA has no propagating graviton, no propagating gravitino, no propagating gauge vector, but still has all the field–theoretic, algebraic, and geometric structures of higher–dimensional supergravity. In Sections 2.8 and 2.9 we add plenty of sugar and uplift our results to $D \geq 4$ rigid SUSY. Finally, in Section 2.10 we add the caffeine, and study the geometrical structures emerging from extended 4D SUGRA.

General references for supergravity include refs. [82], [144], [146], [266], and [289].

2.1 SUSY in diverse dimensions

Supersymmetry is the (only) symmetry that relates bosons to fermions. In fact, it is the only S–matrix quantum symmetry that may connect states of different spins [96, 182]. A classical symmetry between bosons of different spins is usually *not* preserved at the quantum level, unless there is a supersymmetry to protect it. Hence, in most instances, (enough) supersymmetry is required for the classical geometric structures of field theory to make sense at the quantum level.

We assume the reader has some familiarity with the SUSY algebra (the Haag–Łopuszanski–Sohnius theorem [182]) and its general implications, as well as a knowledge of the representations of the algebra [139, 281]. We will not insist on these topics, but just mention them when needed.

The detailed form of (Poincaré) supersymmetry depends on the zoology of spinors existing in the various dimensions (and space–time signatures): Weyl, Majorana, Majorana–Weyl, symplectic–Majorana, etc. However, the general structure is quite universal, so we shall state the properties of the algebra in the different dimensions without elaborating on their derivation.

A word of caution: The SUSY algebra makes sense as an algebra of operators acting on a Hilbert space $\mathcal{H}$, only in rigid supersymmetry. In the case of local supersymmetry, SUGRA, the meaning of the algebra as "operators acting on some Hilbert space" is rather tricky, and we shall return to it when we have enough geometrical tools to study supersymmetry in curved space–times. However, since the variation of a field Φ,[1] under the SUSY transformation of spinorial/Grasmannian parameter ϵ_α, is given by

$$\delta\Phi = -i[\bar{\epsilon}Q, \Phi], \qquad (2.1)$$

when we write the RHS we actually mean the LHS, which is well defined.

2.1.1 Spinors in D dimensions

A spinor is an element of the vector space S on which the universal Dirac matrices Γ_μ act. Its (complex) dimension is $2^{[D/2]}$. In even dimension, $D = 2m$, S is reducible as a representation of $Spin(D, \mathbb{C})$ into the direct sum of two irreducible spin representations $S = S_+ \oplus S_-$, corresponding to the two possible chiralities[2] of the spinor. The elements of the irreducible spaces $S_\pm$ are called *Weyl* spinors.

For the reality properties, we have the standard three possibilities: the $\mathbb{C}$–space S may have a real, or complex, or quaternionic structure, depending on the space–time signature (p, q). We recall the relevant definitions:

Definition 2.1 $E_\mathbb{C}$ a complex vector space. $V_\mathbb{R}$ a real one.

(a) A *real structure* on $E_\mathbb{C}$ is an *anti*–linear map $R\colon E_\mathbb{C} \to E_\mathbb{C}$ with $R^2 = \mathrm{id}$. A vector $v \in E_\mathbb{C}$ is *real* if $Rv = v$, and *purely imaginary* if $Rv = -v$. Let E be the $\mathbb{R}$–subspace of the real vectors: one has $E_\mathbb{C} = \mathbb{C} \otimes_\mathbb{R} E$.

(b) A *quaternionic structure* on $E_\mathbb{C}$ is an *anti*–linear map $J\colon E_\mathbb{C} \to E_\mathbb{C}$ with $J^2 = -\mathrm{id}$. The multiplication (on the left) of a vector $v \in E_\mathbb{C}$ by a quaternion $a + bi + cj + dk$ is defined as $(a + bi + (c + di)J)v$: this makes $E_\mathbb{C}$ into a $\mathbb{H}$–module.

---

[1] Upper case Φ stands for a generic field appearing in the Lagrangian $\mathcal{L}$, bosonic *or* fermionic. Q_α denotes a super–charge (= a generator of SUSY).
[2] Chirality = eigenvalue with respect to the generalized γ_5 matrix: $\Gamma_{[D]} := \Gamma_1 \Gamma_2 \cdots \Gamma_D$.

(c) A *complex structure* on $V_\mathbb{R}$ is an element matrix $I \in \mathrm{End}(V_\mathbb{R})$ with $I^2 = -1$. Multiplication of a vector $v \in V_\mathbb{R}$ by a complex number $(a+bi)$ is defined as $(a+bI)v$. On the complexified space $V_\mathbb{C} \equiv \mathbb{C} \otimes V_\mathbb{R}$ (of double real dimension!) we may diagonalize I, and write $V_\mathbb{C} = V_{(1,0)} \oplus V_{(0,1)}$ with I acting as multiplication by $+i$ (resp. $-i$) on $V_{(1,0)}$ (resp. on $V_{(0,1)} = \overline{V_{(1,0)}}$).

(d) A *quaternionic structure* on $V_\mathbb{R}$ is a pair of elements $I^a \in \mathrm{End}(V_\mathbb{R})$ $(a=1,2)$ satisfying the Clifford algebra

$$I^a I^b + I^b I^a = -2\,\delta^{ab}. \tag{2.2}$$

The (left) multiplication by $a+bi+cj+dk$ on $V_\mathbb{R}$ is defined as the action of $a+bI^1+cI^2+dI^1I^2$. I^1, I^2 and $I^3 := I^1 I^2$ are called *complex structures*.

If $E_\mathbb{C}$ is a complex vector space with a real structure R, we shall write $[[E_\mathbb{C}]]$ for the corresponding real space E. If E_1, E_2 are two complex spaces with quaternionic structures J_1 and J_2, the space $E_1 \otimes E_2$ has a *real* structure given by $R = J_1 \otimes J_2$. We write $[[E_1 \otimes E_2]]$ for the associated real space.

We say that in D dimensions with signature (p, q) there exist *Majorana* spinors, resp. *symplectic–Majorana* spinors, if the corresponding spinor space S admits a real, resp. quaternionic, structure invariant under the action of $Spin(p,q)$. In this case the (pseudo)real elements are called Majorana spinors and, respectively, symplectic–Majorana. If the chiral subspaces $S_\pm$ have an invariant real, resp. quaternionic, structure the corresponding spinors are called *Majorana–Weyl*, and, respectively, *symplectic–Majorana–Weyl* spinors. The *anti*–linear maps of Definition 2.1 are given by

$$\psi \mapsto B^* \psi^*, \qquad B := C\Gamma^0, \tag{2.3}$$

where C is the charge conjugation matrix. The matrix B has the property

$$BB^* = \begin{cases} +1 & \text{Majorana,} \\ -1 & \text{symplectic–Majorana.} \end{cases} \tag{2.4}$$

One shows [262] that the group $Spin(p,q)$ has a *minimal* realization by (pairs of) $2^\alpha \times 2^\alpha$ matrices over the division algebras $\mathbb{R}$, $\mathbb{C}$, or $\mathbb{H}$ as listed in Table 2.1. Hence in Minkowski signature $(D-1, 1)$ the *minimal* spinor is as in Table 2.2; note the periodicity mod 8 in the dimension D (Bott's periodicity). In $D = 4 \mod 4$ is a Weyl spinor and its conjugate may be combined into a (non–minimal) Majorana spinor; in all dimensions D we shall write our fermions in terms of the appropriate minimal spinors.

2.1 SUSY *in diverse dimensions*

Table 2.1 *Spin(p, q) groups*

$p - q$ mod 8		Algebra	2α
0	Pair	$\mathbb{R}$	$p + q - 2$
± 1		$\mathbb{R}$	$p + q - 1$
± 2		$\mathbb{C}$	$p + q - 2$
± 3		$\mathbb{H}$	$p + q - 3$
4	Pair	$\mathbb{H}$	$p + q - 4$

Table 2.2 *Minimal spinors in Minkowski space*

Type of spinor	D
Majorana	$8k + 1, 8k + 3$
Symplectic–Majorana	$8k + 5, 8k + 7$
Weyl	$8k, 8k + 4$
Majorana–Weyl	$8k + 2$
Symplectic Majorana–Weyl	$8k + 6$

2.1.2 Supercharges and superalgebras

To make manifest the automorphism group of the SUSY algebra, it is convenient to write the supercharges in terms of minimal spinors. The SUSY automorphism group may or may not be a symmetry of the physical theory. When it is a symmetry of the given theory, we refer to it as the *R*–symmetry.

In Table 2.3 we present the SUSY generators, Poincaré algebras, and automorphism groups, for the different *D*s (we write one period in *D* mod 8). "CC" stands for *"central charges,"* namely scalar charges generating symmetries commuting with all the other *continuous* symmetries of the theory. Their transformation under the automorphism group can be read directly from the algebra.[3] By the *automorphism group* we mean the compact subgroup preserving the natural metric. In the table, $Sp(\mathcal{N})$ stands for the compact (unitary) form of the symplectic group whose defining representation has dimension $\mathcal{N}$ ($\mathcal{N}$ must be even). The reader may find further details in ref. [281], and also refs. [108, 148, 290].

The last line in the table is very important: it gives the centralizer of the given spin group, namely the endomorphism ring of an *irreducible* representation. It is

---

[3] In Poincaré supersymmetry – in sharp contrast to AdS or conformal SUSY – the central charges Z should be invariant under all the actual (continuous) symmetries of the theory. On the other hand, the SUSY algebra dictates that the Zs are in non–trivial representations of the automorphism group. Hence, in the presence of non–trivial central charges, only the subgroup of the automorphism group that leaves Z invariant may be an *R*–symmetry.

Table 2.3 Supercharges and superalgebras in diverse dimensions

D	3	4	5	6
Supercharges	Majorana	Weyl	Symplectic–Majorana	Symplectic–Majorana–Weyl
Reality	$(Q_\alpha^a)^\dagger = Q_\alpha^a$ (∗)	$(Q_\alpha^a)^\dagger = \bar{Q}_{\dot\alpha a}$	$(Q_\alpha^a)^\dagger = \Omega_{ab} B_\alpha{}^\beta Q_\beta^b$	$(Q_\alpha^a)^\dagger = \bar{Q}_{\alpha a} = \Omega_{ab} B_\alpha{}^\beta Q_\beta^b$
Superalgebra	$\{Q_\alpha^a, Q_\beta^b\} = 2\delta^{ab}(\Gamma^\mu\Gamma_0)_{\alpha\beta} P_\mu + CC$	$\{Q_\alpha^a, \bar{Q}_{\dot\beta b}\} = 2\delta_b^a (\sigma^\mu)_{\alpha\dot\beta} P_\mu$ $\{Q_\alpha^a, Q_\beta^b\} = \epsilon_{\alpha\beta} Z^{ab}$	$\{Q_\alpha^a, Q_\beta^b\} = 2\Omega^{ab}(\Gamma^\mu C)_{\alpha\beta} P_\mu$	$\{Q_\alpha^a, Q_\beta^b\} = 2\Omega^{ab}(\Sigma^\mu)_{\alpha\beta} P_\mu$ $\{Q_\alpha^a, \bar{Q}_{\dot\beta}^b\} = C_{\alpha\dot\beta} Z^{ab}$
Automorphism	$SO(\mathcal{N})$	$U(\mathcal{N})$	$Sp(\mathcal{N})$	$Sp(\mathcal{N}_R) \times Sp(\mathcal{N}_L)$
Centralizer	$\mathbb{R}$	$\mathbb{C}$	$\mathbb{H}$	$\mathbb{H}$

D	7	8	9	10
Supercharges	Symplectic–Majorana	Weyl	Majorana	Majorana–Weyl
Reality	$(Q_\alpha^a)^\dagger = \Omega_{ab} B_\alpha{}^\beta Q_\beta^b$	$(Q_\alpha^a)^\dagger = \bar{Q}_{\dot\alpha a}$	$(Q^a)^\dagger = C(\bar{Q}^a)^t$	$(Q_\alpha^a)^\dagger = Q_\alpha^a$
Superalgebra	$\{Q_\alpha^a, Q_\beta^b\} = 2\Omega^{ab}(\Gamma^\mu C)_{\alpha\beta} P_\mu + CC$	$\{Q_\alpha^a, \bar{Q}_{\dot\beta b}\} = 2\delta_b^a (\Sigma^\mu)_{\alpha\dot\beta} P_\mu$ $\{Q_\alpha^a, Q_\beta^b\} = C_{\alpha\beta} Z^{ab}$	$\{Q_\alpha^a, Q_\beta^b\} = 2\delta^{ab}(\Gamma^\mu C)_{\alpha\beta} P_\mu + CC$	$\{Q_\alpha^a, Q_\beta^b\} = 2\delta^{ab}(\Sigma^\mu)_{\alpha\beta} P_\mu$ $\{Q_\alpha^a, \bar{Q}_{\dot\beta}^b\} = CC$
Automorphism	$Sp(\mathcal{N})$	$U(\mathcal{N})$	$SO(\mathcal{N})$	$SO(\mathcal{N}_R) \times SO(\mathcal{N}_L)$
Centralizer	$\mathbb{H}$	$\mathbb{C}$	$\mathbb{R}$	$\mathbb{R}$

Notes: (1) CC stands for "central charges." (2) (∗) holds in a Majorana rep. In a general rep. we have $(Q^a)^\dagger = BQ^a$, with $B \equiv C\Gamma^0$, where C is the charge–conjugation matrix. The same expression holds for the B matrices in $D = 5, 6, 7$; in the symplectic case we have $B^* B = -1$, in the Majorana one $B^* B = +1$.

Table 2.4 *Physically allowed $\mathcal{N}$s*

D	Allowed extensions $\mathcal{N}$ (resp. $(\mathcal{N}_R, \mathcal{N}_L))^a$
3	1, 2, 3, 4, 5, 6, 8, **9, 10, 12, 16**
4	1, 2, 3, 4, **5, 6, 8**
5	2, 4, **6, 8**
6	(2, 0), (2, 2), (**4, 0**), (**4, 2**), (**6, 0**), (**4, 4**), (**6, 2**), (**8, 0**)
7	2, **4**
8	1, **2**
9	1, **2**
10	(1, 0), (**1, 1**), (**2, 0**)
11	**1**

^{a}For $D = 2(4)$ we have independent left/right supersymmetries (see Table 2.3). Then we have distinct right and left SUSY extension numbers $\mathcal{N}_R, \mathcal{N}_L$. Bold face numbers correspond to SUSY extensions allowed only in the local case (SUGRA).

a division algebra by Schur's lemma. The automorphism group for the minimal supersymmetry ($\mathcal{N} = 1$, except for the symplectic case where $\mathcal{N}_{\min} = 2$) equals the group of elements of unit norm in $\mathbb{R}$, $\mathbb{C}$, and $\mathbb{H}$ respectively: $\mathbb{Z}_2$, $U(1)$, or $Sp(2) \simeq SU(2)$. *This means that the irreducible representation commutes with the identity and, respectively, none, one, or three complex structures which generate the corresponding division algebra* [108]. This corresponds to a very basic fact, especially emphasized by Kugo and Townsend [220]: increasing the space–time dimension D, the physical structures repeat themselves except that they get defined over division algebras of increasing dimension according to the scheme $\mathbb{R} \to \mathbb{C} \to \mathbb{H}$. This is manifest, for instance, in the structures of the Lorentz group for $D = 3$, 4 and 6, which are, respectively,

$$SL(2, \mathbb{R}), \quad SL(2, \mathbb{C}), \quad SL(2, \mathbb{H}), \tag{2.5}$$

as well as the corresponding little groups for massive particles,

$$SU(2, \mathbb{R}), \quad SU(2, \mathbb{C}), \quad SU(2, \mathbb{H}). \tag{2.6}$$

This is also true for the supersymmetric interactions: as we shall see below, the minimal SUSY (scalar) models in $D = 3, 4$, and 6 are based, respectively, on real, complex, and quaternionic differential geometry. Correspondingly, their superalgebras have 2, 4, and 8 supercharges (twice the dimension of the associated division algebra). All this is very beautiful and satisfactory, except that one feels that something is missing. One would like to continue the series to 10 dimensions, in correspondence with SUSY theories with 16 supercharges. On the other hand, we know that there exists a fourth (and last) division algebra, the octonions $\mathbb{O}$. The obvious guess would be a relation like $Spin(9, 1) \sim SL(2, \mathbb{O})$, but this cannot be true at its face value since the octonions are *not* associative, and hence we cannot define a matrix algebra over them. Yet it is true that the 10D Majorana–Weyl spinors are *octonionic* in nature, being related to $\mathbb{O}$ much in the same way as the 3D Majorana ones are related to $\mathbb{R}$, the 4D Weyl ones to $\mathbb{C}$, and the 6D symplectic–Weyl fermions to $\mathbb{H}$.

For our purposes, the important lesson is that there are "special" dimensions, namely $D = 3, 4, 6, 10$, in which the very nature of the spinors induces on a supersymmetric theory (respectively) a real, complex, quaternionic, or (morally) "octonionic" structure (the last one being, more properly, a real structure with a peculiar triality property). The $\mathbb{R}, \mathbb{C}, \mathbb{H}, \mathbb{O}$ sequence is important in many aspects of theoretical physics, especially for superstrings and branes (see e.g., ref. [54]). Then:

General Lesson 2.2 SUSY *in* $D = 3, 4, 6, 10$ *has a natural uniform structure based, respectively, on* $\mathbb{R}$, $\mathbb{C}$, $\mathbb{H}$ *and – in a suitable sense –* $\mathbb{O}$.

Not all extended SUSY algebras may be realized in local field theory with fields having spins ≤ 1 (≤ 2 in the SUGRA case). See Table 2.4 for the list of allowed $\mathcal{N}$s in the different contexts. Below we shall recover this same list from differential geometry.

It is time to leave algebra and study physics (and geometry). We start by considering a baby example, the $D = 2$ case, as a first warm–up.

2.2 A little warm–up: $D = 2$

2.2.1 *The* SUSY *algebra*

In the Weyl–Majorana notation (1 component spinors), the $(\mathcal{N}_+, \mathcal{N}_-)$ SUSY algebra reads

$$\{Q_+^a, Q_+^b\} = 2\,\delta^{ab}\,P_+, \quad \{Q_-^{a'}, Q_-^{b'}\} = 2\,\delta^{a'b'}\,P_-,$$
$$\{Q_+^a, Q_-^{b'}\} = Z^{ab'}, \qquad (2.7)$$
$$a, b = 1, 2, \cdots, \mathcal{N}_+, \quad a', b' = 1, 2, \cdots, \mathcal{N}_-,$$

where $Q_\pm^a$ are Hermitian and Z^{ab} are central charges. From Eq. (2.7) we see that the supercharges $Q_\pm^a$ have mass dimension $1/2$. So, acting on a field, they increase its (engineering) dimension by $1/2$.

2.2.2 $(\mathcal{N}_+, \mathcal{N}_-) = (1,1)$ *models*

For simplicity, we shall limit ourselves to parity–invariant theories; so, in the general Lagrangian (1.20) we set to zero the 2–form coupling that is a peculiar feature of 2D. We also assume that the central charge Z acts *trivially* on the local fields.

Let ϕ^i be the (real) scalar fields appearing in the Lagrangian $\mathcal{L}$ of a $(1,1)$ supersymmetric model. We *define* the Majorana–Weyl fields $\psi_\pm^i$ by

$$\psi_\pm^i = i[Q_\pm, \phi^i]. \tag{2.8}$$

Then the SUSY algebra implies

$$\{Q_\pm, \psi_\pm^i\} = \partial_\pm \phi^i, \tag{2.9}$$

$$\{Q_\mp, \psi_\pm^i\} = F^i, \tag{2.10}$$

where the second equation is the definition of the real scalar F^i, which has (engineering) dimension 1, and hence is non–propagating (an auxiliary field).

From Chapter 1 we know that the ϕ^i are local coordinates in some target manifold $\mathcal{M}$. From their definition, Eq. (2.8), we see that the $\psi_\pm^i$ take values[4] in the tangent bundle $T\mathcal{M}$. In particular the $\psi_\pm^i$ transform as covariant vectors under general field redefinitions, $\phi^i \mapsto \varphi^i(\phi)$,

$$\psi_\pm^i \mapsto (\partial \varphi^i / \partial \phi^j) \psi_\pm^j. \tag{2.11}$$

General Lesson 2.3 *In a supersymmetric theory, the spin–1/2 fermions* which are the SUSY partners of the scalars *live in the (pull–back of the) tangent bundle $T\mathcal{M}$ (twisted by the appropriate space–time spin bundle).*

We added a proviso in the statement since not all the Fermi fields present in the Lagrangian need to appear in the SUSY transformations $\delta \phi^i$. This result is essentially independent of the dimension D of space–time, although in higher dimension one needs to be slightly more precise, see Sections 2.6–2.8.

In view of Chapter 1, the above observation suffices to uniquely fix the kinetic terms of the ψ^i, as well as the couplings of the form $\psi_\pm^i \psi_\pm^j \partial_\mp \phi^k$, in terms of the Riemannian geometry of $\mathcal{M}$. If the theory contains only scalars and fermions, this determines the Lagrangian $\mathcal{L}$ up to the scalar potential $V(\phi)$, the Yukawa couplings $Y_{ij}(\phi) \psi_\pm^i \psi_\mp^j$, and the 4–Fermi interactions (compare with Eq. (1.20)).

We start by considering the SUSY Lagrangians whose bosonic part is a pure σ–model; their action is the one written in Eq. (1.7) plus fermionic terms. This

---

[4] To be *pedantic*: $\psi_\pm^i$ take values in the bundles $\mathcal{V}_\pm := \mathcal{L}^{\pm 1} \otimes \Phi^* T\mathcal{M} \to \Sigma$, where the line bundle $\mathcal{L}$ is obtained by analytic continuation of an Euclidean spin–bundle $\mathcal{L}_E$, i.e., a line–bundle with the property that $\mathcal{L}_E^2 = K$, the canonical bundle. To fix $\mathcal{L}_E$ requires, in general, a choice of spin–structure in the (Euclidean) space–time. To save print, we usually leave *implicit* the space–time part $\mathcal{S}_R$ of the fields' bundles (cf. General Lesson 1.3).

50   *Extended supersymmetry in diverse dimensions*

means that we set, for the moment,[5] $V(\phi) = 0$. In a rigid SUSY theory, $V(\phi) = 0 \Rightarrow Y_{ij}(\phi) = 0$ as well. *Why?* If $V(\phi) = 0$, any constant scalar configuration, $\phi^i = $ const., is a classical vacuum around which we can expand the theory. The scalars' mass^2–matrix in such a vacuum, $(m^2)_i^j$, is proportional to $\partial_i \partial_k V g^{kj} \equiv 0$, identically for all constant ϕ^i. The algebra (2.7) implies that the boson and fermion masses are equal around all SUSY vacua. The fermions' (mass)^2 is[6] $Y_{ik}(\phi) Y^{kj}(\phi)$, so

$$V(\phi) \equiv 0 \;\Rightarrow\; Y_{ik}(\phi) Y^{kj}(\phi) \equiv 0 \;\Rightarrow\; Y_{ij}(\phi) \equiv 0.$$

This also implies that the bosonic part of the auxiliary scalar F^i vanishes.[7] Taking into account their dimension and scaling with the volume of $\mathcal{M}$, we see that on–shell the auxiliary fields should have the form

$$F^i = A^i_{jk}(\phi)\, \psi^j_+ \psi^k_-. \qquad (2.12)$$

The coefficient function $A^i_{jk}(\phi)$ is easily obtained by requiring that the SUSY variation of diff–covariant expressions be diff–covariant. For all 1–forms $f \equiv f_i(\phi)\, d\phi^i \in \Omega^1(\mathcal{M})$, the composite field $f_i(\phi)\, \psi^i \equiv i_\psi f$ transforms as a scalar; the same should hold for its SUSY variation,

$$\begin{aligned}\{Q_\mp, f_i \psi^i_\pm\} &= (\partial_j f_i)\psi^j_\mp \psi^i_\pm + i f_i F^i \\ &= (\nabla_j f_i)\psi^j_\mp \psi^i_\pm + (\Gamma^k_{ji}\, \psi^j_\mp \psi^i_\pm + iF^k) f_k \qquad (2.13) \\ &= (\nabla_j f_i)\psi^j_\mp \psi^i_\pm \quad\Rightarrow\quad F^k = -i\,\Gamma^k_{ji}\, \psi^j_\mp \psi^i_\pm.\end{aligned}$$

In the $V = 0$ case, it remains to determine only the 4–Fermi coupling, $S_{ijkl}(\phi)\, \psi^i_+ \psi^j_+ \psi^k_- \psi^l_-$. One can get it by a two–line computation, but we prefer to argue geometrically, in the spirit of this book. $S_{ijkl}(\phi)$ should be a *covariant* 4–tensor on $\mathcal{M}$ made out of the metric g_{ij} and its derivatives. It should vanish for a flat metric (because, in that case, the theory is free). It should be antisymmetric in the first two indices, as well as in the last two (since the ψ^i anticommute[8]). By parity invariance, it should be invariant under the interchange of the two pairs of indices. Under the scaling $g_{ij} \to \hbar^{-1} g_{ij}$, it should scale like the metric itself, since $1/\hbar$ multiplies the full classical action, 4–fermions included. This last condition – in view of the vanishing of S_{ijkl} in flat space – already says that S_{ijkl} should be linear

---

[5] As already mentioned, the b_{ij} parity–violating terms are set to zero. The logic of the arguments remains valid if $b_{ij} \neq 0$, but then we have two covariant tensors, R_{ijkl} and $H_{ijk} := \partial_{[i} b_{jk]}$, and the geometry is considerably richer. H_{ijk} behaves very much as a torsion [113]. The reader may work out the details of the general case as an exercise.

[6] Indices are raised/lowered with the help of the metrics g_{ij} and g^{ij}.

[7] Indeed, $0 = \delta \mathcal{L}|_{\phi^i = \text{const}} = F^i \frac{\delta \mathcal{L}_F}{\delta \psi^i} \epsilon + 3\text{–fermions}$.

[8] Here $\psi^i_\pm$ are Grassmannian fields (path integral integration variables) *not* operators!

in the Riemann tensor without any covariant derivative (compare the discussion at the end of Section 1.1.1). Then

$$S_{ijkl}(\phi) = c\,R_{ijkl}(\phi) + d(g_{ik}R_{jl} - g_{il}R_{jk} - g_{jk}R_{il} + g_{jl}R_{ik}) \tag{2.14}$$

for some constants c, d. The same argument as in Eq. (1.11) gives $d = 0$. Then

$$S_{ijkl}(\phi) = c\,R_{ijkl}(\phi) \tag{2.15}$$

for some *universal* constant c. As a further check, let us show that the 4–Fermi coupling – whatever it is – should satisfy the second Bianchi identity. Consider the 5–Fermi part of the variation of the Lagrangian $\delta\mathcal{L}|_{5F}$. It receives contributions only from the variation of the 4–Fermi term. Then

$$0 = \delta_+\left(S_{ijkl}\psi_+^i\psi_+^j\psi_-^k\psi_-^l\right)\big|_{5F} = (\nabla_m S_{ijkl})\psi_+^m\psi_+^i\psi_+^j\psi_-^k\psi_-^l \;\Rightarrow\; \nabla_{[m}S_{ij]kl} = 0,$$

which gives another proof of $d = 0$.

The complete Lagrangian and SUSY transformations now read

$$\mathcal{L} = -\frac{1}{2}g_{ij}\partial_\mu\phi^i\partial^\mu\phi^j + \frac{i}{2}g_{ij}\bar\psi^i\gamma^\mu\nabla_\mu\psi^j + c\,R_{ijkl}\,\bar\psi^i\psi^k\,\bar\psi^j\psi^l \tag{2.16}$$

$$\delta\phi^i = \bar\epsilon\psi^i \tag{2.17}$$

$$\delta\psi^i = -i\gamma^\mu\partial_\mu\phi^i\epsilon - \Gamma^i_{jk}\bar\epsilon\,\psi^j\psi^k. \tag{2.18}$$

Our next task is to compute the numerical constant c. As always in this book, we have two choices: either we apply the above SUSY transformations to the Lagrangian, Eq. (2.16), and determine algebraically the unique c that makes it invariant, or we construct a geometric theory predicting its value. We shall follow the second path. Our next subject is the relation between SUSY and the *topology* of the target manifold $\mathcal{M}$. This will explain the geometrical *meaning* of the numerical constant c.

Exercise 2.2.1 Prove Eqs. (2.9) and (2.10).

Exercise 2.2.2 Compute directly the constant c in Eq. (2.16).

2.3 SUSY and the topology of $\mathcal{M}$

2.3.1 The ψs as differential forms

From the previous discussion we see that, geometrically speaking, the $\psi_\pm^i$ behave very much as the differentials of the coordinates, $d\phi^i$, of the target n–fold $\mathcal{M}$. In fact they also satisfy the same anticommuting algebra

$$d\phi^i \wedge d\phi^j = -d\phi^j \wedge d\phi^i \quad \text{(exterior form algebra)}, \quad (2.19)$$

$$\psi_\pm^i \psi_\pm^j = -\psi_\pm^j \psi_\pm^i \quad \text{(Grassmann algebra)}. \quad (2.20)$$

So, an operator containing only fermions of a given chirality, say $+$, may be interpreted as a differential form on $\mathcal{M}$ (twisted, as above, by the appropriate power of the space–time spin bundle) by the rule:

$$A_{i_1 i_2 \cdots i_p}(\phi)\, \psi_+^{i_1} \psi_+^{i_2} \cdots \psi_+^{i_p} \longleftrightarrow A_{i_1 i_2 \cdots i_p}(\phi)\, d\phi^{i_1} \wedge d\phi^{i_2} \wedge \cdots \wedge d\phi^{i_p}. \quad (2.21)$$

To make a long story short, we compactify the model on a circle S^1, keeping only the zero modes; that is, we reduce the theory to one dimension and work in supersymmetric quantum mechanics[9] (SQM). The Lagrangian becomes

$$L = \frac{1}{2} g_{ij} \frac{d\phi^i}{dt} \frac{d\phi^j}{dt} + i g_{ij} \bar{\psi}^i \gamma^0 D_t \psi^j + c R_{ijkl} \bar{\psi}^i \psi^k \bar{\psi}^j \psi^l, \quad (2.22)$$

where c is the constant still to be determined. It is convenient to rewrite L in a basis with γ^0 diagonal:

$$L = \frac{1}{2} g_{ij} \frac{d\phi^i}{dt} \frac{d\phi^j}{dt} + i g_{ij} \psi^{*i} \gamma^0 D_t \psi^j + 3c\, R_{ijkl} \psi^{*i} \psi^{*j} \psi^k \psi^l. \quad (2.23)$$

The fermions now correspond to operators in a Hilbert space with the usual canonical anticommutation relations

$$\{\psi^i, \psi^{*j}\} = g^{ij}(\phi), \qquad \{\psi^i, \psi^j\} = \{\psi^{*i}, \psi^{*j}\} = 0. \quad (2.24)$$

As customarily, we take as reference state the Clifford vacuum $|0\rangle$ defined by the condition $\psi^i |0\rangle = 0$ for all i. Then a generic state in the Hilbert space has the form

$$\left(\Psi(\phi) + \Psi_i(\phi) \psi^{*i} + \cdots + \Psi_{i_1 i_2 \cdots i_n}(\phi) \psi^{*i_1} \cdots \psi^{*i_n} \right) |0\rangle, \quad (2.25)$$

where the coefficient–wavefunctions $\Psi_{i_1 \cdots i_p}(\phi)$ are square–summable and antisymmetric in their indices. They transform under target space diffeomorphisms as the coefficients of the differential form $\Psi_{i_1 \cdots i_p}(\phi)\, d\phi_{i_1} \wedge \cdots \wedge d\phi^{i_p}$.

Hence a state (2.25) is identified with a differential form on $\mathcal{M}$ with *square–summable* (L^2) coefficients, and the Hilbert space $\mathcal{H} \simeq \Omega^\bullet(\mathcal{M}, L^2)$, where the grading $\bullet$ by form–degree corresponds physically to the grading by the Fermi number F of the state. If $\omega \in \Omega^\bullet(\mathcal{M}, L^2)$ is a differential form, we write $|\omega\rangle \in \mathcal{H}$ for

---

[9] We follow ref. [7].

the corresponding state. Equations (2.24) imply

$$\langle \omega_1 | \omega_2 \rangle = \int_{\mathcal{M}} \omega_2 \wedge *\overline{\omega}_1, \qquad (2.26)$$

where $*$ is the Hodge dual (with respect to g_{ij}) and the overbar stands for complex conjugation. We have two supercharges, Q and $Q^\dagger$, which in the Schrödinger picture act as the following differential operators

$$Q|\omega\rangle = |d\omega\rangle, \qquad Q^\dagger |\omega\rangle = |\delta\omega\rangle, \qquad (2.27)$$

where d is the de Rham differential and $\delta \equiv -*d*$ is its Hermitian adjoint (with respect to the inner product (2.26)). The SQM SUSY algebra gives for the Hamiltonian H

$$2H \equiv \{Q, Q^\dagger\} \longrightarrow d\delta + \delta d = \Delta, \qquad (2.28)$$

i.e., the usual Hodge Laplacian [163]. Therefore, the states of zero–energy (the vacua) are precisely the harmonic forms η, that is, the forms satisfying

$$\Delta \eta = 0, \qquad (2.29)$$

having L^2–coefficients. If $\mathcal{M}$ is compact, these are the ordinary harmonic forms $\eta \in \mathbb{H}^\bullet(\mathcal{M})$, and by the Hodge theorem [163] they are in one–to–one correspondence with the de Rham cohomology classes,[10] $H_d^\bullet(\mathcal{M}) \simeq \mathbb{H}^\bullet(\mathcal{M})$.

2.3.2 SUSY and topological index theorems

For any (rigid) supersymmetric model the Witten index [301] is defined as

$$\mathrm{Tr}\big[(-1)^F e^{-\beta H}\big] = \sum_{\text{bosons}} e^{-\beta E_n} - \sum_{\text{fermions}} e^{-\beta E_n}$$

$$= \#\{\text{bosonic } E = 0 \text{ states}\} - \#\{\text{fermionic } E = 0 \text{ states}\}, \qquad (2.30)$$

---

[10] The harmonic form η is just the *unique* form in the given cohomology class $[\eta]$ with the smaller norm $\langle \eta | \eta \rangle^{1/2}$. The proof of this statement is a standard two–line argument:

$$\langle \eta | \Delta | \eta \rangle = \langle \eta | (d\delta + \delta d) | \eta \rangle = \|d|\eta\rangle\|^2 + \|\delta|\eta\rangle\|^2, \quad \text{so} \quad \Delta|\eta\rangle \Leftrightarrow d|\eta\rangle = \delta|\eta\rangle = 0.$$

Let $|\eta + d\xi\rangle$ be another closed form in the same cohomology class:

$$\langle \eta + d\xi | \eta + d\xi \rangle = \langle \eta | \eta \rangle + \langle d\xi | d\xi \rangle + \langle \xi | \delta \eta \rangle + \langle \delta \eta | \xi \rangle = \langle \eta | \eta \rangle + \langle d\xi | d\xi \rangle > \langle \eta | \eta \rangle.$$

where the second line follows from the fact that all states with non–zero energy appear in Bose/Fermi pairs whose contributions cancel each other. From the definition it is clear that the index is an integer independent of β and invariant under continuous supersymmetric deformations of the theory.

Let us compute the Witten index for an SQM σ–model with a compact target space $\mathcal{M}$. From the discussion at the end of Section 2.3.1, we get

$$\text{Tr}\big[(-1)^F e^{-\beta H}\big] = \text{Tr}_{\mathbb{H}}(-1)^F = \sum_{p=0}^{n} (-1)^p B_p = \chi(\mathcal{M}), \qquad (2.31)$$

where $B_p := \dim \mathbb{H}^p(\mathcal{M})$ are the Betti numbers of $\mathcal{M}$, and $\chi(\mathcal{M})$ is its Euler characteristic [163]. On the other hand, the LHS of Eq. (2.31) is equal to the Euclidean path integral with time periodically identified $t \sim t + \beta$ [86]. Since the index is β-independent, we may as well replace the path integral by its $\beta \to 0$ asymptotics. Then, equating the two sides of Eq. (2.31), we get

$$\begin{aligned}
\chi(\mathcal{M}) &= \frac{1}{(2\pi)^{n/2}} \int_{\mathcal{M}} \prod_i d\psi^{*i} d\psi^i \exp\Big(-3c\, R_{ijkl} \psi^{*i} \psi^{*j} \psi^k \psi^l\Big) \\
&= \frac{(-6c)^{n/2}}{(2\pi)^{n/2} (n/2)} \int_{\mathcal{M}} \epsilon_{a_1 a_2 \cdots a_n} R^{a_1 a_2} \wedge \cdots \wedge R^{a_{n-1} a_n} \\
&= (12c)^{n/2} \chi(\mathrm{M}).
\end{aligned} \qquad (2.32)$$

In the second line of Eq. (2.32) we used the fact that only one term in the expansion of the exponential has the right form–degree (i.e., the correct number of fermionic zero–modes) to give a non–zero contribution. We also wrote the Riemann tensor as a 2–form $\frac{1}{2} R_{ij}{}^{ab} d\phi^i \wedge d\phi^j$ taking values in $\mathfrak{so}(T\mathcal{M})$, *alias* with two "flat indices" in the the vielbein formulation [296]. In the third line we used the Gauss–Bonnet formula for the Euler characteristic (compare with Eqs. (10) and (18) of [93]). Thus we predict

$$c = \frac{1}{12}. \qquad (2.33)$$

Of course, this is a silly way to compute the coefficients in the Lagrangian. But computation was not our main motivation. Certainly, we wanted to stress that c is not just *some* coefficient one gets from a boring computation. However the real motivation was to introduce — in a very simple context – the idea that the structure of SUSY is deeply related to algebraic and differential topology. Since the emphasis in this book is on the geometry of the target space $\mathcal{M}$, we could not allow ourselves to ignore the role of the *topology* of $\mathcal{M}$, that is, of its "geometry in the large." Moreover, much of the computability of physical quantities in supersymmetric theories stems from this SUSY–topology connection. During the last quarter of a century, this connection has grown into an impressively powerful tool, namely topological field theory [188, 309, 311]. Already in our simple SQM framework ($D = 1$), we can get all the "classical" index theorems (Hirzebruch, Riemann–Roch–Hirzebruch [186], Atiyah–Singer [20, 21], etc.), just by inserting the suitable operator $\mathcal{O}$ in the path–integral, that is, by replacing

$$\text{Tr}[(-1)^F e^{-\beta H}] \to \text{Tr}[(-1)^F e^{-\beta H} \mathcal{O}]$$

and repeating word–for–word the argument we used above for the Gauss–Bonnet theorem. For details see ref. [7]. Higher-dimensional supersymmetric field theories lead to more sophisticated geometric and topological invariants; just to mention a few: elliptic genera [307], Gromov–Witten invariants [188, 213], Donaldson–Witten invariants [309], etc. See ref. [52] for a review. The topological viewpoint also sets the stage for our next topic.

2.3.3 Adding a scalar potential

The reader may object to our claim that in supersymmetry everything is geometric, by purporting that the geometric viewpoint could not predict the form of the scalar potential $V(\phi)$ and the Yukawa terms. The reader familiar with the superfield approach [157, 297] will already know the structure of these *non–derivative* interactions in terms of the superpotential W. We are going to reproduce those results (and some generalization) from the geometric standpoint. A more intrinsic geometric interpretation of the non–derivative couplings will be given in Chapters 7 and 8.

In the setting of Section 2.3.1, the addition of Yukawa couplings and scalar potential amounts to a deformation of the supercharges which preserves their Fermi–grading and the SUSY algebra,

$$Q \to Q + \Delta Q, \qquad Q^\dagger \to Q^\dagger + \Delta Q^\dagger, \tag{2.34}$$

by operators ΔQ, $\Delta Q^\dagger$ not containing derivatives of the fields (otherwise they will produce new derivative couplings in the Hamiltonian). Without loss of generality, we may study such non–derivative deformations in the dimensionally reduced theory. In the Schrödinger language of Section 2.3.1, we have to deform the de Rham differential d to a new differential operator $\tilde{d}$ which maps even/odd forms into, respectively, odd/even ones, which squares to zero, $\tilde{d}^2 = 0$, and which differs from d by an operator that does not contain derivatives. There are (basically) two possibilities. We can add to d the operator $\xi \wedge$ which acts on a form ω by multiplying it by the odd form ξ, i.e., $\omega \mapsto \xi \wedge \omega$, or we can add the adjoint operator $*\eta*$

which *contracts* the form ω with the odd–degree form η. We *defined* the fermions ψ^i to be the SUSY transformations of the scalars (cf. the begining of Section 2.2.2) which, in the present set-up, requires ξ and η to be 1–forms.

The superpotential and Morse theory

Consider the first possibility: $\tilde{d} = d + \xi$. $\tilde{d}$ must square to zero,

$$0 = (d + \xi)^2 = d\xi, \qquad (2.35)$$

and hence (locally on $\mathcal{M}$) we must have $\xi = dW$ for a certain (real) function W, which we recognize to be the *superpotential*. In this way we recover what we know from a superspace approach. We stress, however, that SUSY does *not* require the existence of a globally defined superpotential: only the d–closed 1–form ξ needs to be globally well defined. Again, the Hamiltonian is given by the deformed Laplacian

$$H = \tilde{\Delta} \equiv \tilde{d}\tilde{\delta} + \tilde{\delta}\tilde{d}, \qquad \text{where } \tilde{\delta} = - * \tilde{d} *, \qquad (2.36)$$

and the supersymmetric vacua are its (normalizable) eigenforms associated to the zero eigenvalue.

Assume $\mathcal{M}$ compact. Since $\tilde{d}^2 = 0$ we can define the $\tilde{d}$–cohomology groups $\tilde{H}^\bullet(\mathcal{M})$ as the $\tilde{d}$–closed •–forms, $\tilde{d}\omega = 0$, modulo the $\tilde{d}$–exact ones, $\omega \sim \omega + \tilde{d}\varrho$. Just as in the de Rham case, the zero–energy states are in one–to–one correspondence with the $\tilde{H}^\bullet(\mathcal{M})$ classes. We can switch on the superpotential continuously, by taking $d \to d + t\, dW$ with t going from 0 to 1. The dimensions of the cohomology groups $\tilde{H}^\bullet(\mathcal{M})_t$ are constant as functions of t (Exercise 2.3.3); for our purpose it suffices to know that these dimensions may possibly jump only by lifting a Bose/Fermi *pair* of vacua to non–zero energy[11] [301]. Therefore,

$$\sum_p (-1)^p \dim \tilde{H}^p(\mathcal{M}) \qquad (2.37)$$

is independent of t, and hence equal to the value at $t = 0$, which is the Euler characteristic $\chi(\mathcal{M}) \equiv$ Witten index, Eq. (2.31).

Can you see the *magic* of this result? As $t \to \infty$ the scalar potential $V(\phi)$ becomes larger and larger, and the zero–energy wave–forms become exponentially small away from the points on $\mathcal{M}$ where $V(\phi)$ vanishes. Since, as you will verify in Exercise 2.3.3,

$$V(\phi) = \frac{1}{2} t^2 g^{ij} \partial_i W \, \partial_j W, \qquad (2.38)$$

---

[11] Recall that the states of non-zero energy are organized into Fermi/Bose pairs of equal energy.

the points where $V(\phi) = 0$ are precisely the *critical points* of the superpotential, namely the loci where $\partial W/\partial \phi^i = 0$. Thus, we are claiming that one can compute the cohomology of a manifold $\mathcal{M}$ by picking an arbitrary smooth function $W\colon \mathcal{M} \to \mathbb{R}$, and doing *local* computations in the vicinity of its critical points. *Sounds magical!* Mathematicians call this magical relation between the behavior of a function at its critical points and the cohomology of $\mathcal{M}$ Morse theory [185, 242]. The reader may find more details on the applications of SUSY to Morse theory in the paper by Witten [302].

Thus, in SUSY also the non–derivative terms have an interesting geometry. Their geometry will be even nicer in the extended case: see Chapters 7 and 8.

The second possibility: equivariant SUSY

The other possibility is more subtle: $\tilde{d} = d + i_v$, where $v = v^i \partial_i$ is some vector field on $\mathcal{M}$. The inner product i_v is a derivation acting on forms as $*e_{\hat{v}}*$, where $e_{\hat{v}}\colon \xi \mapsto \hat{v} \wedge \xi$ stands for the exterior product by the dual[12] form $\hat{v} := g_{ij} v^i d\phi^j$. Now,

$$(d + i_v)^2 = d\, i_v + i_v\, d \equiv \pounds_v \quad \text{(the Lie derivative on forms [325])} \qquad (2.39)$$

is certainly *not* zero. However, assume that v is not a generic vector field, but it is a Killing vector K of $\mathcal{M}$. K generates a $U(1)$ isometry group which – in absence of scalar potential – is a symmetry of the supersymmetric theory (provided one transforms the fermions in the right way, i.e., as elements of $T\mathcal{M}$). This $U(1)$ symmetry is generated by a conserved charge Z which, in the Schrödinger picture, is represented by the differential operator $i\pounds_K$ acting on the wave–forms ω. Then (2.39) is equivalent to

$$\{Q, Q\} = -2iZ, \qquad (2.40)$$

which merely states that Z is a *central charge* of the SUSY algebra.[13] Thus we have found a first example of a SUSY system with non–trivial central charges. Such a model was formulated in 2D SUSY in ref. [8], and analyzed in the TFT context in refs. [222, 310]. In the SQM set–up it was consider by Witten in the second part of ref. [302] and by Alvarez–Gaumé at the end of ref. [7]. The nicer treatment is presented in the paper by Atiyah and Bott ref. [25]; their interest was to formulate the equivariant version of cohomology under the action of some group

---

[12] With respect to the inner product in Eq. (2.26), i_v is the adjoint of $e_{\hat{v}}$.
[13] The SUSY algebra implies the BPS bound $E \geq |Z|$ [139, 281, 298]. The vacua, having $E = 0$, are invariant under Z. The vacua, at the zeroth order in the deformation i_K, are the harmonic forms on $\mathcal{M}$ (assumed compact). Then we have implicitly proved the following Theorem [163, 164, 324]: $\mathcal{M}$ *compact. The harmonic forms on* $\mathcal{M}$ *are invariant under the isometry group* Iso($\mathcal{M}$).

on the given manifold $\mathcal{M}$ and to prove a generalized version of the Lefschetz fixed point formula [23]. See also ref. [178].

The Lefschetz formula is obtained in the following way. From the commutator $\{\widetilde{d}, \widetilde{d}^\dagger\} = 2H$ one computes the potential

$$V(\phi) = \frac{1}{2} g_{ij} K^i K^j. \tag{2.41}$$

Rescaling $K^i \to t K^i$, and taking $t \to \infty$, the vacuum wave-forms concentrate exponentially in the vicinity of the zeros of the Killing vector K^i, that is, on the fixed point of the $U(1)$ isometry. Since the vacuum wave–forms capture the cohomology of $\mathcal{M}$, in this way we end up with computing the cohomology of $\mathcal{M}$ by local analysis around the fixed points. This is precisely what a fixed–point formula is supposed to do. Again, SUSY allows the recovery of a deep mathematical theorem in very easy (and very physical) terms.

Of course the above two mechanisms can be combined together. The most general deformation of the supercharges which produces a potential is $d \to d + i_K + e_{dW}$. This squares to the central charge $\pounds_K$ provided, $i_k dW \equiv \pounds_K W = 0$, i.e., provided W is a $U(1)$ invariant function.

The general Lagrangian in $D = 2$

We are ready to write down the Lagrangian of the general 2D σ–model with potentials (but without gauge coupling and with the parity–violating terms set to zero). The only coupling still to fix is the Yukawa matrix Y_{ij}. By covariance and scaling with the metric, it must be of the general form $a_1 \nabla_i \partial_j W + a_2 \nabla_i K_j$. The two constants are easily found to be 1 (say by checking the free massive case). From this expression we see that the superpotential W contributes to the symmetric part of the Yukawa matrix Y_{ij}, while the Killing vector to the antisymmetric. Thus, reverting to the 2–component notation, we get [8]

$$\mathcal{L} = -\frac{1}{2} g_{ij} \partial_\mu \phi^i \partial^\mu \phi^j + \frac{i}{2} g_{ij} \bar{\psi}^i \gamma^\mu \nabla_\mu \psi^j + \frac{1}{12} R_{ijkl} \bar{\psi}^i \psi^k \bar{\psi}^j \psi^l \\ -\frac{1}{2} g^{ij}(\partial_i W \, \partial_j W + K_i K_j) - \nabla_i \partial_j \bar{\psi}^i \psi^j - \nabla_i K_j \bar{\psi}^i \gamma_5 \psi^j, \tag{2.42}$$

and

$$\delta \phi^i = \bar{\epsilon} \psi^i, \tag{2.43}$$

$$\delta \psi^i = -i \gamma^\mu \partial_\mu \psi^i \epsilon - \Gamma^i_{jk} \bar{\epsilon} \psi^j \psi^k - g^{ij} \partial_j W \epsilon - K^i \gamma_5 \epsilon. \tag{2.44}$$

The β–function of the metric g_{ij}. The arguments of Section 1.1.1 apply also to the SUSY case. The metric β–function, β_{ij}, is, at each order in $\hbar$, a polynomial in the $\nabla_{i_1} \cdots \nabla_{i_s} R^j{}_{klm}$. Supersymmetry puts severe restrictions on these polynomials. Again to the one–loop order we get $c_1 R_{ij}$ by the arguments of Section 1.1.1. We can compute by purely geometrical means the numerical value[14] of c_1, without making any reference to Feymann diagrams. A possible strategy is: since c_1 is independent of the manifold $\mathcal{M}$, choose it to be the two–dimensional sphere S^2. Then the correct value of c_1, $1/2\pi$, is the unique one which makes the Kallan–Symanzik equation of RG consistent with the index theorem for the Dirac operator $\slashed{D}$, see, e.g., [188]. The index theorem is an exact statement, not just a "one–loop" result. This means that the geometric approach leads to *exact* computations of some β–functions in SUSY QFTs. Results of this kind are known as *supersymmetric non–renormalization theorems*. Although they are often proved using other techniques (as superspace [157]), they are most easily understood via geometry and topology; see [188].

Exercise 2.3.1 Starting from Eq. (2.36), work out the detailed form of the scalar potential and Yukawa couplings for a given W.

Exercise 2.3.2 Prove that the dimensions of the cohomology groups $\tilde{H}^p(\mathcal{M})$ of $\tilde{d} = d + t\, dW \wedge$ are independent of t.

Exercise 2.3.3 Compute directly the Yukawa couplings in Eq. (2.42).

Exercise 2.3.4 Compute c_1 both directly and geometrically.

2.4 Extended supersymmetry in 3D

Our next task is to generalize the findings of Section 2.2.2 to $\mathcal{N} > 1$. It would be natural to continue to work in 2D, since this is the dimension in which the arguments below were originally formulated [9, 10, 12] and the dimension in which the supersymmetric σ–models are power–counting renormalizable (in fact *finite* for $\mathcal{N}$ large enough). We shall work, instead, in 3D. The reason is twofold. Aesthetically, *three* is the dimension associated with $\mathbb{R}$. Pedagogically, we wish to treat rigid and local SUSY on the same footing, since the two theories are more conveniently developed together, emphasizing their common properties and contrasting their structural differences. For supergravity, 2D is a tricky, exceptional dimension, since the gravitino ψ_μ reduces, in a conformal gauge, to a pure gamma–trace

---

[14] $c_1|_{\text{SUSY}} \neq c_1|_{\text{bosonic}}$ since the first gets contributions also from the fermionic loop.

$\gamma_\mu \psi$, as is well known from the quantization of the fermionic string `a la Polyakov [260]. In 3D, on the contrary, SUGRA looks very much like in higher dimensions, although it is still "trivial" since neither the metric nor the gravitino fields propagate local degrees of freedom. In the *rigid* case, 2D left–right symmetric models with (1, 1) SUSY have the same geometrical structure as their 3D, $\mathcal{N} = 1$ counterparts; therefore, if readers prefer, they may rephrase the discussion below in 2D, by a straightforward dimensional reduction.

We work with Lagrangians in canonical form in the sense of Section 1.6. In this set-up all bosonic propagating degrees of freedom are represented by scalars ϕ^i, and the spin–1/2 fields χ^i are in one-to-one correspondence with the ϕ^i.[15] In addition to the ϕ^i, and χ^j the models possibly have gauge vectors A_μ^a, Rarita–Schwinger fields ψ_μ^A, and a metric $g_{\mu\nu}$ which do not propagate local degrees of freedom; in particular, the derivatives of the A_μ^a enter the Lagrangian only through Chern–Simons terms: see Sections 1.6.2 and 2.6. The 3D supersymmetric models with Chern–Simons terms are described in full detail and generality in Chapter 8.

The 3D algebra is[16]

$$\{Q_\alpha^A, Q_\beta^B\} = 2\, \delta^{AB} (\gamma^\mu \gamma^0)_{\alpha\beta} P_\mu + Z^{AB} \epsilon_{\alpha\beta}, \qquad (2.45)$$
$$A, B = 1, \ldots, \mathcal{N}, \qquad \alpha, \beta = 1, 2.$$

This algebra has an automorphism group $SO(\mathcal{N})_R$ under which the supercharges Q_α^A transform in the vector representation and Z^{AB} in the adjoint. $SO(\mathcal{N})_R$ need *not* to be a symmetry of the physical theory. If a subgroup $\mathcal{R} \subset SO(\mathcal{N})_R$ is actually a symmetry, we refer to it as the *R*–symmetry.

2.4.1 Rigid SUSY and parallel complex structures

In $\mathcal{N} = 1$ SUSY the fermions are tangent vectors to the scalar manifold $\mathcal{M}$. Indeed, as in Section 2.2.2, we can always *define* the Fermi fields χ^i so that the SUSY transformation of the physical scalars reads

$$\delta \phi^i = \bar{\epsilon} \chi^i. \qquad (2.46)$$

Assuming the scalar kinetic terms have the standard form, $g_{ij} \partial_\mu \phi^i \partial^\mu \phi^j / 2$, the Noether theorem applied to the symmetry (2.46) gives for the corresponding

---

[15] We mean the *propagating* spin–1/2. In certain formalisms there may be additional *auxiliary* spin–1/2 fields. We assume such auxiliary fields to have been integrated out.
[16] We adopt the Majorana representation $\gamma^0 = i\sigma_2$, $\gamma^1 = \sigma_1$, $\gamma^2 = \sigma_3$, $C = \gamma^0$, with $\bar{\psi} := \psi^t \gamma^0$.

2.4 Extended supersymmetry in 3D

conserved super–current $J_{\mu\alpha}$ and supercharge Q_α,

$$J_{\mu\alpha} = g_{ij}(\gamma^\nu \gamma_\mu \chi^i)_\alpha \, \partial_\nu \phi^j + \text{non derivative terms}, \qquad (2.47)$$

$$Q_\alpha = \int J_{0\alpha} \, d^2 x. \qquad (2.48)$$

Rigid $\mathcal{N} > 1$ supersymmetry is a special case of $\mathcal{N} = 1$ SUSY. Indeed, one can always focus one's attention on *one* supercharge, say Q_α^1, forgetting the others. This is *not* true in the local case, since $\mathcal{N}$–extended supergravity requires the presence of $\mathcal{N}$ species of gravitini, ψ_μ^A, $A = 1, 2, \ldots, \mathcal{N}$.

The parallel complex structures

Given a rigid $\mathcal{N}$–extended SUSY model, we apply the above remark, fix our attention on Q_α^1, and use Eq. (2.46) to define a preferred basis,[17] χ^i, for the spin–1/2 fields.

Next, we consider the other $\mathcal{N} - 1$ supercharges[18] Q_α^a. For each of them we have an equation like Eq. (2.46),

$$\delta^a \phi^i = \bar{\epsilon}_a \chi^{ai} \qquad \text{(no summation over } a\text{!)}. \qquad (2.49)$$

Since the χ^i form a basis of the spin–1/2 fields, we must have $\chi^{ai} = f^{ai}{}_j \chi^j$ for certain coefficients $f^{ai}{}_j(\phi)$, in general ϕ–dependent. Geometrically the $f^{ai}{}_j$ are $(1,1)$ tensor fields on $\mathcal{M}$, that is, sections of $T\mathcal{M} \otimes T^*\mathcal{M}$.

The SUSY algebra puts severe restrictions on the $f^{ai}{}_j$. From

$$\{Q_\alpha^A, J_{0\beta}^B\} = \delta^{AB} \gamma_{\alpha\beta}^\mu g_{ij} \partial_0 \phi^i \partial_\mu \phi^j + \text{at most 1 derivative}, \qquad (2.50)$$

using Eqs. (2.46), (2.47) and (2.49) we infer

$$(f^a)^t g + g f^a = 0, \qquad (2.51)$$

$$(f^a)^t g f^b + (f^b)^t g f^a = 2 \delta^{ab} g; \qquad (2.52)$$

that is, the tensors f^a obey the Clifford algebra

$$f^a f^b + f^b f^a = -2 \delta^{ab}, \qquad (2.53)$$

and $f_{ij}^a \equiv g_{ik} f^{ak}{}_j$ are antisymmetric in the i,j indices. On the other hand, we could have chosen another supercharge, say Q_α^2, as the reference one to define

---

[17] This is the point where we need a Lagrangian in canonical form, that is, all propagating vectors have been dualized to scalars; otherwise some fermions would be SUSY partners of the bosonic degrees of freedom associated with the vectors and χ^i would not be a complete basis.
[18] In this section $A, B, C, \ldots$ take values $1, 2, \ldots, \mathcal{N}$, while $a, b, c, \ldots$ take values $2, \ldots, \mathcal{N}$.

the Fermi fields' basis χ^i. The Lagrangian cannot depend on our arbitrary choice of a reference supercharge. Hence we must have

$$\forall a: \quad g_{ij} \bar{\chi}^i \gamma^\mu D_\mu \chi^j = g_{ij} \bar{\chi}^{ai} \gamma^\mu D_\mu \chi^{aj} \quad \text{not summed over } a \quad (2.54)$$

which, in view of Eq. (2.52), implies

$$0 = D_\mu f^{ai}{}_j \equiv \partial_\mu \phi^k \big(\partial_k f^{ai}{}_j + \Gamma^i{}_{kl} f^{al}{}_j - \Gamma^l{}_{kj} f^{ai}{}_l \big). \quad (2.55)$$

That is, *the tensors $f^{ai}{}_j$ are parallel*[19] for the Levi–Civita connection on $\mathcal{M}$.

General Lesson 2.4 *To have $\mathcal{N}$ extended* SUSY, *on the scalar manifold $\mathcal{M}$ there must exist $\mathcal{N} - 1$ parallel 2–forms f^a_{ij} such that the associated tensors $f^a \in \mathrm{End}(T\mathcal{M})$ satisfy the Clifford algebra*

$$f^a f^b + f^b f^a = -2\delta^{ab}. \quad (2.56)$$

In particular, $\mathcal{N}$–extended SUSY *implies*

$$\dim_{\mathbb{R}} \mathcal{M} = \mathbf{N}(\mathcal{N}) m \quad \text{where } m \in \mathbb{N} \text{ and}$$

$$\mathbf{N}(k) := \begin{cases} \mathbf{N}(k+8) = 16\,\mathbf{N}(k), \\ \mathbf{N}(1) = 1, \ \mathbf{N}(2) = 2, \ \mathbf{N}(3) = \mathbf{N}(4) = 4, \\ \mathbf{N}(5) = \mathbf{N}(6) = \mathbf{N}(7) = \mathbf{N}(8) = 8. \end{cases} \quad (2.57)$$

The last statement is just the formula for the allowed dimensions of a module of the Clifford algebra (2.56). The tensors $f^{ai}{}_j \in \mathrm{End}(T\mathcal{M})$ are called *complex structures*, since[20] they have square -1 (cf. Definition 2.1).

The existence of a set of parallel complex structures implies quite strong constraints on the geometry of $\mathcal{M}$. The identity

$$0 = [D_l, D_m] f^{ai}{}_j = R_{lm}{}^i{}_k f^{ak}{}_j - R_{lm}{}^k{}_j f^{ai}{}_k \quad (2.58)$$

shows that each f^a puts an algebraic constraint on the curvature of the manifold $\mathcal{M}$. Therefore *only certain target geometries are compatible with $\mathcal{N}$–extended supersymmetry.* As we increase $\mathcal{N}$ we get geometries that are more and more "special." Understanding the geometries that appear at given $\mathcal{N}$, and their physical meaning, is the main goal of the present book.

---

[19] A tensor or a spinor is called *parallel* iff its covariant derivative vanishes.
[20] According to the mathematical terminology, we should call the f^a *almost* complex structures. We shall show in Part II that they are *true* complex structures. In this chapter we shall be sloppy with language, and refer to them simply as "complex structures."

2.4 Extended supersymmetry in 3D

The automorphism group $SO(\mathcal{N})_R$

In rigid supersymmetry $SO(\mathcal{N})_R$ acts as a global group that *cannot* be gauged.[21] The $\mathcal{N}(\mathcal{N}-1)/2$ matrices Σ^{AB},

$$\Sigma^{1a} = -\Sigma^{a1} := f^a,$$
$$\Sigma^{ab} := \frac{1}{2}\left(f^a f^b - f^b f^a\right) \qquad a, b, = 2, 3, \cdots, \mathcal{N}, \tag{2.59}$$

generate its Lie algebra $\mathfrak{spin}(\mathcal{N})$. The fermions χ^i are in a spinorial representation of this $\mathfrak{spin}(\mathcal{N})$ algebra which, for $\mathcal{N}$ even, has definite chirality. The representation is reducible, in general. $Spin(\mathcal{N})$ acts on the fermions as

$$\chi \mapsto \exp\left(\Lambda_{AB}\Sigma^{AB}\right)\chi, \tag{2.60}$$

and may or may not be a symmetry of the Lagrangian $\mathcal{L}$. Even when we have an R-symmetry, the actual invariance may correspond to a combination of transformations of the form (2.60) and isometries of $\mathcal{M}$ whose infinitesimal action on the fields has the form

$$\delta\phi^i = \Lambda_m K^{mi}, \tag{2.61}$$
$$\delta\chi^i = \pounds_{\Lambda_m K^m}\psi^i + \Lambda_m \Theta^m_{AB}(\Sigma^{AB})^i{}_j\chi^j, \tag{2.62}$$

where K^m are Killing vectors on $\mathcal{M}$ and Θ^m_{AB} are suitable coefficients restricted by the condition of closure of the algebra (Exercise 2.4.2).

Since the tensors f^a are parallel, so are the tensors Σ^{AB}. In view of the fact that the R-symmetry is *not gauged,* this is merely an application of General Lesson 1.4.

Algebraic properties of the complex structures

We need to understand better the algebraic properties of the complex structures $f^{ai}{}_j$, since they are crucial for constructing SUSY/SUGRA models. We denote by $\mathbb{C}l(m)$ the universal Clifford algebra generated over $\mathbb{R}$ by elements Γ^a ($a = 1, 2, \ldots, m$) satisfying

$$\Gamma^a \Gamma^b + \Gamma^b \Gamma^a = -2\,\delta^{ab}, \tag{2.63}$$

and write $\mathbb{A}(n)$ for the algebra of $n \times n$ matrices with entries in the algebra $\mathbb{A}$. One has [262] the following theorem.

---

[21] Otherwise the commutator $[\Lambda(x)_{AB}R^{AB}, \bar{\epsilon}_C Q^C] = (\Lambda_{AC}(x)\bar{\epsilon}_A)Q^C$ of a local R-transformation and a global supersymmetry would produce a local supersymmetry.

Theorem 2.5 *The following are isomorphisms of $\mathbb{R}$–algebras:*

$$Cl(8k) \simeq \mathbb{R}(2^{4k}) \qquad Cl(8k+1) \simeq \mathbb{C}(2^{4k})$$
$$Cl(8k+2) \simeq \mathbb{H}(2^{4k}) \qquad Cl(8k+3) \simeq \mathbb{H}(2^{4k}) \oplus \mathbb{H}(2^{4k})$$
$$Cl(8k+4) \simeq \mathbb{H}(2^{4k+1}) \qquad Cl(8k+5) \simeq \mathbb{C}(2^{4k+2})$$
$$Cl(8k+6) \simeq \mathbb{R}(2^{4k+3}) \qquad Cl(8k+7) \simeq \mathbb{R}(2^{4k+3}) \oplus \mathbb{R}(2^{4k+3})$$

Here there is a subtlety which should be stressed: for $m = 8k+3$ and $m = 8k+7$, corresponding to $\mathcal{N} = 4n$, the Clifford algebra $Cl(m)$ is *not* isomorphic to a matrix algebra but rather to the direct sum of two matrix algebras. From the Theorem we see that the algebra $Cl(m+8)$ is Morita equivalent to $Cl(m)$ (Bott's periodicity). One also shows [262] the following theorem.

Theorem 2.6 *Let $Cl^0(\mathcal{N}) \subset Cl(\mathcal{N})$ be the subalgebra of the elements* even *under the involution $\Gamma^a \mapsto -\Gamma^a$. Then*

$$Cl^0(\mathcal{N}) \simeq Cl(\mathcal{N}-1). \tag{2.64}$$

In view of these results, our General Lesson 2.4 can be restated as:

General Lesson 2.7 *Let $\mathcal{M}$ be the scalar manifold of a 3D $\mathcal{N}$–extended supersymmetric theory. Then $T\mathcal{M}$ is an ungraded module of $Cl^0(\mathcal{N})$.*

The modules of $Cl^0(\mathcal{N})$ are described in Atiyah, Bott and Shapiro ref. [26]; see their Table 2. The dimension of an irreducible module is given by $\mathbf{N}(\mathcal{N})$, where $\mathbf{N}(k)$ is the function defined in Eq. (2.57). As a consequence of the above subtlety, a general $Cl^0(\mathcal{N})$ module has the following structure:

$$\underbrace{M \oplus M \oplus \cdots \oplus M}_{p \text{ times}} \qquad \text{for } \mathcal{N} \not\equiv 0 \mod 4, \tag{2.65}$$

$$\underbrace{M \oplus M \oplus \cdots \oplus M}_{p \text{ times}} \oplus \underbrace{\widetilde{M} \oplus \widetilde{M} \oplus \cdots \oplus \widetilde{M}}_{q \text{ times}} \qquad \text{for } \mathcal{N} \equiv 0 \mod 4, \tag{2.66}$$

where $M, \widetilde{M}$ are non–isomorphic irreducible $Cl^0(\mathcal{N})$ modules. Indeed, precisely for $\mathcal{N} = 4n$ there are *two* such inequivalent irreducible modules, whereas in the other dimensions there is just one isoclass. $\widetilde{M}$ is the *twisted* module. If we view M and $\widetilde{M}$ as *graded* $Cl(\mathcal{N}-1)$–modules,

$$M = M^0 \oplus M^1, \tag{2.67}$$

the twisted module $\widetilde{M}$ is obtained simply by interchanging the two summands ([26] **proposition [5.5]**),

$$\widetilde{M} = M^1 \oplus M^0. \tag{2.68}$$

2.4 Extended supersymmetry in 3D

This means (for $\mathcal{N} \neq 4n$, say) that we may introduce a frame $\{e^i_{\sigma\,r}, \partial_i\}$ in $T\mathcal{M}$, (here $\sigma = 1, 2, \ldots, \mathbf{N}(\mathcal{N})$ and $r = 1, 2, \ldots, \dim \mathcal{M}/\mathbf{N}(\mathcal{N})$), such that

$$f^a\, e_{\sigma\,r} = (\Gamma^a)_\sigma{}^\tau\, e_{\tau\,r}, \qquad (2.69)$$

where the Γ^a are the ordinary (numerical) Dirac matrices of $\mathbb{C}l(\mathcal{N}-1)$. Since the f^a are Levi–Civita parallel, the frame may be taken to be orthonormal. Then, as in Section 1.2.2, the frame $e_{\sigma\,r}$ may be seen as a *bundle isomorphism*

$$T\mathcal{M} \simeq S \otimes \mathcal{U}, \qquad (2.70)$$

where S is a rank $\mathbf{N}(\mathcal{N})$ vector bundle with structure group $Spin(\mathcal{N})_R$, acting as in Eq. (2.60), and $\mathcal{U}$ is the vector bundle associated with the index r, whose structure group is the centralizer C of $Spin(\mathcal{N})$ in $SO(\dim \mathcal{M})$. Note that the frame $\{e_{\sigma\,r}\}$ is unique up to local $Spin(\mathcal{N}) \times C$ rotations.

Thus our General Lesson may be stated, in the spirit of Section 1.2.2, as:

General Lesson 2.8 *$\mathcal{M}$ the manifold of a 3D $\mathcal{N}$–extended* SUSY *model:*

- *If $\mathcal{N} \not\equiv 0 \mod 4$, on $\mathcal{M}$ there are two vector bundles S and $\mathcal{U}$, of respective ranks $\mathbf{N}(\mathcal{N})$ and $(\dim \mathcal{M})/\mathbf{N}(\mathcal{N})$, such that*

$$T\mathcal{M} \simeq S \otimes \mathcal{U}. \qquad (2.71)$$

- *If $\mathcal{N} \equiv 0 \mod 4$, we have two rank $\mathbf{N}(\mathcal{N})$ vector bundles, S and $\widetilde{S}$, and two vector bundles $\mathcal{U}, \widetilde{\mathcal{U}}$, with $\mathrm{rank}\,\mathcal{U} + \mathrm{rank}\,\widetilde{\mathcal{U}} = (\dim \mathcal{M})/\mathbf{N}(\mathcal{N})$, such that*

$$T\mathcal{M} \simeq S \otimes \mathcal{U} \oplus \widetilde{S} \otimes \widetilde{\mathcal{U}}. \qquad (2.72)$$

- *The bundles S, $\widetilde{S}$ have structure group $Spin(\mathcal{N})_R$. In rigid* SUSY *the $Spin(\mathcal{N})_R$ connection – and hence the bundles S, $\widetilde{S}$ – are flat. See Section 2.6 for the corresponding statement in* local SUSY.

The splitting of $T\mathcal{M}$ into two distinct summands for $\mathcal{N} \equiv 0 \mod 4$ is a direct consequence of the subtlety in the structure of the Clifford algebras. This subtlety is well understood in $\mathcal{N} = 4$ SUSY/SUGRA. It corresponds to the statement that the scalar manifold – in general – splits in a product space $\mathcal{M} \times \widetilde{\mathcal{M}}$, with $T\mathcal{M} \simeq S \otimes \mathcal{U}$ and $T\widetilde{\mathcal{M}} \simeq \widetilde{S} \otimes \widetilde{\mathcal{U}}$. The scalars living in the two spaces have different properties under SUSY (as the discussion above implies); according to the standard jargon, the scalars in $\mathcal{M}$ are said to belong to *hypermultiplets* and those of $\widetilde{\mathcal{M}}$ to *twisted hypermultiplets*.

For the cognoscenti. Assume that our 3D SUGRA is obtained by compactifying Type IIA (or Type IIB) superstring on $Y_3 \times S^1$ with Y_3 a Calabi–Yau 3–fold. Then $\mathcal{M}$ parametrizes the complex structures of Y_3 and $\widetilde{\mathcal{M}}$ its complexified Kähler moduli: the interchange of the two $\mathbb{H}$s in $\mathbb{C}l(3) \simeq \mathbb{H} \oplus \mathbb{H}$ thus corresponds to the replacement of Y_3 with its mirror $\widetilde{Y}_3$ [91, 188].

2.4.2 An example: 3D σ–models

For σ–models with only "metric" interactions (i.e., in the absence of superpotential and gauge interactions), General Lesson 2.4 has an inverse:

Proposition 2.9 (SUSY enhancement) *If the metric g_{ij} on the target manifold $\mathcal{M}$ admits $\mathcal{N} - 1$ parallel complex structures which generate the Clifford algebra (2.53), the $\mathcal{N} = 1$ supersymmetry of the corresponding σ–model*

$$\mathcal{L} = -\frac{1}{2} g_{ij} \partial^\mu \phi^i \partial_\mu \phi^j + \frac{i}{2} g_{ij} \bar{\psi}^i \gamma^\mu D_\mu \psi^j - \frac{1}{24} R_{ijkl} \bar{\psi}^i \gamma_\mu \psi^j \bar{\psi}^k \gamma^\mu \psi^l \qquad (2.73)$$

is enhanced to $\mathcal{N}$–extended SUSY. *Moreover $\mathcal{N} = 3 \Rightarrow \mathcal{N} = 4$. If $\mathcal{M}$ is irreducible,[22] the R–symmetry is $SO(2)$ for $\mathcal{N} = 2$ and $SO(3)$ for $\mathcal{N} = 3, 4$. In the presence of a superpotential $W(\phi)$, the condition for the enhancement of supersymmetry is that there exist $\mathcal{N} - 1$ functions $W^a(\phi)$ such that*

$$\partial_i W = f^{a\,j}{}_i \, \partial_j W^a \qquad \text{not summed over } a \qquad (2.74)$$

(the generalized Cauchy–Riemann conditions).

The proof in elementary. The replacement $\psi^i \leftrightarrow f^{a\,i}{}_j \psi^j$ leaves invariant the metric and hence all couplings in (2.73). Hence if $J_\mu = \gamma^\nu \gamma_\mu \psi^i \, g_{ij} \, \partial_\nu \phi^j + \cdots$ is a conserved supercharge, so is $J^a_\mu = \gamma^\nu \gamma_\mu (f^a \psi)^i \, g_{ij} \, \partial_\nu \phi^j + \cdots$. Now assume we have two anticommuting complex structures f^1, f^2. It follows from the theory of the Pauli matrices that $f^3 = f^1 f^2$ is a third complex structure which anticommutes with the f^1 and f^2. $SO(\mathcal{N})_R$ acting on the fermions as in Eq. (2.60) is a symmetry. For even $\mathcal{N}$, the ψs transform as *Weyl* spinors of $Spin(\mathcal{N})$. In particular, for $\mathcal{N} = 4$ – assuming $\mathcal{M}$ irreducible – they transform according to the irreducible representation $(\mathbf{2}, \mathbf{1})$ of $Spin(4) \simeq SU(2) \times SU(2)$; thus the second $SU(2)$ acts trivially on all fields. The effective R–symmetry[23] is then $SU(2)$.

The last statement in proposition 2.9 is the condition under which the replacement $\psi^i \leftrightarrow f^{a\,i}{}_j \psi^j$ does not spoil the SUSY invariance. Written in terms of the new

---

[22] By *irreducible* we mean that the universal cover of $\mathcal{M}$ is not a metric product $\mathcal{M}_1 \times \mathcal{M}_2$.
[23] This is true for a *generic* $\mathcal{N} = 4$ metric. If the given metric has isometries with certain special properties, the R–symmetry may be enhanced to a larger group.

fermions $\psi^{ai} \equiv f^{ai}{}_j \psi^j$ the Lagrangian should have the standard $\mathcal{N} = 1$ SUSY form for some superpotential W^a; this gives (2.74). In the case of $\mathcal{N} = 2$, i.e., just one complex structure, this condition reduces to the usual Cauchy–Riemann equation whose general solution is $W = \operatorname{Re} h(\phi)$, $W^1 = \operatorname{Im} h(\phi)$ with $h(\phi)$ holomorphic.[24] For $\mathcal{N} = 3, 4$ the condition is very restrictive.

Proposition 2.9 is *not* true in presence of Chern–Simons couplings. Generically an $\mathcal{N} = 3$ CS theory is not enhanced to $\mathcal{N} = 4$. In fact, it happens only under very special conditions. Moreover an $\mathcal{N} = 4$ CS model may have the full $Spin(4)$ R–symmetry acting non–trivially on the fields. The conditions under which SUSY enhancement *does* happen are known, and we shall return to them in Chapter 8.

Exercise 2.4.1 Write down the condition on Θ (Eq. (2.62)) from the closure of the R–symmetry algebra. Hint: The Lie derivative with respect to a Killing vector of a parallel 2–form is again a parallel 2–form.

2.5 The language of G–structures. Flat (G_1, G_2)–structures

The results of Section 2.4 are best stated in the language of G–structures [62, 209, 214, 280]. Let G be a closed subgroup of $GL(n, \mathbb{R})$; as reviewed in the Appendix, a *G–structure* on a smooth n–fold $\mathcal{M}$ is a smooth subbundle P_G of the bundle $L(\mathcal{M})$ of the linear frames over $\mathcal{M}$ having structure group G.

Let $\mathcal{M}$ be the scalar manifold of an $\mathcal{N}$–extended 3D SUSY model. The frames $\{e_{\sigma\, r}\}$ introduced around Eq. (2.69) endow $\mathcal{M}$ with a $Spin(\mathcal{N}) \times C$–structure,[25] the groups C being listed in Table 2.5. The case $\mathcal{N} = 4$ is special, since the Lie algebra $\mathfrak{spin}(4) = \mathfrak{su}(2) \oplus \mathfrak{su}(2)$ is not simple, and the indecomposable spinor representations $(\mathbf{2}, \mathbf{1})$ and $(\mathbf{1}, \mathbf{2})$ transform under only one of the two $SU(2)$s. Comparing with the Appendix, we see that $\mathcal{M}$ is endowed with a $(Spin(\mathcal{N}) \times C)$–structure (for $\mathcal{N} = 4$ we may also have $SU(2) \times C$, if only one kind of Clifford module is present). The connection on the $Spin(\mathcal{N}) \times C$–structure is the Levi–Civita one, and hence *torsionless*. The forms Σ^{AB} transform the adjoint of $Spin(\mathcal{N})$ which is non–trivial for $\mathcal{N} \geq 3$. Since Σ^{AB} are parallel, $[\nabla_i, \nabla_j]\Sigma^{AB} = 0$, that is, in $\mathcal{N} \geq 3$ rigid SUSY, the $\mathfrak{spin}(\mathcal{N})$ part of Levi–Civita connection is *flat* (in fact trivial).

---

[24] Here we are cheating a bit: we have not proven yet (but it is true) that all parallel complex structures are integrable. Granted this, the argument is perfectly rigorous.
[25] The *actual* structure may be further reduced to a closed subgroup $G \subset Spin(\mathcal{N}) \times C$; for large $\mathcal{N}$ this is an automatic consequence of the torsionless condition: see Chapters 3,4.

Table 2.5 *Torsionless G–structures on $\mathcal{M}$ for 3D $\mathcal{N}$–extended* SUSY *(rigid or local)*

$\mathcal{N}$	G–structure[a]
$8k$	$Spin(\mathcal{N}) \times SO(n_1/\mathbf{N}(\mathcal{N})) \times SO(n_2/\mathbf{N}(\mathcal{N}))$
$8k \pm 1$	$Spin(\mathcal{N}) \times SO(n/\mathbf{N}(\mathcal{N}))$
$8k \pm 2$	$Spin(\mathcal{N}) \times U(n/\mathbf{N}(\mathcal{N}))$
$8k \pm 3$	$Spin(\mathcal{N}) \times Sp(2n/\mathbf{N}(\mathcal{N}))$
$8k + 4$	$Spin(\mathcal{N}) \times Sp(2n_1/\mathbf{N}(\mathcal{N})) \times Sp(2n_2/\mathbf{N}(\mathcal{N}))$

[a]$\dim_{\mathbb{R}} \mathcal{M} = n$; n_1, n_2 are integers ≥ 0 so that $n_1 + n_2 = n$.

We shall need an extension of the G–structure idea. Consider a sequence of two closed subgroups

$$G_2 \subsetneq G_1 \subset G_0 \equiv GL(n, \mathbb{R}), \tag{2.75}$$

with Lie algebras $\mathfrak{g}_2 \subsetneq \mathfrak{g}_1$ such that there is a G_2–invariant decomposition[26] $\mathfrak{g}_1 = \mathfrak{g}_2 \oplus \mathfrak{m}$, i.e., $[\mathfrak{g}_2, \mathfrak{m}] \subset \mathfrak{m}$. Suppose we have a corresponding sequence of principal subbundles on $\mathcal{M}$

$$P_{G_2} \subset P_{G_1} \subset P_{G_0} \equiv L(\mathcal{M}), \tag{2.76}$$

where P_{G_ℓ} has structure group G_ℓ. This is a pair of G_ℓ–structures $\{P_{G_\ell}\}$ which satisfy mutual compatibility conditions: namely, the invariant tensor/spinor fields for the G_1–structure should also be invariant for the G_2–structures. By itself, this just gives a G_2–structure (which may always be prolonged to a G_1 one). However, suppose in addition that P_{G_1} admits a connection form A_1 which is *flat* and *torsionless*. Let $\varrho \colon \mathfrak{g}_1 \to \mathfrak{g}_2$ be the natural projection. Then we have a natural connection A_2 on P_{G_2}:

$$A_2 = \varrho(A_1)\big|_{P_{G_2}}, \qquad A_1\big|_{P_{G_2}} = A_2 + \Phi. \tag{2.77}$$

The curvature of A_2 has the *minus a commutator* form,

$$F_2 \equiv dA_2 + \frac{1}{2}[A_2, A_2] = -\frac{1}{2}[\Phi, \Phi], \tag{2.78}$$

while the $\mathfrak{m}$–valued 1–form Φ is *covariantly closed*,

$$D\Phi \equiv d\Phi + [A_2, \Phi] = 0. \tag{2.79}$$

[26] This is equivalent to requiring the coset G_1/G_2 to be *reductive*. The geometry of reductive homogeneous spaces is discussed in Chapter 11.

Equations (2.78) and (2.79) are typical of tt^* [87], Higgs bundles [77], and many other geometrical constructions. They will appear again and again in this book in different contexts. We shall refer to this situation as a *flat (G_1, G_2)-structure*, We say that a (flat) (G_1, G_2)-structure is *torsionless* if the underlying G_2-structure is torsionless. In particular, if G_2 is compact and torsionless, A_2 is the Levi-Civita connection of the associated Riemannian metric, and F_2 is the Riemann tensor which takes values in $\Lambda^2(\mathfrak{g}_2)$ and has the special form in Eq. (2.78). Such Riemannian geometries are called *special*.

2.6 Local extended supersymmetry in 3D

In supergravity, SUSY is a local symmetry; that is, the spinorial Grassmann parameter $\epsilon_\alpha^A(x)$ depends on the space–time coordinate x. Each SUSY parameter is associated with a vector–spinor gauge field $\psi_{\mu\alpha}^A$. Its SUSY transformation must have the form[27]

$$\delta \psi_\mu^A = \widehat{\mathcal{D}}_\mu \epsilon^A := D_\mu \epsilon^A + \mathcal{Q}_{\mu B}^A \epsilon^B + \mathcal{M}_{\mu\nu B}^A \gamma^\nu \epsilon^B, \qquad (2.80)$$

where D_μ is the standard covariant derivative acting on spinors, and $\mathcal{Q}_{\mu B}^A$ and $\mathcal{M}_{\mu\nu B}^A$ are some covariant expressions depending on the various fields. Local supersymmetry implies general covariance, since the anticommutator of two x–dependent supersymmetries is an x–dependent translation, i.e., a general reparametrization of the space–time coordinates. In particular, the metric $h_{\mu\nu}(x)$ must be one of the fields in the theory.

3D is special since the metric $h_{\mu\nu}$, the gravitini ψ_μ^A, and the vector fields (with Chern–Simons kinetic terms) A_μ^a do not propagate independent local degrees of freedom. In fact, the Hilbert–Einstein action $\int \sqrt{-h}\, R$, the Rarita–Schwinger one $\int \bar{\psi}_\mu \gamma^{\mu\nu\rho} D_\nu \psi_\rho$, and the Chern–Simons term lead to equations of motion of the form

$$\left. \begin{array}{c} R_{\mu\nu\rho\lambda} \\ D_{[\mu} \psi_{\nu]} \\ F_{\mu\nu} \end{array} \right\} = \text{source currents.} \qquad (2.81)$$

Hence the fields $h_{\mu\nu}, \psi_\mu^A, A_\mu^a$ are pure–gauge away from their sources, which means that the source fields are the only gauge–invariant degrees of freedom.

---

[27] We use the fact that in 3D the four matrices $1, \gamma_\mu$ form a basis of the Clifford algebra, .

2.6.1 Pure 3D $\mathcal{N}$-supergravity

Pure supergravity in 3D exists for *all* $\mathcal{N}$s. The field content is the metric vielbein[28] $e_\mu{}^a$, the spin–connection $\omega_\mu^a \equiv \epsilon^{abc}\omega_{\mu bc}/2$, and $\mathcal{N}$ Majorana gravitini ψ_μ^A ($A = 1, 2 \ldots, \mathcal{N}$). The Lagrangian is the obvious one, Einstein–Hilbert plus Rarita–Schwinger,

$$\mathcal{L}_{SG} = -\frac{1}{2}\epsilon^{\mu\nu\rho}\{e_\mu{}^a R_{\nu\rho a}(\omega) + i\bar{\psi}_\mu^A D_\nu \psi_\rho^A\}, \tag{2.82}$$

where

$$R_{\mu\nu}{}^a = \partial_\mu \omega_\nu{}^a - \partial_\nu \omega_\mu{}^a + \epsilon^{abc}\omega_{\mu b}\omega_{\nu c}, \tag{2.83}$$

$$D_\mu \psi = \left(\partial_\mu + \frac{1}{2}\omega_\mu{}^a \gamma_a\right)\psi, \tag{2.84}$$

$$D_\mu e_\nu{}^a = \partial_\mu e_\nu{}^a + \epsilon^{abc}\omega_{\mu b} e_{\nu c}. \tag{2.85}$$

Here the spin connection $\omega_\mu{}^a$ is meant to be eliminated through its algebraic equations of motion, which just state that the torsion is a bilinear in the gravitini

$$D_{[\mu} e_{\nu]}^a = \frac{i}{4}\bar{\psi}_\mu^A \gamma^a \psi_\nu^A. \tag{2.86}$$

The Lagrangian is invariant under

$$\delta e_\mu{}^a = \frac{i}{2}\bar{\epsilon}^A \gamma^a \psi_\mu^A, \qquad \delta\psi_\mu^A = D_\mu \epsilon^A. \tag{2.87}$$

Indeed,[29]

$$\delta\mathcal{L} = -\frac{i}{4}\epsilon^{\mu\nu\rho}\{\bar{\epsilon}^A \gamma^a \psi_\mu^A R_{\nu\rho a} + 4\bar{\psi}_\mu^A D_\nu D_\rho \epsilon^A\}$$
$$= -\frac{i}{4}\epsilon^{\mu\nu\rho}\{\bar{\epsilon}^A \gamma^a \psi_\mu^A R_{\nu\rho a} + R_{\nu\rho a} \bar{\psi}_\mu^A \gamma^a \epsilon^A\} \equiv 0, \tag{2.88}$$

where we have used the the Ricci identity

$$[D_\mu, D_\nu]\epsilon = \frac{1}{2}R_{\mu\nu a}\gamma^a \epsilon. \tag{2.89}$$

---

[28] The space–time metric is $h_{\mu\nu} = e_\mu{}^a e_\nu{}^b \eta_{ab}$, with $\eta_{ab} = \text{diag}(-1, +1, +1)$.
[29] One has

$$\delta\mathcal{L} = \frac{\delta\mathcal{L}}{\delta e_\mu{}^a}\delta e_\mu{}^a + \frac{\delta\mathcal{L}}{\delta\omega_\mu{}^a}\delta\omega_\mu{}^a + \frac{\delta\mathcal{L}}{\delta\psi_\mu^A}\delta\psi_\mu^A,$$

but, since we are assuming that $\omega_{\mu a}$ satisfies its equation of motion, $\delta\mathcal{L}/\delta\omega_{\mu a} \equiv 0$, we do not need to vary the spin–connection in $\mathcal{L}$.

2.6.2 Coupling 3D SUGRA to a non–linear σ–model

We couple the pure SUGRA to a SUSY σ–model with scalars ϕ^i parametrizing some manifold $\mathcal{M}$ and their spin–1/2 superpartners χ^i. The ϕ^i and the χ^i are the true propagating degrees of freedom.[30]

Up to higher–order couplings, we expect a Lagrangian having the general structure

$$\mathcal{L} = \mathcal{L}_{SG} + \mathcal{L}_{\sigma-\text{model}} + \mathcal{L}_{\text{inter.}} \tag{2.90}$$

At the linearized level in the gravitino fields the interaction Lagrangian, $\mathcal{L}_{\text{inter.}}$, should have the Noether form

$$\sqrt{-h}\,(\text{gauge field})_\mu\,(\text{current})^\mu, \tag{2.91}$$

that is, $\sqrt{-h}\,\bar{\psi}^A_\mu J^{\mu A}$ where J^A_μ is the σ–model supercurrent. In fact, this extends to the full non–linear coupling by the standard rule

$$\frac{\delta \mathcal{L}_{\text{inter.}}}{\delta \bar{\psi}^A_\mu} = \text{properly \underline{supercovariantized} supercurrent.} \tag{2.92}$$

Thus, to have a consistent theory, the matter σ–model must have (at least) $\mathcal{N}$ supersymmetries, i.e., $\mathcal{N}$ conserved supercurrents at the zero order in the supergravity fields. From the SUSY variation $2\,\delta\phi^i = \bar{\epsilon}^A f^{Ai}{}_j \chi^j$, which we take as the definition[31] of both the χ^i and the $f^{Ai}{}_j$, we learn that on $\mathcal{M}$ there should be $(\mathcal{N}-1)$ complex structures,[32] $f^{ai}{}_j$, generating a Clifford algebra as in Eqs. (2.51)–(2.53). To produce more symmetric formulae we set

$$f^{Ai}{}_j := \begin{cases} \delta^i{}_j & A = 1, \\ f^{ai}{}_j & A = 2, 3, \ldots, \mathcal{N}. \end{cases} \tag{2.93}$$

The situation is much the same as in the rigid case, Section 2.4.1, *with two major differences:* the one mentioned in the discussion following Eq. (2.47), and that, in

---

[30] Non–propagating scalar and spin–1/2 fields, if any, have been integrated away.
[31] We have changed the normalization of the spin–1/2 fields, $\chi^i \mapsto \frac{1}{2}\chi^i$ in order to make easier the comparison with the existing literature, which in 3D SUGRA has such a factor of 1/2.
[32] They should be called *almost*–complex structures according to the mathematical terminology.

SUGRA, the automorphism group, $SO(\mathcal{N})_R$, may be gauged, and indeed *should* be gauged. This is already manifest from Eq. (2.80): as we shall show presently, $\mathcal{Q}^A_{\mu B}$ is a composite connection for $SO(\mathcal{N})_R$.

The composite $SO(\mathcal{N})_R$ connection

The general form of $\mathcal{Q}_\mu{}^{AB}$ is[33]

$$\mathcal{Q}_\mu{}^{AB} = \partial_\mu \phi^i \, Q_i^{AB}(\phi) + \text{fermions}. \tag{2.94}$$

The antisymmetric part of Q_i^{AB} is locally on $\mathcal{M}$ a 1–form with coefficients in $\mathfrak{so}(\mathcal{N})_R$ which, in view of Eq. (2.80), transforms as a connection under ϕ–dependent $SO(\mathcal{N})_R$ transformations. It remains to show that the symmetric part of Q_i^{AB} vanishes. Indeed, in order to cancel in the SUSY variation of the Lagrangian all terms proportional to

$$\epsilon^{\mu\nu\rho}\, \partial_\mu \phi^i\, \bar\epsilon^A D_\nu \psi_\rho^B,$$

the same tensor Q_i^{AB} should appear both in the SUSY transformation of the gravitini, Eq. (2.80), and in their kinetic terms

$$\mathcal{L}_{RS} = -\frac{i}{2}\epsilon^{\mu\nu\rho}\, \bar\psi_\mu^A \mathcal{D}_\nu \psi_\rho^A, \tag{2.95}$$

$$\mathcal{D}_\mu \psi_\nu^A := D_\mu \psi_\nu^A + \partial_\mu \phi^i\, Q_i^{AB}\, \psi_\nu^B. \tag{2.96}$$

where D_μ, as before, is the spin–connection covariant derivative. The symmetric part $Q_i^{(AB)}$ decouples from $\mathcal{L}_{RS}$ since $\epsilon^{\mu\nu\rho}\bar\psi_\mu^A \psi_\rho^B$ is antisymmetric in $A \leftrightarrow B$; hence SUSY invariance implies

$$Q_i^{AB} = -Q_i^{BA}, \tag{2.97}$$

and Q_i^{AB} is a $\mathfrak{so}(\mathcal{N})_R$ connection on the scalar manifold $\mathcal{M}$. This interpretation is confirmed by the fact that the proper $SO(\mathcal{N})_R$–covariant derivative $\mathcal{D}_\mu$ enters into the Lagrangian. A crucial aspect of local SUSY is that the $\mathfrak{so}(\mathcal{N})_R$ connection Q_i^{AB} *cannot be flat*, as we shall prove presently.

The connection Q_i^{AB} defines a principal $SO(\mathcal{N})$–bundle $\mathcal{P} \to \mathcal{M}$. We denote by $V(\mathcal{P})$, $\text{Adj}(\mathcal{P})$, and $S_\pm(\mathcal{P})$ the vector bundles associated to $\mathcal{P}$ through, respectively, the vector, adjoint, and $\pm$ chirality spinor representations[34] of $SO(\mathcal{N})$. The SUSY parameters $\epsilon^A(x)$ rotate under local $SO(\mathcal{N})$ transformations. Since $2\,\delta\phi^i = \bar\epsilon_A(f^A\chi)^i$, so do the f^A. Then the $(f^A\chi)^i$ are sections of the vector bundle

$$T\mathcal{M} \otimes V(\mathcal{P}) \to \mathcal{M}. \tag{2.98}$$

---
[33] $SO(\mathcal{N})_R$ indices $A, B, \ldots$ are raised/lowered with the invariant metric δ_{AB}.
[34] Physical consistency requires that the $SO(\mathcal{N})$ connection may be uplifted to a *Spin($\mathcal{N}$)* connection.

We stress that the objects $(f^A\chi)^i$ are the $SO(\mathcal{N})_R$ covariant fields; to define the Fermi fields χ^i we have to choose what we mean by the "first" supercharge and this is a non–covariant $SO(\mathcal{N})_R$ gauge–choice.

Since $V(\mathcal{P})$ is not flat, the f^A cannot be covariantly constant as tensors on $\mathcal{M}$. To understand the property that replaces in local SUSY the covariant constancy of the rigid case, we have to study the χs kinetic terms. But before that we compute the curvature of $\mathcal{P}$.

The $SO(\mathcal{N})_R$ curvature

From the definition (2.96) we have (suppressing $\mathfrak{so}(\mathcal{N})$ matrix indices)

$$\mathcal{D}_\mu \mathcal{D}_\nu = D_\mu D_\nu + (D_\mu \partial_\nu \phi^i) Q_i + \partial_\nu \phi^i \partial_\mu \phi^j (\nabla_j Q_i) \\ + (\partial_\nu \phi^i Q_i D_\mu + \partial_\mu \phi^i Q_i D_\nu) + \partial_\mu \phi^i \partial_\nu \phi^j\, Q_i Q_j, \tag{2.99}$$

so

$$[\mathcal{D}_\mu, \mathcal{D}_\nu] = \mathcal{R}_{\mu\nu} + \partial_\mu \phi^i \partial_\nu \phi^j \big(\partial_i Q_j - \partial_j Q_i + Q_i Q_j - Q_j Q_i\big), \tag{2.100}$$

where $\mathcal{R}_{\mu\nu}$ is the Riemann tensor of $\mathcal{M}$ seen as a 2–form taking values in $\mathfrak{so}(\dim\mathcal{M})$. The term in the parentheses multiplying $\partial_\mu\phi^i\partial_\nu\phi^j$ is precisely the curvature of the $\mathfrak{so}(\mathcal{N})_R$ connection Q_i^{AB}. We write P_{ij}^{AB} for this curvature. Let $\mathcal{L}_{EH}$ and $\mathcal{L}_{RS}$ be the Einstein–Hilbert and gravitino kinetic terms, respectively. The computation in Section 2.6.1 implies (cf. Eq. (2.88))

$$\delta(\mathcal{L}_{EH} + \mathcal{L}_{RS}) = -\frac{i}{2}\epsilon^{\mu\nu\rho}\, \partial_\nu\phi^i \partial_\rho\phi^j\, P_{ij}{}^{AB}\, \bar\psi_\mu^A \epsilon^B. \tag{2.101}$$

To leading order, this term can be canceled only by the variation of χ^i in the Nother coupling $\mathcal{L}_{\text{inter.}}$. In view of Eq. (2.47), one has[35]

$$\mathcal{L}_{\text{inter.}} = \frac{i}{2} e\, g_{ij}\, \bar\psi_\mu^A \gamma^\nu \partial_\nu\phi^i\, \gamma^\mu f^{Aj}{}_k \chi^k + 4\text{-fermions}, \tag{2.102}$$

while the variation of χ^i is

$$\delta\chi^i = \frac{1}{2} g^{ij} g_{kl}\, \gamma^\mu \partial_\mu\phi^k\, f^{A\,l}{}_j \epsilon^A + \text{covariantizing fermions}. \tag{2.103}$$

Thus the variation of χ^i in $\mathcal{L}_{\text{inter.}}$ produces the term

$$\frac{i}{4} e (\bar\psi_\mu^A \gamma^\nu \gamma^\mu \gamma^\rho \epsilon^B)\, g_{ij} g^{kh} g_{lm}\, \partial_\nu\phi^i \partial_\rho\phi^l\, f^{Aj}{}_k f^{Bm}{}_h. \tag{2.104}$$

---

[35] $e \stackrel{\text{def}}{=} \det e_\mu^a \equiv \sqrt{-h}$.

Replacing $\gamma^\nu \gamma^\mu \gamma^\rho = \epsilon^{\nu\mu\rho} +$ traces, we see that the variation (2.104) cancels the one in Eq. (2.101) provided the curvature of the bundle $V(\mathcal{P}) \to \mathcal{M}$ is

$$P_{ij}^{AB} = \frac{1}{2} \Sigma^{AB}_{ij}, \qquad (2.105)$$

where

$$\Sigma^{AB} := -\frac{1}{2}\left(gf^A g^{-1}(f^B)^t g - gf^B g^{-1}(f^A)^t g\right) \qquad (2.106)$$

is the tensor defined in Eq. (2.59), a section of $\wedge^2 T^*\mathcal{M} \otimes \mathrm{Adj}(\mathcal{P})$. Thus,

$$P_{ij}^{AB} = -\frac{1}{4}\left(gf^A g^{-1}(f^B)^t g - gf^B g^{-1}(f^A)^t g\right), \qquad (2.107)$$

a formula which is somehow reminiscent of the structure discussed in Section 2.5, with a curvature that is minus a commutator.

The Bianchi identity is

$$\mathcal{D}_{[i} \Sigma^{AB}{}_{jk]} = 0, \qquad (2.108)$$

where the derivative $\mathcal{D}_i$ is both Levi–Civita and $\mathfrak{so}(\mathcal{N})_R$ covariant.

χ^i kinetic terms

In the discussion around Eq. (2.98), we learned that an $SO(\mathcal{N})_R$ covariant kinetic term for the χs must have the form

$$-\frac{i}{2\mathcal{N}} e\, g_{ij} (f^A \bar{\chi})^i \gamma^\mu \mathcal{D}_\mu (f^A \chi)^j, \qquad (2.109)$$

where $\mathcal{D}_\mu$ is spin–connection, target Levi–Civita, and $\mathfrak{so}(\mathcal{N})_R$ covariant

$$\mathcal{D}_\mu (f^A \chi)^i = \left(\partial_\mu + \frac{1}{2}\omega_\mu^m \gamma_m\right)(f^A \chi)^i + \partial_\mu \phi^j \{\Gamma^i_{jk}(f^A \chi)^k + Q^{AB}_i (f^B \chi)^i\}. \qquad (2.110)$$

Consistency of Eq. (2.109) with Eq. (2.90) requires

$$\mathcal{D}_\mu (f^A \chi)^i \equiv f^A \mathcal{D}_\mu (f^1 \chi)^i, \qquad (2.111)$$

which leads to [119, 120]

$$\nabla_i f^A + Q^{AB}_i f^B \equiv Q^{1b}_i f^A f^b + Q^{A1} f^1. \qquad (2.112)$$

The $A = a$ component of this equation can be recast in the form

$$\nabla_i \Sigma^{1a} + Q^{ab}_i \Sigma^{1b} + Q^{1b}_i \Sigma^{ba} = 0. \qquad (2.113)$$

2.6 Local extended supersymmetry in 3D

This is just one component of a generally covariant equation. So we must have, in full generality,

$$\mathcal{D}_i \Sigma^{AB} \equiv \nabla_i \Sigma^{AB} + Q_i^{AC} \Sigma^{CB} + Q_i^{BC} \Sigma^{AC} = 0; \qquad (2.114)$$

that is, the Σ_{ij}^{AB} – and hence the curvature P_{ij}^{AB} – *are covariantly constant with respect to the Levi–Civita and SO($\mathcal{N}$)$_R$ connections*. In particular, the Bianchi identity (2.108) is trivially satisfied.

In the local case, the condition

$$\mathcal{D}_i \Sigma_{jk}^{AB} = 0 \qquad (2.115)$$

replaces the rigid SUSY requirement that the f^A, and hence *a fortiori* the Σ^{AB}, are *parallel* $D_i \Sigma_{jk}^{AB} = 0$. As a consequence, all tensors constructed out of the Σ^{AB} that are SO($\mathcal{N}$)$_R$ singlets are *parallel* for the Levi–Civita connection. This remark applies in particular to the central tensor $\mathbf{K}_{ijkl} := \epsilon_{ABCD} \Sigma_{ij}^{AB} \Sigma_{kl}^{CD}$ of $\mathcal{N} = 4$ SUGRA.

Finally, we note that the χ^i kinetic terms may be written, canonically but less covariantly, in the form

$$-\frac{i}{2} e\, g_{ij}\, \bar\chi^i \gamma^\mu \widetilde{\mathcal{D}}_\mu \chi^j, \qquad (2.116)$$

where

$$\widetilde{\mathcal{D}}_\mu \chi^i = D_\mu \chi^i + \partial_\mu \phi^j (\Gamma^i_{jk} \chi^k + Q_j^{1a} f^{ai}{}_k) \chi^k. \qquad (2.117)$$

The 4–Fermi coupling

The first source of 4–Fermi couplings is the elimination of the spin connection $\omega_\mu{}^a$ through its equations of motion as in Section 2.6.1. For the other terms we repeat in the 3D context the argument of Section 2.2.2 for the 2D σ–model. The most general form of the coupling of 4 spin–1/2 fields is[36]

$$A_{ijkl}[g]\, \bar\chi^i \gamma^\mu \chi^k\, \bar\chi^j \gamma_\mu \chi^l \qquad (2.118)$$

for some covariant tensor $A_{ijkl}[g]$ on $\mathcal{M}$, depending only on the metric g and its derivatives, and having the following properties:

$$A_{ijkl} = -A_{jikl} = -A_{ijlk} = A_{klij}, \qquad (2.119)$$

$$\nabla_i A_{jklm} + \nabla_j A_{kilm} + \nabla_k A_{ijlm} = 0. \qquad (2.120)$$

---

[36] The simplest way to get Eqs. (2.118), (2.119) and (2.120) is by dimensionally reducing from 3D to 2D; then the arguments of Section 2.2.2 apply directly.

We replace the metric by $g_{ij} \to \lambda g_{ij}$ and take the limit $\lambda \to 0$, that is, infinite σ–model coupling. In this limit the gravitational interactions are negligible, and we should recover (2.73). Recalling the scaling properties of the various polynomials in the curvatures and their covariant derivatives, Section 1.1.1, we see that the general form of the coupling of the four χ^i should be

$$A_{ijkl}[g] = -\frac{1}{4} e R_{ijkl} + a e (g_{ik} g_{jl} - g_{il} g_{jk}), \qquad (2.121)$$

for some universal constant a. The value of a will depend on the precise normalization of the metric. In Exercise 2.6.3 you will show that in a certain natural normalization $a = 1/16$ [119].

The other couplings

To complete the computation of $\mathcal{L}$, it remains to determine a few couplings. They belong to $\mathcal{L}_{\text{inter.}}$, contain at least one gravitino ψ_μ^A, and can be obtained from integrating Eq. (2.92). To do this, define the supercovariant derivative[37] of ϕ^i to be

$$\widehat{\partial}_\mu \phi^i := \partial_\mu \phi^i - \frac{1}{2} \bar{\psi}_\mu^A (f^A \chi)^i. \qquad (2.122)$$

The *"properly supercovariantized supercurrent"* on the RHS of Eq. (2.92) is, quite obviously, the rigid supercurrent with the replacement $\partial_\mu \phi^i \to \widehat{\partial}_\mu \phi^i$. Then the integral is

$$\mathcal{L}_{\text{inter}} = \frac{1}{4} e g_{ij} (f^A \bar{\chi})^i \gamma^\mu \gamma^\nu (\partial_\nu \phi^j + \widehat{\partial}_\nu \phi^j) \psi_\mu^A. \qquad (2.123)$$

This last contribution completes the Lagrangian [119].

2.6.3 Conclusions: the geometry of 3D SUGRA

Let us summarize our results for *local* extended SUSY in 3D. We wrote the most general Lagrangian in the absence of vector fields and scalar potential. However, from the absolute generality of the arguments, it is obvious that our findings remain valid even in the presence of vectors (after they have been dualized as in Section 1.6.2), with the proviso that — in the presence of gauge couplings – the gauge group G should be realized on the scalar manifold $\mathcal{M}$ as a group of isometries. The gauge coupling is then obtained by adding the gauge connection in the covariant derivatives, through the corresponding Killing vectors K^i, as discussed

---

[37] This is the derivative that commutes with local SUSY transformations.

in Section 1.1.2. For $\mathcal{N} > 1$ the SUSY completion of the gauge interactions also requires specific Yukawa couplings, a scalar potential $V(\phi)$ and, of course, the Chern–Simons terms.

The kinetic terms and the 4–χ^i interactions are not affected by the gaugings (Exercise 2.6.3); for the moment we focus on those couplings which are differential–geometric in nature. This does *not* mean that in the gauge and Yukawa couplings there is no interesting geometry: on the contrary, the interplay between the Killing vectors K^i generating the gauge group and the "parallel" tensors f^A and Σ^{AB} is one of the deepest aspects of the geometry and is crucial to understand which gaugings, Yukawas, and potentials are consistent with $\mathcal{N}$–extended SUSY/SUGRA. We shall return to these topics in Chapters 7 and 8 after having developed weapons powerful enough.

Here is what we learned in this section:

General Lesson 2.10 *Let $\mathcal{M}$ be the scalar manifold of a 3D theory with $\mathcal{N}$–extended local supersymmetry in its "canonical" form. Then*

1. *There exists a $\mathrm{Spin}(\mathcal{N})_R$ principal bundle $\mathcal{P} \to \mathcal{M}$, with a covariantly constant curvature, P^{AB}_{ij}*

$$\mathcal{D}_i P^{AB}_{jk} = 0. \tag{2.124}$$

2. *The endomorphisms*

$$\mathbf{1} := \partial_i \otimes d\phi^i, \quad \text{and} \quad L^{AB} := -P^{ABj}_i \partial_j \otimes d\phi^i \in \mathrm{End}(T\mathcal{M}) \tag{2.125}$$

generate a subalgebra of $\mathrm{End}(T\mathcal{M})$ isomorphic to the even Clifford subalgebra $\mathbb{C}l^0(\mathcal{N})$.

3. *Under this isomorphism the curvature endomorphisms L^{AB} are mapped to the standard generators $\tfrac{1}{2}\Gamma^{AB}$ of $\mathfrak{spin}(\mathcal{N})_R$. Hence*

$$T\mathcal{M} \simeq \begin{cases} S(\mathcal{P}) \otimes \mathcal{U} & \mathcal{N} \not\equiv 0 \mod 4, \\ S(\mathcal{P}) \otimes \mathcal{U} \oplus \widetilde{S}(\mathcal{P}) \otimes \widetilde{\mathcal{U}} & \mathcal{N} \equiv 0 \mod 4, \end{cases} \tag{2.126}$$

and this decomposition is preserved by the parallel transport on $\mathcal{M}$. Equivalently, we have the G–structures in Table 2.5.

4. *For $\mathcal{N} \equiv 0 \mod 4$, the two $\mathrm{Spin}(\mathcal{N})$ bundles have opposite Euler classes*

$$e(S(\mathcal{P})) = -e(\widetilde{S}(\mathcal{P})). \tag{2.127}$$

5. *Finally, the 4–χ^i coupling has the form*

$$A_{ijkl}[g] = -\frac{1}{4} e\, R_{ijkl} + \frac{1}{16} e\, (g_{ik}g_{jl} - g_{il}g_{jk}). \tag{2.128}$$

78 *Extended supersymmetry in diverse dimensions*

Exercise 2.6.1 Predict from geometry the coefficient a in Eq. (2.121). HINT: take the target space $P\mathbb{C}^N$ with standard metric canonically normalized as $\text{Vol}(P\mathbb{C}^N) = (4\pi)^N/N$.

Exercise 2.6.2 Prove that the kinetic terms and the 4–Fermi interactions are not affected by gaugings, provided the gauge group acts by isometries of the scalar metric.

Exercise 2.6.3 Prove Eq. (2.127). HINT: use ref. [26], proposition (5.5).

2.7 Connections with algebraic geometry

To illustrate how deep and powerful General Lesson 2.10 is, we state without proof (which will be given in Part II[38] and further elaborated in Part III) what it says in the simplest possible case, namely $\mathcal{N} = 2$ with $\mathcal{M}$ compact. The result is true, word–for–word, also in 4D $\mathcal{N} = 1$ SUGRA [318].

Corollary 2.11 *Let $\mathcal{M}$ be the scalar manifold of a (canonical) 3D $\mathcal{N} = 2$ SUGRA. Assume $\mathcal{M}$ compact. Then $\mathcal{M}$ is a projective algebraic variety over $\mathbb{C}$. Conversely, all projective varieties are target spaces of a $\mathcal{N} = 2$ SUGRA.*

Corollary 2.11 says that for some $M \in \mathbb{N}$ there is an embedding

$$\kappa : \mathcal{M} \hookrightarrow P\mathbb{C}^M, \qquad (2.129)$$

and the metric on $\mathcal{M}$ is the pull–back of twice the canonical one on $P\mathbb{C}^N$. The physical interpretation of this result is that the Newton constant is quantized (for $\mathcal{M}$ compact). See ref. [318] in the context of 4D, $\mathcal{N} = 1$ SUGRA. This in turn gives a natural normalization of the metric, which is the one we used to say that the 4–Fermi coupling a is $1/16$. This quantization is, in fact, Dirac's monopole charge quantization. The curvature $P \equiv P^{12}$ may be seen as an Abelian field–strength in target space $\mathcal{M}$, and its flux on a closed surface $\mathcal{C} \subset \mathcal{M}$ measures the number of monopoles trapped inside $\mathcal{C}$. Then the Corollary 2.11 describes "magnetic monopoles" in target space.

---

[38] Anticipation for the cognoscenti: Let $f^i{}_j$ be the complex structure predicted by $\mathcal{N} = 2$. The 2–form $g_{ik}f^k{}_j$ is proportional to the curvature P, hence closed, and therefore a Kähler form. It is proportional to the curvature of the $SO(2)$ bundle $V(\mathcal{P})$, and hence $\mathcal{M}$ is a Hodge manifold. Then the corollary follows from Kodaira embedding theorem [171, 216].

The map κ is an analogue to the map μ defined in Eq. (1.88). In fact, we have

$$\Sigma \xrightarrow{\Phi} \mathcal{M} \xrightarrow{\kappa} \frac{SU(M+1)}{U(1) \times SU(M)} \simeq P\mathbb{C}^M, \qquad (2.130)$$

and both κ and μ map the space $\mathcal{M}$ in some symmetric space G/H. Metrics, connections, bundles, etc. on $\mathcal{M}$ are obtained from those of G/H via pull–back. The geometry of G/H is completely specified in terms of the Lie groups G and H, so to write the Lagrangian $\mathcal{L}$ one has only to understand which maps κ, μ are allowed. This boils down to the compatibility conditions between the geometric structures pulled–back by κ and, respectively, μ.

Generalization to $\mathcal{N} = 3, 4$: the Swann bundle. A similar structure is present also in the $\mathcal{N} = 3, 4$ cases. The Hodge bundle is replaced in these cases by the Swann bundle over $\mathcal{M}$ [282].

2.8 Supersymmetry in $D = 4$ and 6 dimensions

Although 3D is a nice theoretical laboratory, we are mainly interested in $D \geq 4$ physics. To justify the approach advocated in this book, we have to show that the structures we found in 3D apply, with minor modifications, in all dimensions. From the properties of the supercharges in diverse dimensions, Section 2.1.2, we see that the most interesting dimensions are $D = 3, 4, 6$, and 10. Of course, we are interested in other dimensions as well, but we focus first on these four. From the discussion in Section 2.1.2, we expect that SUSY in 4 and 6 dimensions behaves much like in 3D, except that the underlying geometry is complex, or respectively quaternionic, in nature. This is particularly evident in the σ–models (models with only scalar and spin–1/2 fields). We begin our discussion with these theories, then in Section 2.9 we add the vector fields and analyze the geometry of the SUSY gauge theories.

From SUSY representation theory [281] we know that SUSY σ–models exist in 4D only for $\mathcal{N} = 1, 2$, in 6D only for $(\mathcal{N}_L, \mathcal{N}_R) = (2, 0)$,[39] and in 10D never; in all other cases the minimal supermultiplet contains higher spins. However, here we pretend ignorance and study general $\mathcal{N}$'s. It is for the geometry to say, *a posteriori*, which $\mathcal{N}$ may be realized in a given dimension.

---

[39] To avoid all misunderstanding we stress that $(2, 0)$ is the *minimal* supersymmetry in 6D (eight supercharges) called $(1, 0)$ by other authors. In our conventions $\mathcal{N}_{L,R}$ are even.

2.8.1 Six–dimensional σ–models.

The basic spinors in 6D are symplectic–Majorana–Weyl. To make the SUSY automorphism group $Sp(\mathcal{N})_R$ manifest, we choose a symplectic basis for the right spin–1/2 fields. In such a basis the ψ_α^m ($m = 1, 2, \ldots, 2M$) satisfy the reality condition

$$(\psi_\alpha^m)^\dagger \stackrel{\text{def}}{=} \bar{\psi}_{\alpha\, m} = \Omega_{mn} B_\alpha{}^\beta \psi_\beta^n, \tag{2.131}$$

where Ω_{mn} is a constant $2M \times 2M$ symplectic matrix and B is as in Eq. (2.3). The Fermi kinetics terms then take the canonical form

$$-\frac{i}{2} \Omega_{mn} \psi_\alpha^m \partial^{\alpha\beta} \psi_\beta^n. \tag{2.132}$$

We also take a symplectic basis for the supercharges and SUSY parameters

$$(\epsilon_\alpha^A)^\dagger \stackrel{\text{def}}{=} \bar{\epsilon}_{\alpha\, A} = \epsilon_{AB} B_\alpha{}^{\dot\beta} \epsilon_{\dot\beta}^B. \tag{2.133}$$

Let ϕ^i be a set of (real) scalar fields which parametrize (locally) $\mathcal{M}$. We must have[40]

$$\delta\phi^i = \gamma^i_{A\, m}(\phi)\, \bar{\epsilon}^A \psi^m \tag{2.134}$$

for certain field–dependent coefficients $\gamma^i_{A\, m}$. The reality condition implies

$$(\gamma^i_{A\, m})^* = \epsilon^{AB} \Omega^{mn} \gamma^i_{B\, n}. \tag{2.135}$$

From Eq. (2.134) we see that the supercurrent has the form

$$\Gamma^\nu \Gamma_\mu \psi^m \gamma^i_{A\, m}\, g_{ij}\, \partial_\nu \phi^j + \cdots \tag{2.136}$$

where the ellipsis stand for terms without $\partial_\nu \phi^i$. By the same argument we used in 3D, the SUSY variation of the supercurrent then implies the Clifford–like property

$$\gamma^i_{A\, m}\, g_{ij}\, \gamma^j_{B\, n} + \gamma^i_{B\, m}\, g_{ij}\, \gamma^j_{A\, n} = 2\, \epsilon^{AB}\, \Omega_{mn}. \tag{2.137}$$

The variation $\delta\phi^i$ transforms under a diffeomorphism as a section of $T\mathcal{M}$. Then, geometrically, $\gamma^i_{A\, m}$ is a bundle map

$$\mathcal{S} \otimes \mathcal{U} \to T\mathcal{M}, \tag{2.138}$$

where $\mathcal{S}$ is a rank $\mathcal{N}$ vector bundle related to the index A, and $\mathcal{U}$ is a rank $2M$ vector bundle associated with the index m. The reality condition requires the structure

---

[40] One raises and lowers the indices with the rules $\chi^A = \epsilon^{AB} \chi_B$, $\chi_A = \epsilon_{BA} \chi^B$, $\epsilon_{AB} \epsilon^{BC} = -\delta_A{}^C$ and the same rules for the indices m, n using the symplectic matrix Ω_{mn}.

2.8 Supersymmetry in D = 4 and 6 dimensions

groups of these bundles to be contained in $Sp(\mathcal{N})$ and $Sp(2M)$, respectively. The SUSY generators carry the index A; thus the structure group $Sp(\mathcal{N})$ of $\mathcal{S}$ is, in fact, the automorphism group of the 6D, *left* SUSY algebra. Equation (2.137) says that this map has rank $2\mathcal{N}M$. This is a contradiction, unless $\mathcal{N} = 2$, since each pair of symplectic–Majorana–Weyl spinors is equivalent to four real degrees of freedom, so equality of Bose/Fermi degrees of freedom requires $\dim \mathcal{M} = 4M$ and hence $\mathcal{N} = 2$. Then (2.138) becomes an isomorphism

$$T\mathcal{M} \simeq \mathcal{S} \otimes \mathcal{U}. \tag{2.139}$$

$\gamma^i_{A\,m}$ is like a vielbein in General Relativity, converting the "curved" index i into the "flat" bi–index $A\,m$, except that here the bundle structure group is reduced from $SO(4M)$ to the subgroup $Sp(2) \times Sp(2M)$, as is manifest from the fact that the metric in flat indices is (cf. Eq. (2.137)) [31]

$$g_{ij}\, \gamma^i_{A\,m}\gamma^j_{B\,n} = \epsilon_{AB}\, \Omega_{mn}, \qquad \epsilon^{AB}\, \Omega^{mn}\, \gamma^i_{A\,m}\gamma^j_{B\,n} = g^{ij}. \tag{2.140}$$

In the language of Section 2.5 this is *a $Sp(2) \times Sp(2M)$–structure on $\mathcal{M}$*.

In rigid SUSY the R–group $Sp(\mathcal{N})$ cannot be gauged. Hence, by General Lesson 1.4, $\mathcal{S}$ should correspond to a *parallel structure*. The $Sp(2)$ connection is then flat, and all flat–index tensors of the form (say)

$$T_{A_1 A_2 \cdots A_k}\, \Omega_{m_1 n_1} \cdots \Omega_{m_r n_r} \qquad (T_{A_1 A_2 \cdots A_k} \text{ a numerical tensor}),$$

are covariantly constant. This, in particular, applies to the tensors $L^a_{AB}\, \Omega_{mn}$, where $L^a_{AB} = L^a_{BA}$ are the generators of $\mathfrak{sp}(2) \simeq \mathfrak{su}(2)$ in the defining representation. Converting to curved indices with the help of the γs, we construct three *covariantly constant*, hence closed, 2–forms

$$(\omega^a)_{ij} = \gamma_i^{A\,m}\, \gamma_j^{A\,m}\, L^a_{AB}\, \Omega_{mn}. \tag{2.141}$$

The situation is very much like the one in 3D, the parallel 2–forms L^a associated with the generators of $\mathfrak{sp}(2)$ replacing the parallel 2–forms Σ^{AB} corresponding to the generators of $\mathfrak{so}(\mathcal{N})$.

Except for the limitation to $\mathcal{N}_L = 2$, the structure we find is exactly the same as in 3D with the $SO(\mathcal{N})$ geometric structures replaced by $Sp(\mathcal{N}_L)$ ones, which precisely corresponds to the philosophical principle that we pass from 3D to 6D by a change of the "ground (skew) field" $\mathbb{R} \to \mathbb{H}$. The limitation to $\mathcal{N}_L = 2$ follows from dimensional reduction from 6D to 3D: after reduction, the $Sp(\mathcal{N}_L)$ structures should be reinterpreted as $SO(\mathcal{N})$ ones, and this is possible only for $\mathcal{N}_L = 2$ thanks to the identity $\mathfrak{sp}(2) = \mathfrak{so}(3)$.

2.8.2 Four dimensions

The general pattern should now be clear. In 4D we have complex structures replacing the real ones: Weyl fermions and $U(\mathcal{N})$ automorphism groups instead of Majoranas and $SO(\mathcal{N})$s. Indeed, 4D corresponds to the "ground field" $\mathbb{C}$. The isomorphism in General Lesson 2.10 is replaced by

$$T\mathcal{M} \otimes \mathbb{C} \simeq \mathcal{S} \otimes \mathcal{U} \oplus \overline{\mathcal{S}} \otimes \overline{\mathcal{U}}, \qquad (2.142)$$

where $\mathcal{S}$ is a *flat* $U(\mathcal{N})_R$ bundle and $\overline{\mathcal{S}}$ is its complex conjugate. Correspondingly, we expect a set of $\mathcal{N}^2$ *parallel* 2–forms, $\omega_{ij}^a d\phi^i \wedge d\phi^j$, with the property that the associated endomorphisms $\omega^a{}^i{}_j$ generate the Lie algebra $\mathfrak{u}(\mathcal{N}) \subset \text{End}(\mathcal{S})$. The story is slightly subtler, however, since the actual tangent bundle is not $T_{\mathbb{C}}\mathcal{M} := T\mathcal{M} \otimes \mathbb{C}$ but a real subspace. By construction, $T_{\mathbb{C}}\mathcal{M}$ is a complex vector bundle with a real structure. By Definition 2.1 a real structure is an *anti*linear involution $R: T_{\mathbb{C}}\mathcal{M} \to T_{\mathbb{C}}\mathcal{M}$. The real tangent space $T\mathcal{M} \equiv [[T_{\mathbb{C}}\mathcal{M}]]$ is the R invariant subspace of $T_{\mathbb{C}}\mathcal{M}$. The parallel 2–forms then correspond to the R–even generators of $\mathfrak{u}(\mathcal{N})$, $R^*\omega^a = \omega^a$.

Example. Consider an $\mathcal{N} = 2$ σ–model in 4D obtained from a 6D one by dimensional reduction. Since (for a σ–model) $\mathcal{M}$ is invariant under dimensional reduction, from Section 2.8.1 we learn that the target space has a parallel $Sp(2)$ structure. We have, in general, only three parallel forms, transforming in the adjoint of $Sp(2)$. Given $Sp(2) \simeq SU(2)$ we recover the correct 4D structure, but there is no parallel form associated with the $\mathfrak{u}(1)$ generator: it has been projected out by the reality condition R, which in this case is the quaternionic one, Eq. (2.131).

Consistency under dimensional reduction. As the previous example illustrates, the parallel structures of a σ–model should be invariant under dimensional reduction. In (Q)FT dimensional reduction is a "structure transporting map," like functors in categories. In the case at hand, by pushing down the 4D models to 3D, we get that the $U(\mathcal{N})$ structures should be equivalent to $SO(2\mathcal{N})$ ones. The embedding $U(\mathcal{N}) \hookrightarrow SO(2\mathcal{N})$ is not enough since, in 3D, the flat 2–forms should make a *complete* representation of $SO(2\mathcal{N})$. This constraint has only *two solutions:* $\mathcal{N} = 1$, thanks to the isomorphism $U(1) \simeq SO(2)$, and $\mathcal{N} = 2$. In the second case the three parallel forms coming from 4D make the $(\mathbf{3}, \mathbf{1})$ representation of $SO(4) \simeq SU(2) \times SU(2)$. This result implies that in 4D SUSY σ–models exist only for $\mathcal{N} = 1, 2$.

2.9 4D SUSY gauge theories. Special Kähler geometry

Theories with vectors introduce additional geometric structures, as we saw already at the bosonic level in Section 1.4.1. In this chapter we limit ourselves to the most interesting case, namely 4D. It is convenient to analyze the basic geometric structures with the gauge coupling constants set to zero (that is, all vectors are Abelian). Then to switch on the Yang–Mills coupling g will be easy since the gauge interactions are naturally described by geometrical objects of the $g = 0$ theory, as we saw in Section 1.1.2.

The field content of the models is: scalars ϕ^i ($i = 1, 2, \ldots, n$), spin–1/2 fermions ψ^m ($m = 1, 2, \ldots, F$), and (Abelian) vectors A_μ^u ($u = 1, 2, \ldots, V$). Equality of Bose/Fermi degrees of freedom requires $2F = 2V + n$.

2.9.1 Gauge–matter splitting

As the reader already knows from the superspace approach, or from the theory of SUSY representations on fields, in general there may be two kinds of spin–1/2 field: those in the same supermultiplet with the vectors *(gaugini)*, and those belonging to supermultiplets containing only spin–0 and spin–1/2 fields. In geometric language, the vectors' SUSY transformations, $\delta A_\mu^u = \bar{\epsilon}_A \gamma_\mu \lambda^{Au}$, split the Fermi bundle $\mathcal{V} \to \mathcal{M}$ into the direct sum $\Lambda \oplus \Lambda^\perp$, where Λ is the bundle of the fermions λ^{Au} appearing in the δA_μ^u, whereas $\Lambda^\perp$ corresponds to the fermions which are orthogonal to the λ^{Au} (with respect to the metric defined by the kinetic terms). As in 3D, we fix our attention on an $\mathcal{N} = 1$ subalgebra. The scalars' transformation under this subalgebra, $\delta \phi^i = \bar{\epsilon}_1 \psi^i$, defines a bundle *monomorphim*

$$T\mathcal{M} \xrightarrow{j} \mathcal{V} \simeq \Lambda \oplus \Lambda^\perp \quad \Rightarrow \quad T\mathcal{M} \simeq (\Lambda/\mathrm{coker}\,j) \oplus \Lambda^\perp, \tag{2.143}$$

where we used the fact that $\Lambda^\perp \subset j(T\mathcal{M})$ by the positivity of the kinetic terms of the $\psi \in \Lambda^\perp$ (in down–to–earth language: these fermions are not the superpartners of a vector, hence they should be the superpartners of a scalar). We claim that the splitting in Eq. (2.143) is preserved by parallel transport on $\mathcal{M}$ (Exercise 2.9.4). We shall show in Chapter 3 that this condition implies that the manifold $\mathcal{M}$ is a Riemannian product

$$\mathcal{M} = \mathcal{M}_{\text{gauge}} \times \mathcal{M}_{\text{matter}}, \tag{2.144}$$

with a direct sum metric

$$ds^2 = g^{(1)}_{mn}(x)\,dx^m dx^n + g^{(2)}_{ab}(y)\,dy^a dy^b, \tag{2.145}$$

where $g^{(1)}$ (resp. $g^{(2)}$) is a metric on $\mathcal{M}_{\text{gauge}}$ (resp. $\mathcal{M}_{\text{matter}}$). The geometry of $\mathcal{M}_{\text{matter}}$ is, of course, that of an $\mathcal{N}$–extended σ–model, already discussed in Section 2.8. It remains to analyze $\mathcal{M}_{\text{gauge}}$. As we know from representation theory, SUSY gauge theories exist in 4D for three values of $\mathcal{N}$, namely 1, 2, and 4. The

$\mathcal{N} = 1$ gauge supermultiplet has no scalars, so $\mathcal{M}_{\text{gauge}} =$ (a point), and there is no geometry to study.[41] We start with the $\mathcal{N} = 4$ case.

2.9.2 $\mathcal{N} = 4$ gauge theories

The $\mathcal{N} = 4$ geometry is so constrained that we do not have much work to determine it. Schematically, the SUSY tansformations have the generic form

$$\delta A^u_{\alpha\dot{\alpha}} = \bar{\epsilon}_{\dot{\alpha}A} \lambda^{Au}_\alpha + \text{H.c.}, \qquad (2.146)$$

$$\delta \lambda^{Au}_\alpha = E_i^{ABu}(\phi) \partial^{\dot{\alpha}}_\alpha \phi^i \, \bar{\epsilon}_{\dot{\alpha}B} + F^u_{\alpha\beta} \epsilon^{\beta\gamma} \epsilon^A_\gamma, \qquad (2.147)$$

where $E_i^{ABu}(\phi)$ is the analogue of γ_i^{Am} in Section 2.8.1. From the SUSY algebra and the reality conditions[42] one gets

$$E_i^{ABu} = -E_i^{BAu}, \qquad (2.148)$$

$$E_{iAB}{}^u \stackrel{\text{def}}{=} (E_i^{ABu})^* = -\frac{1}{2}\epsilon_{ABCD} E_i^{CDu}. \qquad (2.149)$$

Just as in Section 2.8.1, we interpret this result as a bundle isomorphism:

$$T\mathcal{M}_{\text{gauge}} \simeq \mathcal{S}_6 \otimes \mathcal{U}, \qquad (2.150)$$

where $\mathcal{S}_6$ is the vector bundle associated with the representation **6** of $SU(4)_R \sim SO(6)$ and $\mathcal{U}$ is a real bundle. Again, $\mathcal{S}_6$ is flat (otherwise we would be forced to gauge $SU(4)_R$, and that would contradict rigid SUSY) and hence for all constant symmetric 6×6 matrices $S_{AB\,CD}$ the symmetric tensor[43]

$$\hat{S}_{ij} := E_i^{ABu} E_j^{CDv} \mathcal{N}_{uv} S_{AB\,CD} \qquad (2.151)$$

should be *parallel*. This is a very strong constraint on the geometry of $\mathcal{M}$, and – as we shall see in Chapter 3 – it implies its metric to be flat. Hence, in this case, there is no need for a more in–depth analysis.

---

[41] This is *not* correct: although the metric geometry of $\mathcal{M}$ is just the one of an $\mathcal{N} = 1$ σ–model, the vector kinetic terms define a map $\mu: \mathcal{M} \to Sp(2V, \mathbb{R})/U(V) \simeq \mathfrak{H}_V$ and one has to describe the geometric conditions on the map μ for the model to be supersymmetric. We know that $T_{\mathbb{C}}\mathcal{M} \simeq T \oplus \bar{T}$; the same argument (applied to the action of $U(V)$ on $T\mathfrak{H}_V$) gives $T\mathfrak{H}_V = \mathfrak{h} \oplus \bar{\mathfrak{h}}$. The condition will turn out to be $\mu_*T \subset \mathfrak{h}$ (μ is holomorphic).
[42] In the linearized theory the scalars $\phi^{ABu} \sim E_i^{ABu} \phi^i$ transform in the real **6** rep. of $SU(4)$. This fixes the properties of E_i^{ABu}.
[43] As in Section 1.4.2 $\mathcal{N}_{uv}$ is the matrix appearing in the vector kinetic terms.

2.9.3 $\mathcal{N} = 2$ gauge theories in 4D

Geometrically speaking, the rigid $\mathcal{N} = 2$ gauge theories in 4D are among the most interesting field theories. The discussion is less elementary than the rest of the chapter, and the reader may prefer to skip it.

There are two fundamental structures. The first one is easily obtained by the "forgetful trick": Forget half supercharges, and consider the model just an $\mathcal{N} = 1$ supersymmetric theory. The result of Section 2.8.2. applies; hence

$$T_{\mathbb{C}}\mathcal{M} \simeq F \oplus \overline{F} \qquad \text{(where } F \simeq \mathcal{S} \otimes \mathcal{U}\text{),} \tag{2.152}$$

that is, on $\mathcal{M}$ there is <u>one</u> *parallel* complex structure and then a torsionless $U(V)$–structure (cf. Table 2.5). Correspondingly, the cotangent bundle also splits $T^*_{\mathbb{C}}\mathcal{M} \simeq F^\vee \oplus \overline{F}^\vee$. Given a (complex) 1–form ξ, that is, a section of $T^*_{\mathbb{C}}\mathcal{M}$, we can decompose it as

$$\xi = \xi|_{F^\vee} + \xi|_{\overline{F}^\vee}. \tag{2.153}$$

$\xi|_{F^\vee}$ (resp. $\xi|_{\overline{F}^\vee}$) is called the $(1, 0)$ (resp. $(0, 1)$) part of the form ξ. We apologize for anticipating here some results from Parts II and III (expecially Chapter 9): there we show that *parallel* (almost) complex structures are *integrable*; that is, there exist on $\mathcal{M}$ holomorphic functions z^i ($i = 1, 2, \ldots, \dim \mathcal{M}/2$) such that their differentials dz^i (resp. $d\bar{z}^{\bar{i}}$) span $F^\vee$ (resp. $\overline{F}^\vee$), i.e., they form a basis for the $(1, 0)$ (resp. $(0, 1)$) differential forms. More generally, a differential form Ξ, of degree $n = p + q$, is called a form of *type* (p, q) if, in holomorphic coordinates, it takes the form

$$\Xi_{i_1 i_2 \cdots i_p \, \bar{j}_1 \bar{j}_2 \cdots \bar{j}_q} \, dz^{i_1} \wedge dz^{i_2} \wedge \cdots \wedge dz^{i_p} \wedge d\bar{z}^{\bar{j}_1} \wedge \cdots \wedge d\bar{z}^{\bar{j}_q}. \tag{2.154}$$

The second structure is subtler. It is a *real* structure and is best described from the real point of view. Thus we revert to Majorana (= Hermitian) supercharges.[44] The formula[45]

$$-i\epsilon_{AB} \, C^{\alpha\beta} \, \{Q^A_\alpha, [Q^B_\beta, A^u_\mu]\} = Q^u_i(\phi) \, \partial_\nu \phi^i \tag{2.155}$$

defines a set of V real 1–forms, $\theta^u \equiv Q^u_i(\phi) \, d\phi^i$ on $\mathcal{M}$ ($a = 1, 2, \ldots, V$).

In Section 1.4.1 we learned that the theory has a dual formulation in terms of the dual vectors B_v (defined by $G_v = dB_v$). The dual of Eq. (2.155) reads

$$-i\epsilon_{AB} \, C^{\alpha\beta} \, \{Q^A_\alpha, [Q^B_\beta, B_{v\,\mu}]\} = P_{v\,i}(\phi) \, \partial_\nu \phi^i, \tag{2.156}$$

---

[44] C stands for the Majorana representation charge conjugation matrix. It satisfies $C^t = -C$.
[45] In the linearized theory the RHS is proportional the gradient of a scalar belonging to the vector supermultiplet; in the curved theory we should allow for a vielbein–like tensor to convert "curved" indices into "flat" ones, exactly as we did in Section 2.8 for the σ models.

with $\widetilde{\theta}_v \equiv P_{vi} d\phi^i$ *another* set of V real 1–forms on $\mathcal{M}$. By linearity of Eqs. (2.155) and (2.156), the vector of 1–forms $\Theta^U \equiv \begin{pmatrix} \theta^u \\ \widetilde{\theta}_v \end{pmatrix}$ transforms canonically under $Sp(2V, \mathbb{R})$:

$$\begin{pmatrix} \theta^u \\ \widetilde{\theta}_v \end{pmatrix} \longrightarrow \mathbf{S} \begin{pmatrix} \theta^u \\ \widetilde{\theta}_v \end{pmatrix}, \qquad \text{where } \mathbf{S} \in Sp(2V, \mathbb{R}). \tag{2.157}$$

Consider the associated (real) symplectic 2–form

$$\omega := \theta^u \wedge \widetilde{\theta}_u. \tag{2.158}$$

As we saw in Chapter 1, this symplectic structure cannot be degenerate, otherwise the energy–momentum tensor would be singular. Then $\omega^V \neq 0$. Since $\dim_\mathbb{R} T^*\mathcal{M} = 2V$, the 1–forms Θ^U ($U = 1, 2, \ldots, 2V$) must span the cotangent bundle $T^*\mathcal{M}$. Thus Θ^U defines a bundle isomorphism

$$T^*\mathcal{M} \simeq \mathcal{S}, \tag{2.159}$$

where $\mathcal{S}$ is a vector bundle with structure group $Sp(2V, \mathbb{R})$. In other words: *on $\mathcal{M}$ there is an $Sp(2V, \mathbb{R})$–structure*. $\mathcal{S}$ is a very special kind of bundle: it is a *flat* one; that is, the curvature of its connection vanishes identically. This follows from the fact that only *rigid* $Sp(2V, \mathbb{R})$ rotations are allowed, i.e., the matrix $\mathbf{S}$ in Eq. (2.157) should be locally independent of the point[46] $\phi \in \mathcal{M}$; therefore the $Sp(2V, \mathbb{R})$ holonomy, $P \exp\bigl(-\int_C A_{Sp(2V,\mathbb{R})}\bigr)$, cannot change under a continuous deformation of the path C. Then, if C is contractible, its holonomy equals 1, which means that $A_{Sp(2V,\mathbb{R})}$ is locally *pure gauge* (= no curvature).

We write ∇_i for the $Sp(2V, \mathbb{R})$ covariant derivative, and $\nabla \equiv d\phi^i \nabla_i$ for the associated exterior derivative. The flatness condition is

$$\nabla^2 = 0. \tag{2.160}$$

The coframe Θ^U is the symplectic analogue of the usual orthogonal coframe (vielbein forms) of General Relativity, the usual orthogonal (symmetric) pairing δ_{ab} being replaced by the symplectic (skew–symmetric) one Ω_{UT}, and the $\mathfrak{so}(n)$ spin–connection 1–form by an $\mathfrak{sp}(2V, \mathbb{R})$ connection 1–form $\varpi^U{}_T \in \Omega^1(\mathfrak{sp}(2V, \mathbb{R}))$ (this holds in a *general* symplectic coframe; in our particular coframe Θ^U the connection $\varpi^X{}_Y$ vanishes by construction). Then we have the symplectic versions of the Cartan's curvature and structural equations:

$$d\Theta^U + \varpi^U{}_T \wedge \Theta^T = T^U \qquad \text{(symplectic torsion)}, \tag{2.161}$$

$$d\varpi^U{}_T + \varpi^U{}_S \wedge \varpi^S{}_T = R^U{}_T \qquad \text{(symplectic curvature)}. \tag{2.162}$$

---

[46] We can, or better we should, have non–trivial monodromies at the global level.

2.9 4D SUSY gauge theories. Special Kähler geometry

We already know that $R^U{}_T = 0$. What about the symplectic *torsion?*

Lemma 2.12 *The symplectic torsion vanishes identically $T^U \equiv 0$.*

There are many ways of seeing this. By far the simplest method is to use $\mathcal{N} = 2$ superfields. However, let us sketch a poor man's argument, showing that the vanishing of the symplectic torsion on $\mathcal{M}$ is just the push–forward to the target space of gauge–invariance in space–time Σ. Again, $\Phi \colon \Sigma \to \mathcal{M}$ is the scalars' field configuration, seen as a map.

Equation (2.155) may be written as ($A^u \equiv A^u_\mu \, dx^\mu$)

$$\Phi^* \theta^u = -i\epsilon_{AB} C^{\alpha\beta} \{Q^A_\alpha, [Q^B_\beta, A^u]\}. \tag{2.163}$$

The symmetry of the index contractions allows to rewrite the RHS, with the help of the super–Jacobi identity, in the form

$$\Phi^* \theta^u = -\frac{i}{2} [\epsilon_{AB} C^{\alpha\beta} \{Q^A_\alpha, Q^B_\beta\}, A^u]. \tag{2.164}$$

The RHS is formally zero, since the anticommutator inside the bracket is a central charge, and a massless gauge vector cannot be charged under a central charge. The apparent paradox is easily solved: the SUSY algebra is realized only up to gauge transformations, that is, only on *gauge–invariant* operators. But then

$$\Phi^* d\theta^u \equiv d\Phi^* \theta^u = -\frac{i}{2} [\epsilon_{AB} C^{\alpha\beta} \{Q^A_\alpha, Q^B_\beta\}, F^u] \equiv 0, \tag{2.165}$$

since $F^u = dA^u$ is gauge–invariant. Then $d\theta^u = 0$, and, by $Sp(2V, \mathbb{R})$ covariance,

$$\mathbf{T}^U \equiv d\Theta^U \equiv d\begin{pmatrix} \theta^u \\ \tilde{\theta}_v \end{pmatrix} = 0. \tag{2.166}$$

The compatibility conditions. 4D $\mathcal{N} = 2$ SUSY predicts that the manifold $\mathcal{M}_{\text{gauge}}$ has a $U(V)$–structure as well as an $Sp(2V, \mathbb{R})$–structure (both torsionless). These two structures should be compatible with each other. First of all, the parallel symplectic 2–form of the $Sp(2V, \mathbb{R})$–structure, ω (see Eq. (2.158)), and the parallel symplectic 2–form of the $U(V)$–structure, $g_{ik} f^k{}_j d\phi^i \wedge d\phi^j$, should coincide (if properly normalized). We can see this in two different ways: geometrically, we cannot have two non–proportional *commuting* symplectic forms on an irreducible manifold. Physically, the constraints on the fermion kinetic terms we get by relating them to the scalars should match those we obtained by relating them to the vectors. In particular, it is easy to see directly that the canonical symplectic 2–form ω should be of type $(1, 1)$ with respect to the parallel complex structure: in fact, ω should remain invariant under the action of the structure subgroup $U(1)_R$

of the bundle F in Eq. (2.152) because the vectors A_μ^u have zero R–charge. Alternatively, one may extract this result from the consistency of the canonical form of the energy–momentum tensor with SUSY.

The second consistency condition may be stated as the requirement

$$\nabla\left(\Theta^U|_{(1,0)}\right) = 0, \qquad (2.167)$$

known as the *special Kähler condition* [50, 145].

Corollary 2.13 (to **Lemma 2.12**)   *The special Kähler condition holds.*

Proof   From Eq. (2.163), we have

$$\Phi^*\left(\theta^u|_{(1,0)}\right) = -i\,\epsilon_{AB}\,\epsilon^{\alpha\beta}\,\{Q_\alpha^A, [Q_\beta^B, A^u]\}, \qquad (2.168)$$

where now Q_α^A denotes *Weyl* (2–component) supercharges. Again, the anticommutator inside the bracket is a central charge, and hence

$$\nabla\left(\theta^u|_{(1,0)}\right) = d\left(\theta^u|_{(1,0)}\right) = 0. \qquad (2.169)$$

$\square$

Remark. The $(1,0)$ form $\Theta^U|_{(1,0)}$ is ∇–closed but *not* ∇_i–parallel.

The geometry resulting from *compatible U(V)– and Sp(2V, $\mathbb{R}$)–structures* is called *special Kähler geometry*.

The prepotential

Let us show that the above description is equivalent to the usual one in terms of a prepotential. The fact that ∇_i is both flat and torsionless implies that there exist (local) Darboux coordinates, i.e., local functions (q^x, p_y) such that $\theta^x = dq^x$, $\widetilde{\theta}_y = dp_y$ and

$$\omega = dq^x \wedge dp_x, \qquad (2.170)$$

while we may set $\nabla = d$. On the other hand, let z^i be holomorphic coordinates for $\mathcal{M}$. We must have $dq^x|_{(1,0)} = A_i^x\,dz^i$, $dp_y|_{(1,0)} = B_{iy}\,dz^i$, for certain functions A_i^x and B_{iy}. The special Kähler condition requires $\nabla dq^x|_{(1,0)} = \nabla dp_y|_{(1,0)} = 0$; decomposing into type we get

$$\overline{\partial}A_i^x = \overline{\partial}B_{iy} = 0, \qquad (2.171)$$

$$\partial(A_i^x\,dz^i) = \partial(B_{iy}\,dz^i) = 0. \qquad (2.172)$$

$A_i^x \, dz^i$ and $B_{iy} dz^i$ are ∂–closed *holomorphic* forms; by the Poincaré lemma, locally on $\mathcal{M}$, there exist holomorphic functions $Z^x(z)$, $W_y(z)$ such that

$$dq^x|_{(1,0)} = \frac{1}{2} dZ^x \Rightarrow dq^x = \mathrm{Re}(dZ^x), \tag{2.173}$$

$$dp_y|_{(1,0)} = -\frac{1}{2} dW_y \Rightarrow dp_y = -\mathrm{Re}(dW_y). \tag{2.174}$$

By the first compatibility condition, the symplectic form

$$\omega = -\frac{1}{4}(dZ^x + d\overline{Z}^x) \wedge (dW_x + d\overline{W}_x) \tag{2.175}$$

has pure type $(1,1)$; hence its $(2,0)$ part vanishes,

$$dW_x \wedge dZ^x = 0 \Rightarrow d(W_x \, dZ^x) = 0 \Rightarrow W_x \, dZ^x = d\mathcal{F}, \tag{2.176}$$

for some holomorphic $\mathcal{F}$ function (again the Poincaré lemma). $\mathcal{F}$ is known as *the superpotential*. Choosing the Z^x as the (local) holomorphic coordinates, we write

$$W_x = \frac{\partial \mathcal{F}}{\partial Z^x}, \tag{2.177}$$

and ω is

$$\omega = \frac{i}{2} \mathrm{Im}\left(\frac{\partial \mathcal{F}}{\partial Z^x \partial Z^y}\right) dZ^x \wedge d\overline{Z}^y. \tag{2.178}$$

2.9.4 Special Kähler geometry as a (G_1, G_2)-structure

In the language of Section 2.5, a special Kähler geometry is just a *torsionless* $(Sp(2n,\mathbb{R}), U(n))$–structure on the $2n$–fold $\mathcal{M}$. Indeed, the first compatibility condition says that the symplectic form of the $U(n)$–structure corresponding to the Kähler geometry is the same as the symplectic form of the flat $Sp(2n,\mathbb{R})$ structure, which is just the same as saying that $P_{U(n)}$ is a subbundle of the flat $P_{Sp(2n,\mathbb{R})}$ principal bundle. The second compatibility condition (2.167) is equivalent to stating that the connection $A_{U(n)}$ of Eq. (2.77) is *torsionless*. Let us show that (2.167) implies that the $(Sp(2n,\mathbb{R}), U(n))$–structure is torsionless. Indeed, we saw above that (2.167) implies

$$\Theta|_{(1,0)} = \begin{pmatrix} dz^i \\ dw_j \end{pmatrix}, \qquad i,j = 1,2,\ldots,n, \tag{2.179}$$

where z^i and w_j are two sets of local complex coordinates on $\mathcal{M}$. The flat $Sp(2n,\mathbb{R})$ connection has the general form

$$\nabla_i dz^j = A^j_{ik} dz^k + A^j_{i\bar{k}} d\bar{z}^k, \qquad \nabla_{\bar{i}} dz^j = A^j_{\bar{i}k} dz^k + A^j_{\bar{i}\bar{k}} d\bar{z}^k. \tag{2.180}$$

The subalgebra $\mathfrak{u}(n) \subset \mathfrak{sp}(2n,\mathbb{R})$ is given by the elements commuting with the complex structure; the induced $U(n)$ connection $\nabla = d + A_{\mathfrak{u}(n)}$ is obtained, as in Section 2.5, by projecting ∇ on $\mathfrak{u}(n)$, that is, by keeping only the $(1,0)$ part of the RHS of (2.180):

$$\nabla_i dz^j = A^j_{ik} dz^k, \qquad \nabla_{\bar{i}} dz^j = A^j_{\bar{i}k} dz^k. \tag{2.181}$$

∇ is torsionless if and only if

$$A^j_{ik} = A^j_{ki}, \qquad A^j_{\bar{i}k} = 0, \tag{2.182}$$

while Eq. (2.167), $\nabla dz^j \equiv 0$, yields

$$A^j_{ik} = A^j_{ki}, \qquad A^j_{i\bar{k}} = A^j_{\bar{i}k} = 0, \qquad A^j_{\bar{i}\bar{k}} = A^j_{\bar{k}\bar{i}}, \tag{2.183}$$

which implies (2.182). By construction, ∇ is both metric and torsionless, and hence is the Levi–Civita connection of the underlying Kähler metric

$$A^j_{ik} = \Gamma^j_{ik}, \qquad A^j_{\bar{i}k} = 0. \tag{2.184}$$

Since, in the Θ coframe, $\nabla = d$,

$$0 = \nabla_i(dz^j + d\bar{z}^j) = \Gamma^j_{ik} dz^k + \left(A^{\bar{j}}_{i\bar{k}}\right)^* dz^k \quad \Rightarrow \quad A^{\bar{j}}_{i\bar{k}} = -\left(\Gamma^j_{ik}\right)^*. \tag{2.185}$$

Hence $A_{\mathfrak{u}(n)}$ torsionless implies the special Kähler condition (2.183).

Remark. For any G_1-structure P_{G_1} and closed subgroup $G_2 \subset G_1$, let P_{G_1}/G_2 be the bundle with fiber G_1/G_2 associated with the bundle P_{G_1}.

Proposition 2.14 ([215] proposition 5.6) *The principal subbundles $P_{G_2} \subset P_{G_1}$ with group G_2 are in one-to-one correspondence with the sections*

$$\mathcal{M} \to P_{G_1}/G_2. \tag{2.186}$$

For the $(Sp(2n,\mathbb{R}), U(n))$-structure on $\mathcal{M}$, Proposition 2.1.4 says that the $U(n)$-substructure of $P_{Sp(2n,\mathbb{R})}$ is *locally* equivalent to a map

$$\mu \colon \mathcal{M} \to Sp(2n,\mathbb{R})/U(n), \tag{2.187}$$

that is, to specifying the couplings of the Abelian vectors (Section 1.4). The section of (2.186) is just the vielbein $\mathcal{E}(\phi)^{-1}$ of Chapter 1. Then $A_{\mathfrak{u}(n)} = (\mathcal{E} d\mathcal{E}^{-1})_{\mathfrak{u}(n)}$.

The structural equations and Lax pair

We write D for the $(1,0)$ part of the Levi–Civita connection ∇, and $\mathfrak{sp}(2n,\mathbb{R}) = \mathfrak{u}(n) \oplus \mathfrak{m}$. The difference of two connections is a tensor field. We define C as the

2.9 4D SUSY gauge theories. Special Kähler geometry

$\mathfrak{m}$–valued $(1,0)$–form

$$C = (\nabla - \overline{\nabla})|_{(1,0)}, \qquad (2.188)$$

so that

$$\boldsymbol{\nabla} = D + \overline{D} + C + \overline{C}. \qquad (2.189)$$

$\boldsymbol{\nabla}$ is *flat*. Since $\boldsymbol{\nabla}$ is Kählerian, decomposing into type $\boldsymbol{\nabla}^2 = 0$ yields

$$D^2 = \overline{D}^2 = 0, \qquad (2.190)$$

$$DC = \overline{D}\,\overline{C} = 0, \qquad (2.191)$$

$$\left(D\overline{D} + \overline{D}D + C \wedge \overline{C} + \overline{C} \wedge C\right) + \left(D\overline{C} + \overline{D}C\right) = 0. \qquad (2.192)$$

The terms in the first parenthesis of Eq. (2.192) belong to $\mathfrak{u}(n)$ while those in the second one belong to $\mathfrak{m}$; hence the two expressions vanish separately. The first gives for the curvature of the special Kähler metric

$$R_{i\bar{j}} = -[C_i, \overline{C}_{\bar{j}}], \qquad (2.193)$$

which has the *minus a commutator* form. The two terms in the second parentheses of (2.192) belong to distinct subspaces of $\mathfrak{sp}(2n, \mathbb{R})$ [their index structure is $(D\overline{C})^{\bar{j}}{}_i$ and $(\overline{D}C)^j{}_{\bar{i}}$] and vanish independently:

$$D\overline{C} = \overline{D}C = 0. \qquad (2.194)$$

Finally, the torsionless condition gives

$$C_{ijk} := g_{i\bar{l}}(C_j)^{\bar{l}}{}_k = \text{totally symmetric in } i, j, k. \qquad (2.195)$$

The equations (2.195), (2.190), (2.191), (2.193) and (2.194) are typical of the tt^* geometry [87, 130]. However, special Kähler geometry is *not* a particular instance of tt^* geometry, since the groups are different in the two cases. Nevertheless, the Lax pair representation of tt^* geometry extends to the special Kähler geometry: it says that the connection

$$\boldsymbol{\nabla}^{(\lambda)} = \boldsymbol{\nabla} + \lambda C + \lambda^{-1} \overline{C} \qquad (2.196)$$

is flat for *all* values $\lambda \in \mathbb{C}^\times$ of the spectral parameter. Then any special Kähler manifold corresponds to an integrable system. For alternative points of view on integrability see refs. [129] and [145].

Exercise 2.9.1 Prove the claim stated after Eq. (2.143).

Exercise 2.9.2 Prove that a Ricci–flat special Kähler manifold is flat.

Exercise 2.9.3 Show that the Ricci curvature of a special Kähler manifold is non–negative.

2.10 4D supergravity

We close this introductory chapter by presenting some geometrical aspects of 4D extended supergravity. The structures are similar to the ones we found in 3D, but now the graviton, the gravitini, and the gauge vectors do propagate physical states. At first sight, the theory looks quite a mess; however, once one has understood its internal logic, it appears to be rather elegant and simple. This section is meant as a mere appetizer. We shall return to 4D SUGRA after the development of the necessary tools in Parts II and III. Our immediate goal is to extract from the physics of local SUSY enough geometric data to uniquely determine the relevant couplings in the Lagrangian $\mathcal{L}$ of any $\mathcal{N}$–extended supergravity.

The field content is the General Relativity vielbein e_μ^a (related to the space–time metric by the usual relation $h_{\mu\nu} = e_\mu^a e_\nu^b \eta_{ab}$); the spin–3/2 gravitini, ψ_μ^A, ($A = 1, 2, \ldots, \mathcal{N}$); the vector fields, A_μ^u, ($u = 1, 2, \ldots, V$); the spin–1/2 fermions χ^m ($m = 1, 2, \ldots, M$); and the scalars ϕ^i ($i = 1, 2, \ldots, n$) parametrizing the n–fold $\mathcal{M}$. As usual in 4D SUGRA, we adopt the following convention [106, 107, 291]: fermions with upper indices have chirality $+1$, while those with lower indices have chirality -1, namely

$$\gamma_5 \psi_\mu^A = \psi_\mu^A, \qquad \gamma_5 \psi_{A\mu} = -\psi_{A\mu},$$
$$\gamma_5 \chi = \chi^m, \qquad \gamma_5 \chi_m = -\chi_m, \qquad (2.197)$$
$$\bar{\psi}_{A\mu} = (\psi_\mu^A)^\dagger \gamma^0, \qquad \bar{\chi}_m = (\chi^m)^\dagger \gamma^0.$$

Without loss of generality, we assume that the fermions correspond to a "flat" (i.e., unitary) basis, that is, their kinetic terms have the form

$$-i e \, \bar{\chi}_m \slashed{D} \chi^m + \cdots. \qquad (2.198)$$

The vectors have Yang–Mills–like kinetic terms

$$-\frac{1}{4} i e \, \mathcal{N}_{uv} h^{\mu\rho} h^{\nu\sigma} F_{\mu\nu}^{u+} F_{\rho\sigma}^{v+} + \text{H.c.}, \qquad (2.199)$$

2.10 4D supergravity

and the transformation of the gravitini has the general form

$$\delta\psi_\mu^A = D_\mu \epsilon^A + \mathcal{Q}^A_{\mu B} \epsilon^B \mathcal{T}^{AB}_{\rho\sigma} \gamma^{\rho\sigma} \gamma_\mu \epsilon_B \\ + \mathcal{M}^{AB} \gamma_\mu \epsilon_B + \mathcal{K}^A_{\nu B} \gamma^\nu \gamma_\mu \epsilon^B + \mathcal{J}_\mu^{\nu AB} \gamma_\nu \epsilon_B, \quad (2.200)$$

where, in analogy with Eq. (2.197), we adopted the convention [106, 107, 291] that $\mathcal{T}^{AB}_{\rho\sigma}$ with *upper* indices refers to the *anti*self–dual component, while we write the self–dual part with *lower* ones, $\mathcal{T}_{\rho\sigma\, AB} \equiv (\mathcal{T}^{AB}_{\rho\sigma})^*$.

In this section we consider three basic geometric structures:

G–structures: the $U(\mathcal{N})_R \times H$–structure on $\mathcal{M}$ and its curvatures;
Duality: the $Sp(2V, \mathbb{R})$–bundles, connections, and curvatures;
Z–map: the "central charge" geometry and flag manifolds.

In Chapter 7 we shall address two other structures: the one connected with the *gauging* of a subgroup $\subset \text{Iso}(\mathcal{M})$, and the one related to the scalar potential. These structures are more than sufficient to determine all couplings. In fact, the study of all three structures is already pleonastic: the geometry of $\mathcal{M}$ is uniquely determined by the G–structure alone.

2.10.1 The $U(\mathcal{N})_R$–bundles

The analysis in Section 2.4.1 and Section 2.6 for the 3D case may be repeated here with similar conclusions. According to the philosophy of Section 2.1.2, the geometric structures defined over the ground field $\mathbb{R}$ are replaced by the corresponding ones based on $\mathbb{C}$. In particular, on the scalar manifold, $\mathcal{M}$, we have bundles with structure group $U(\mathcal{N})_R$, the automorphism group of 4D SUSY.

Tangent bundle isomorphisms

As in Section 2.6, the gravitino SUSY transformations, Eq. (2.200), define a $U(\mathcal{N})_R$ connection on $\mathcal{M}$, $\mathcal{Q}^A_{i\,B}(\phi)$, by the rule

$$\mathcal{Q}^A_{\mu B} = \mathcal{Q}^A_{i\,B} \partial_\mu \phi^i + \cdots. \quad (2.201)$$

This gives a $U(1)_R$–principal bundle $\mathcal{P}_1 \to \mathcal{M}$ and an $SU(\mathcal{N})_R$–principal bundle $\mathcal{P}_\mathcal{N} \to \mathcal{M}$. From these two bundles we construct vector bundles associated with any given representation of $U(1)_R \times SU(\mathcal{N})_R$.

As in Section 2.6, we are particularly interested in the *bundle isomorphisms* from the tangent bundle $T\mathcal{M}$ to bundles with $U(1)_R \times SU(\mathcal{N})_R \times H$ structure, where $H \subset C$ with C the centralizer of the image of $U(\mathcal{N})_R$ in $SO(n)$. By such a *structure* we mean vector bundles equipped with metric and compatible torsionless

$U(1)_R \times SU(\mathcal{N})_R \times H$ connection, and *isomorphism* always means isomorphism of the full structure. There are two ways of determining the bundle isomorphism associated with a given SUGRA model: either we go through the detailed analysis, as we did in 3D, or we use a short cut – by the "target space equivalence principle" (Corollary 1.2): the linear structure on $T_\phi \mathcal{M}$ is the one dictated by the SUSY representations at the *linear* level; that is, *the decomposition of $T\mathcal{M}$ into irreducible representations of $U(\mathcal{N})_R$ is given by the R–symmetry representations of the scalars in the linearized* SUSY *multiplets* [281].

The linear analysis is quite easy: at the linearized level a given SUGRA model contains a *free* graviton supermultiplet and a set of *free* matter supermultiplets whose scalars have the same $U(\mathcal{N})_R$ quantum numbers as in rigid SUSY (cf. Sections 2.8.2, 2.9). It remains to determine the $U(\mathcal{N})_R$ representation content of the scalars in the free graviton supermultiplet. To construct the free supermultiplet, one starts from the helicity $+2$ graviton state, which is a $U(\mathcal{N})_R$ singlet, $|+2, \mathbf{1}\rangle$, and acts repeatedly with the supercharges Q^A which lower the helicity by $1/2$. Since the Q^A anticommute, the free state

$$Q^{A_1} Q^{A_2} \cdots Q^{A_k} |+2, 1\rangle \equiv |2 - k/2, [A_1 A_2 \cdots A_k]\rangle \qquad (2.202)$$

has helicity $2 - k/2$ and belongs to the k–index antisymmetric representation of $U(\mathcal{N})$ of dimension $\binom{\mathcal{N}}{k}$. The scalars correspond to $k = 4$. For $\mathcal{N} = 8$ the supermultiplet so constructed is PCT self–conjugate; in all other cases we should add the PCT conjugate states, in particular the scalars in the $(\mathcal{N} - 4)$–index antisymmetric representation. In this way we get the list of representations in the fourth column of Table 2.6. In the table, n is the number of matter chiral multiplets for $\mathcal{N} = 1$ and vector–multiplets for $\mathcal{N} = 3, 4$, while v (resp. h) is the number of $\mathcal{N} = 2$ vector–multiplets (resp. hypermultiplets).

The target space equivalence implies that this free–theory analysis is an *exact* statement in the full interacting SUGRA *provided* we interpret it as a statement about the G–structures on the scalar manifold $\mathcal{M}$, i.e., as giving an isomorphism of $T\mathcal{M}$ with vector bundles of structure group $U(\mathcal{N})_R \times H$. In Table 2.6 we present the list of G–structures for all 4D SUGRA.

We add a few remarks on the table. The first issue is the splitting of $T\mathcal{M}$ into the direct sum of two *real* bundles, which (as we shall prove in the next chapter) means that the scalar manifold is a *metric product* $\mathcal{M}_1 \times \mathcal{M}_2$. This happens precisely if the action of the group G on $T\mathcal{M}$ is reducible. This is the case for $\mathcal{N} = 2$ and $\mathcal{N} = 4$. Geometrically, the splitting of $\mathcal{M}$ arises in the following way. Consider the curvature $(P_{kl}^{(1)})^i{}_j$ of the $U(1)_R$ connection; it is a 2–form with values in $\mathfrak{u}(1)_R \subset \mathrm{End}(T\mathcal{M})$. Let

$$\Pi^i{}_j = (P^{(1)\,kl})^i{}_m (P^{(1)}_{kl})^m{}_j. \qquad (2.203)$$

Table 2.6 G–structures on the scalar manifold of 4D $\mathcal{N}$–extended SUGRA with matter

$\mathcal{N}$	Splitting of $\mathcal{M}$	G_s–structures on $\mathcal{M}_s$	$T\mathcal{M}_s$ as a representation of G_s	$T\mathcal{M} \simeq ?^a$
1	$\mathcal{M}$	$G = U(1) \times SU(n)$	$(\boldsymbol{n})_{+1} \oplus (\boldsymbol{\overline{n}})_{-1}$	$[[\mathcal{L} \otimes B \oplus \overline{\mathcal{L}} \otimes \overline{B}]]$
2	$\mathcal{M}_1 \times \mathcal{M}_2$	$G_1 = U(1)_R \times SU(v)$ $G_2 = SU(2)_R \times Sp((2h)$	$(\boldsymbol{v})_{+2} \oplus (\boldsymbol{\overline{v}})_{-2}$ $(\boldsymbol{2}, \boldsymbol{2h})$	$[[\mathcal{L}^2 \otimes B \oplus \overline{\mathcal{L}}^2 \otimes \overline{B}]] \oplus$ $\oplus [[\mathcal{V} \otimes \mathcal{U} \oplus \overline{\mathcal{V}} \otimes \overline{\mathcal{U}}]]$
3	$\mathcal{M}$	$G = U(1)_R \times SU(3)_R \times SU(n)$	$(\boldsymbol{3}, \boldsymbol{\overline{n}})_{+2} \oplus (\boldsymbol{\overline{3}}, \boldsymbol{n})_{-2}$	$[[\mathcal{L}^2 \otimes \wedge^2 \mathcal{V} \otimes B \oplus \overline{\mathcal{L}}^2 \otimes \wedge^2 \overline{\mathcal{V}} \otimes \overline{B}]]$
4	$\mathcal{M}_1 \times \mathcal{M}_2$	$G_1 = U(1)_R$ $G_2 = SU(4)_R \otimes SO(n)$	$(\boldsymbol{1})_{+4} \oplus (\boldsymbol{1})_{-4}$ $(\boldsymbol{6}, \boldsymbol{n})$	$[[\mathcal{L}^4 \oplus \overline{\mathcal{L}}^4]] \oplus$ $\oplus [[\wedge^4 \mathcal{V} \otimes B]]$
5	$\mathcal{M}$	$SU(5)_R \times U(1)_R$	$(\boldsymbol{\overline{5}})_{+4} \oplus (\boldsymbol{5})_{-4}$	$[[\mathcal{L}^4 \otimes \wedge^4 \mathcal{V} \oplus \overline{\mathcal{L}}^4 \otimes \wedge^4 \overline{\mathcal{V}}]]$
6	$\mathcal{M}$	$G = U(1)_R \times SU(6)_R$	$(\boldsymbol{\overline{15}})_{+4} \oplus (\boldsymbol{15})_{-4}$	$[[\mathcal{L}^4 \otimes \wedge^4 \mathcal{V} \oplus \overline{\mathcal{L}}^4 \otimes \wedge^4 \overline{\mathcal{V}}]]$
7		$G = U(1)_R \times SU(7)_R$	$(\boldsymbol{\overline{35}})_{+4} \oplus (\boldsymbol{35})_{-4}$	<u>**IMPOSSIBLE !**</u>
8	$\mathcal{M}$	$G = SU(8)_R$	70	$[[\wedge^4 \mathcal{V}]]$

[a] $\mathcal{L}$ is the $U(1)_R$ line–bundle and $\mathcal{V}$ is the $SU(\mathcal{N})_R$ vector bundle associated with the fundamental representation.

Since $(P^{(1)}_{kl})_{ij}$ is *anti*symmetric in $i \leftrightarrow j$, Π is symmetric and may be diagonalized; then we have an orthogonal splitting of $T\mathcal{M}$ into eigenbundles[47]

$$T\mathcal{M} = \bigoplus (T\mathcal{M})_\lambda, \qquad (T\mathcal{M})_\lambda \equiv \mathrm{Ker}(\Pi - \lambda), \qquad \lambda \in \mathbb{R}. \tag{2.204}$$

The curvature computations for 3D imply, in that case, that $(T\mathcal{M})_\lambda$ is trivial for $\lambda \neq 0, 1$; the same result applies to all Ds. Then the splitting takes the form

$$T\mathcal{M} \simeq \mathrm{Ker}\,\Pi \oplus \mathrm{Im}\,\Pi. \tag{2.205}$$

Thus, for $\mathcal{N} = 2, 4$, $T\mathcal{M}$ has the structure

$$T\mathcal{M} \simeq [[\mathcal{L} \otimes \mathcal{U}]] \oplus [[\Theta \otimes \mathcal{B}]]. \tag{2.206}$$

Here $\mathcal{L}$ is the complex line bundle[48] associated with the charge $+1$ representation of $U(1)_R$, and $\mathcal{U} \simeq \mathcal{L}^{-1} \otimes \mathrm{Im}\,\Pi$ is a complex bundle with structure group *contained* in the semi–simple part of the centralizer of $U(1)_R$ in $SO(\dim \mathrm{Im}\,\Pi)$, i.e., $SU(\dim \mathrm{Im}\,\Pi/2)$. Θ is an $SU(\mathcal{N})_R$ bundle associated with the *real* representation **6** for $\mathcal{N} = 4$, and to the *quaternionic* one **2** for $\mathcal{N} = 2$ (cf. the rigid case Section 2.8.2).

For $\mathcal{N} = 3, 5, 6$, the structure of the tangent bundle is simpler:

$$T\mathcal{M} \otimes \mathbb{C} = \mathcal{L} \otimes \mathcal{U} \oplus \mathcal{L}^{-1} \otimes \mathcal{U}^\vee, \tag{2.207}$$

with $\mathcal{L}$ the $U(1)_R$ line bundle of the appropriate charge, and $\mathcal{U}$ a complex bundle with structure group contained in the semi–simple part of the centralizer of $U(1)_R$ in $SO(\dim \mathcal{M})$. This structure group is

$$SU(3)_R \times SU(\dim \mathcal{M}/6) \qquad \text{for } \mathcal{N} = 3 \tag{2.208}$$

$$SU(\mathcal{N})_R \qquad \text{for } \mathcal{N} = 5, 6. \tag{2.209}$$

$SU(\mathcal{N})_R$ acts on $\mathcal{U}$ for $\mathcal{N} = 3, 5, 6$ in the $\overline{\mathbf{3}}$, $\overline{\mathbf{5}}$, and $\overline{\mathbf{15}}$, respectively.

The $\mathcal{N} = 7$ case is special. At face value, it looks similar to the previous ones. Naively one would expect to get $T\mathcal{M} \simeq [[\mathcal{L} \otimes \mathcal{U}]]$, with $\mathcal{U}$ a $SU(7)_R$ bundle associated with the representation $\overline{\mathbf{35}}$. However, we shall see in Part II that *no Riemannian manifold $\mathcal{M}$ may have a tangent bundle isomorphic to a vector bundle with such a structure group and representation*. Therefore, geometry leaves us with only one possible conclusion: *there is no such a thing as 4D $\mathcal{N} = 7$* SUGRA.

---

[47] Recall that if $\sigma: \mathcal{A} \to \mathcal{B}$ is a bundle map, $\mathrm{Ker}\,\sigma \subset \mathcal{A}$ and $\mathrm{Im}\,\sigma \subset \mathcal{B}$ are *subbundles*. See Section 0.5 of ref. [171].

[48] *Definition:* Complex line bundle = vector bundle with fiber isomorphic to $\mathbb{C}$.

2.10 4D supergravity

In $\mathcal{N} = 8$ SUSY, we must have the isomorphism

$$T\mathcal{M} \simeq \Theta, \tag{2.210}$$

where Θ is a vector bundle with structure group $SU(8)_R$ associated with the *real* irreducible representation of dimension **70**.

We shall see in Part II that it is a basic fact of Riemannian geometry that there are precisely *three* simply connected manifolds with the property (2.210): they correspond, respectively, to positive, negative and zero scalar curvature. The same statement holds for the $T\mathcal{M}$–isomorphisms we got for $\mathcal{N} = 3, 4, 5, 6$; in all cases there are precisely three Riemannian manifolds compatible with the given tangent bundle isomorphisms. Again they correspond to the three possible signs of the curvature. Since, as we are going to see, the $SU(\mathcal{N})_R$ curvature should be *negative*, we remain with a *single* (simply connected) Riemannian space with the given properties for $\mathcal{N} \geq 3$.

Computing the $U(\mathcal{N})_R$ curvatures

We compute the curvature of the R–symmetry connection along the lines we followed in Section 2.6 for the 3D case. As we mentioned at the end of the previous subsection, strictly speaking we need only the *sign* of the curvature. In fact, we already know even that: the sign is easily determined by reducing to 3D. The SUSY transformations of scalars have the form

$$\delta \phi^i = K^{iA}{}_m(\phi) \bar{\epsilon}_A \chi^m + \text{H. c.} \tag{2.211}$$

for some tensor K^{iA}_m; we saw above that this tensor is a kind of "vielbein;" in particular it is covariantly constant (in the appropriate sense). In geometric terms this means that the $U(\mathcal{N})_R \times H$ structure on $\mathcal{M}$ is *torsionless*.

We set

$$K^m_{iA} \overset{\text{def}}{=} g_{ik} K^{km}{}_A \overset{\text{def}}{=} g_{ik} (K^{kA}{}_m)^*. \tag{2.212}$$

Equation (2.211) requires the (chiral) supercurrent to have the form

$$J^A_\mu = \gamma^\rho \gamma_\mu \chi^m \, g_{ij}(\phi) K^{iA}{}_m(\phi) \, \partial_\rho \phi^j + \cdots. \tag{2.213}$$

Thus the fermion SUSY transformation should contain the term

$$\delta \chi_m = \gamma^\mu \epsilon_A \, K^{iA}_m \, g_{ij} \, \partial_\mu \phi^j + \cdots. \tag{2.214}$$

This form of the SUSY transformations, exactly as in 3D and in rigid SUSY, implies a Clifford–like property

$$[K^A (K^B)^\dagger + (K^B)^* (K^A)^t]_{ij} = 2 \delta^A{}_B \, g_{ij}. \tag{2.215}$$

The gravitino equation of motion,[49]

$$i\gamma^{\mu\nu\rho} \mathcal{D}_\nu \psi^A_\rho = J^{A\mu}, \tag{2.216}$$

must transform in the right way under SUSY. The LHS of Eq. (2.216) gives $i\gamma^{\mu\nu\rho}\mathcal{D}_{[\nu}\mathcal{D}_{\rho]}\epsilon^A$, which contains a term linear in the $U(\mathcal{N})_R$ curvature $P_{ij}{}^A{}_B$

$$\cdots + i\gamma^{\mu\nu\rho} \partial_{[\nu}\phi^i \partial_{\rho]}\phi^j P_{ij}{}^A{}_B \epsilon^B + \cdots. \tag{2.217}$$

The term proportional to the $\gamma^{\mu\nu\rho}\partial_{[\nu}\phi^i\partial_{\rho]}\phi^j$ in the RHS of Eq. (2.216) reads

$$\gamma^{\mu\nu\rho}\partial_{[\nu}\phi^i\partial_{\rho]}\phi^j \left(-\frac{1}{2}\left[g_{ik}K^{kA}{}_m K^m{}_{jB} - g_{jk}K^{kA}{}_m K^m{}_{iB}\right]\right). \tag{2.218}$$

Thus

$$P_{ij}{}^A{}_B = -\frac{1}{2}\left[K^A (K^B)^\dagger - (K^B)^*(K^A)^t\right]_{ij}. \tag{2.219}$$

From this equation we infer

$$P_{ij}{}^A{}_B = -P_{ji}{}^A{}_B, \tag{2.220}$$

$$(P_{ij}{}^A{}_B)^* = -P_{ij}{}^B{}_A, \tag{2.221}$$

that is, $P \in \Lambda^2(\mathfrak{u}(\mathcal{N})_R)$, a Lie algebra valued 2–form. In particular, the curvature is *not* zero. In fact it is *negative*.

General Lesson 2.15 *In $D = 4$ SUGRA the tangent bundle $T\mathcal{M}$ has the reduced structure group listed in Table 2.6. The $U(\mathcal{N})_R$ curvature is given by Eq. (2.219). The curvature tensor $P_{ij}{}^A{}_B$ is* PARALLEL.

2.10.2 Duality bundles and connections

If our SUGRA models has V vector fields, we have a map

$$\mu: \mathcal{M} \to \frac{Sp(2V, \mathbb{R})}{U(V)} \simeq \mathfrak{H}_V, \tag{2.222}$$

describing the scalar–vector couplings (Section 1.4). Over $\mathfrak{H}_V$ there are natural bundles: the flat $Sp(2V, \mathbb{R})$ bundle $\mathcal{F}$, the tautological bundle $\mathcal{T}$, and the quotient bundle $\mathcal{F}/\mathcal{T}$. All these bundles may be pulled back to $\mathcal{M}$. As in rigid special Kähler geometry, SUSY gives geometric relations between the metric on $\mathcal{M}$ and the connections pulled back by μ. We shall not pursue this subject further here.

---

[49] The derivative $\mathcal{D}_\mu$ is covariant with respect to all the relevant local symmetries.

2.10 4D supergravity

For $2 \leq \mathcal{N} \leq 4$ SUGRA the "duality geometry" is subsumed in the much more powerful Z–map geometry.

It remains to discuss the case $\mathcal{N} = 1$. If $V \neq 0$, we have a map μ as in Eq. (2.222). We need to determine its properties, namely the compatibily condition with the geometric structures on $\mathcal{M}$. In fact, on $\mathcal{M}$ we have only the (almost) complex structure f, which – as in Section 2.7 – is related to the curvature of a complex line bundle $\mathcal{L}$. Also the coset $Sp(2V, \mathbb{R})/U(V)$, being the Siegel's upper half–space, is a complex manifold with a complex structure I. We claim that the compatibility condition is

$$\mu_* f = I \, \mu_*, \quad \text{or dually,} \quad \mu^* \, I^t = f^t \, \mu^*, \tag{2.223}$$

i.e., complex structure goes to complex structure. In fact, both complex structures are manifestations of the $U(1)_R$ automorphism group, and both ways of implementing a $U(1)_R$ transformation should give the same answer.

2.10.3 Z–map

In General Relativity one can define the Poincaré algebra generators P_μ, $M_{\mu\nu}$ only for space–times which are asymptotic to flat Minkowski space; in such a geometry one constructs the ADM momentum 4–vector P_μ in terms of an integral at spatial infinity (see Chapter 6). Under certain circumstances, which are also reviewed in Chapter 6, the same holds [205, 247, 300] for the other generators of the *super*Poincaré algebra: supercharges and central charges. For central charges this boils down to the following simple rule: take the gravitino transformations, Eq. (2.200), and consider the complex two form $\mathcal{T}^{AB}_{\rho\sigma}$. Then set

$$Z^{AB} = \int_{\text{spatial } \infty} \mathcal{T}^{AB}_{\mu\nu} \, dx^\mu \wedge dx^\nu. \tag{2.224}$$

SUSY implies that Z^{AB}, when defined, is the central charge. We have

$$\mathcal{T}^{AB}_{\mu\nu} = H^{AB}_u(\phi) \, F^{u\,-}_{\mu\nu} + \text{(fermions)} \tag{2.225}$$

for a certain tensor $H^{AB}_u(\phi)$. Then the central charge Z^{AB} is a linear combination of the Abelian charges $Q^u \equiv \int_\infty F^{u\,-}$ with coefficients $H^{AB}_u(\phi_\infty)$, where ϕ^i_∞ is the (constant) value of the scalars at spatial infinity.

We may describe the situation in the following geometric terms: over $\mathcal{M}$ we have the flat $Sp(2V, \mathbb{R})$ vector bundle $\mathcal{F}$ on which the field–strength $\binom{F^{u\,-}}{G^-_v}$ takes values. Inside $\mathcal{F}$ there is a subbundle, $\mathcal{Z}$, of rank $\mathcal{N}(\mathcal{N}-1)/2$, whose fiber $\mathcal{Z}_\phi$

over the point $\phi \in \mathcal{M}$ is the subspace spanned by the $\mathcal{T}_{\mu\nu}^{AB}$. Concretely, the coefficients $H_\mu^{AB}(\phi)$ define a bundle map $H\colon \mathcal{F}^\vee \to \mathcal{Z}^\vee$ dual to the embedding map $H^\vee\colon \mathcal{Z} \to \mathcal{F}$.

The subbundle $\mathcal{Z} \subset \mathcal{F}$ describes how the linear space $\mathcal{Z}_\phi$ moves inside the fiber $\mathcal{F}_\phi$ as we move our base point $\phi \in \mathcal{M}$. Physically this has the following interpretation: we are expanding the theory around the classical "vacuum," $\phi = \phi_\infty$, and asking which linear combinations of the vector fields A_μ^u play the role, in this particular vacuum, of the "graviphotons"[50] and which play that of the "matter" vectors. The question is meaningful only for $2 \leq \mathcal{N} \leq 4$, because for $\mathcal{N} = 1$ there are no graviphotons and for $\mathcal{N} \geq 5$ there is no matter.

Replacing $\mathcal{M}$ by its universal cover, the flat $Sp(2V, \mathbb{R})$ bundle $\mathcal{F}$ may be identified with the trivial bundle with fiber $\mathbb{C}^{2V}$ equipped with two additional structures: (i) a real structure R and (ii) a symplectic pairing Ω. At a given point $\phi \in \mathcal{M}$, the fiber $\mathcal{Z}_\phi$ is an $\mathcal{N}(\mathcal{N}-1)/2$–dimensional complex subspace of $\mathbb{C}^{2V}$; physical consistency requires the subspace $\mathcal{Z}_\phi \subset \mathbb{C}^{2V}$ to have specific properties under the real and symplectic structures: see Chapters 4, 11, and 12. It turns out that the group $Sp(2V, \mathbb{R})$ acts transitively on the subspaces having these properties. Hence the subbundle $\mathcal{Z}$ may be seen as a map (the *Z–map*)

$$\zeta\colon \mathcal{M} \to Sp(2V, \mathbb{R})/I, \qquad (2.226)$$

where $I \subset U(V)$ is the isotropy subgroup of $\mathcal{Z}_\phi$ (see Chapter 11). Over the coset in the RHS we have God–given homogeneous bundles equipped with canonical metrics and connections. Their pull–backs by ζ define natural geometric structures on $\mathcal{M}$ which, for $2 \leq \mathcal{N} \leq 4$, fully determine the Lagrangian of (ungauged) SUGRA. The detailed analysis will be presented in Chapters 4, 11, and 12.

Both maps μ and ζ, Eqs. (2.222) and (2.226), may be uplifted to maps

$$\hat{\mu}, \hat{\zeta}\colon \mathcal{M} \to Sp(2V, \mathbb{R}). \qquad (2.227)$$

We will see that the two uplifts may be chosen to be equal $\hat{\mu} = \hat{\zeta}$. This is a general compatibility condition between the two geometric structures.

Exercise 2.10.1 From Table 1 of ref. [281] construct the table of G–structures for 5D SUGRA.

Exercise 2.10.2 Same as previous exercise but for 7D SUGRA.

---

[50] A *graviphoton* is a vector belonging to the supermultiplet containing the graviton. It is not an invariant notion, as the discussion in the text indicates; the concept of *graviphoton* makes sense in the linearized theory, but depends on the configuration around which we linearize.

Part II

Geometry and extended SUSY:
more than eight supercharges

3
Parallel structures and holonomy

In Chapter 2 we saw that $\mathcal{N}$–extended SUSY/SUGRA requires several *parallel* tensors on the target manifold $\mathcal{M}$. In 3D rigid SUSY all tensors in the even Clifford algebra $\mathbb{C}\ell^0(\mathcal{N}) \subset \text{End}(T\mathcal{M})$ must be parallel. Likewise, in 3D SUGRA[1] $\mathcal{P}(\Sigma)$ must be a parallel $2k$–form for *all* degree–k Ad–invariant polynomial $\mathcal{P}\colon \odot^k \mathfrak{spin}(\mathcal{N}) \to \mathbb{C}$. The same story holds in $D = 4$ or 6 replacing $Spin(\mathcal{N})$ with $U(\mathcal{N})$ or, respectively, $Sp(\mathcal{N})$. In the general case, the parallel tensors on $\mathcal{M}$ are described by the G–structure appropriate for the given $\mathcal{N}$ and D. Indeed, a fundamental principle of differential geometry (see Section 3.1) states that the existence of parallel tensors is equivalent to the reduction of the holonomy group of $T\mathcal{M}$, that is, to a torsionless G–structure with $G \subsetneq SO(n)$.

In this chapter we study the implications for the metric geometry of $\mathcal{M}$ of the existence of several parallel structures [42, 51, 208, 209, 267].

3.1 The holonomy group

In this section $\mathcal{M}$ is a connected Riemannian manifold equipped with the metric g and the Levi–Civita connection ∇, the unique connection which is both metric and torsionless.

3.1.1 Definitions

Let $\phi \in \mathcal{M}$ be a point, and C a piecewise smooth loop starting and ending at ϕ. We denote by $W(C)$ the *parallel transport* along C (i.e., the Wilson line of the

---

[1] Here $\Sigma \in \Lambda^2(\mathfrak{spin}(\mathcal{N}))$ is the 2–form in Eq. (2.105). The parallel form $\mathcal{P}(\Sigma)$ represents a Chern–Weil characteristic class [94, 186, 235] of the gravitino $Spin(\mathcal{N})$–bundle.

connection ∇). Since the connection is metric, $W(C)$ is an element of the orthogonal group[2] $O(T_\phi)$. Given two such loops, C_1 and C_2, we can define their product, which is also a loop,

$$C_1 \cdot C_2(t) = \begin{cases} C_2(2t) & 0 \le t \le 1/2, \\ C_1(2t-1) & 1/2 \le t \le 1, \end{cases} \tag{3.1}$$

and $W(C_1 \cdot C_2) = W(C_1) \cdot W(C_2)$ as elements of $O(T_\phi)$.

Definition 3.1 The *holonomy group* of $\mathcal{M}$ at ϕ, $\mathrm{Hol}(\phi)$, is the subgroup of $O(T_\phi)$ generated by all the $W(C)$s, where C runs through all piecewise smooth loops on $\mathcal{M}$ based at ϕ. The *restricted holonomy group*, $\mathrm{Hol}^0(\phi)$ is the subgroup generated by the $W(C_0)$s, with C_0 a contractible loop.

Let us change the base point from ϕ to $\widetilde{\phi}$. Fix a path $C_{\widetilde{\phi}\phi}$ from ϕ to $\widetilde{\phi}$. Then

$$\mathrm{Hol}(\widetilde{\phi}) = W(C_{\widetilde{\phi}\phi}) \, \mathrm{Hol}(\phi) \, W(C_{\widetilde{\phi}\phi})^{-1}, \tag{3.2}$$

$$\mathrm{Hol}^0(\widetilde{\phi}) = W(C_{\widetilde{\phi}\phi}) \, \mathrm{Hol}^0(\phi) \, W(C_{\widetilde{\phi}\phi})^{-1}, \tag{3.3}$$

so the holonomy groups are independent of the point ϕ up to isomorphism. Then we speak of *the* holonomy groups of $\mathcal{M}$ and denote them as $\mathrm{Hol}(g)$, $\mathrm{Hol}^0(g)$, respectively. The following proposition is obvious from the definitions.

Proposition 3.2 *Saying that an n–fold $\mathcal{M}$ has a Riemannian metric with holonomy group $\mathrm{Hol}(g) = G$ is equivalent to saying that $\mathcal{M}$ admits a* TORSIONLESS *G–structure with G a closed (hence compact) subgroup of $O(n)$.*

Here we are interested only in the (simpler) group $\mathrm{Hol}^0(g)$. For a simply connected manifold $\mathrm{Hol}(g) \equiv \mathrm{Hol}^0(g)$. Hence, replacing $\mathcal{M}$ with its universal Riemannian covering (if necessary), we assume $\mathcal{M}$ to be simply connected. In this case

$$\mathrm{Hol}(g) \subset SO(\dim \mathcal{M}). \tag{3.4}$$

Suppose on $\mathcal{M}$ there is a tensor field $K \in \Gamma(\mathcal{M}, \otimes^k T\mathcal{M} \otimes^l T^*\mathcal{M})$ which is invariant under parallel transport, i.e., for every $\phi_1, \phi_2 \in \mathcal{M}$ and every path $C_{\phi_2\phi_1}$ from ϕ_1 to ϕ_2 one has

$$W^*(C_{\phi_2\phi_1}) K(\phi_1) = K(\phi_2), \tag{3.5}$$

---

[2] To save print, we write simply T_ϕ for the fiber $T_\phi \mathcal{M}$ of the tangent bundle at the point ϕ. $O(T_\phi)$ is the group of endomorphisms of T_ϕ which preserve the fiber metric.

where $W^*(C)$ denotes the parallel transport in the representation appropriate for the tensor K. Then, by Definition 3.1, the tensor $K(\phi)$ at ϕ is invariant under the group $\mathrm{Hol}(g) \subset O(T_\phi)$ (acting in the appropriate representation). Conversely, given a tensor K_0 at ϕ_0 which is invariant under $\mathrm{Hol}(\phi_0)$, we may construct a tensor field K which is invariant by parallel transport defining $K(\phi)$, for all $\phi \in \mathcal{M}$, by the formula (3.5). Obviously $K(\phi)$ is independent of the choice of path $C_{\phi\phi_0}$. Moreover, we known from general Riemannian geometry that a tensor K is invariant under parallel transport if and only if its covariant derivative vanishes, $\nabla_i K = 0$. Hence,

Fundamental Principle 3.3 *Let $\mathcal{M}, g$ be a Riemannian manifold. The following three properties are equivalent:*

(a) *There exists on $\mathcal{M}$ a tensor field K of type (k, l) which has zero covariant derivative $\nabla_i K = 0$.*
(b) *There exists on $\mathcal{M}$ a tensor field K of type (k, l) which is invariant under parallel transport.*
(c) *There esists a point $\phi \in \mathcal{M}$ and a tensor $K_0 \in (T_\phi)^{\otimes k} \otimes (T_\phi^\vee)^{\otimes l}$ which is invariant under the (appropriate) action of $\mathrm{Hol}(\phi)$.*

The geometric problem of finding all the parallel structures (tensors with vanishing covariant derivative) on a given manifold $\mathcal{M}$ – which is our basic concern – is thus equivalent to the algebraic problem of finding the invariants of the holonomy group. Then our program reduces to:

1. Find all Lie groups G and representations R_G, such that G is the holonomy group of some Riemannian manifold $\mathcal{M}$, with G acting on $T\mathcal{M}$ according to the R_G representation.
2. Classify all invariants ($\equiv$ parallel structures) for given (G, R_G), and see for which pairs (G, R_G) these parallel structures coincide with the ones required by SUSY/SUGRA (cf. Chapter 2).
3. Determine the metric geometries corresponding to each such (G, R_G) and understand their properties.

We are left with two basic questions:

(A) Which Lie groups G may be holonomy groups of Riemannian manifolds? Which representations R_G are allowed?
(B) Given the holonomy group $\mathrm{Hol}(g)$, what can we say about the metric geometry of $\mathcal{M}$?

We start with (A). As a first step, we consider the holonomy of the "trivial" situations, namely the Riemannian products $\mathcal{M}_1 \times \mathcal{M}_2$, the target spaces of decoupled σ–models.

3.1.2 Riemannian products

Let $\mathcal{M}_1$ and $\mathcal{M}_2$ be two Riemannian manifolds with metrics g_1 and g_2, respectively. By the *Riemannian product* of the two manifolds we mean the product manifold $\mathcal{M}_1 \times \mathcal{M}_2$ equipped with the metric $g_1 \oplus g_2$; that is, in local coordinates,

$$ds^2_{g_1 \oplus g_2} = g_1(x)_{ij}\, dx^i\, dx^j + g_2(y)_{ab}\, dy^a\, dy^b. \tag{3.6}$$

The subbundles of $T(\mathcal{M}_1 \times \mathcal{M}_2)$ of vectors tangent, respectively, to $\mathcal{M}_1$ and $\mathcal{M}_2$ are *involutive*.[3] The corresponding integral submanifolds have the form $\{\phi_1\} \times \mathcal{M}_2$ and $\mathcal{M}_1 \times \{\phi_2\}$ and are *totally geodesic*.[4]

Definition 3.4 A Riemann manifold $\mathcal{M}$ is *reducible* (resp. *locally reducible*) if it is isometric (resp. locally isometric) to a Riemann product.

The two tangent subbundles $T\mathcal{M}_{1,2} \subset T(\mathcal{M}_1 \times \mathcal{M}_2)$ are manifestly invariant under the holonomy group $\mathrm{Hol}(g_1 \oplus g_2)$. By Fundamental Principle 3.3, $\mathrm{Hol}(g_1 \oplus g_2)$ is a subgroup of[5] $SO(n_1 + n_2)$ leaving invariant the two complementary and orthogonal subspaces $T\mathcal{M}_{1,2}$. Thus $\mathrm{Hol}(g_1 \oplus g_2) \subseteq SO(n_1) \times SO(n_2) \subset SO(n_1 + n_2)$. In fact,

$$\mathrm{Hol}(g_1 \oplus g_2) = \mathrm{Hol}(g_1) \times \mathrm{Hol}(g_2), \tag{3.7}$$

since $W(C) = W(\pi_1 C)_1 \times W(\pi_2 C)_2$, where $\pi_{1,2} \colon \mathcal{M}_1 \times \mathcal{M}_2 \to \mathcal{M}_{1,2}$ are the natural projections. This result has a local converse:

Theorem 3.5 *Let $\mathcal{M}, g$ be a Riemannian manifold and $\phi \in \mathcal{M}$ a point. Let $T_0 \subset T_\phi \mathcal{M}$ be the subspace on which $\mathrm{Hol}(\phi)$ acts trivially, and let $T_0^\perp$ be its orthogonal complement. Since $\mathrm{Hol}(\phi) \subset O(T_\phi)$, $T_0^\perp$ can be decomposed into the orthogonal sum of irreducible representations of the holonomy group,*

$$T_0^\perp = T_1 \oplus T_2 \oplus \cdots \oplus T_k. \tag{3.8}$$

Then $\mathcal{M}, g$ is locally a Riemannian product $g_0 \oplus g_1 \oplus \cdots \oplus g_k$ where g_0 is flat. Moreover $\mathrm{Hol}(\phi)$ is a direct sum of representations

$$\mathrm{Hol}(\phi) = H_1 \times H_2 \times \cdots \times H_k \tag{3.9}$$

---

[3] A subbundle $\mathcal{D} \subset T\mathcal{M}$ is said to be *involutive* if, given any two vector fields in $\mathcal{D}$, $X = X^i \partial_i$ and $Y = Y^i \partial_i$, their Lie bracket $[X, Y] \equiv \pounds_X Y \in \mathcal{D}$. A subbundle $\mathcal{D} \subset T\mathcal{M}$ is said to be *integrable* if for each point $\phi \in \mathcal{M}$ there is a submanifold $\iota \colon \mathcal{S} \to \mathcal{M}$, with $\phi \in \mathcal{S}$, such that $\mathcal{D}|_\mathcal{S} = T\mathcal{S}$. The Frobenius theorem [224, 225] states that $\mathcal{D}$ *is integrable if and only if it is involutive*. For the dual statement in terms of differential forms, see, e.g., ref. [67].

[4] A Riemann submanifold $\iota \colon \mathcal{S} \to \mathcal{M}$ (equipped with the metric $\iota^* g$ induced by the immersion ι) is a *totally geodesic submanifold* if all the geodesics of $\mathcal{S}$ are also geodesics of the ambient space $\mathcal{M}$. Equivalently, the *second fundamental form* of $\mathcal{S}$ vanishes [263].

[5] We set $n_{1,2} = \dim \mathcal{M}_{1,2}$.

with $H_l \subseteq O(T_l)$ acting irreducibly on T_l and trivially on T_j, $j \neq l$.

Proof Consider the orthogonal splitting of the tangent bundle, $TM \simeq \oplus_{l=0}^{k} \mathcal{T}_l$, where $\mathcal{T}_l$ is obtained by parallel transporting T_l on $\mathcal{M}$. Each subbundle is involutive: in fact if $X \in \mathcal{T}_l$, also $\nabla_Y X \in \mathcal{T}_l$, since $\mathcal{T}_l$ is preserved by parallel transport; hence $X, Y \in \mathcal{T}_l \Rightarrow [X, Y] \equiv \nabla_X Y - \nabla_Y X \in \mathcal{T}_l$. By Frobenius's theorem,[6] the $\mathcal{T}_l$ are integrable; then $\mathcal{M}$ is *locally* diffeomorphic to $\mathcal{F}_0 \times \mathcal{F}_1 \times \cdots \times \mathcal{F}_k$, where $T\mathcal{F}_l \simeq \mathcal{T}_l$. Since $T\mathcal{F}_l \perp T\mathcal{F}_j$ for $l \neq j$, the metric g has the "block–diagonal" form $g_0 \oplus g_1 \oplus \cdots \oplus g_k$. Let $x^{(l,\alpha)}$ be local coordinates on $\mathcal{F}_l$. Invariance under parallel transport of the $T\mathcal{F}_l$s requires

$$\Gamma^{(l,\alpha)}_{(j,\beta)(k,\gamma)} = 0 \quad \text{unless } l = j = k. \tag{3.10}$$

Now, for $j \neq l$,

$$\partial_{(j,\alpha)} g_{(l,\beta)(l,\gamma)} = g_{(l,\gamma)(l,\delta)} \Gamma^{(l,\delta)}_{(j,\alpha)(l,\beta)} + (\beta \leftrightarrow \gamma) = 0, \tag{3.11}$$

and hence the metric–block g_l is a function of the $x^{(l,\alpha)}$ only. $\square$

This local result has a global version:

Theorem 3.6 (de Rham) *If a Riemannian manifold $\mathcal{M}, g$ is complete,[7] simply connected, and its holonomy acts reducibly on TM, then $\mathcal{M}, g$ is a Riemannian product.*

3.1.3 Holonomy and curvature

The variation of a vector $v \in T_\phi \mathcal{M}$ under parallel transport along the infinitesimal parallelogram spanned by ϵx^i and ϵy^i is given by

$$\Delta v^i = -\epsilon^2 x^k y^l R_{kl}{}^i{}_j v^j \equiv -\epsilon^2 R(x,y)^i{}_j v^j, \tag{3.12}$$

where $R_{kl}{}^i{}_j$ is the Riemann tensor. Hence the Lie algebra $\mathfrak{hol}(\phi)$ of $\text{Hol}(\phi) \subset O(T_\phi)$ certainly contains all the curvature endomorphisms $R(x,y) \in \text{End}(T_\phi)$ for all $x, y \in T_\phi$. Let ϕ, ϕ' be two points in $\mathcal{M}$, and $C_{\phi'\phi}$ a path connecting them. We construct the following loop $\widetilde{C}$ starting and ending at ϕ. First we go to ϕ' along $C_{\phi'\phi}$, then we follow an infinitesimal parallelogram of sizes ϵx^i, ϵy^j, and then we return back

---

[6] See footnote 3 of this chapter.
[7] The equivalence between the different notions of *completeness* is the following:

Theorem (Hopf–Rinow) *For a Riemannian $\mathcal{M}$ the following are equivalent: (i) $\mathcal{M}$ is geodesically complete (i.e., each maximal geodesic $\gamma(\tau)$ is defined for τ on the entire real axis $\mathbb{R}$); (ii) $\mathcal{M}$ is complete as a metric space (i.e., the Cauchy sequences converge); (iii) the bounded subset of $\mathcal{M}$ are relatively compact. If one (hence all) of these conditions holds, given two points $\phi, \phi' \in \mathcal{M}$ there exists at least one geodesic connecting them.*

to ϕ through $C_{\phi'\phi}^{-1}$. The variation of the vector $v^i \in T_\phi$ under the parallel transport along the circuit $\tilde{C}$ is given by

$$\Delta v^i = -\epsilon^2 \, (W(C_{\phi'\phi}))^{-1} \, R(x,y)' \, W(C_{\phi'\phi}))^i{}_j \, v^j, \tag{3.13}$$

where $R(x,y)' = x^k y^l R_{kl}(\phi') \in \text{End}(T_{\phi'})$ is the curvature endomorphism at the point ϕ'. Hence

$$\begin{aligned}W(C_{\phi'\phi})^{-1} \, R(x,y)' \, W(C_{\phi'\phi}) &\in \mathfrak{hol}(\phi), \\ \forall \, \phi' \in \mathcal{M}, \; \forall \, C_{\phi'\phi}, \; \text{and} \; \forall \, x, y \in T_{\phi'},&\end{aligned} \tag{3.14}$$

which, in particular, implies the following theorem.

Theorem 3.7 (Nijenhuis [251]) $R_{ijk}{}^l \in \mathfrak{hol}$ and $\nabla_m R_{ijk}{}^l \in \mathfrak{hol}$.

Hence the Lie algebra $\mathfrak{hol}(g)$ contains all the curvature endomorphisms at every point of $\mathcal{M}$, for all pairs of tangent vectors, everything pulled back to a fixed point ϕ along all possible parallel transportations. It is quite a big deal, but it is everything:

Theorem 3.8 (Ambrose–Singer [13]) *The Lie algebra $\mathfrak{hol}(\phi)$ is <u>exactly</u> the subalgebra of $\mathfrak{so}(T_\phi \mathcal{M})$ generated by all the elements in Eq. (3.14).*

This theorem is true for any connection, not only for a Riemannian one.

3.2 Symmetric Riemannian spaces

At face value, question (A) of Section 3.1.1 has a very simple answer:

<u>All</u> compact groups G are the holonomy group of some Riemannian manifold.

Indeed, for *G simple* take as Riemannian space G itself with the (unique up to overall normalization) metric which is invariant under both right and left translations.[8] Then, as you will show in Exercise 3.2.4, $\text{Hol}(G) = G$. However, this is quite a special example. All connections, curvatures, etc. are constructed using the commutators in the Lie algebra $\mathfrak{g}$ of G. In particular,

$$\nabla_i R_{jkl}{}^m = 0, \tag{3.15}$$

and the Riemann tensor *itself* is PARALLEL.

---

[8] Such a metric always exists for G compact: see, e.g., [263], proposition 26.2.

3.2 Symmetric Riemannian spaces

To get an interesting classification theorem for the Riemannian holonomy groups (and representations) we have to leave aside such cheap algebraic examples, namely the manifolds whose curvature is, essentially, a constant numerical tensor. Therefore as a preliminary we study and classify the manifolds in which the full Riemann tensor is parallel in order to separate these special cases from the rest.

3.2.1 Definitions

We begin by reviewing some geometrical facts.

Geodesic symmetries

Let $\mathcal{M}$ be a Riemannian manifold and $\phi_0 \in \mathcal{M}$ a point. The *exponential map*

$$\exp_{\phi_0} : T_{\phi_0}\mathcal{M} \to \mathcal{M} \tag{3.16}$$

associates with each vector $X \in T_{\phi_0}\mathcal{M}$ the point $\gamma_X(1) \in \mathcal{M}$, where $\gamma_X(\tau)$ is the unique geodesic starting at ϕ_0 and having initial velocity

$$\left.\frac{d\gamma_X}{d\tau}\right|_{\tau=0} = X \in T_{\phi_0}\mathcal{M}. \tag{3.17}$$

We write $\exp(X)$ for the image of the exponential map. It is a *local* diffeomorphism from a neighborhood of the origin in $T_{\phi_0}\mathcal{M}$ to a neighborhood of ϕ_0 in $\mathcal{M}$. A priori, $\exp(\,\cdot\,)$ is defined only locally near ϕ_0. If $\mathcal{M}$ is *complete*, the exponential map is defined globally but it is not a diffeomorphism, since for $\|X\| \geq \rho$ (the injectivity radius [239]) $\exp(X)$ is not injective.

Definition 3.9 By the *geodesic symmetry* s_{ϕ_0} at the point ϕ_0 we mean the map (defined only locally, unless $\mathcal{M}$ is *complete*)

$$s_{\phi_0} : \exp(X) \mapsto \exp(-X), \tag{3.18}$$

i.e., the exponential image of a "parity" transformation in the tangent space.

Definition 3.10 A Riemannian manifold $\mathcal{M}$ is *locally symmetric* if for each point ϕ the geodesic symmetry s_ϕ (which is defined only locally) is an isometry. $\mathcal{M}$ is *symmetric* if for each point ϕ the geodesic symmetry is a globally defined isometry. Equivalently, $\mathcal{M}$ is symmetric if for all $\phi \in \mathcal{M}$ there is an isometry s_ϕ of $\mathcal{M}$ such that

$$s_\phi(\phi) = \phi, \quad s_{\phi*}|_\phi = -\mathrm{Id}_{T_\phi\mathcal{M}}. \tag{3.19}$$

A *complete, connected, simply connected* locally symmetric manifold is globally symmetric ([263] theorem 5.1).

Parallel curvatures

The relevance of the symmetric spaces for us stems from the following proposition.

Proposition 3.11 *A Riemannian manifold $\mathcal{M}$ is locally symmetric* if and only if *its Riemann tensor is* PARALLEL:

$$\nabla_i R_{jkl}{}^m = 0. \tag{3.20}$$

Proof Let $\mathcal{M}$ be symmetric. The Riemann tensor R is invariant under any isometry,[9] so $R = s_\phi^* R$. By Eq. (3.19),

$$s_\phi^* \nabla R|_\phi = -\nabla R|_\phi = 0. \tag{3.21}$$

Conversely, let R be covariantly constant. Fix a point ϕ. The *normal coordinates* x^i centered at ϕ (see [12, 43, 134, 232, 257]) are defined by taking an orthonormal basis e_i on $T_\phi \mathcal{M}$ and parametrizing (locally) $\mathcal{M}$ in the form $\exp_\phi [x^i e_i]$. In these coordinates the metric[10] has a Taylor expansion with coefficients given by covariant expressions in the Riemann tensor and its covariant derivatives computed at the base point ϕ,

$$\begin{aligned}g_{ij}(x) = \delta_{ij} &- \frac{1}{3} R_{ikjl} x^k x^l + \frac{1}{6} \nabla_s R_{iklj} x^k x^l x^s \\ &+ \left(\frac{1}{20} \nabla_s \nabla_t R_{iklj} + \frac{2}{45} R_{ikl}{}^m R_{istm}\right) x^k x^l x^s x^t + \cdots.\end{aligned} \tag{3.22}$$

If R_{ijkl} is parallel, the coefficients in the RHS are products of Riemann tensors, $R_{i_1 j_1 k_1 l_1} R_{i_1 j_1 k_1 l_1} \cdots R_{i_1 j_1 k_1 l_1}$, with the indices contracted either with δ^{kl} or with x^l. By index counting, all terms have an *even* number of x^l. Hence from Eq. (3.22)

$$g_{ij}(-x) = g_{ij}(x), \tag{3.23}$$

that is, the geodesic symmetry at ϕ is a (local) isometry. Since ϕ was arbitrary, $\mathcal{M}$ is (locally) symmetric. □

3.2.2 Cartan theorem

So, for a complete, simply connected manifold the conditions of *being symmetric* and *having a parallel Riemann tensor* are equivalent. But there is a lot more. We have the following wonderful theorem.

---

[9] We omit the indices in the Riemann tensor and covariant derivative.
[10] 2nd CARTAN THEOREM (proposition E.III.7 of ref. [43]): *The coefficients of the Taylor expansion of* $\exp_\phi^* g(x)$ *near $x = 0 \in T_\phi \mathcal{M}$ are* UNIVERSAL *polynomials in the covariant derivatives of the curvature tensor at the point ϕ.*

3.2 Symmetric Riemannian spaces

Theorem 3.12 (Cartan) $\mathcal{M}$ *a complete Riemannian manifold. $\mathcal{M}$ is (globally) symmetric if and only if it is isometric to a homogeneous space G/H, with G a connected Lie group, H a compact subgroup of G, and there is an involutive automorphism σ of the group G for which, if S denotes the fixed set of σ and S_e its connected component of the identity, one has $S_e \subset H \subset S$. The symmetric metric of G/H is invariant under G.*

Remark. Not all *homogeneous* spaces G/H are symmetric. Symmetry requires the existence of an involutive automorphism σ of G having the properties stated in the theorem. σ is part of the definition of the symmetric structure on the manifold G/H, and an isomorphism of symmetric manifolds is defined to be an isometry preserving σ. Certain manifolds can be written as homogeneous spaces in more than one way. For example:

$$S^{2n+1} = \frac{SO(2n+2)}{SO(2n+1)} = \frac{U(n+1)}{U(n)}, \quad S^6 = \frac{SO(7)}{SO(6)} = \frac{G_2}{SU(3)}, \quad etc. \quad (3.24)$$

In these cases, the *symmetric* space G/H representation (if it exists) is *unique*. This will be evident from the proof.

Before proving theorem 3.13, we establish two simple Lemmas. Consider the isometry group $\text{Iso}(\mathcal{M})$. It is a Lie group.[11]

Lemma 3.13 *$\mathcal{M}$ complete and symmetric. The isometry group $\text{Iso}(\mathcal{M})$ acts* TRANSITIVELY *on $\mathcal{M}$. [That is, given two points $\phi, \phi' \in \mathcal{M}$, there exists $g \in \text{Iso}(\mathcal{M})$ such that $\phi' = g \cdot \phi$.]*

Proof By Hopf–Rinow (cf. footnote 7) there is a geodesic passing through the two points. Let $\tilde{\phi}$ be the middle point on the geodesic arc between ϕ and ϕ'. $s_{\tilde{\phi}}$ is an isometry mapping ϕ to ϕ' (and vice versa). □

Lemma 3.14 *Each smooth manifold $\mathcal{X}$ on which a Lie group G acts smoothly and transitively is diffeomorphic to the quotient manifold G/H, where H is the stabilizer of an arbitrary point $p_0 \in \mathcal{X}$. The diffeomorphism*

$$\varphi: G/H \to \mathcal{X} \quad (3.25)$$

is given by

$$\varphi(gH) = g \cdot p_0, \quad g \in G. \quad (3.26)$$

Proof Obvious (see ref. [263] lemma 9.3). □

---

[11] This is the content of the *Myers–Steenrod* theorem [234].

Proof (of Cartan theorem) (1) Let G be a Lie group, H a Lie subgroup, and σ an involutive automorphism of G such that

$$\text{Fix}(\sigma)_e \subset H \subset \text{Fix}(\sigma). \tag{3.27}$$

We have to show that G/H equipped with σ is a *symmetric space*. Since the quotient is smooth, it is enough to show that for all points $p \in G/H$ there is a geodesic isometry s_p. A point in the coset has the form gH for $g \in G$. Let $p_i = g_i H$, $i = 1, 2$. We set

$$s_{p_1} p_2 = g_1 \sigma(g_1^{-1} g_2) H. \tag{3.28}$$

It is easy to check that $s_{p_1}^2 = 1$. It is an isometry, since both multiplication by an element $g \in G$ (on the left) and σ are isometries.[12]

(2) Conversely, assume $\mathcal{M}$ is complete and symmetric. Let $G = \text{Iso}(\mathcal{M})$. Fix a point $\phi_0 \in \mathcal{M}$, and let $H \subset G$ be the stabilizer of ϕ_0. We define a map $\sigma: G \to G$ by

$$\sigma(g) = s_{\phi_0} g s_{\phi_0}. \tag{3.29}$$

σ is an involutive automorphism of G. For any automorphism $\rho: \mathcal{M} \to \mathcal{M}$, and any point $\phi \in \mathcal{M}$, we have

$$\rho \circ s_\phi \circ \rho^{-1} = s_{\rho(\phi)}. \tag{3.30}$$

Hence for $\forall\, h \in H$, we have $h s_{\phi_0} h^{-1} = s_{h \cdot \phi_0} = s_{\phi_0}$, so

$$\sigma(h) = s_{\phi_0} h s_{\phi_0} = s_{\phi_0}^2 h = h \quad \Rightarrow \quad H \subset \text{Fix}(\sigma). \tag{3.31}$$

Let $A \in T_e(\text{Fix}(\sigma))$. $\sigma_* A = A$, and therefore $\sigma(\exp \tau A) = \exp \tau A$, for all $\tau \in \mathbb{R}$. Now

$$s_{\phi_0}((\exp \tau A)\phi_0) = (s_{\phi_0} \exp \tau A)\phi_0 = (\exp \tau A\, s_{\phi_0})\phi_0 = (\exp \tau A)\phi_0, \tag{3.32}$$

that is, the point $(\exp \tau A)\phi_0$ is a fixed point of s_{ϕ_0}. But ϕ_0 is an isolated fixed point, and hence $(\exp \tau A)\phi_0 = \phi_0$. So $\exp \tau A \in H$, which means $\text{Fix}(\sigma)_e \subset H$. Thus,

$$\text{Fix}(\sigma)_e \subset H \subset \text{Fix}(\sigma). \tag{3.33}$$

The map in Eq. (3.26),

$$\varphi: G/H \to \mathcal{M}, \tag{3.34}$$

---

[12] The homogeneous coset metric on G/H is constructed using the Lie algebras of G and H, and hence all algebraic automorphisms are isometries. See Chapter 5 where the invariant metrics are discussed in detail.

3.2 Symmetric Riemannian spaces

is a diffeomorphism. Now, from Eqs. (3.28) and (3.29), for all $g_i \in G$,

$$\varphi(s_{g_1} g_2 H) = \varphi(g_1 \sigma(g_1^{-1} g_2) H) = g_1 \sigma(g_1^{-1} g_2) \cdot \phi_0$$
$$= g_1 s_{\phi_0} g_1^{-1} g_2 s_{\phi_0} \phi_0 = s_{g_1 \phi_0} g_2 s_{\phi_0} \phi_0 \qquad (3.35)$$
$$= s_{g_1 \phi_0}(g_2 \phi_0),$$

so the diffeomorphism φ is an automorphism of symmetric spaces. The argument also shows that the representation G/H is unique. □

Lie algebraic constructions

We can prove Cartan's theorem from a different viewpoint, perhaps more useful. We ask ourselves: *Given a Riemannian manifold $\mathcal{M}$ with a* PARALLEL CURVATURE, *can we find directly two groups G and H such that $\mathcal{M} \simeq G/H$?*

Yes, we can. Fix a point $\phi_0 \in \mathcal{M}$. Let $\mathfrak{m} = T_{\phi_0}\mathcal{M}$. For $x, y \in \mathfrak{m}$, $R(x, y) \in \mathfrak{so}(\mathfrak{m}) \subset \text{End}(\mathfrak{m})$. Let $\mathfrak{h}$ be the Lie subalgebra of $\mathfrak{so}(\mathfrak{m})$ generated by the $R(x, y)$ for all $x, y \in \mathfrak{m}$. The direct sum $\mathfrak{g} = \mathfrak{h} \oplus \mathfrak{m}$ has a natural structure of Lie algebra precisely iff the curvature is parallel. In fact, define the bracket

$$[x, y] = R(x, y) \in \mathfrak{h}, \qquad x, y \in \mathfrak{m}, \qquad (3.36)$$

$$[r, x] = r \cdot x \in \mathfrak{m}, \qquad x \in \mathfrak{m},\ r \in \mathfrak{h} \subset \text{End}(\mathfrak{m}), \qquad (3.37)$$

$$[r, s] = r \cdot s - s \cdot r \in \mathfrak{h} \subset \text{End}(\mathfrak{m}), \qquad r, s \in \mathfrak{h}. \qquad (3.38)$$

To show that it is a Lie algebra, we have to check the Jacobi identities:

$$([[x, y], z] + [[y, z], x] + [[z, x], y])^i$$
$$= -(R_{klm}{}^i + R_{lmk}{}^i + R_{mkl}{}^i) x^k y^l z^m \equiv 0, \qquad (3.39)$$

$$[[x, r], s] + [[r, s], x] + [[s, x], r] \equiv [r, s] \cdot x + (s \cdot r - r \cdot s) \cdot x \equiv 0, \qquad (3.40)$$

$$[[x, y], r] + [[y, r], x] + [[r, x], y]$$
$$= R(x, y) r - r R(x, y) - R(r \cdot y, x) + R(r \cdot x, y) \stackrel{?}{=} 0. \qquad (3.41)$$

Only the last one is non–trivial. r is a linear combination of endomorphisms of $\mathfrak{m}$ of the form $R(w, z)$, so the Jacobi identity (3.41) is equivalent to

$$x^i y^j w^k z^l (R_{ij}{}^m{}_p R_{kl}{}^p{}_n - R_{kl}{}^m{}_p R_{ij}{}^p{}_n - R_{kl}{}^p{}_j R_{pi}{}^m{}_n + R_{kl}{}^p{}_i R_{pj}{}^m{}_n)$$
$$\equiv x^i y^j w^k z^l [\nabla_k, \nabla_l] R_{ij}{}^m{}_n = 0 \qquad (3.42)$$

since the Riemann tensor is parallel. $\mathfrak{g}$ and $\mathfrak{h}$ are the Lie algebras of G and H, respectively. The involution σ_* acts on the Lie algebra $\mathfrak{g}$ as the automorphism

$$\sigma_*|_\mathfrak{h} = \text{Id}, \qquad \sigma_*|_\mathfrak{m} = -\text{Id}. \qquad (3.43)$$

The exponential map of $(\mathfrak{g} \bmod \mathfrak{h}) \simeq T_e(G/H)$ and that of $T_{\phi_0}\mathcal{M}$ coincide and the metric in normal coordinates is the same for G/H and $\mathcal{M}$. Indeed, the Riemann tensors at the points e, ϕ_0 are equal by construction, and the coefficients in the expansion of $g(x)$ are universal polynomials in this covariantly constant tensor. Hence the two manifolds are (locally) isometric.

We shall discuss symmetric spaces in greater detail in Chapter 5: models, realizations, metrics, connections, etc. Here we are only interested in them for their peculiar role in the general theory of holonomy groups.

3.2.3 The holonomy group of a symmetric space

As before, for simplicity we replace all manifolds with their universal covers, i.e., we assume $\mathcal{M}$ to be simply connected. By Theorem 3.6 we can limit ourselves to *irreducible* manifolds without loss of information.

The following result describes the holonomy of a *symmetric* manifold.

Proposition 3.15 *The holonomy group* $\mathrm{Hol}(G/H)$ *of an irreducible, simply connected symmetric space G/H is equal to H, and its action on $T(G/H)$ is induced by the adjoint representation of G.*

Proof By Theorem 3.8, $\mathfrak{hol}(G/H)$ is generated by all parallel transports of the endomorphism $R(x, y)$. Since the Riemann tensor of a symmetric space is invariant under parallel transport, $\mathfrak{hol}(G/H)$ is generated by the curvature at a fixed point, say at eH; that is, it is generated by the endomorphism of $x^k y^l R_{kl}{}^i{}_j$ of $T_e(G/H) \simeq \mathfrak{m}$, cf. Section 3.2.2. As shown there, $\mathfrak{h} \subset \mathrm{End}(\mathfrak{m})$ is precisely the span of the endomorphisms $R(x, y)$. Hence $\mathfrak{h} = \mathfrak{hol}(G/H)$. □

3.2.4 Rank and transitive actions on spheres

Let $\mathfrak{g} = \mathfrak{h} \oplus \mathfrak{m}$ be as in Section 3.2.2. Consider the maximal Abelian subalgebras of $\mathfrak{m}$. By a theorem of Cartan [69], two such algebras are conjugated under the adjoint action of H. Their common dimension is called the *rank* of the symmetric space. For our present purposes, the following result is of interest.

Proposition 3.16 *A symmetric space G/H has rank 1 if and only if the action of H on the unit tangent sphere in $T_e(G/H) \simeq \mathfrak{m}$ is* TRANSITIVE.

Proof Let $x \in \mathfrak{m}$, and $H \cdot x$ be its H–orbit in $\mathfrak{m}$. The tangent space to the orbit in x is, by definition, given by the elements $[\mathfrak{h}, x]$. Suppose the action of H is *not* transitive. Then there is a vector, y, not proportional to x, which is orthogonal to

3.4 Parallel forms on $\mathcal{M}$

It is time to study the issue systematically. We ask: what are the parallel tensors (apart from the trivial ones constructed out of g and the Levi–Civita tensor ϵ) for each holonomy group allowed by the Berger theorem?

We already know from Corollary 3.19 that for a non–symmetric manifold the (non–trivial) parallel tensors cannot be symmetric.

Proposition 3.22 *Let $\mathcal{M}$ be an irreducible non–symmetric Riemannian n–fold with holonomy group Hol^0. The parallel forms of degree $1 \leq \ell \leq (n-1)$ on $\mathcal{M}$ are:*

(a) $\mathrm{Hol}^0 = SO(n)$: *none;*
(b) $\mathrm{Hol}^0 = U(m)$: *the 2–form $\omega := g_{ik}f^k{}_j d\phi^i \wedge d\phi^j$ (called the Kähler form) and its exterior powers ω^k;*
(c) $\mathrm{Hol}^0 = SU(m)$: *the Kähler form, its powers, a type $(m,0)$ complex volume form ε and its type $(0,m)$ conjugate $\bar{\varepsilon}$;*
(d) $\mathrm{Hol}^0 = Sp(2m)$: *the three Kähler 2–forms $\omega^a := g_{ik}f^{ak}{}_j d\phi^i \wedge d\phi^j$ and the polynomial algebra generated by them[14];*
(e) $\mathrm{Hol}^0 = Sp(2) \times Sp(2m)$: *the 4–form $\Theta := \omega^a \wedge \omega^a$ and its powers;*
(f) $\mathrm{Hol}^0 = G_2$: *a 3 form ϕ and its dual 4–form $*\phi$;*
(g) $\mathrm{Hol}^0 = Spin(7)$: *a self–dual 4 form ϕ.*

Proof Fundamental Principle 3.3 reduces the statement to elementary group theory (Exercise 3.4). □

Relation with Clifford algebras. We see that the structure of a parallel form is quite constrained. For instance, assume that we have a parallel 2–form on an irreducible non–symmetric manifold $\mathcal{M}$. We wish to show that this form comes from a parallel *complex structure* $f \in \mathrm{End}(T\mathcal{M})$, $f^2 = -1$, through the formula $\omega_{ij} = g_{ik}f^k_j$. Indeed, consider $L_{ij} := \omega_{ik}\omega_{jl}g^{kl}$; it is symmetric and non-vanishing (its trace is $\|\omega\|^2$). Hence, by Corollary 3.19 L_{ij} should be proportional to g_{ij}. Normalize ω so that the constant of proportionality is 1. Then $\omega g^{-1}\omega^t = g$, and since $\omega^t = -\omega$, $(g^{-1}\omega)^2 = -1$, i.e., $f \equiv g^{-1}\omega$ is a complex structure. By the same argument, if on $\mathcal{M}$ there are n linearly independent parallel 2–forms, the corresponding skew–symmetric endomorphisms f^a should generate a $\mathbb{C}l(n)$ algebra as in Chapter 2.

Exercise 3.4.1 Fill in the details of the proof of Proposition 3.22.

---

[14] Of course, there are relations in this algebra [176].

3.5 Parallel spinors and holonomy

The above results for parallel tensors may be generalized to parallel spinors. In this section, $\mathcal{M}$ is a Riemannian n–fold with a spin structure, and $\mathcal{S}_\pm$ are the corresponding spin bundles (of given chirality for n even). As we already mentioned, they are the vector bundles associated to the principal $Spin(n)$ bundle through the fundamental spinorial representations (the spin structure being precisely the uplift of the usual Riemannian $SO(n)$–principal bundle given by the connection ∇ to a principal bundle with fiber its double cover $Spin(n)$; this uplift is always possible provided a certain $\mathbb{Z}_2$ cohomology class, the *second Stiefel–Whitney class*, vanishes [201, 235]). The Levi–Civita connection on $T\mathcal{M}$ induces a connection on $\mathcal{S}_\pm$, called the *spin connection*

$$D^{\mathcal{S}} = \partial + \frac{1}{4}\omega_{ij}\,\Gamma^{ij}, \qquad (3.47)$$

where ω_{ij} is the $SO(n)$ connection form and $\Gamma^{ij} = \Gamma^{[i}\Gamma^{j]}$ are the Dirac matrices generating $Spin(n)$. $\mathrm{Hol}(D^{\mathcal{S}})$ is either equal to $\mathrm{Hol}(\nabla)$ or to a double cover.

A smooth section ψ of $\mathcal{S}_\pm$ is called a *parallel spinor* if

$$D^{\mathcal{S}}\psi = 0. \qquad (3.48)$$

Again we have an integrability condition, $0 = (D^{\mathcal{S}})^2\psi \propto R_{ij}\Gamma^{ij}\psi$, which, as in Fundamental Principle 3.3, says that ψ *is parallel if and only if it is invariant under* $\mathrm{Hol}(g) \subset Spin(n)$.

To find the number of (linearly independent) parallel spinors $N_\pm(G)$ for a given holonomy group $\mathrm{Hol}(g) = G$ (and, for n even, a given chirality $\pm$), we decompose the spinor representations of $Spin(n)$ into irreducible representations of the subgroup G and count how many times we get the trivial representation; that is,

$$N_\pm(G) = \int_G \mathrm{Tr}_{\mathcal{S}_\pm}(h)\,dh, \qquad (3.49)$$

where dh is the Haar measure [263, 69] on the compact group G normalized so that the total volume is 1. We have the following:

Theorem 3.23 (Wang [295]) *Let $\mathcal{M}$ be a (simply connected) irreducible spin Riemannian n–fold with $n \geq 3$. Let $N_\pm$ be the dimensions of the spaces of parallel spinors in $\mathcal{S}_\pm$. $N_+ + N_- \geq 1$ if and only if the holonomy group $\mathrm{Hol}(\mathcal{M})$ is one of the five in Table 3.4.*

Before going to the proof, let us state a crucial lemma:

Lemma 3.24 *Assume that on the Riemannian manifold $\mathcal{M}$ there is a non–vanishing parallel spinor, $D_i^{\mathcal{S}}\psi = 0$. Then $\mathcal{M}$ is Ricci–flat $R_{ij} = 0$.*

3.5 Parallel spinors and holonomy

Table 3.4 *Holonomy groups and parallel spinors*

n	$\text{Hol}(\mathcal{M})$	N_+, N_-^a
$4m$	$SU(2m)$	2, 0
$4m$	$Sp(2m)$	$m+1, 0$
$4m+2$	$SU(2m+1)$	1, 1
7	G_2	$N = 1$
8	$Spin(7)$	0, 1

[a] For a choice of orientation; for the opposite one $N_+ \leftrightarrow N_-$. For n odd there is no distinction of chirality; we have a single number N of parallel spinors.

Proof We take the integrability condition

$$0 = 4 [D_i^{\mathcal{S}}, D_j^{\mathcal{S}}] \psi = R_{ijkl} \Gamma^{kl} \psi \qquad (3.50)$$

and contract it with Γ^j. Using the identity $\Gamma^j \Gamma^{kl} = \Gamma^{jkl} - \delta^{jk}\Gamma^l + \delta^{jl}\Gamma^k$, we get

$$R_{ijkl} \Gamma^{jkl} \psi = -2 R_{ij} \Gamma^j \psi. \qquad (3.51)$$

The LHS vanishes by the first Bianchi identity. So

$$0 = g^{ij} R_{ik} \Gamma^k R_{jl} \Gamma^l \psi = g^{ij} g^{kl} R_{ik} R_{jl} \psi, \qquad (3.52)$$

and hence $R_{ij} = 0$. □

Proof (of **Theorem 3.23**) The argument goes through several steps:

(1) If $\mathcal{M}$ admits a non–zero parallel spinor it cannot be an irreducible symmetric space. In fact, by Cartan's classification of these spaces[15] [51, 184], the only (complete, simply connected, irreducible) Ricci–flat symmetric manifold is $\mathbb{R}$, which has dimension 1, while we assume $n \geq 3$. Then the holonomy group $\text{Hol}(\mathcal{M})$ should be in the Berger's list.

(2) If $\text{Hol} = SO(n)$ we have no parallel spinors since $\mathcal{S}_\pm$ are *irreducible* representations of $\mathfrak{so}(n)$.

(3) Consider $\text{Hol} = U(1) \times SU(m)$. In this case we can split the $Spin(n)$ gamma matrices into $(1,0) \oplus (0,1)$ type, namely Γ^i and $\Gamma^{\bar i} \equiv (\Gamma^i)^\dagger$, so that the Clifford algebra reads

$$\Gamma^i \Gamma^{\bar j} + \Gamma^{\bar j}\Gamma^i = 2\delta^{i\bar j}, \qquad (3.53)$$

$$\Gamma^i \Gamma^j + \Gamma^j \Gamma^i = 0, \quad i,j = 1, 2, \ldots, m. \qquad (3.54)$$

---

[15] For a proof see Section 5.7.2

122   *Parallel structures and holonomy*

The representation spaces are constructed by picking a Clifford vacuum ψ_0 that satisfies $\Gamma^{\bar{i}} \psi_0 = 0$ for all $\bar{i}$. The spinor spaces S_+ and S_- are spanned by the vectors

$$\Gamma^{i_1 i_2 \cdots i_l} \psi_0, \tag{3.55}$$

with l even and odd, respectively. The $U(1)$ charge of the vector in Eq. (3.55) is $(l - m/2)$ (by "PCT symmetry"). There are only two vectors that are invariant under $SU(m)$, namely ψ_0 and $\Gamma^{123\cdots m} \psi_0$. Their $U(1)$ charges are $\pm m/2$, so no component of $\mathcal{S}$ in invariant under the full $U(m)$.

(4) However, the previous calculation shows that we have two parallel spinors if Hol $= SU(m)$, namely ψ_0 and $\Gamma^{123\cdots m} \psi_0$. They have the same chirality if m is even, and opposite chirality if m is odd.

(5) We computed the $Sp(2)$ curvature of a proper Quaternionic–Kähler manifold under the name, say, of the $Spin(3)_R$–curvature for an $\mathcal{N} = 3$ SUGRA in 3D. The explicit formula, Eq. (2.105), implies that an irreducible *proper* Quaternionic–Kähler manifold is *never* Ricci–flat. So Hol $= Sp(2) \times Sp(2m)$ is ruled out[16] by Lemma 3.24.

(6) Hol $= Sp(2m) \subset SU(2m)$. Equation (3.55) still holds, but the Ω_{ij}–traces now are $Sp(2m)$–invariant (here Ω_{ij} is the symplectic matrix). Calling $\omega := \Omega_{ij} \Gamma^{ij}$, we have the following list of $Sp(2m)$ singlets:

$$\psi_0, \; \omega \psi_0, \; \omega^2 \psi_0, \; \cdots \; \omega^m \psi_0. \tag{3.56}$$

In total we have $m + 1$ singlets, all of the same chirality.

(7) Hol $= G_2 \subset Spin(7)$. The **8** of $Spin(7)$ decomposes as $\mathbf{7} \oplus \mathbf{1}$ under G_2, so $N = 1$.

(8) Hol $= Spin(7) \subset Spin(8)$. $\mathcal{S}_+ = \mathcal{S}|_{Spin(7)}$, whereas $\mathcal{S}_- = \mathbf{7} \oplus \mathbf{1}$. So $N_+ = 0$ and $N_- = 1$. □

The proof is even more interesting than the result itself. In particular, we have proved the following useful theorem.

Theorem 3.25 *Let $\mathcal{M}$ be a Calabi–Yau m–fold (that is Hol $\subseteq SU(m)$). Denote by $\Omega^{(*,0)} = \oplus_{p=0}^{m} \Omega^{(p,0)}$ the graded ring of smooth $(p, 0)$ forms, $0 \leq p \leq m$, and $\mathcal{S}$ the space of smooth spinor fields (sections of the spin bundle). Let $\psi_0 \in \mathcal{S}$ be a parallel spinor as in the previous theorem (unique up to multiplication by $\lambda \in \mathbb{C}$). Then*

$$\Omega^{(*,0)} \simeq \mathcal{S} \quad \text{isomorphic as } \Omega^{(*,0)} \text{ modules}, \tag{3.57}$$

---

[16] Alternatively, to rule out the existence of parallel spinors in manifolds of holonomy $Sp(2) \times Sp(2m) \subset Spin(4m)$, look at the explicit construction of the $Sp(2m) \subset Spin(4m)$ invariant spinors in step (**6**) of the proof. The $(m+1)$ $Sp(2m)$–invariant spinors belong to the *irreducible* spin $m/2$ representation of the $Sp(2) \simeq SU(2) \equiv$ centralizer of $Sp(2m)$ in $Spin(4m)$. Since the representation is *irreducible*, it does not contain any singlet.

$$\phi_{i_1 i_2 \cdots i_p} dz^{i_1} \wedge dz^{i_2} \wedge \cdots dz^{i_p} \mapsto \phi_{i_1 i_2 \cdots i_p} \Gamma^{i_1 i_2 \cdots i_p} \psi_0. \tag{3.58}$$

Under the above isomorphism, the Dirac operator $\not{D}$ is mapped into the Kähler–Dirac operator

$$\not{D} \leftrightarrow \partial + \bar{\partial}, \tag{3.59}$$

where $\bar{\partial}$ is the adjoint of the Dolbeault operator ∂. In particular,

$$(\text{parallel } (p,0) \text{ forms}) \leftrightarrow (\text{parallel spinors}).$$

Proof The Dirac operator is $\Gamma^i D_i + (\Gamma^i D_i)^\dagger$. One has

$$\Gamma^i D_i (\phi_{i_1 i_2 \cdots i_p} \Gamma^{i_1 i_2 \cdots i_p} \psi_0) =$$
$$= (D_i \phi_{i_1 i_2 \cdots i_p}) \Gamma^{i \, i_1 i_2 \cdots i_p} \psi_0 + \phi_{i_1 i_2 \cdots i_p} \Gamma^{i \, i_1 i_2 \cdots i_p} D_i \psi_0 = \tag{3.60}$$
$$= (\partial \phi)_{i_0 i_1 i_2 \cdots i_p} \Gamma^{i_0 i_1 i_2 \cdots i_p} \psi_0,$$

so $\partial \leftrightarrow \Gamma^i D_i$. On the other hand, the antilinear map $\phi \mapsto *\phi^*$ corresponds on spinors to $\psi \mapsto \psi^\dagger$, so the two notions of adjoint coincide and

$$\bar{\partial} = (\partial)^\dagger \leftrightarrow (\Gamma^i D_i)^\dagger. \tag{3.61}$$

$\square$

For the phenomenological implications of this result see Witten, ref. [306].

Exercise 3.5.1 Let $\Omega^{(*)}$ be the space of *all* (smooth) differential forms on a manifold $\mathcal{M}$ (not necessarily Calabi–Yau). Show that $\Omega^{(*)} \simeq \mathcal{S} \otimes \bar{\mathcal{S}}$. Use this isomorphism to deduce from the classification of the parallel spinors Proposition 3.22 for the classification of parallel forms.

3.6 *G–structures and Spencer cohomology*

Berger's holonomy theorem is a special case of a more general theorem (largely due to Berger himself) which classifies irreducible *torsionless* affine holonomy groups. Restricting to *metric* connections we get back the result of Section 3.3, but we may consider more general (non–metric) connections on $T\mathcal{M}$.

It is convenient to use the language of G–structures $P_G \subset L(\mathcal{M})$ and compatible connections ∇ (a connection ∇ on $T\mathcal{M}$ is *compatible with the G–structure P_G* iff $\text{Hol}(\nabla) \subseteq G$; see the Appendix). If G is compact, the G–structure is metric; in this section we drop the compactness condition.

One shows that all closed subgroups $G \subset GL(n, \mathbb{R})$ may be realized as the holonomy of some compatible connection on $T\mathcal{M}$. An interesting classification theorem is obtained by asking *for which closed subgroups $G \subset GL(n, \mathbb{R})$ there exists a* TORSIONLESS *compatible connection ∇ with a non–parallel curvature $\nabla R \neq 0$*. The question is relevant since the natural G–structures are precisely the torsionless ones; see the examples in the Appendix. Of course, if we limit ourselves to irreducible (closed) subgroups of $SO(n)$ the answer is just Berger's list of Riemannian holonomy groups.

One may also ask how many torsion–free connections are compatible with the given G-structure P_G. If ∇ and ∇' are two connections of P_G, their difference $\alpha = \nabla' - \nabla \in C^\infty(\mathrm{adj}(P_G) \otimes T^*\mathcal{M})$ is a tensor $\alpha_{ij}{}^k$ and

$$T(\nabla')_{ij}{}^k = T(\nabla)_{ij}{}^k - \alpha_{ij}{}^k + \alpha_{ji}{}^k. \tag{3.62}$$

Hence, if a torsionless connection on P_G exists, all the others are in 1–1 correspondence with the $\alpha \in C^\infty(\mathrm{adj}(P_G) \otimes T^*\mathcal{M}) \cap C^\infty(\odot^2 T^*\mathcal{M} \otimes T\mathcal{M})$. We may rephrase the situation as a cohomology problem (Spencer cohomology).

Spencer cohomology

Let V be a vector space and $\mathfrak{g}$ a Lie subalgebra of $\mathfrak{gl}(V) := V \otimes V^\vee$. Define recursively the following $\mathfrak{g}$–modules:

$$\begin{aligned} \mathfrak{g}^{(-1)} &= V, \\ \mathfrak{g}^{(0)} &= \mathfrak{g}, \\ \mathfrak{g}^{(k)} &= [\mathfrak{g}^{(k-1)} \otimes V^\vee] \cap [V \otimes \odot^{k+1} V^\vee], \quad k = 1, 2, 3, \ldots, \end{aligned} \tag{3.63}$$

and define the map

$$\partial \colon \mathfrak{g}^{(k)} \otimes \wedge^{l-1} V^\vee \to \mathfrak{g}^{(k-1)} \otimes \wedge^l V^\vee \tag{3.64}$$

as antisymmetrization on the last l indices.[17] Since $\partial^2 = 0$, we can define the *Spencer cohomology groups* $H^{k,l}(\mathfrak{g})$ as

$$H^{k,l}(\mathfrak{g}) = \ker \partial|_{\mathfrak{g}^{(k-1)} \otimes \wedge^l V^\vee} \big/ \operatorname{Im} \partial|_{\mathfrak{g}^{(k)} \otimes \wedge^{l-1} V^\vee}. \tag{3.65}$$

Given a connection ∇ on P_G, its torsion $T(\nabla)$ is an element of $T\mathcal{M} \otimes \wedge^2 T^*\mathcal{M}$. Under the identification $V = T_\phi\mathcal{M}$, $V^\vee = T_\phi^*\mathcal{M}$, $T(\nabla)|_\phi \in \mathfrak{g}^{-1} \otimes \wedge^2 V^\vee$, where $\mathfrak{g} = \mathrm{Lie}(G) \subset \mathfrak{gl}(V)$ By Eq. (3.62) the difference between the torsion of two such connections is

$$\big(T(\nabla') - T(\nabla)\big)\big|_\phi \in \partial(\mathfrak{g} \otimes V^\vee). \tag{3.66}$$

---

[17] We identify the vector bundles and the corresponding sheaf of C^∞ sections.

Hence the class $[T(\nabla)] \in H^{0,2}(\mathfrak{g})$ is independent of the particular connection, and depends only on the principal bundle P_G. The class $[T(\nabla)] \in H^{0,2}(\mathfrak{g})$ is called the INTRINSIC TORSION *of the principal G–bundle P_G.* Thus:

Proposition 3.26 *The G–structure P_G admits a torsionless connection ∇ if and only if its Spencer class $[T(\nabla)] \in H^{0,2}(\mathfrak{g})$ vanishes. In this case, the space of torsionless compatible connections is isomorphic to $(\mathfrak{g} \otimes V^\vee)/\partial \mathfrak{g}^{(1)}$.*

We say that ∇ is a a *symmetric G–connection* iff its curvature is covariantly constant. We ask again for the classification of irreducible holonomy representations of torsion–free non–symmetric G–compatible connections. If $G \subseteq O(n)$, the connection is metric and we get back the Berger classification. The possibility of non–metric connections opens for us a whole new world. The general classification was initiated by Berger, who wrote down a list of groups and representations, stating that it contained all possible holonomies *but* a finite number of *exotic* ones. Finally, after contributions by many people, the list was completed by S. Merkulov and L. Schwachhöfer in 1998 [233]; the number of *exotic holonomies* turned out to be infinite. Technically, one has to compute the relevant Spencer cohomology groups; Merkulov and Schwachhöfer do this using a deep generalization of Simons's "transitivity on the unit sphere" idea, reintepreting the relevant group representations in the light of the Bott–Borel–Weil theorem [59].

3.7 The holonomy groups of Lorentzian manifolds

In Chapter 1 we stressed that the *same* geometrical structures appear in the physical space–time and in the scalars' target space $\mathcal{M}$. This holds, in particular, for the holonomy groups. The space–time holonomy groups are crucial for problems like the SUSY–preserving compactifications of superstrings [74, 124] and rigid SUSY in curved spaces ([133] and references therein): see Chapter 6. However, in most physical applications space–time is not a Riemmanian manifold but rather a pseudo–Riemannian one of signature $(D-1, 1)$. The holonomy groups in indefinite signature (p, q) present new phenomena which require new definitions.

Definition 3.27 A connected pseudo–Riemannian manifold $M^{p,q}$ of signature (p, q) is called *irreducible* if its holonomy group $\text{Hol}(M^{p,q})$ acts irreducibly on the tangent space $T_\phi M^{p,q} \sim \mathbb{R}^{p,q}$. It is called *indecomposable* if its holonomy group $\text{Hol}(M^{p,q}) \subset O(p, q)$ does not leave invariant any proper *non–degenerate* subspace of $T_\phi M^{p,q} \sim \mathbb{R}^{p,q}$.

In the Riemannian case (positive signature) *indecomposable* ⇔ *irreducible*. In the indefinite signature case *irreducible* ⇒ *indecomposable*, but the opposite implication is *false*. Indeed, consider the general Lorentzian metric in 2D. It can be put in the form $ds^2 = g(x^+, x^-) \, dx^+ \, dx^-$. The holonomy, acting by local Lorentz transformations, should leave invariant the two one–dimensional subspaces of the tangent bundle generated by $\partial/\partial x^+$ and $\partial/\partial x^-$ which transform into multiples of themselves. Leaving invariant the two light–cone directions, the holonomy does not act irreducibly; yet the manifold is *not* a product. However, the inner product restricted to each invariant subspace is identically zero. Thus the invariant subspaces are degenerate, and this action is indecomposable according to our definition. From this example we see that, in general signature (p, q), the de Rham Theorem 3.6 should be replaced by the following theorem.

Theorem 3.28 (de Rham, Wu [37, 321, 322]) *Let $M^{p,q}$ be a simply connected complete pseudo–Riemannian manifold. $M^{p,q}$ is isometric to the product of a flat space times the product of simply connected complete indecomposable pseudo–Riemannian manifolds.*

The classification of the holonomies of *irreducible* pseudo–Riemannian manifolds is covered by the Merkulov–Schwachhöfer theorem [233]. In Lorentzian signature $(n - 1, 1)$ this leaves only one possibility, namely [35]

$$\text{Hol}^0(M) = SO_0(n - 1, 1). \tag{3.67}$$

Thus *in Lorentzian signature <u>all</u> indecomposable special holonomy groups arise from manifolds which are indecomposable but not irreducible.* If M is decomposable, $M = M^{d-1,1} \times M^{D-d,0}$, with $M^{D-d,0}$ a Riemannian $(D - d)$–fold which may have special holonomy as described in previous sections.

Example: 3D Lorentzian holonomies. To give a feeling of the indecomposable but non–irreducible Lorentzian spaces, we consider the 3D case. The Lorentz group is $SO_0(2, 1) \simeq SL(2, \mathbb{R})/ \pm 1$. The two subgroups which have indecomposable non–irreducible actions are

$$A^1(\mathbb{R}) = \left\{ \begin{pmatrix} 1 & b \\ 0 & 1 \end{pmatrix}, \ b \in \mathbb{R} \right\}, \tag{3.68}$$

$$A^2(\mathbb{R}) = \left\{ \begin{pmatrix} a & b \\ 0 & a^{-1} \end{pmatrix}, \ a \in \mathbb{R}^\times, b \in \mathbb{R} \right\}. \tag{3.69}$$

Examples of metrics with these holonomy groups are, respectively,

$$A^1(\mathbb{R}): \quad ds^2 = 2 \, dx^- \, dx^+ + f(x^+)^2 \, dy^2, \tag{3.70}$$

$$A^2(\mathbb{R}): \quad ds^2 = 2 \, dx^- \, dx^+ + f(x^+)^2 \, dy^2 + g(x^+)^2 (dx^+)^2. \tag{3.71}$$

Physicists call these spaces *pp–waves*. A particular instance is the Cahen–Wallach space which is *symmetric* [70].

3.7 The holonomy groups of Lorentzian manifolds

Classification. The possible holonomies of indecomposable non–irreducible Lorentzian manifolds had been classified [39, 153, 227, 228]. Lorentz signature, $(n-1, 1)$, is "easy": let $W \subset T_\phi$ be a Hol–invariant non–zero subspace; $W \cap W^\perp$ is a Hol–invariant light–like *line*. The subalgebra of $\mathfrak{so}(n-1, 1)$ preserving this line, $(\mathbb{R} \oplus \mathfrak{so}(n-2)) \ltimes \mathbb{R}^{n-2}$, consists of matrices of the form[18]

$$\left\{ \begin{pmatrix} 0 & a & X^t \\ a & 0 & X^t \\ X & -X & A \end{pmatrix} : a \in \mathbb{R}, X \in \mathbb{R}^{n-2}, A \in \mathfrak{so}(n-2) \right\}. \tag{3.72}$$

The *indecomposable* subalgebras are obtained by putting suitable restrictions on a, X, and A. One restriction on A is obvious: it should belong to a Lie subalgebra $\mathfrak{g}$ of $\mathfrak{so}(n-2)$; we write $\mathfrak{g} = \mathfrak{z} \oplus \mathfrak{g}'$ (center $\oplus$ semi–simple part).

Definition 3.29 A Lorentzian space N is called a *Brinkmann space* if $\mathfrak{hol}(N)$ preserves a *null* vector, that is, it is a subalgebra of the algebra of matrices of the form (3.72) with $a = 0$.

More generally, one shows [36, 39, 153] that there are four types of indecomposable subalgebras of the algebra of matrices of the form (3.72):

(1) $(\mathbb{R} \oplus \mathfrak{g}) \ltimes \mathbb{R}^{n-2}$.
(2) $\mathfrak{g} \ltimes \mathbb{R}^{n-2}$.
(3) $(\mathrm{graph}(\phi) \oplus \mathfrak{g}') \ltimes \mathbb{R}^{n-2}$ with $\phi: \mathfrak{z} \to \mathbb{R}$ linear.
(4) $(\mathfrak{g} \oplus \mathrm{graph}(\psi)) \ltimes \mathbb{R}^r$, where $0 < r < n - 2$, $\mathfrak{g}' \subset \mathfrak{so}(r)$ and $\psi: \mathfrak{z} \to \mathbb{R}^{n-2-r}$ is linear and surjective.

An indecomposable Lorentzian manifold N is a Brinkmann space iff $\mathfrak{hol}(N)$ is of type (2) or (4). From the above list we see that the important datum in the holonomy of an indecomposable but not–irreducible Lorentzian holonomy is the Lie subalgebra $\mathfrak{g} \subseteq \mathfrak{so}(n-2)$. We have the following elegant theorem.

Theorem 3.30 (Leistner [227, 228]) *An indecomposable subalgebra of*

$$(\mathbb{R} \oplus \mathfrak{so}(n-2)) \ltimes \mathbb{R}^{n-2} \tag{3.73}$$

of type (1)–(4) is the holonomy algebra of a Lorentzian manifold N if and only if $G = \exp \mathfrak{g}$ is a product of Berger's holonomy groups.

One says that an indecomposable non–irreducible Lorentzian manifold N is a *pp-wave* if $\mathfrak{g} = 0$, while if $\mathfrak{g} \neq 0$ one says that *it has a $\mathfrak{g}$–flag*. N is said to have a *Kähler flag* (resp. *Calabi–Yau flag, Quaternionic–Kähler flag*, etc.) iff

---

[18] We write the matrix in a base for which $\eta_{\alpha\beta} = \mathrm{diag}(-1, 1, 1, \ldots, 1)$.

$\mathfrak{g} \subseteq \mathfrak{u}(m)$, $m \equiv [(n-2)/2]$ (resp. $\subseteq \mathfrak{su}(m)$, $\subseteq \mathfrak{sp}(2) \oplus \mathfrak{sp}(m)$, etc.). An indecomposable non–irreducible manifold N admitting a non–zero parallel spinor is called a *Brinkmann–Leistner space*. For the classification of the Brinkmann–Leistner spaces see Exercise 3.7.2.

Exercise 3.7.1 Give an elementary argument why a Lorentzian manifold with a parallel spinor is necessarily reducible.

Exercise 3.7.2 N is an indecomposable non–irreducible Lorentzian manifold. Show that N has parallel spinors $\Leftrightarrow$ is a Brinkmann space and G is trivial or a product of groups $SU(k)$, $Sp(2\ell)$, G_2, or $Spin(7)$.

Exercise 3.7.3 Compute the number of linearly independent parallel spinors on a Brinkmann–Leistner space with given $\mathfrak{g}$–flag.

Exercise 3.7.4 Show that a Brinkmann–Leistner space has a nowhere–vanishing *null* Killing vector.

Exercise 3.7.5 State and prove the Lorentzian version of Lemma 3.24.

Exercise 3.7.6 Give a physicist–style proof of Eq. (3.67).

4
SUSY/SUGRA Lagrangians and U–duality

The geometrical results of Chapter 3 fully determine the Lagrangian for any (ungauged) SUSY/SUGRA model. In this chapter we present the general picture, valid for all $\mathcal{N}$s and Ds, and discuss in detail the cases $D = 3, 4$.

4.1 Determination of the scalar manifold $\mathcal{M}$

In Chapter 2 we found that the scalar manifold tangent bundle $T\mathcal{M}$ of a SUSY/SUGRA theory is isomorphic to a (direct sum of) product(s) $\mathcal{S} \otimes \mathcal{U}$, where the bundle $\mathcal{S}$ has structure group Aut_R, the SUSY automorphism group. The product structure is preserved by parallel transport. We also found that the Aut_R curvature vanishes in rigid SUSY, while it is a commutator in SUGRA. The details of Aut_R depend on the particular D and $\mathcal{N}$, but otherwise the situation is pretty universal. These facts may equivalently be stated as conditions on the holonomy group $\text{Hol}^0(\mathcal{M})$

$$\text{Hol}^0(\mathcal{M}) \subseteq \begin{cases} \alpha(\text{Aut}_R) \times \mathcal{C}(\alpha(\text{Aut}_R)) & \text{SUGRA}, \\ \mathcal{C}(\alpha(\text{Aut}_R)) & \text{rigid SUSY}, \end{cases} \qquad (4.1)$$

where $\text{Aut}_R \xrightarrow{\alpha} O(\dim \mathcal{M})$ is the representation of Aut_R on $T\mathcal{M}$, i.e., $\alpha(\text{Aut}_R)$ is the quotient of Aut_R acting *faithfully* on the scalars, and $\mathcal{C}(\alpha(\text{Aut}_R))$ is its *centralizer* in $O(\dim \mathcal{M})$. The decompositions of $T\mathcal{M}$ in irreducible representations of Aut_R are listed in Chapter 2. In all cases $\alpha(\mathfrak{aut}_R) \neq 0$ but for 3D $\mathcal{N} = 1$, where $\mathfrak{aut}_R \equiv 0$.

4.1.1 Rigid supersymmetry

We discuss the manifolds $\mathcal{M}$ for the dimensions in the $\mathbb{R} \leftrightarrow \mathbb{C} \leftrightarrow \mathbb{H} \leftrightarrow \mathbb{O}$ sequence, i.e., $D = 3, 4, 6, 10$, leaving the other Ds as an exercise. In the 6D case we assume that no chiral 2–forms are present; see Exercise 4.1.2.

Table 4.1 *The scalars" manifolds in rigid* SUSY *with $N_S \leq 8$ supercharges*

Riemannian	$D = 3, \mathcal{N} = 1$
Kähler	$D = 3, \mathcal{N} = 2$
	$D = 4, \mathcal{N} = 1$
	$\mathcal{M}_{\text{gauge}}$ for $D = 4, \mathcal{N} = 2$ [a]
hyperKähler	$D = 3, \mathcal{N} = 3, 4$ [b]
	$\mathcal{M}_{\text{matter}}$ for $D = 4, \mathcal{N} = 2$
	$D = 6, (\mathcal{N}_L, \mathcal{N}_R) = (2, 0), (0, 2)$ [c]

[a] $\mathcal{M}_{\text{gauge}}$ is *special* Kähler: see Section 2.9.4
[b] For $\mathcal{N} = 4$ $\mathcal{M}$ is a "twisted" product: see Section 2.4.1.
[c] This holds in absence of chiral two forms.

Corollary 4.1 *In rigid* SUSY, $\mathcal{M}$ *irreducible and non–symmetric is possible only if the* TOTAL NUMBER *of supercharges N_S is ≤ 8. In this case $\mathcal{M}$ is as in Table 4.1. Whenever $N_S \geq 9$, $\mathcal{M}$ is* FLAT.

Proof If $\mathcal{M}$ is irreducible and non–symmetric, its holonomy group should be in the Berger list. Their centralizers $\mathcal{C}(\text{Hol}^0)$ are listed in Table 3.3. Let $\mathfrak{c}$ be the Lie algebra of $\mathcal{C}(\text{Hol}^0)$. $\mathfrak{c}$ is $\mathfrak{u}(1)$ for Kähler, $\mathfrak{sp}(2)$ for hyperKähler, and 0 otherwise. Equation (4.1) implies that $\alpha(\mathfrak{aut}_R) \subseteq \mathfrak{c}$. Except for 3D $\mathcal{N} = 1$, $\alpha(\mathfrak{aut}_R)$ is a non–zero quotient algebra of $\mathfrak{aut}_R$. Comparing Table 3.3 with the list of $\mathfrak{aut}_R$'s in Table 2.3 we get the first two statements. In particular, from Section 2.4.1 we see that 3D $\mathcal{N} = 4$ is special since $\mathfrak{aut}_R \simeq \mathfrak{su}(2)_l \oplus \mathfrak{su}(2)_r$ is not simple, and we may have a product $\mathcal{M} = \mathcal{M}_l \times \mathcal{M}_r$ so that

$$\alpha(\mathfrak{aut}_R)|_{\mathcal{M}_l} \simeq \mathfrak{su}(2)_l, \qquad \alpha(\mathfrak{aut}_R)|_{\mathcal{M}_r} \simeq \mathfrak{su}(2)_r, \qquad (4.2)$$

both factor manifolds being hyperKähler but with different actions of Aut_R. Likewise, from Section 2.9.1 4D $\mathcal{N} = 2$ is special, with $\mathcal{M} = \mathcal{M}_{\text{gauge}} \times \mathcal{M}_{\text{hyper}}$ and

$$\alpha(\mathfrak{aut}_R)|_{\mathcal{M}_{\text{gauge}}} \simeq \mathfrak{u}(1), \qquad \alpha(\mathfrak{aut}_R)|_{\mathcal{M}_{\text{hyper}}} \simeq \mathfrak{su}(2). \qquad (4.3)$$

For $N_S \geq 9$, $\mathcal{M}$ cannot be irreducible and non–symmetric. Assume $\mathcal{M}$ to be *irreducible symmetric*. By Proposition 3.21, $\mathfrak{c}$ is

$$\mathfrak{c} = \begin{cases} \mathfrak{u}(1) & \text{if } H = U(1) \times K, \\ 0 & \text{otherwise.} \end{cases} \qquad (4.4)$$

Since $\alpha(\mathfrak{aut}_R) \subseteq \mathfrak{c} \subset \mathfrak{u}(1)$, in rigid SUSY an irreducible symmetric space is possible only for those Ds and $\mathcal{N}$s having *trivial* or *Abelian* $\alpha(\text{Aut}_R)$. The first case corresponds to $D = 3\;\mathcal{N} = 1$: *all* Riemannian manifolds are allowed, in particular the symmetric ones. Comparing with Table 2.3, the second case corresponds to $D = 3\;\mathcal{N} = 2$ and $D = 4,\;\mathcal{N} = 1$, which have $N_S \leq 4$. By (4.4) the allowed symmetric spaces have the form $G/[U(1) \times K]$, i.e., are Hermitian symmetric spaces, which are a particular instance of Kähler spaces;[1] we do not get any new space besides those listed in Table 4.1. The reader may wonder about the case of $\mathcal{M}_{\text{gauge}}$ in 4D $\mathcal{M} = 2$, which is special Kählerian; one shows that a special Kähler manifold of *rigid* 4D $\mathcal{N} = 2$ SUSY is symmetric if and only if it is flat.

Thus, for $N_S \geq 9$, we remain with only one possibility: $\mathcal{M}$ is *reducible*. By de Rham's theorem, the universal cover $\widetilde{\mathcal{M}}$ of $\mathcal{M}$ has the form

$$\widetilde{\mathcal{M}} = (\text{flat}) \times \mathcal{M}_1 \times \cdots \times \mathcal{M}_k \tag{4.5}$$

with the $\mathcal{M}_l$ non–flat irreducible. Since the holonomy acts separately on each factor space, the previous arguments apply to each irreducible $\mathcal{M}_l$, which then should be trivial. It remains the flat factor. □

In particular, $\mathcal{M}$ should be *flat* for $(\mathcal{N}_L, \mathcal{N}_R) = (2, 2)$ in 6D, $\mathcal{N} = 4$ in 4D, and $5 \leq \mathcal{N} \leq 8$ in 3D. Corollary 4.1 contains only the conditions on $\mathcal{M}$ following from the Aut_R–structure of $T\mathcal{M}$. In general there may be other restrictions on $\mathcal{M}$. In fact, the above corollary fully characterizes the allowed $\mathcal{M}$'s with only one exception, $\mathcal{N} = 2\;D = 4$, where we have in addition the restrictions coming from "special geometry," see Section 2.9.4. However, even this case is completely determined by the above corollary by requiring consistency with reduction to 3D. The allowed Kähler manifolds for $\mathcal{M}_{\text{gauge}}$ in 4D $\mathcal{N} = 2$ are precisely those which become hyperKähler in 3D after dualizing vectors into scalars [91].

4.1.2 $\widetilde{\mathcal{M}}$ in supergravity

Proposition 4.2 *Let $\widetilde{\mathcal{M}}$ be the universal covering space of the scalar manifold of $\mathcal{N}$-extended* SUGRA *in $D = 3, 4, 6, 10$ dimensions. Then:*

(a) $\widetilde{\mathcal{M}}$ is IRREDUCIBLE *except possibly for:*
  (i) $\mathcal{N} = 1, 2, 4$ in 3D;
  (ii) $\mathcal{N} = 1, 2, 4$ in 4D;
  (iii) $(2, 0), (2, 2)$ in 6D;

(b) $\mathcal{M}$ is SYMMETRIC *except for $N_S \leq 8$ in which case Table 4.1 applies with* "HYPERKÄHLER" *replaced by* "QUATERNIONIC–KÄHLER" *for $D \neq 6$. In $D = 6$, $\widehat{\mathcal{M}} = T \times H$ where H is "Quaternionic-Kähler" and T is*

---
[1] Cf. discussion at the end of Section 3.3

parametrized by the scalars of the tensor supermultiplets (see Section 4.8). The Quaternionic–Kähler spaces have NEGATIVE *Ricci curvature.*
(c) if $N_S \geq 9$, $\widetilde{\mathcal{M}}$ *is a* SYMMETRIC *manifold of the form*

$$\frac{G}{\mathrm{Aut}_R \times K} \quad \text{with} \quad \mathfrak{g} = \mathfrak{m} \oplus \mathfrak{aut}_R \oplus \mathfrak{k} \text{ and } \mathfrak{m} \simeq T_e\mathcal{M},$$

and the adjoint action of $\mathfrak{aut}_R$ *on* $\mathfrak{m}$,

$$[R^a, X^j] = f^{aj}{}_k X^k \qquad R^a \in \mathfrak{aut}_R, \ X^i \in \mathfrak{m},$$

is given by the representations in Table 2.6. Therefore, $\widetilde{\mathcal{M}}$ *is the symmetric space* G/H *listed for the diverse* D *and* $\mathcal{N}$ *in Table 4.2.*

Note that for $N_S \geq 9$ $\widetilde{\mathcal{M}}$ is IRREDUCIBLE SYMMETRIC except for 4D $\mathcal{N} = 4$ and 6D $(2, 2)$. In Table 4.2 we also give a reference to the historical papers where the geometry of the coset G/H was first deduced for the given SUGRA. Usually in the SUGRA literature the flat space $\mathbb{R}$ is identified with the group manifold $SO(1, 1)$. The note "no graviton" refers to would–be SUSY theories, whose existence is not forbidden on purely differential geometric grounds, that cannot be supergravities (since there is no massless representation of the corresponding SUSY algebra containing the graviton) and that were never constructed [266].

Proof (a). We have only to compare the isomorphisms of Chapter 2 with de Rham's Theorem 3.6. In 3D the holonomy group is $Spin(\mathcal{N})_R \times K$, with $Spin(\mathcal{N})_R$ acting on $T\mathcal{M}$ in a spinorial representation. If $Spin(\mathcal{N})$ is simple, de Rham's theorem implies that $\mathcal{M}$ is irreducible. The only exceptions are $Spin(1) = \{e\}$, $Spin(2) = U(1)$, and $Spin(4) \simeq SU(2)_1 \times SU(2)_2$. In the last case we may have $\mathcal{M} = \mathcal{M}_1 \times \mathcal{M}_2$ with $\mathrm{Hol}(\mathcal{M}_i) = SU(2)_i \times K_i$. Physically, one manifold is parametrized by hypermultiplets and the other by twisted hypermultiplets. In 4D the same story applies: $\mathrm{Aut}_R = U(1) \times SU(\mathcal{N})$, and the only possible factorization is in a manifold of $U(1) \times K_1$ holonomy and one of $SU(\mathcal{N}) \times K_2$; thus the holonomy representation on $T\mathcal{M}$ may decompose only if the representation of $SU(\mathcal{N})_R$ on $T\mathcal{M}$ is *real* or *pseudoreal* (quaternionic). Comparing with Table 2.6, the first case happens for $\mathcal{N} = 4$ and the second for $\mathcal{N} = 2$. For $\mathcal{N} = 1$ $SU(1)_R$ is trivial and hence we have no *local* restriction on the Kähler manifold $\mathcal{M}$. We leave 6D as an exercise.

(b) Just compare the list of G–structures (or, equivalently, bundle isomorphisms) of Table 2.6, and their generalizations, with Berger's theorem and Definition 3.20. There is one special case, namely the scalars of the 6D $(2, 0)$ tensor supermultiplet, which are inert under Aut_R [281]. This leaves undetermined the factor manifold T, whose geometry will be described in Section 4.8. As a side remark, notice that $\mathfrak{hol}(\mathcal{M}) = \mathfrak{aut}_R$ implies $\mathcal{M}$ to be (locally) *symmetric*. Indeed, from the curvature

Table 4.2 *Symmetric manifolds for $N_S > 8$ extended* SUGRA

D	$\mathcal{N}$	G/H [a]	Notes
3	5	$Sp(4, 2k)/(Sp(4) \times Sp(k))$	[119]
3	6	$SU(4, k)/(SU(4) \times U(k))$	[119]
3	7	none	cannot exist
3	8	$SO(8, k)/(SO(8) \times SO(k))$	[119]
3	9	$F_{4(-20)}/SO(9)$	[119]
3	10	$E_{6(-14)}/(SO(10) \times SO(2))$	[119]
3	11	none	cannot exist
3	12	$E_{7(-14)}/(SO(12) \times SO(3))$	[119]
3	13, 14, 15	none	cannot exist
3	16	$E_{8(+8)}/SO(16)$	[119]
3	≥ 17	none	cannot exist
4	3	$SU(3, k)/(SU(3) \times U(k))$	[80]
4	4	$\frac{SU(1,1)}{U(1)} \times \frac{SO(6,k)}{SO(6) \times SO(k)}$	[45, 104]
4	5	$SU(5, 1)/U(5)$	[98]
4	6	$SO^*(12)/U(6) \equiv SO(6, \mathbb{H})/U(6)$	[98]
4	7	none	cannot exist
4	8	$E_{7(7)}/SU(8)$	[98]
4	≥ 9	none	cannot exist
6	(2, 2)	$\mathbb{R} \times \frac{SO(4,k)}{SO(4) \times SO(k)}$	[27]
6	(4, 0)	$SO(5, k)/(SO(5) \times SO(k))$	[27, 283]
6	(4, 2)	$SO(5, 1)/SO(5)$	
6	(6, 0)	$SU^*(6)/Sp(6) \equiv SL(3, \mathbb{H})/Sp(6)$	no graviton
6	(4, 4)	$SO(5, 5)/(SO(5) \times SO(5))$	[283]
6	(6, 2)	$F_{4(4)}/(Sp(6) \times SU(2))$	no graviton

Table 4.2 *(Cont.)*

D	$\mathcal{N}$	$G/H\ ^a$	Notes
6	(8, 0)	$E_{6(6)}/Sp(8)$	no graviton
10	(1, 0)	$\mathbb{R}$	[92]
10	(1, 1)	$\mathbb{R}$	[72, 158]
10	(2, 0)	$SU(1,1)/U(1)$	[191]

^{a}The small numbers in parentheses appended to certain groups in the table refer to the specific real form: the number in parentheses is the signature of the invariant Killing quadratic form for the given real form.

computations in Chapter 2, we know that the $\mathfrak{aut}_R$ part of the curvature, $R|_{\mathfrak{aut}_R}$, is covariantly constant. Then $R \equiv R|_{\mathfrak{aut}_R}$ implies $DR = 0$, i.e., $\mathcal{M}$ symmetric. Thus, for instance, in 3D, $\mathcal{N} = 7$, the possibility $\mathrm{Hol}^0(\mathcal{M}) = Spin(7)$ and $\mathcal{M}$ a $Spin(7)$–manifold is ruled out.[2]

(c) The first statement follows from the explicit description of the holonomy for a symmetric manifold G/H and $H \equiv \mathrm{Hol}(\mathcal{M}) = \mathrm{Aut}_R \times K$. The second one is the relation between the holonomy representation of G/H and the bundle isomorphisms (G–structures) of Chapter 2.

Table 4.2 is then constructed as follows. The irreducible symmetric manifolds were classified by Cartan. Once we know that the relevant manifolds are irreducible symmetric, we take the Cartan list [51, 69, 184], and look for cosets G/H with the right subgroup H and holonomy representation. We implement the procedure first for 3D, and then for all $D \geq 4$.

Three dimensions. In 3D with $\mathcal{N} \geq 5$, $\mathcal{M}$ is irreducible symmetric. It is a manifold of dimension $k \mathbf{N}(\mathcal{N})$, $k \in \mathbb{N}$ (cf. General Lesson 2.4) which has a holonomy of the form $Spin(\mathcal{N}) \times K$, where K acts orthogonally, unitarily, or symplectically if $Spin(\mathcal{N})$ is (respectively) real, complex, or quaternionic, see Table 2.1.

The crucial point is that $T\mathcal{M}$ belongs to a *spinorial representation* of $Spin(\mathcal{N})$. In the Cartan classification, there are very few symmetric spaces whose holonomy acts in a spinorial representation. Indeed, for a symmetric space G/H, we have (see Section 3.2.2) $T_e G/H \simeq \mathfrak{m} := \mathfrak{g} \ominus \mathfrak{h}$, and the holonomy representation $\mathfrak{m}_{\mathfrak{hol}}$ is given by the decomposition of the adjoint representation of $\mathfrak{g}$ into representations

---

[2] There are other reasons why a $Spin(7)$–manifold is excluded: for instance, a $Spin(7)$ manifold is necessarily *Ricci–flat*, whereas $\mathcal{N} \geq 3$ SUGRA requires $\mathcal{M}$ to be Einstein with a specific, *negative*, cosmological constant.

of the subalgebra $\mathfrak{h}$,

$$\mathfrak{g} = \text{adj}(\mathfrak{h}) \oplus \mathfrak{m}_{\mathfrak{hol}}. \tag{4.6}$$

The decomposition of the Lie algebra of a *classical group* never produces spinorial representations. Thus holonomy groups acting *via* spinorial representations can arise only in two ways. First, for small $\mathcal{N}$'s through the Lie algebra isomorphisms:

$$\begin{aligned}
&\mathfrak{spin}(1) \simeq 0, &&\mathfrak{spin}(2) \simeq \mathfrak{u}(1), \\
&\mathfrak{spin}(3) \simeq \mathfrak{su}(2) \simeq \mathfrak{sp}(2), &&\mathfrak{spin}(4) \simeq \mathfrak{su}(2) \oplus \mathfrak{su}(2), \\
&\mathfrak{spin}(5) \simeq \mathfrak{sp}(4), &&\mathfrak{spin}(6) \simeq \mathfrak{su}(4),
\end{aligned} \tag{4.7}$$

as well as the triality automorphism of $\mathfrak{spin}(8)$,

$$\mathfrak{spin}(8) \simeq \mathfrak{so}(8), \tag{4.8}$$

which allow us to *reinterpret* classical representations as spinorial ones. These isomorphisms: (i) fix the geometry of the 3D SUGRAs with $N_S \leq 8$ to be as in Proposition 4.2, (ii) produce the 3D $\mathcal{N} = 5, 6, 8$ entries in Table 4.2, and (iii) show that, in the presence of propagating matter, *no* $\mathcal{N} = 7$ SUGRA may exist in 3D.

Second, G may be an *exceptional* Lie group. In this case we have the four esoteric *projective planes* based on the octonions[3] [28]:

$$\begin{array}{ll}
\mathbb{O}P^2, & (\mathbb{C} \otimes \mathbb{O})P^2, \\
(\mathbb{H} \otimes \mathbb{O})P^2, & (\mathbb{O} \otimes \mathbb{O})P^2,
\end{array} \tag{4.9}$$

which correspond to four pairs of symmetric spaces whose holonomy representations are genuinely spinorial, namely [51, 69, 184]

$$\frac{F_4}{Spin(9)}, \quad \frac{E_6}{Spin(10) \times SO(2)}, \quad \frac{E_7}{Spin(12) \times SU(2)}, \quad \frac{E_8}{Spin(16)}, \tag{4.10}$$

and their non–compact (= negatively curved) duals.

In Chapter 2 we computed the $Spin(\mathcal{N})$ curvature in 3D SUGRA: it was *negative*. Hence we should keep the non–compact versions. The four non–compact octonionic spaces then should be the $\mathcal{M}$s for $\mathcal{N} = 9, 10, 12, 16$. No other physical 3D SUGRA may exist! In particular, any 3D SUGRA, having local propagating degrees of freedom, must have $\mathcal{N} \leq 16$. This constraint stems, ultimately, from the Hurwitz theorem, which states that the octonions $\mathbb{O}$ are the last normed division algebra [28].

The $D \geq 4$ case. In order to work out all Ds at once, we take the tables of *linear* SUSY representations in ref. [281], look for the quantum numbers of the

---

[3] Only projective *planes* and *lines* are well defined for octonions.

scalar fields under Aut_R, invoke our blessed "target space equivalence principle" to identify them as holonomy representations, and check which – if any – G/H space in Cartan's list has *that* holonomy representation. In doing this we use the Lie algebra isomorphisms (4.7). The method works smoothly even when $\mathcal{M}$ is a *reducible* symmetric manifold. For each $\mathcal{N}$ and D one finds either *two* G/H spaces with the right representation, or none. The second case corresponds (magically?) to situations where physical intuition predicts that *no* SUGRA exists. In the other cases, the two solutions form a dual pair of Cartan spaces: a compact one (with positive curvature), called a *space of Type I* in Helgason's classification [184], and its non–compact negatively–curved companion, a *space of Type III* in Helgason's classification (see Chapter 5 for details). Since we know that the curvature of the Aut_R–connection is negative, we opt for the Type III space. In this way everything gets determined, up to a few scalars which split from $T\mathcal{M}$ because they are singlets of $(\text{Aut}_R)_{\text{semi-simple}}$. For $N_S \geq 9$, they are either a single real scalar – which corresponds to the trivial symmetric space $\mathbb{R}$ – or a complex one charged under $U(1)_R$. If the $U(1)$ curvature is constant and negative,[4] the complex scalars should parametrize $SU(1,1)/U(1)$, i.e., the upper half–plane. In this way one writes down Table 4.2 in less than two minutes. It is remarkable than one *never* finds more than one solution! □

On the face of it, in the following SUGRAs:

$$3D\ \mathcal{N} \geq 5, \qquad\qquad 4D\ \mathcal{N} \geq 3,$$
$$6D\ \mathcal{N}_R + \mathcal{N}_L \geq 4, \qquad\qquad 10D\ \mathcal{N}_R + \mathcal{N}_L \geq 2,$$

the scalar sector is invariant under a big non–compact symmetry group G, namely the isometry group $G \equiv \text{Iso}(G/H)$, with G as in Table 4.2. We have still to see how these isometries act on the non–scalar sectors, and notably on the gauge forms. We anticipate that they are symmetries of the full <u>ungauged</u> SUGRA. However, Dirac quantization of fluxes will reduce the actual physical symmetry to an arithmetic subgroup of G; see Section 4.9. Historically, the symmetry G was known as the *hidden symmetry* of the SUGRA model.

Global geometry of $\mathcal{M}$

Proposition 4.2 describes the metric geometry of the universal cover of $\widetilde{\mathcal{M}}$. Then the most general scalar manifold *classically* consistent with extended local supersymmetry has the form

$$\mathcal{M} = \Gamma \backslash \widetilde{\mathcal{M}}, \tag{4.11}$$

---

[4] These statements will be justified in Sections 4.4–4.7.

4.1 Determination of the scalar manifold $\mathcal{M}$

Table 4.3 *Symmetric manifolds for $N_S > 8$, 5D SUGRA*

$\mathcal{N}$	G/H
4	$\mathbb{R} \times \big(SO(5,k)/SO(5) \times SO(k)\big)$
6	$SU^*(6)/Sp(6)$
8	$E_{6(6)}/Sp(8)$

where $\Gamma \subset \text{Iso}(\widetilde{\mathcal{M}})$ is a discrete subgroup. In particular, for $N_S > 8$ $\mathcal{M}$ is the *double coset*

$$\mathcal{M} = \Gamma\backslash G/H, \qquad \Gamma \subset G \text{ discrete}, \tag{4.12}$$

with G and H as in Table 4.2. At the classical level one would require that Γ acts freely in order to garantee that $\mathcal{M}$ is smooth.

We will see in Section 4.9 below that quantum consistency puts severe constraints on the discrete group Γ. At the non–perturbative level it seems that non–smooth $\mathcal{M}$s are also allowed. However, we will see that all consistent $\mathcal{M}$ have a *finite* smooth cover.

Exercise 4.1.1 Write down the scalars' manifolds for rigid extended SUSY in $D = 5, 7, 8$.

Exercise 4.1.2 Show that in rigid $\mathcal{N} = 2$ 4D SUSY $\mathcal{M}_{\text{gauge}}$ is symmetric only if it is flat.

Exercise 4.1.3 Consider a 6D rigid $(2, 0)$ SUSY model with k chiral 2–forms (i.e., k $(2,0)$ tensor supermultiplets [252, 281]). Show that $\widetilde{M} = \mathbb{R}^k \times H$ with H hyperKähler.

Exercise 4.1.4 Work out the details of the proof of part (a) of Proposition 4.2 in 6D. HINT: use the push–forward to $\widetilde{M}$ of the 2–form gauge symmetry in analogy with Section 2.9.3.

Exercise 4.1.5 Check the manifold list for $D = 5$ $\mathcal{N} \geq 4$ SUGRA, given in Table 4.3. Write down the analogous tables for $D = 7, 8$.

4.2 Dimensional reduction and totally geodesic submanifolds

Given a SUSY/SUGRA model in $D \geq 4$ dimensions, we can dimensionally reduce it to 3D by assuming that the fields depend only on three coordinates, and rewriting all couplings in a manifestly 3D covariant way. In general the 3D scalar manifold, $\mathcal{M}_3$, has a larger dimension than the original one, $\mathcal{M}_D$, since we get new scalars from the internal components of tensor fields; moreover, to put the theory in canonical form, we have to dualize the various form–fields into scalars, increasing the dimension of $\mathcal{M}_3$. Thus, in general, the original scalar manifold $\mathcal{M}_D$ is just a *submanifold* immersed in $\mathcal{M}_3$,

$$\mathcal{M}_D \hookrightarrow \mathcal{M}_3, \tag{4.13}$$

equipped with the induced metric. We ask: *what is the geometrical relation between the manifolds $\mathcal{M}_D$ and $\mathcal{M}_3$ or, more generally, the scalar manifold $\mathcal{M}_d$ of the model reduced to $d < D$–dimensions?*

To answer the question, we perform a *gedanken experiment*. Consider our SUGRA theory (ungauged and with vanishing scalars' potential[5]) in D dimensions. Look for a solution of the equations of motion in which all fields are set to zero, except the scalars and the metric. We take the scalars to depend only on time t, $\phi^i = \phi^i(t)$. Moreover, we consider the adiabatic limit; that is, we write $\phi^i = \phi^i(\epsilon\, t)$, and take $\epsilon \to 0$. In this limit we can forget about the gravitational back–reaction (since $T^{\mu\nu} = O(\epsilon^2)$) and take the metric to be flat. The equations of motion reduce to

$$-\frac{d}{dt}\left(G_{ij}\,\dot{\phi}^j\right) + \Gamma^l_{ij}\, G_{lk}\, \dot{\phi}^j\, \dot{\phi}^k + O(\epsilon^3) = 0, \tag{4.14}$$

so the solutions are simply the geodesics on $\mathcal{M}_D$. Reduce to d dimensions by requiring the fields not to depend on $D-d$ spatial coordinates. The above *gedanken* solutions should be also solutions of the d–dimensional theory (in the same adiabatic limit). But, in d–dimensional theory the adiabatic–limit solutions are precisely the geodesics of $\mathcal{M}_d$. Hence under the embedding $\mathcal{M}_D \hookrightarrow \mathcal{M}_d$ geodesics go to geodesics. This is precisely the definition of a *totally geodesic submanifold* [51, 263] Hence:

---

[5] Is this a consistent assumption? Yes, it is. If, in some SUGRA model, a certain gauging, or potential, was needed for consistency in D dimension, we would have the corresponding condition on the gauging and potential down in $D = 3$ and this is not the case: see Chapter 2.

General Lesson 4.3 Let $\mathcal{M}_D$ and $\mathcal{M}_d$ be, respectively, the scalars' manifolds of a (Q)FT in D space–time dimensions and of its (trivial) reduction to d dimensions. Then

$$\varrho_{D,s}: \mathcal{M}_D \hookrightarrow \mathcal{M}_d \tag{4.15}$$

is a TOTALLY GEODESIC SUBMANIFOLD.

In the above *gedanken experiment* we assumed that $V(\phi) \equiv 0$. But, of course, the relation between $\mathcal{M}_D$ and $\mathcal{M}_d$ cannot depend on the scalar potential; thus our conclusion is fairly general.

The pull–back $\varrho^*_{D,d}$ is so important as to deserve a name: it is called *group disintegration*; see ref. [210]. The name stems from the following theorem.

Theorem 4.4 *Let G/H be a symmetric Riemannian manifold with Cartan involution σ, and $X \hookrightarrow G/H$ a totally geodesic submanifold. Then X is also a symmetric space G'/H' with $G' \subset G$ and involution $\sigma' = \sigma|_{G'}$.*

Proof The restriction of the geodesic symmetry σ to a totally geodesic submanifold is still a symmetry. □

Then suppose we have a D–dimensional SUGRA with $N_S > 8$, so that $\mathcal{M}_3 = G_3/H_3$ is symmetric. From the theorem we have $\mathcal{M}_d = G_d/H_d$ for $D \geq d \geq 3$ and a chain of subgroups

$$G_D \subset G_{D-1} \subset \cdots \subset G_4 \subset G_3, \tag{4.16}$$

which is the *disintegration* of the Lie group G_3.

Exercise 4.2.1 Show that the 10D (2,0) group disintegration is

$$SL(2,\mathbb{R}) \subset SL(2,\mathbb{R}) \times SL(2,\mathbb{R}) \subset SL(3,\mathbb{R}) \times SL(2,\mathbb{R}) \subset$$
$$\subset SL(5,\mathbb{R}) \subset SO(5,5) \subset E_{6(6)} \subset E_{7(7)} \subset E_{8(8)}.$$

Exercise 4.2.2 Construct the group disintegration chain of the 6D (2,2) SUGRA with $G_6 = SO(1,1) \times SO(4,k)$.

4.3 Four–fermion couplings vs. holonomy

The results of Section 4.1 may be obtained more directly from 4–fermion couplings.

4.3.1 3D

From the Ambrose–Singer theorem, we know that $\mathrm{Hol}^0(\mathcal{M})$ is generated by the curvature tensors. These tensors can be read directly in the 4–χ couplings which read (schematically)

$$a R_{ijkl}\, \bar\chi^i \gamma^\mu \chi^j\, \bar\chi^k \gamma_\mu \chi^l \qquad \text{(rigid)}, \tag{4.17}$$

$$\left(a R_{ijkl} - b\, g_{ik} g_{jl}\right) \bar\chi^i \gamma^\mu \chi^j\, \bar\chi^k \gamma_\mu \chi^l \qquad \text{(local)}. \tag{4.18}$$

Rigid supersymmetry

In rigid SUSY, if the scalar potential[6] $V(\phi) \equiv 0$, the group Aut_R is actually a symmetry, by an argument presented in Chapter 2. Aut_R acts on the fields as an isometry of $\mathcal{M}$ and an Aut_R rotation of the fermions, say

$$\chi \mapsto \exp\left[\tfrac{1}{2}\alpha_{AB}\, \Sigma^{AB}\right] \chi, \tag{4.19}$$

in 3D, or the analogous formula in $D > 3$. If Aut_R is a symmetry of $\mathcal{L}$, it should be — in particular – a symmetry of the 4–Fermi coupling, i.e., the Riemann tensor must satisfy

$$R_{ijkl} = S_i^{\,m} S_i^{\,n} S_i^{\,p} S_i^{\,q} R_{mnpq} \qquad S_i^{\,j} \in \alpha(\mathrm{Aut}_R). \tag{4.20}$$

This means that the Lie algebra $\mathfrak{hol}$ spanned by $R(x, y) \in \mathrm{End}(T\mathcal{M})$ is invariant under the adjoint action of $\alpha(\mathfrak{aut}_R)$. Since the $S_i^{\,m}$ in the above equation are *constant matrices* (General Lesson 1.4), all covariant derivatives $D_{m_1} \cdots D_{m_r} R_{ijkl}$ are invariant, and hence by an adaption of the Ambrose–Singer argument, the Aut_R transformations *commute* with $\mathfrak{hol}$ and

$$\alpha(\mathfrak{aut}_R) \subseteq \mathfrak{c}. \tag{4.21}$$

We have got back the previous result.

Supergravity

More interesting is the local case. Still Eq. (4.20) holds, but now the $S_i^{\,j}$ are ϕ–dependent matrices. Besides, Eq. (4.20) is true in *full generality,* because Aut_R, being a *gauge* symmetry, should be always *exact*.

By definition, a manifold $\mathcal{M}$ is (locally) symmetric if and only if Eq. (4.20) holds for all $S \in \mathrm{Hol}^0(\mathcal{M})$. Thus in 3D, $\mathcal{N} \geq 9$ SUGRA, where (Table 2.6)

$$\mathrm{Hol}^0(\mathcal{M}) = \mathrm{Aut}_R, \tag{4.22}$$

---

[6] $V \equiv 0$ in the ungauged theory for $\mathcal{N} \geq 3$: see Chapter 7.

we immediately conclude that $\mathcal{M}$ is a symmetric space, *without any need for the Berger theorem:* it is a direct consequence of Eq. (4.20), i.e., of the gauge invariance of $\mathcal{L}$.

For general $\mathcal{N}$, however, we have

$$\mathfrak{hol} = \alpha(\mathfrak{aut}_R) \oplus \mathfrak{k}, \tag{4.23}$$

and we can only conclude that the projection of the curvature endomorphism on $\mathfrak{aut}_R$ is covariantly constant

$$D_i R_{ijkl}|_{\mathfrak{aut}_R} = 0 \tag{4.24}$$

(as we saw in Chapter 2 by direct computation of the curvatures), and also *invariant* under the adjoint action of Aut_R. By Ambrose–Singer, the Riemann tensor is an element of

$$\odot^2 \mathfrak{hol} \equiv \odot^2(\alpha(\mathfrak{aut}_R) \oplus \mathfrak{k}). \tag{4.25}$$

Let $\mathfrak{s} \subset \alpha(\mathfrak{aut}_R)$ be a *simple* Lie subalgebra. The Riemann tensor decomposes, in general, into the following representations of $\mathfrak{s}$

$$\odot^2 \mathfrak{s} \oplus \mathfrak{s} \otimes V_1 \oplus V_2 \tag{4.26}$$

(here V_1, V_2 stand for trivial $\mathfrak{s}$–modules). By Schur's lemma, (4.26) may be invariant under $\mathfrak{s}$ only if (i) $V_1 = 0$, and (ii) the first term in the RHS is proportional to the quadratic Casimir with a *constant* coefficient.[7] Hence, in local SUSY, the Riemann tensor is a sum of the quadratic Casimirs of the simple factors of Aut_R, plus a term in $\odot^2(\mathfrak{k} \oplus \mathfrak{aut}_R|_{\mathrm{Abelian}})$. If $\mathfrak{aut}_R$ is Abelian (the Kähler case) this is an empty condition; in all other cases we get a non–trivial restriction on the geometry. First of all $R|_\mathfrak{s}$ is covariantly constant; this condition inserted in the Bianchi identity forces *all* components of the curvature to be covariantly constant, except in the case $\mathfrak{s} = \mathfrak{sp}(2)$ since the Bianchi identity requires antisymmetrization with respect to three indices, and the $\mathfrak{sp}(2)$ index takes only two values. In this case the Bianchi identities give no further conditions beyond $DR|_{\mathfrak{sp}(2)} = 0$. Then we have the following theorem:

Theorem 4.5 (Salamon [267]) *(1) Let $\mathcal{M}$ be Quaternionic–Kähler with $T\mathcal{M} \simeq \mathcal{S} \otimes \Lambda$ ($\mathcal{S}, \Lambda$ have structure groups $Sp(2)$ and $Sp(2m)$, respectively) and dimension $4n \geq 8$. Then the Riemann tensor belongs to the space*

$$\mathbb{R} R_0 \oplus \odot^4 \Lambda \tag{4.27}$$

---

[7] That the coefficient α is a constant follows from the Bianchi identity: $0 = DR|_\mathfrak{s} = d\log\alpha \wedge R|_\mathfrak{s}$, where $R|_\mathfrak{s} = \alpha\, C_2 \in \odot^2 \mathfrak{s}$.

where R_0 is the curvature tensor of the canonical projective space $P\mathbb{H}^n$.

(2) $\mathcal{M}$ is Einstein (i.e., $R_{ij} = \lambda g_{ij}$) and it is Ricci–flat if and only if it is hyperKähler.

Proof (1) is the condition of invariance under the gauge symmetry $Sp(2)_R$. (2) is the fact that $\odot^4\Lambda$ component of the Riemann tensor does not contain $\wedge^2\Lambda$ (unless $m = 1$), and hence does not contribute to the Ricci tensor which is $\lambda \in \mathbb{R}$ times that of $P\mathbb{H}^n$. Finally, if $\mathcal{M}$ is Ricci flat, $\lambda = 0$, and $\mathfrak{hol} \subset \mathfrak{sp}(2m)$, which is the definition of hyperKähler. □

Remark. The restriction to real dimension ≥ 8 stems from the fact that the general holonomy of a 4–fold $O(4) \simeq Sp(2) \times Sp(2)$, so the restriction to "quaternionic–Kähler" is empty in dimension 4; in the SUGRA literature one adopts the convention that a "Quaternionic–Kähler 4–fold" is a manifold for which Eq. (4.27) holds. This are the manifolds which may be coupled to $\mathcal{N} = 2$ SUGRA [31]. With this convention one has:

General Lesson 4.6 *Comparing Eq.* (4.27) *with the computation of the* $Spin(\mathcal{N})$ *curvature in Chapter 2, we see that, given any <u>negatively curved</u> Quaternionic–Kähler metric g, there is a (unique) re–scaling, $g \to g' \equiv \lambda g$, such that g' is the target space metric of an $\mathcal{N} = 4$* SUGRA *theory, for an appropriate scale λ. This should be contrasted with the $\mathcal{N} = 2$ case, where the analogous argument shows that <u>not all</u> Kähler manifolds are allowed as target spaces, but only the subclass of the Hodge ones (see Section 2.7).*

In Exercise 2.6.3 you were asked to compute, by geometric arguments, the coefficients a and b in Eq. (4.18). One may check that the linear combination is precisely right so that the tensor in parenthesis which couples to χ^4 is $\odot^4\Lambda$; that is, the effect of the SUGRA correction to the rigid 4–χ coupling is to project out the $Sp(2)_R$ contribution to the curvature. The formula is exactly as in the rigid case, except that R_{ijkl} is replaced by Ω_{mnpq}, the $Sp(2m)$ curvature (which is the only one present in the rigid case!). The $Sp(2m)$ curvature is $Sp(2)_R$ invariant by definition.

4.3.2 $D \geq 4$

We write χs for the fermions taking value in $T\mathcal{M}_D$ in D dimensions. From the 3D viewpoint, the D–dimensional 4–χ coupling looks like

$$\bar{\chi}^i \gamma^\mu \chi^j \bar{\chi}^k \gamma_\mu \chi^l \left(a\, g_{ik} g_{jl} - b\, R^{(3)}_{ijkl}\right)\Big|_{T\mathcal{M}_D \subset T\mathcal{M}_3}, \qquad (4.28)$$

where $R^{(3)}$ is the curvature computed in $\mathcal{M}_3$. By General Lesson 4.3, $\mathcal{M}_D \hookrightarrow \mathcal{M}_3$ is totally geodesic. The main property of a totally geodesic submanifold is precisely that the curvature tensor computed using the (induced) metric on $\mathcal{M}_D$ is equal to the restriction of the curvature computed using the metric on $\mathcal{M}_3$. Thus, in (4.28) we may drop the superscript (3), and learn that the formula (4.18) is true (for the fermions spanning $T\mathcal{M}$) for SUGRA in *any* dimension. Therefore our 3D argument relating the 4–Fermi coupling to holonomy, and hence metric geometry, holds in all dimensions. What about the other 4–Fermi couplings involving gaugini, dilatini, and spin–3/2 gravitini? Do they have a geometrical description?

They do. Consider the dimensional reduction embedding map

$$\varrho_{D,3}: \mathcal{M}_D \hookrightarrow \mathcal{M}_3. \qquad (4.29)$$

The D–dimensional fermion vector bundle $\mathcal{F} \to \mathcal{M}_D$ is just $\varrho_{D,3}^* T\mathcal{M}_3$. The bundle $\mathcal{F}$ contains all D–dimensional fermions, even the physical polarizations of the gravitini ψ_μ^A, $\mu = 3, 4, \ldots, D-1$. Hence the full 4–Fermi coupling is just the (centralizer component of) curvature of the $\varrho_{D,3}^* T\mathcal{M}_3$ bundle. Computing the 4–Fermi terms by actually constructing the map $\varrho_{D,3}$ may be cumbersome, but once we know that they are *curvatures of certain natural bundles,* we have only to identify these bundles over $\mathcal{M}_D$ by comparing their Aut_R quantum numbers, using "target space equivalence." This is particularly easy for $\mathcal{N} \geq 3$ where everything is Lie–algebraic, and the homogeneous bundles are uniquely fixed by their representation content. Thus:

General Lesson 4.7 *The 4–Fermi couplings are also uniquely determined by the holonomy group* $\mathrm{Hol}(\mathcal{M})$.

This $\varrho_{D,3}$ game may be played with the other couplings as well.

4.4 Vector couplings in 4D SUGRA

4.4.1 First example: $\mathcal{N} = 8$ SUGRA

To fix the ideas, let us start by considering the example of $\mathcal{N} = 8$ SUGRA [98]. From linear representation theory, we know that there are 28 vectors; then, by Section 2.10.2, their field–strengths take value in a flat $Sp(56, \mathbb{R})$–bundle $\mathcal{F}_{56} \to \mathcal{M} \equiv E_{7(7)}/SU(8)$. Moreover, our pet "equivalence principle" gives

$$\mathcal{F}_{56} \simeq \wedge^2 \mathcal{S}_8 \oplus \wedge^2 \mathcal{S}_8^\vee, \qquad (4.30)$$

where $\mathcal{S}_8$ is the $SU(8)$ gravitino bundle which, by Section 4.1, is associated *via* the **8** representation to the tautological principal $SU(8)$–bundle

$$E_{7(7)} \xrightarrow{\pi} E_{7(7)}/SU(8). \qquad (4.31)$$

Equation (4.30) means that there is an invertible vielbein

$$(U_u^{[AB]}, V_{u\,[AB]}), \qquad u = 1, 2, \ldots, 56;\ A, B = 1, 2, \ldots, 8, \qquad (4.32)$$

converting the curved $SU(8)$ indices, $A, B, \ldots$, into flat $Sp(56, \mathbb{R})$ indices, $u, v, \ldots$, which is covariantly constant under the combined $Sp(56, \mathbb{R})$ and $SU(8)$ connections. We may choose our frame $(U_u^{[AB]}, V_{u\,[AB]})$ so that the $Sp(56, \mathbb{R})$ connection vanishes. Then $(U_u^{[AB]}, V_{u\,[AB]})$, $u = 1, \ldots, 56$, are 56 linearly independent covariantly constant sections of the $SU(8)$–bundle

$$\wedge^2 S_8 \oplus \wedge^2 S_8^\vee \longrightarrow E_{7(7)}/SU(8).$$

Left multiplication by elements of $E_{7(7)}$ maps covariantly constant sections into covariantly constant ones, and hence the 56 covariantly constant sections $(U_u^{[AB]}, V_{u\,[AB]})$ should form a definite representation[8] ρ of $E_{7(7)}$. This representation should be *real symplectic*, by consistency with $Sp(56, \mathbb{R})$. Thus the representation ρ on the 56 covariantly constant sections is in fact a subgroup embedding

$$\rho : E_{7(7)} \to Sp(56, \mathbb{R}) \tag{4.33}$$

given by the 56×56 matrix $(U_u^{[AB]}, V_{u\,[AB]})$, which is precisely the vielbein $\mathcal{E}^{-1}$ we discussed at length in Chapter 1 in the context of general dualities,

$$(U_u^{[AB]}, V_{u\,[AB]}) \longleftrightarrow \mathcal{E}^{-1} \in Sp(56, \mathbb{R}). \tag{4.34}$$

Then we are exactly in the situation of the *prototypical example* of page 26: we have only to check that

$$\rho(E_{7(7)}) \cap U(28) = SU(8) \tag{4.35}$$

and then the commutative diagram (1.96),

$$\begin{array}{ccc} E_{7(7)} & \xrightarrow{\rho} & Sp(56, \mathbb{R}) \\ \pi_{\text{can}} \downarrow & & \downarrow \pi_{\text{can}} \\ E_{7(7)}/SU(8) & \xrightarrow{\mu} & Sp(56, \mathbb{R})/U(28) \end{array}, \tag{4.36}$$

will determine the vector coupling μ for us. Equation (4.35) is obvious: the image under ρ of the maximal compact subgroup $SU(8)$ is a compact subgroup, and hence is contained in a $U(28)$ maximal compact subgroup of $Sp(56, \mathbb{R})$.

From now on we identify the elements of the group $E_{7(7)}$ with the corresponding 56×56 matrices in the faithful **56** representation. We also identify the point $\phi \in E_{7(7)}/SU(8)$ with the corresponding vielbein, $\mathcal{E}^{-1} \in E_{7(7)}$, with the proviso that

---

[8] This is a particular instance of the general construction of homogeneous bundles over G/H. See Chapter 11.

4.4 Vector couplings in 4D SUGRA

two vielbeins $\mathcal{E}$ and $\mathcal{E}'$ correspond to the same point if there exists an $U \in SU(8)$ such that

$$\mathcal{E}' = \begin{pmatrix} \rho_{28}(U) & \\ & \rho_{28}^*(U) \end{pmatrix} \mathcal{E}. \tag{4.37}$$

The scalar–vector couplings are directly determined by the vielbein (Chapter 1), that is, by the commutative diagram (4.36). So also the vector couplings are elegantly predicted by geometry.

4.4.2 Generalization to all $\mathcal{N}$

$\mathcal{N} \geq 3$ SUGRAs

Consider first the case $\mathcal{N} \geq 3$, in which – as we saw in Section 4.1.2 – the scalar manifold has the form G/H. In all such models the isometry group G acts on the field–strengths through a real symplectic representation F of dimension $2n$ (n being the number of vector fields). The representation F embeds $G \hookrightarrow Sp(2n, \mathbb{R})$, and we identify G with its image. The subgroup $H \subset G$, being compact, is mapped in $U(n)$, and again the commuting diagram (1.96) does its job. So all scalar–vector couplings get determined (and are guaranteed to be G–invariant): the map μ of General Lesson 1.7 is induced by the representation map $\rho_F \colon G \hookrightarrow Sp(2n, \mathbb{R}) \subset \mathrm{End}(\mathbb{R}^{2n})$,

$$\begin{array}{ccc} G & \xrightarrow{\rho_F} & Sp(2n, \mathbb{R})) \\ {\scriptstyle \pi_{\mathrm{can}}}\downarrow & & \downarrow{\scriptstyle \pi_{\mathrm{can}}} \\ G/H & \xrightarrow{\mu} & Sp(2n, \mathbb{R})/U(n) \end{array}, \tag{4.38}$$

and thus the couplings are determined once the representation content of F is known. This is simplicity itself: we list the representations F under G for the various $\mathcal{N}$ in Table 4.4 (k is the number of matter vector multiplets).

$\mathcal{N} = 1, 2$ SUGRAs

The case of $\mathcal{N} = 1$ already has been treated in Chapter 2; all *holomorphic* maps $\mu \colon \mathcal{M} \to Sp(2V, \mathbb{R})/U(V)$ are allowed.

The case $\mathcal{N} = 2$ is more interesting. Here we have the SUGRA version of special geometry[9] called *projective special Kähler geometry* [145]. In the particular case in which we have "enough" symmetries – that is the group $\mathrm{Iso}(\mathcal{M})$ is transitive – the argument we used above for the $\mathcal{N} \geq 3$

---
[9] The rigid version is discussed in Section 2.9.3.

Table 4.4 *Representations of the 4D duality groups on the field–strengths*

$\mathcal{N}$	G	Representation F
3	$SU(3,k)$	$[[(\mathbf{3}+\mathbf{k}) \oplus \overline{(\mathbf{3}+\mathbf{k})}]]$
4	$SU(1,1) \times SO(6,k)$	$(\mathbf{2}, \mathbf{6}+\mathbf{k})$
5	$SU(5,1)$	**20** (self–dual part of $\wedge^3 V_\mathbf{6}$)
6	$SO^*(12)$	**32** (the chiral spinor)
8	$E_{7(7)}$	**56** (the fundamental rep.)

case goes through word–for–word for $\mathcal{N} = 2$ too, and all couplings are fixed (and easy to compute explicitly) once we know in which representations of Iso($\mathcal{M}$) the field–strengths $\mathcal{F}$ are [91]. The general case is similar, and again uniquely determined by geometrical considerations. However, for the sake of order and clarity, we prefer to postpone the discussion of the general $\mathcal{N} = 2$ model until after the foundation of the relevant geometric theory in Chapter 12.

4.5 The gauge point of view

We have seen that in 4D $\mathcal{N} \geq 3$ SUGRA the scalar fields live on a coset manifold G/H. The number of scalar degrees of freedom is $\dim G - \dim H$. A value of the scalar fields ϕ^i is identified with the *class* of the vielbein $\mathcal{E}(\phi)$ in the coset G/H. This is the mathematicians' language. Physicists think differently. They introduce a full set of $\dim G$ scalar fields, parametrizing the *group G*, a much easier object than the coset G/H. Now the entries of the matrix $\mathcal{E}$ *are* the scalar fields. However, in this way one gets $\dim H$ more scalar fields than the number of physical degrees of freedom; therefore one introduces a *gauge* symmetry with $\dim H$ generators "to eat" the unwanted degrees of freedom. This is easily done. As gauge group one takes H itself, and the gauge transformations act on the scalars by

$$\mathcal{E}^{-1} \to \mathcal{E}^{-1} U(x)^{-1} \qquad U(x) \in H. \tag{4.39}$$

The theory with target space being the group manifold G with the subgroup of right multiplication by H *gauged* is physically equivalent to the original one with target space the coset G/H. Indeed, the coset theory is just the gauged one in the "unitary gauge," and all H–gauge–invariant observables are manifestly the same in the two formulations.

The gauge formulation, however, has formidable advantages. It is more intuitive, and has more symmetry: the formalism now has an automorphism group which is $G_{\text{GLOBAL}} \times H_{\text{LOCAL}}$ acting on $\mathcal{E}^{-1}$ as follows:

$$\mathcal{E}^{-1} \to g\,\mathcal{E}^{-1}\,U(x)^{-1}, \qquad g \in G_{\text{GLOBAL}}, \quad U(x) \in H_{\text{LOCAL}}. \tag{4.40}$$

Having a formalism with automorphism group so large is a tremendous technical asset. We shall adhere to it. G_{GLOBAL} may, or may not, be a symmetry of the full Lagrangian. On the contrary, H_{LOCAL} is *always* exact, since it was "artificially" constructed to be a symmetry. The discrete group Γ in Eq. (4.12) should also be seen as a *gauge* symmetry.

4.6 The complete 4D Lagrangian: *U*–duality

In Section 4.4 we have not just computed the full non–linear scalar–vector couplings. We have done a lot more: we have proved that the G isometry of the scalar sector is also a symmetry of the vector sector. On the other hand, the arguments of Section 4.3 imply that the 4–Fermi couplings are also G–invariant, as are the Fermi kinetics terms, since their H–bundles are homogeneous bundles[10] over G/H, and hence G–covariant.

It appears that the full Lagrangian $\mathcal{L}^{\mathcal{N}\geq 3}$ of ungauged $\mathcal{N} \geq 3$ supergravity is invariant under the *hidden symmetry* group G. To establish this result we have to prove invariance of the last two classes of couplings present in $\mathcal{L}^{\mathcal{N}\geq 3}$:

(1) couplings proportional to $\bar\psi^A_\mu \gamma^\nu \gamma^\mu \chi_m \partial_\nu \phi^i$;
(2) Pauli couplings with two fermions and a vector field–strength.

We start by giving an *a priori* argument for their G–invariance. These couplings are sections of some bundles that are obtained by pulling–back with $\rho^*_{D,3}$ certain sections of the appropriate bundles of the 3D theory. In Chapter 2 we constructed the full 3D ungauged Lagrangian, and found it to be invariant under all the isometries of $\mathcal{M}$. Then the couplings pulled back to 4D are automatically invariant by *group disintegration*.

Let us be explicit. For simplicity we assume $\mathcal{N} \geq 5$, leaving the other cases as an exercise. The couplings in (1), linear in the gravitini, are just the Noether term, coupling the "superconnection" ψ^a_μ to the supercurrent, which is proportional to

$$\gamma^\nu \gamma_\mu \chi_{BCD}\, P_i^{ABCD}\, \partial_\nu \phi^i + \text{higher–order in fermions}, \tag{4.41}$$

---

[10] The geometry of homogeneous bundles over reductive cosets G/H is described in detail in Chapter 11.

where, as shown in Chapter 2, $P_i^{ABCD}(\phi)$ is the vielbein realizing the bundle isomorphisms of Table 2.6,

$$TM \otimes \mathbb{C} \simeq \wedge^4 S_\mathcal{N} \oplus \wedge^4 \overline{S}_\mathcal{N}. \tag{4.42}$$

P_i^{ABCD} plays exactly the same role as γ_i^{Am} in the rigid 6D $\mathcal{N} = 2$ σ–model, Section 2.8.1, and it generates a Clifford–like subalgebra of $\text{End}(TM)$ for reasons already reviewed many times (for the $\mathcal{N} = 8$ case, cf. Eq. (4.16) of ref. [106]). In particular, the G–invariant metric on G/H is

$$g_{ij} = P_i^{ABCD} (P_j^{ABCD})^* + (i \leftrightarrow j), \tag{4.43}$$

and thus the coupling

$$\overline{\psi}_{A\mu} \gamma^\nu \gamma^\mu \chi_{BCD} P_i^{ABCD} \partial_\nu \phi^i \tag{4.44}$$

is automatically invariant under *any* isometry of g_{ij}.

Let us now show explicitly that the Pauli couplings (2) are G–invariant. We return to Eq. (1.75), but now we replace the generic locution *"other fields"* in the RHS with the appropriate term, bilinear in the fermions, which enters in $G_{u\mu\nu} \equiv *(\partial \mathcal{L}/\partial F_{\mu\nu}^u)$ as a consequence of the Pauli couplings. In the $Sp(2V, \mathbb{R})$–covariant formalism, these terms are rewritten in the form of a **2V**–vector of 2–forms $\mathcal{K}_{\mu\nu}$. The constraint (1.75) now reads[11]

$$\frac{1}{2}(1 - i\Omega)\mathcal{E} \mathcal{F}^+ = \mathcal{K}^+, \tag{4.45}$$

while the vector equation of motion is simply $d\mathcal{F} = 0$. Since $(1 - i\Omega)/2$ is a projector, consistency requires $(1 - i\Omega)\mathcal{K}^+ = 2\mathcal{K}^+$, so only half the components of $\mathcal{K}$ are independent:

$$\mathcal{K}^+ = \begin{pmatrix} K^+ \\ iK^+ \end{pmatrix}. \tag{4.46}$$

The H gauge symmetry acts on the LHS of Eq. (4.45) as in Eq. (4.40). Since $H \subset U(V)$, $U(x) \in H$ commutes with Ω. Then gauge invariance requires the RHS of (4.45) to transform in the same way

$$\mathcal{K}^+ \to U(x)\mathcal{K}^+. \tag{4.47}$$

To be more explicit, we write the matrix $\mathcal{E}$ as in Eq. (1.76):

$$\mathcal{E} = \begin{pmatrix} \mathcal{A} & \mathcal{B} \\ \mathcal{C} & \mathcal{D} \end{pmatrix}. \tag{4.48}$$

---

[11] I omit space–time, G, and H indices. I hope this will not cause confusion.

The entries of $\mathcal{E}$ are related by various algebraic identities, reflecting the fact that the matrix $\mathcal{E}$ is an element of the group G. A matrix of the form (4.48) belongs to $H \equiv G \cap U(V)$ iff it satisfies the algebraic equations defining the group G and, in addition, has the form

$$\begin{pmatrix} A & B \\ -B & A \end{pmatrix} \quad \text{with} \quad (A + iB) \in U(V), \tag{4.49}$$

so

$$\begin{pmatrix} K^{\pm} \\ \pm iK^{\pm} \end{pmatrix} \to \begin{pmatrix} A & B \\ -B & A \end{pmatrix} \begin{pmatrix} K^{\pm} \\ \pm iK^{\pm} \end{pmatrix} \equiv \begin{pmatrix} (A \pm iB)K^{\pm} \\ \pm i(A \pm iB)K^{\pm} \end{pmatrix}, \tag{4.50}$$

and K^+, K^- transform in conjugate representations of the gauge symmetry $H \subset U(V)$. Define the H–gauge–invariant **2V**–vector of 2–forms

$$\widehat{\mathcal{K}}_{\mu\nu} = \mathcal{E}^{-1} \mathcal{K}_{\mu\nu} \tag{4.51}$$

and the *improved* field–strengths $\widehat{\mathcal{F}}_{\mu\nu} \equiv \mathcal{F}_{\mu\nu} - \widehat{\mathcal{K}}_{\mu\nu}$. The constraint now reads

$$(1 - i\Omega) \mathcal{E} \widehat{\mathcal{F}}^+ = 0, \tag{4.52}$$

which is manifestly G–invariant under

$$\widehat{\mathcal{F}}^+ \to g \widehat{\mathcal{F}}^+, \qquad \mathcal{E}^{-1} \to g \mathcal{E}^{-1}, \quad g \in G. \tag{4.53}$$

We see that the global G symmetry is induced by local H–invariance. Hence the vector equations of motion and the Pauli couplings are G–invariant.

General Lesson 4.8 *In $\mathcal{N} \geq 3$, $D = 4$ ungauged* SUGRA *the complete equations of motion are invariant under the non–compact "hidden symmetry" G. G acts on the field–strengths $\mathcal{F}$ through a real symplectic representation F by dualities, and all couplings are uniquely determined by the G–representations F which are listed in Table 4.4.*

The fact that the classical equations of motion are G–invariant does not mean that the *physics* is G–invariant: we have still to take into account the subtleties implied by Dirac's quantization of fluxes, Section 1.5.3, whose effect is to break down the symmetry to an arithmetic subgroup $G_{\mathbb{Z}} \subset G$. The resulting discrete symmetries are called U–dualities [199]. These dualities are quite important in the context of superstrings/M–theory, where they are exact *quantum* symmetries.

The reader may have the feeling that our *explicit* Lagrangian $\mathcal{L}^{\mathcal{N}}$ for $\mathcal{N} \geq 3$ is not explicit enough for his/her taste. Don't worry! In the next chapter we shall give very explicit expressions.

Exercise 4.6.1 Show the G–invariance of the full Lagrangian for $\mathcal{N} = 3, 4$.

4.7 U–duality, Z–map, and Grassmannians

This is the right place to discuss the central charge geometry we introduced in Section 2.10.3. By far the most interesting case is $\mathcal{N} = 2$ SUGRA. We defer this more sophisticated case to Part III, and here limit ourselves to the $\mathcal{N} \geq 3$ SUGRAs which we can study by elementary Lie–algebraic techniques.

To discuss central charges, we have to be slightly more explicit in our treatment of U–duality. To get simpler and nicer formulae, we rewrite the real matrix $\mathcal{E}$ in the complex Cayley rotated basis where the action of the compact subgroup $H \subset U(V)$ is diagonal; see Section 1.5.1 for details.

4.7.1 The H–covariant field–strength $\mathfrak{F}$

Using the machinery Section 1.5.1, we define two projections of the $2V$–vector $\mathcal{F}^+$:

$$P_+ \mathcal{E}\, \mathcal{F}^+ = \begin{pmatrix} K^+ \\ 0 \end{pmatrix}, \qquad (4.54)$$

$$P_- \mathcal{E}\, \mathcal{F}^+ = \begin{pmatrix} 0 \\ \mathfrak{F}^+ \end{pmatrix}. \qquad (4.55)$$

The first equation is just (4.45) in the Cayley basis. The second equation *defines* $\mathfrak{F}^+$, which is called the *H–covariant vector field–strength*. $\mathfrak{F}^+$ transforms in the opposite representation with respect to K^+, that is, in the same representation as its conjugate K^-. In the old SUGRA jargon, in passing from $\mathcal{F}^+$ to $\mathfrak{F}^+$, we have converted the "curved" $G \subset Sp(2V, \mathbb{R})$ indices into "flat"[12] $H \subset U(V)$ indices using the vielbein $\mathcal{E}$.

Writing $\mathcal{E}$ as in Eq. (1.111), we solve Eqs. (4.54) and (4.55) in the form

$$\begin{aligned} 2\,\mathfrak{F}^+ =& \{V^* + U^* + (U^* - V^*)(U - V)^{-1}(U + V)\} F^+ \\ & + 2(V^* - U^*)(U - V)^{-1} K^+ \end{aligned} \qquad (4.56)$$

where F^+ is the (complex) self–dual part of the Abelian field–strength dA (all indices suppressed from the notation). From Eqs. (1.116), we get

$$V^* = (U^t)^{-1} V^\dagger U, \qquad U^* = (U^t)^{-1} V^\dagger V, \qquad (4.57)$$

which inserted into the first line of (4.56) give

$$\mathfrak{F}^+ = (U^t)^{-1}(U - V)^{-1} U F^+ + \cdots \qquad (4.58)$$

---

[12] Note that in the SUGRA jargon the indices of the *flat* G–bundle are called "curved," while those of the *curved* H–bundle are called "flat." This terminological twist is due to analogy with the language one uses for General Relativity.

for the relation between the H–invariant field–strengths $\mathfrak{F}$ and the Maxwell ones F^+. $\mathfrak{F}^+$ has manifestly the right transformation under H, Eq. (1.112),

$$\mathfrak{F}^+ \mapsto (h^t)^{-1}\mathfrak{F}^+ \equiv h^*\mathfrak{F}^+. \tag{4.59}$$

In SUGRA the local symmetry group, H, has the form

$$H \equiv \mathrm{Aut}_R \times \widetilde{H} = \begin{cases} U(1)_R \times SU(\mathcal{N})_R \times \widetilde{H} & \mathcal{N} \neq 8, \\ SU(8)_R & \mathcal{N} = 8 \end{cases} \tag{4.60}$$

and $K^+, \mathfrak{F}^-$ belong to the representations[13] (for $\mathcal{N} < 8$),

$$\left(\binom{\mathcal{N}}{2}, \mathbf{1}\right)_{+2} \oplus \left(\overline{\binom{\mathcal{N}}{6}}, \mathbf{1}\right)_{-6} \oplus (\mathbf{1}, k)_0, \tag{4.61}$$

where k is the number of matter vectors ($k = 0$ for $\mathcal{N} \geq 5$).

4.7.2 Pauli couplings and central charges

Local H symmetry requires the Pauli couplings to be proportional to

$$e\,(K^+_{\mu\nu})^t\,\mathfrak{F}^{+\,\mu\nu} + \mathrm{H.\,c.}, \tag{4.62}$$

which can be also be written in the more suggestive form $e\mathcal{K}^t_{\mu\nu}\Omega\mathcal{E}\mathcal{F}^{\mu\nu}$. The equation $G^+ = i\partial\mathcal{L}/\partial F^+$ fixes the overall coefficient to be $+1$. Now the "target space equivalence principle" implies that K^+ is a bilinear in the fermions without any scalar; K^+ is exactly given by the formula one would get from the linear theory [109]. We are especially interested in the term bilinear in the gravitini ψ^A_μ. Gauge invariance, Fermi statistics, and covariance determine this term up to a numerical coefficient

$$K^{+\,AB}_{\mu\nu} \propto \bar{\psi}^A_\rho \gamma^{[\rho}\gamma_{\mu\nu}\gamma^{\sigma]}\psi^B_\sigma + \cdots. \tag{4.63}$$

The variation of the action with respect to ψ^A_μ gives the supercurrent, from which one reads the SUSY transformation of the fields. Covariance fixes the term in $\delta\psi^A_\mu$ containing the vector field–strength to the form

$$\delta\psi^A_\mu = \mathcal{D}_\mu\epsilon^A + c\,\mathfrak{F}^{-\,AB}_{\rho\sigma}\gamma^{\rho\sigma}\gamma_\mu\epsilon_B + \cdots, \tag{4.64}$$

for a certain (convention–dependent) constant c. This formula is uniquely singled out by $U(\mathcal{N})_R \subset H$ covariance. In particular, $\mathfrak{F}^-$ is the unique object linear in

---

[13] In a convention where the $U(1)_R$ charge of the left–handed gravitino is 1.

the field–strengths with precisely the right $U(\mathcal{N})_R$ properties to lead to a locally covariant formula. Comparing with our discussion in Section 2.10.3, for $2 \leq \mathcal{N} \leq 4$ we conclude that

$$Z^{AB} = c \int_{\text{spatial } \infty} \mathfrak{F}^{-AB}$$

$$= c \left((U^{-1})^\dagger (U^* - V^*)^{-1} U^*\right)^{AB}_{u} \int_{\text{spatial } \infty} F^{-u}. \tag{4.65}$$

that is, the central charges are linear combinations of the Maxwell electric and magnetic fluxes at infinity with coefficients which are given by the value at infinity[14] of the matrix in front of the integral in the second line.

Equation (4.65) may be interpreted as saying that the complex central charge Z^{AB} defines an $\mathcal{N}(\mathcal{N}-1)/2$–dimensional linear subspace $\mathcal{Z}$ of the $2V$–dimensional space of (complexified) electric and magnetic charges $\Gamma_\mathbb{C} \equiv \Gamma \otimes \mathbb{C}$ (Γ being the Dirac lattice of quantized electric/magnetic charges). $\Gamma_\mathbb{C}$ is endowed with a real structure[15] R and a symplectic pairing Ω (the Dirac pairing). The subspace $\mathcal{Z}$ is isotropic for Ω, while the Hermitian form $\Omega(\cdot, \bar{\cdot})|_\mathcal{Z}$ should be non–degenerate. In Chapter 11 we show that such subspaces are classified by a *polarized flag manifold* whose detailed properties depend on $\mathcal{N}$ and k. Then the central charges defines a Z–map

$$\zeta : \mathcal{M} \to \text{(suitable polarized flag manifold)}, \tag{4.66}$$

from which we may read all the couplings in the Lagrangian. For the case $\mathcal{N} = 3, 4$ the situation simplifies; for $\mathcal{N} = 3$ because the duality representation is reducible over $\mathbb{C}$ (cf. Table 4.4), and in $\mathcal{N} = 4$ for the peculiar form of the reality condition (the matter vector multiplet is PCT self–conjugate). In these cases ζ reduces to a map from $\mathcal{M}$ to the Grassmannian manifold parametrizing, respectively, positive–definite complex 3–planes in $\mathbb{C}^{3,k}$ (i.e., $\mathbb{C}^{3+k}$ endowed with a Hermitian form of signature $(3, k)$) and positive–definite real 6–planes in $\mathbb{R}^{6,k}$:

$$\zeta : \mathcal{M} \to \begin{cases} \text{Gr}(3, \mathbb{C}^{3,k}) \equiv SU(3,k)/[SU(3) \times U(k)], & \mathcal{N} = 3, \\ \text{Gr}(6, \mathbb{R}^{6,k}) \equiv SO(6,k)/[SO(6) \times SO(k)], & \mathcal{N} = 4. \end{cases} \tag{4.67}$$

For $\mathcal{N} = 3$, ζ is just the identity map

$$\mathcal{M} \equiv \text{Gr}(3, \mathbb{C}^{3,k}) \xrightarrow{\zeta \equiv \text{Id}} \text{Gr}(3, \mathbb{C}^{3,k}), \tag{4.68}$$

---

[14] The value at infinity of the scalars is assumed to be constant, otherwise the charges make no sense.
[15] In fact a stronger *integral* structure. The overbar is complex conjugation with respect to R.

while for $\mathcal{N}=4$ it is the projection π_2 on the second factor space:

$$\mathcal{M} \equiv SU(1,1)/U(1) \times \text{Gr}(6, \mathbb{R}^{6,k}) \xrightarrow{\zeta \equiv \pi_2} \text{Gr}(6, \mathbb{R}^{6,k}). \tag{4.69}$$

This gives an alternative explanation of why $\mathcal{M}$ is the symmetric space we found in Section 2.1. We discuss the $\mathcal{N}=4$ example in detail.

4.7.3 Example: $\mathcal{N}=4$

From Table 4.4 we see that for $\mathcal{N}=4$ the vielbein matrix $\mathcal{E}$ is the tensor product of an $SU(1,1)$ matrix and an $SO(6,k)$ matrix.

$$\mathcal{E} = \begin{pmatrix} \phi_1 & \phi_2 \\ \phi_2^* & \phi_1^* \end{pmatrix} \otimes \left(L^{AB}{}_u, L^m{}_u \right) \tag{4.70}$$

where

$$A, B = 1, \ldots, 6, \qquad m = 1, \ldots, k, \qquad u = 1, 2, \ldots k+6, \tag{4.71}$$

$$|\phi_1|^2 - |\phi_2|^2 = 1. \tag{4.72}$$

The $SO(6,k)$ matrix L has the properties

$$L^{AB}{}_u = (L_{AB\,u})^* = \frac{1}{2}\epsilon^{ABCD} L_{CD\,u} \quad \text{(reality condition)},$$

$$-L^{AB}{}_u L_{AB\,v} + L^m{}_u L^m{}_v = \eta_{uv} \quad \text{("vielbein" property)}, \tag{4.73}$$

where;

$$\eta_{xy} = \text{diag}(\underbrace{-,-,-,-,-,-}_{6 \text{ times}}, \underbrace{+,+,\cdots,+}_{k \text{ times}}). \tag{4.74}$$

The matrix $\mathcal{E}$ is well defined *up to multiplication* on the left by $H = U(1) \times SO(6) \times SO(k)$ (gauge invariance).

Applying the formula (4.58), we get

$$\mathfrak{F}^{-\,AB} = (\phi_1^* - \phi_2^*)^{-1} L^{AB}{}_u F^{-u}, \tag{4.75}$$

so the image of the central–charge map ζ of Eq. (4.67) is the complex subspace of $\mathbb{C}^{6+k}$ spanned by the 6 vectors $L^{AB}{}_u E^u$ (where E^u is a canonical basis of $\mathbb{C}^{6+k}$), which – using the real structure of $\mathcal{N}=4$ SUGRA – is naturally identified with a *real 6–plane* in $\mathbb{R}^{6+k}$.

We may be more specific. We have the vector space $\mathbb{R}^{6,k}$, endowed with an inner product, $(\cdot,\cdot)$, of signature $(6,k)$ given by $-\eta$, Eq. (4.74). The central charges define a *positive–definite* 6–plane in $\mathbb{R}^{6,k}$, that is, a linear subspace $W \subset \mathbb{R}^{6,k}$ of dimensions six such that $(\cdot,\cdot)|_W$ is *positive–definite*. The space of all such positive–definite planes is the non–compact Grassmannian [51]

$$\mathrm{Gr}(6, \mathbb{R}^{6,k}) = \frac{SO(6,k)}{SO(6) \times SO(k)}, \quad (4.76)$$

and the central–charge map ζ,

$$\mathcal{M} \equiv \frac{SU(1,1)}{U(1)} \times \frac{SO(6,k)}{SO(6) \times SO(k)} \xrightarrow{\zeta} \mathrm{Gr}(6, \mathbb{R}^{6,k}) \equiv \frac{SO(6,k)}{SO(6) \times SO(k)}, \quad (4.77)$$

is – in the $\mathcal{N}=4$ case – simply the projection into the second factor. This gives an alternative explanation of why, in the $\mathcal{N}=4$ case, the target space $\mathcal{M}$ turns out to be that specific symmetric space.

Exercise 4.7.1 Show that the identity map $\mathrm{Gr}(1, \mathbb{C}^{1,k}) \xrightarrow{\zeta} \mathrm{Gr}(1, \mathbb{C}^{1,k})$ defines a *particular* $\mathcal{N}=2$ 4D SUGRA with transitive duality group $U(1,k)$ acting reducibly (over $\mathbb{C}$) on the field–strengths.

Exercise 4.7.2 Give the analogue Z–map/Grassmannian construction of 3D $\mathcal{N}=5,6,8$ SUGRAs.

4.8 6D chiral forms

In Section 4.1 we postponed the description of the geometry of the factor manifold T parametrized by the scalars φ^I of the 6D $(2,0)$ tensor supermultiplets.[16] It is time to fill the gap. The bosonic fields of a $(2,0)$ tensor supermultiplet [252] are a 2–form $B^-_{\mu\nu}$ with *anti*–self–dual field–strength and a scalar φ. We consider the coupling of n such tensor multiplets to 6D $(2,0)$ SUGRA. For a generic matter (and representation) content such chiral supergravities suffer gravitational anomalies at the quantum level [11, 286]; under favorable circumstances the anomalies get cancelled by a Green–Schwarz mechanism [169], see refs. [221, 272] for a detailed and deep analysis. Our discussion in this section is purely classical/formal.

The gravity supermultiplet contains a 2–form field $B^+_{\mu\nu}$ with a self–dual field–strength [252]. Hence, in total, the 2–form content is 1 self–dual, $B^+_{\mu\nu}$, and n anti–self–dual ones, $B^{-I}_{\mu\nu}$. By General Lesson 1.7 we know that the coupling of the

---

[16] Recall that $(2,0)$ is the *minimal* SUSY in 6D (8 supercharges). Most authors call it $(1,0)$.

scalars φ^I to the two forms defines a map

$$\mu: T \to SO(1,n)/SO(n). \tag{4.78}$$

The 6D analogue of the 4D "pushing–forward gauge invariance to target space" (cf. Section 2.9.3) shows that the differential μ_* has maximal rank. Since both spaces, T and $SO(1,n)/SO(n)$, have dimension n, μ is locally a diffeomorphism, and we may identify the universal cover $\widetilde{T}$ with the hyperbolic symmetric space $SO(1,n)/SO(n)$. In fact, μ should be an *isometry*. This may be seen in different ways: (i) we may mimic the argument we used for special geometry in 4D, a flat $SO(1,n)$–bundle replacing the $Sp(2n,\mathbb{R})$ one, or (ii) just realize that SUSY requires $SO(1,n)$ to be a symmetry of the full classical equations of motion. Alternatively, (iii) one may consider consistency with rigid SUSY in the decoupling limit of supergravity, or else (iv) consider the 4–Fermi interactions. In conclusion, for 6D $(2,0)$ SUGRA [265],

$$\widetilde{T} = SO(1,n)/SO(n) \equiv H_n \equiv \operatorname{Gr}(1,\mathbb{R}^{1,n}), \tag{4.79}$$

where H_n is the hyperbolic n–space.

We may use the map μ to get an alternative derivation of the scalar manifold for the 6D $(4,0)$ SUGRA coupled to n $(4,0)$ tensor supermultiplets. The gravitational supermultiplet of 6D $(4,0)$ contains 5 self–dual 2–forms and no scalars, while each $(4,0)$ tensor supermultiplet contains an anti–self–dual 2–form and 5 real scalars [265]. Hence $\dim \widetilde{T} = 5n$, and μ is a map:

$$\mu: \widetilde{T} \to SO(5,n)/SO(5) \times SO(n) \equiv \operatorname{Gr}(5,\mathbb{R}^{5,n}). \tag{4.80}$$

The two spaces have equal dimension, so μ is an isometry. Thus we identify $\widetilde{T} \equiv SO(5,n)/SO(5) \times SO(n)$, recovering the 6D $(4,0)$ entry of Table 4.2.

4.9 Arithmetics of U–duality. Global geometry of $\mathcal{M}$

For $N_S > 8$ the analysis of the previous sections completely determines the *local* geometry of the scalar manifold $\mathcal{M}$ but leaves open two *global* questions: (i) the discrete subgroup $\Gamma \subset \operatorname{Iso}(\widetilde{\mathcal{M}})$ in Eq. (4.11), and (ii) the breaking of the duality group G due to Dirac quantization of fluxes, Section 1.5.3. These two issues are strictly related in any SUGRA that is the low–energy effective description of a *consistent* quantum theory of gravity. A consistent–looking low–energy theory that cannot arise from a consistent theory of quantum gravity is said to belong to the *swampland*, while the fully consistent ones form the *landscape*.

We discuss the 4D case. *Mutatis mutandis,* the arguments apply to all D.

4.9.1 The landscape and the swampland

The low–energy theory of any consistent quantum gravity is believed to satisfy certain conditions expressed by the Banks–Seiberg conjectures [32]:

BS 1 There are no global (i.e. ungauged) symmetries. A slightly weaker statement follows from black–hole physics [32]: the order of the discrete global symmetry group is bounded by a universal *finite* constant.
BS 2 All continuous gauge groups are *compact*.
BS 3 The charges form a complete set consistent with Dirac quantization.

In Section 1.5.3 we saw that when the Abelian gauge group is *compact*, as required by **BS 2**, at most an arithmetic subgroup[17]

$$L \subseteq G_{\mathbb{Z}} \equiv Sp(2n, \mathbb{Z}) \cap G \tag{4.81}$$

of the classical duality group $G \subset Sp(2n, \mathbb{R})$ may correspond to *physical* symmetries. L is the *U–duality group* [199]. **BS 3** is consistent with (and strongly suggests) the equality $L = G_{\mathbb{Z}}$, which, in particular, implies that the discrete group L is infinite. This contradicts **BS 1** unless "almost all" of $G_{\mathbb{Z}}$ is gauged. More precisely, the gauged subgroup $\Gamma \subset G_{\mathbb{Z}}$ should have *finite index* in $G_{\mathbb{Z}}$. By definition, the gauged subgroup $\Gamma \subset G$ coincides with the discrete isometry group Γ in Eq. (4.12). As we are going to show, the condition $[G_{\mathbb{Z}} : \Gamma] < \infty$ is equivalent to the geometric statement that the scalar manifold $\mathcal{M}$ has *finite volume*

$$\mathrm{Vol}(\mathcal{M}) \equiv \mathrm{Vol}(\Gamma \backslash G / H) < \infty. \tag{4.82}$$

This result is consistent with other conjectures by Vafa and Ooguri [255, 287]

VO 1 Let $\mathcal{M}$ be the scalar manifold of the low–energy effective theory of a consistent quantum gravity. Then the Riemannian manifold $\mathcal{M}$ is *non–compact, complete,* and has *finite volume*.
VO 2 Consider the theory around the vacuum $\langle \phi \rangle \in \mathcal{M}$ where the point $\langle \phi \rangle$ is a distance T from some reference point ϕ_0. As $T \to \infty$, there emerges an infinite tower of light particles with mass of the order $\exp(-\alpha T)$ for some $\alpha > 0$.
VO 3 There is no non–trivial 1–cycle with minimal length within a given homotopy class in $\mathcal{M}$.

---
[17] Here we see G as a matrix subgroup of $SL(2n, \mathbb{R})$ *via* the representations listed in Table 4.4. Then G is realized as a certain subvariety of the algebraic space of $2n \times 2n$ unimodular matrices defined by polynomial equations with rational coefficients (that is, G is an algebraic subvariety of $SL(2n, \mathbb{R})$ defined over $\mathbb{Q}$). This fact is crucial to make sense of Eq. (4.81).

4.9 Arithmetics of U–duality. Global geometry of $\mathcal{M}$

We claim that (at least) when the low–energy theory has enough supersymmetries, $N_S > 8$, the conjectures **BS 1**–**BS 3** are equivalent to **VO 1**–**VO 3** and in fact to **VO 1** from which the other two conjectures follow. Indeed, for $N_S > 8$ we have $\mathcal{M} = \Gamma \backslash G/H$ for some non–compact Lie group $G \subset Sp(2n, \mathbb{R})$ (cf. Table 4.2). G is the product of simple non–compact pairwise non–isotypic groups,[18] and H a maximal compact subgroup. Then $\mathrm{Vol}(\mathcal{M}) < \infty$ if and only if $\mathrm{Vol}(\Gamma \backslash G) < \infty$. A discrete subgroup $\Gamma \subset G$ which satisfies the finite covolume condition is called a *lattice in* G [230, 241]. Moreover, $\Gamma \backslash G/H$ is non–compact if and only if the $\mathbb{Q}$–rank of Γ is > 0 [207, 241]. Thus **VO 1** requires Γ to be a lattice in G of positive $\mathbb{Q}$–rank. In a Lie group $G = G_1 \times \cdots \times G_s$ with G_ℓ simple, non–compact, and pairwise non–isotypic, all lattices Γ such that $\Gamma \backslash G$ is *not* compact have the form $\Gamma_1 \times \cdots \times \Gamma_s$ with Γ_ℓ an *irreducible* lattice in G_ℓ. A *non*–cocompact lattice Γ in a Lie group having the stated properties is said to be *arithmetic* if Γ is isogenous[19] to $G_\mathbb{Z}$ (up to isomorphism).

From Eq. (4.81) we see that the Banks–Seiberg conjectures require Γ to be isogenous to $G_\mathbb{Z}$. This condition implies **VO 1** on account of the following theorem.

Theorem 4.9 *$G_\mathbb{Z}$ is a lattice in G. Moreover,*

$$\mathbb{Q}\text{--rank}(G_\mathbb{Z}) = \mathbb{R}\text{--rank}(G) = \mathrm{rank}(G/H) \geq 1. \tag{4.83}$$

Proof For the first statement see [207, 241]. The second one follows from the definitions and the footnote on page 156 (here implicitly we use the standard rational embedding $G \subset Sp(2n, \mathbb{R})$, implied by **BS 3**). □

The inverse implication, **VO 1** $\Rightarrow$ (Γ isogenous to $G_\mathbb{Z}$), is given by the next theorem:

Theorem 4.10 (Margulis arithmeticity theorem [230, 241]) *G is a simple and non–compact Lie group which is not isogenous to $SO(1, n)$ or $SU(1, n)$. Then all lattices $\Gamma \subset G$ are arithmetic.*

This settles the question for all SUGRAs whose group G (Table 4.2) does not contain $SO(1, n)$ or $SU(1, n)$ as a factor. Notice that these last groups do have non–arithmetic lattices: for instance, by the the uniformization theorem [136] all compact (hence finite volume) Riemann surfaces of genus ≥ 2 can be written in the form $\Gamma \backslash \mathfrak{H} \equiv \Gamma \backslash SU(1, 1)/U(1)$ for some Fuchsian group $\Gamma \in SU(1, 1) \equiv SL(2, \mathbb{R})$; then all such Fuchsian groups are lattices in $SU(1, 1)$, but only a zero–measure subset of them may be arithmetic.

---

[18] We say that two simple Lie groups G_1, G_2 are *isotypic* if their Lie algebras satisfy $\mathfrak{g}_1 \otimes \mathbb{C} = \mathfrak{g}_2 \otimes \mathbb{C}$.
[19] Two subgroups Γ_1, Γ_2 are *isogenous* if the common subgroup $\Gamma_1 \cap \Gamma_2$ has finite index in both.

Comparing with Table 4.2 we see that the theorem does not cover three cases: (i) $\mathcal{N} = 5$, (ii) $\mathcal{N} = 4$, and (iii) $\mathcal{N} = 3$ with $k = 1$. However, they may also be settled by requiring consistency with the (inverse) group disintegration. One reduces the 4D SUGRA to 3D and uses the consistency of flux quantization under dimensional reduction along the lines of ref. [272]. The integers k in Table 4.2, which count the number of *matter* supermultiplets, for the 4D and the dimensional reduced 3D theories are related as

$$k_{3D} = k_{4D} + \begin{cases} 1 & \text{for } N_S = 12, \\ 2 & \text{for } N_S = 16, \end{cases} \tag{4.84}$$

and thus the group G for a 3D SUGRA with $N_S > 8$ obtained by dimensional reduction of a 4D one is *never* isogenous to $SO(1,n)$ or $SU(1,n)$.

To relate **VO 2**, **VO 3** to **VO 1**, we recall the following proposition:

Proposition 4.11 (1) If $\mathrm{Vol}(\Gamma \backslash G) < \infty$ while $\Gamma \backslash G$ is not compact, Γ contains a non–trivial unipotent element u [207, 241].

(2) (Jacobson–Morosov [241]) If $G \subset SL(\ell, \mathbb{R})$ and $u \in G$ is a non–trivial unipotent element, there is a polynomial homomorphism $\phi \colon SL(2, \mathbb{R}) \to G$ (defined over $\mathbb{R}$) such that[20] $\phi(T) = u$.

Then ϕ embeds $\mathfrak{h} \equiv SL(2, \mathbb{R})/U(1) \hookrightarrow G/H$ as a totally geodesic submanifold and, as $y \to +\infty$ the point $z = iy \in \mathfrak{h}$ is mapped at infinity in $\Gamma \backslash G/H$. Modding out by the subgroup $\langle u \rangle \subset \Gamma$, we see that, restricting on the submanifold $\phi(SL(2, \mathbb{Z}) \backslash \mathfrak{h})$, the asymptotic geometry behaves as the geometry around the cusp of the fundamental modular domain $SL(2, \mathbb{Z}) \backslash SL(2, \mathbb{R})/U(1)$, which was precisely the example that motivated Ooguri and Vafa to propose conjectures **VO 2,3**. As we go to infinity in $\phi(SL(2, \mathbb{Z}) \backslash \mathfrak{h}) \subset \mathcal{M}$ the central charges, Eq. (4.58), vanish exponentially, so all BPS states have exponentially small masses in the limit. Moreover, the BPS bound is saturated by an infinite tower of states; indeed the monodromy at infinity u is unipotent, and hence maps BPS states into BPS states of higher charge, exactly as the monodromy at infinity of the Coulomb branch in the Seiberg–Witten solution of pure $SU(2)$ $\mathcal{N} = 2$ SYM [273, 274], which generates a tower of dyons as a consequence of the Witten effect [299].

---

[20] T is the usual modular matrix corresponding to $\tau \to \tau + 1$.

4.9.2 Volume of $G_{\mathbb{Z}}\backslash G/H$

The volume of the $G_{\mathbb{Z}}\backslash G/H$ can be computed in the form

$$\text{Vol}(G_{\mathbb{Z}}\backslash G/H) = \sharp \mathbf{Z}(H) \frac{\text{Vol}(G_{\mathbb{Z}}\backslash G)}{\text{Vol}(H)}, \tag{4.85}$$

where $\sharp \mathbf{Z}(H)$ is the order of the center of H. If G is defined over $\mathbb{Q}$ and is $\mathbb{Q}$–split (i.e., $\mathbb{Q}$–rank(G) = rank G), the difficult factor in the RHS, namely $\text{Vol}(G_{\mathbb{Z}}\backslash G)$, was computed by Langlands in ref. [226] (see also [284, 156]). It is given by the following very elegant formula:

$$\text{Vol}(G_{\mathbb{Z}}\backslash G) = \sharp(\pi_1(G)) \prod_{k=1}^{l} \zeta(a_i), \tag{4.86}$$

where $\zeta(s)$ is the Riemann zeta function, l = rank G, and $\{a_1 = 2, a_2, \cdots, a_l\}$ are the degrees of the basic Casimir invariants of $G_{\mathbb{C}}$ (equal to the Coxeter exponents plus 1). Thus, for instance, the volume of $Sp(2n, \mathbb{Z})\backslash \mathfrak{H}_n$ is

$$\text{Vol}(Sp(2n, \mathbb{Z})\backslash \mathfrak{H}_n) = \frac{n}{(2\pi)^{n(n+1)/2}} \prod_{k=1}^{n} k \zeta(2k), \tag{4.87}$$

where we have used [240]

$$\text{Vol}(U(n)) = (2\pi)^{n(n+1)/2} \prod_{k=1}^{n} \frac{1}{k}. \tag{4.88}$$

For the volume of $E_{7(7)\mathbb{Z}}\backslash E_{7(7)}/SU(8)$ one gets

$$\frac{\zeta(2)\zeta(6)\zeta(8)\zeta(10)\zeta(12)\zeta(14)\zeta(18)}{2^{32} c^{63} \pi^{35}} \prod_{k=1}^{8} k, \tag{4.89}$$

where $c = \sqrt{6}$ is the rescaling of the $SU(8)$ metric. Equation (4.89) is easily evaluated using

$$\zeta(2k) = \frac{2^{2k-1}}{(2k)} B_k \pi^{2k}, \tag{4.90}$$

where B_k are Bernoulli numbers ([276] proposition VII.7).

Exercise 4.9.1 Verify that the asymptotic behavior at the infinite ends of $\Gamma\backslash G/H$ satisfies **VO 2,3**.

5
σ–Models and symmetric spaces

To make our formulation of $N_S > 8$ SUGRA totally concrete we have to give explicit expressions for the G–invariant metrics, bundles, and connections over the symmetric spaces G/H of Table 4.2. This requires studying the geometry of these remarkable Riemannian spaces. From the gauge point of view (Section 4.5) it is clear that we have to start from G–invariant geometric structures on the group manifold G itself, and then "gauge away" the spurious H part. Hence we begin with the differential geometry of a Lie group.

5.1 Cartan connections on G

5.1.1 Left–invariant vector fields

A Lie group G is a group which has a differentiable structure so that the group operations

$$G \times G \to G \qquad (g, h) \mapsto gh, \tag{5.1}$$

$$G \to G \qquad g \mapsto g^{-1}, \tag{5.2}$$

are smooth maps. In particular, a Lie group G is a smooth manifold. To be concrete, we see G as a group of matrices, $G \subset \mathbb{K}(n)$ with $\mathbb{K} = \mathbb{R}$ or $\mathbb{C}$, and take the differential structure induced on G by that of $\mathbb{K}(n) \simeq \mathbb{K}^{n^2}$. As a matrix, an element of $g \in G$ has the form

$$g = \exp \mathsf{X} \equiv \sum_{k=1}^{\infty} \frac{\mathsf{X}^k}{k}, \qquad \text{with } \mathsf{X} \in \mathfrak{g} \subset \mathbb{K}(n), \tag{5.3}$$

where $\mathfrak{g}$ is the Lie algebra of G, seen as an algebra of $n \times n$ matrices,[1] and is identified with the tangent space at the origin $T_e G$.

---

[1] The Lie bracket is given by the matrix commutator.

5.1 Cartan connections on G

By Eq. (5.1) each element $h \in G$ defines a diffeomorphism

$$L_h: G \to G \qquad (5.4)$$
$$g \mapsto hg$$

called *left translation by h*. Let $\mathsf{X} \in T_e G \simeq \mathfrak{g}$ be an element of the tangent space at the identity $e \in G$, and consider the vector field on G

$$X(g) := (L_g)_* \mathsf{X} \in T_g G.$$

By construction,

$$X(hg) = (L_h)_* X(g), \qquad (5.5)$$

so the vector field $X(g)$ is invariant under left translation or, as we shall say, *left–invariant*. Conversely, any left–invariant vector field arises in this way. Hence the space of left–invariant vectors on G is naturally isomorphic to $T_e G$, that is, to the Lie algebra $\mathfrak{g}$. The integral line of a left–invariant vector field,[2] $X(g)$, is called a *one–parameter subgroup*. The commutator of two left–invariant vector fields, $X(g)$ and $Y(g)$, is again a left–invariant vector field. Hence, if $X_i(g)$ ($i = 1, 2, \ldots, \dim G$) is a basis of such vector fields,[3]

$$[X_i(g), X_j(g)] = f_{ij}{}^k X_k(g), \qquad (5.6)$$

where $f_{ij}{}^k$ are constants (the Lie group structure constants) satisfying the Jacobi identity.

Analogously, we can define *right–invariant* vector fields $\widetilde{X}_i$. The inversion map $i: g \mapsto g^{-1}$ interchanges the left– and right–invariant vector fields,

$$i_* X_i = \widetilde{X}_i. \qquad (5.7)$$

5.1.2 Left–invariant and Cartan connections

A connection D_i on a Lie group G is *left–invariant* if, given any two left–invariant vector fields, X and Y, the vector field $X^i D_i Y^j$ is also left–invariant.

A left–invariant connection defines a multiplication on the Lie algebra $\mathfrak{g}$:

$$\alpha: \mathfrak{g} \times \mathfrak{g} \to \mathfrak{g} \qquad \alpha(X, Y) = X^i D_i Y^j. \qquad (5.8)$$

---

[2] That is, the solution $g(t)$ to the differential equation

$$\frac{d}{dt} g(t) = X(g(t)), \qquad g(0) = e.$$

[3] The sign in the RHS of Eq. (5.6) is *tricky*. It depends on the way we interpret the action of the group. We shall adhere to the viewpoint of ref. [263].

Given a connection D_i, we have a notion of parallel transport, and hence of geodesic curves. It would be desirable that the geodesics (passing through e) and the one–parameter subgroups be the *same* curves; this is equivalent to the validity of the following natural–looking equation:

$$\exp_e(tX) = \exp(tX), \tag{5.9}$$

where the LHS is the exponential in the sense of differential geometry [239, 263] and the RHS is the matrix exponential as in Eq. (5.3).

Definition 5.1 A left–invariant connection on a Lie group G is called a *Cartan connection* if, for all vectors $X \in T_e G$, the one–parameter subgroup $\exp(tX) \subset G$ and the geodesic $\gamma_{e,X}$ coincide.

In particular, a Lie group G is always *complete* with respect to any Cartan connection. The equation of the geodesic $\gamma_{e,X}$ then reads

$$0 = \left(\frac{d\gamma_{e,X}}{dt}\right)^i D_i \left(\frac{d\gamma_{e,X}}{dt}\right)^j = X^i(g(t)) D_i X^j((g(t)) = \alpha(X,X)^j, \tag{5.10}$$

so, for a Cartan connection, the product $\alpha(\cdot,\cdot)$ defined in Eq. (5.8) is anti–commutative; that is,

$$D_X Y = -D_Y X, \tag{5.11}$$

for all left–invariant vectors X, Y.

Proposition 5.2 *On G there is a unique torsionless Cartan connection*

$$D_X Y = \frac{1}{2}[X,Y]. \tag{5.12}$$

Proof By the definition of torsion,

$$0 = T(X,Y) = D_X Y - D_Y X - [X,Y]$$
$$= 2 D_X Y - [X,Y]. \qquad \square$$

When G is simple, the general Cartan connection is given by $D_X Y = \lambda[X,Y]$, $\lambda \in \mathbb{R}$. Its torsion is given by $T(X,Y) = (2\lambda - 1)[X,Y]$.

5.1.3 Curvature of a Cartan connection

Let X, Y, Z be three left–invariant vector fields on the simple Lie group G. Let us compute the curvature of a Cartan connection

$$\begin{aligned} R(X,Y)Z &= D_X D_Y Z - D_Y D_X Z - D_{[X,Y]} Z \\ &= \lambda^2 ([X,[Y,Z]] - [Y,[X,Z]]) - \lambda\,[[X,Y],Z] \\ &= (\lambda^2 - \lambda)\,[[X,Y],Z]. \end{aligned} \quad (5.13)$$

Thus we have two Cartan connections (with torsion) which are *flat*, namely $\lambda = 0, 1$. The corresponding parallel sections are, respectively, the left–invariant vectors X, and the right–invariant ones $\widetilde{X}$.

Theorem 5.3 *The curvature of the torsionless Cartan connection is*

$$R(X,Y) = -\frac{1}{4}[X,Y] \quad (5.14)$$

acting on $\mathfrak{g} \simeq T_e G$ *by the adjoint representation. In particular, the Riemann tensor is covariantly constant.*

Again the curvature is minus a commutator.

5.1.4 Invariant metrics

By the fundamental theorem of differential geometry, the above torsionless Cartan connection should be the Christoffel connection for any *left–invariant* metric on G. Since the left–invariant fields span TG, a left–invariant metric is defined by its *constant* values on any basis $\{X_i\}$ of left–invariant vectors (that is, any basis $\{X_i\}$ of $\mathfrak{g}$)

$$g(X_i, X_j) = g_{ij}. \quad (5.15)$$

A left–invariant metric $g(\cdot, \cdot)$ on G is so identified with a metric on $\mathfrak{g}$.

One has

$$0 = D_k\, g(X_i, X_j) = \frac{1}{2} g([X_k, X_i], X_j) + \frac{1}{2} g(X_i, [X_k, X_j]), \quad (5.16)$$

so $g(\cdot, \cdot)$ should be an *invariant metric* on the Lie algebra $\mathfrak{g}$. Unfortunately, not all groups have invariant metrics. By the Cartan–Killing criterion and Schur's lemma, a simple Lie algebra $\mathfrak{g}_{\text{simple}}$ admits one and only one (up to normalization) invariant bilinear pairing $\mathfrak{g}_{\text{simple}} \times \mathfrak{g}_{\text{simple}} \to \mathbb{K}$, namely the *Cartan–Killing form*, but this form is not always a real positive–definite inner–product. Its signature depends on the specific real form $\mathfrak{g}$ of the abstract complex Lie algebra $\mathfrak{g}_\mathbb{C}$. See Section 5.3 below.

Corollary 5.4 *The left–invariant vectors X_i are Killing vectors for any left–invariant metric g_{ij} on the group manifold G.*

Proof Let Y be a left–invariant vector. Since $\pounds_Y$ is a derivation,

$$(\pounds_Y g)(X_i, X_j) = \pounds_Y\bigl(g(X_i, X_j)\bigr) - g\bigl(\pounds_Y X_i, X_j\bigr) - g\bigl(X_i, \pounds_Y X_j\bigr)$$
$$= Y^a \partial_a g_{ij} - g([Y, X_i], X_j) - g(X_i, [Y, X_j]) = 0. \qquad \square$$

Exercise 5.1.1 Prove the last statement in Theorem 5.3.

5.2 Maurier–Cartan forms

5.2.1 Left–invariant forms

Given a basis $\{X_i\}$ of left–invariant vector fields, we construct a dual basis $\{\omega^i\}$ of left–invariant *1–forms* by defining

$$\omega^i(X_j) = \delta^i{}_j. \tag{5.17}$$

It is convenient to write ω as a 1–form taking values in the Lie algebra $\mathfrak{g}$. Let $\{\mathsf{X}_i\}$ be the set of generators of $\mathfrak{g}$ corresponding to the X_i. We write

$$\omega = \omega^i \otimes \mathsf{X}_i \in \Lambda^1(G) \otimes \mathfrak{g}. \tag{5.18}$$

By construction, ω is the *unique* left–invariant element of $\Lambda^1(G) \otimes \mathfrak{g}$ which, at the origin e, corresponds to the identity $\mathrm{Id}_\mathfrak{g}$ in the Lie algebra $\mathfrak{g}$ under the isomorphism

$$\Lambda^1(G)\bigr|_e \otimes \mathfrak{g} \equiv \mathfrak{g} \otimes T_e^*(G) \simeq \mathfrak{g} \otimes \mathfrak{g}^\vee \simeq \mathrm{End}(\mathfrak{g}), \tag{5.19}$$

that is,

$$\omega(X)|_e = \mathsf{X}. \tag{5.20}$$

Definition 5.5 ω *is called the (left–invariant)* Maurier–Cartan form.

The explicit formula for the Maurier–Cartan forms follows easily from Eq. (5.20). Again, we consider $G \subset GL(n, \mathbb{K})$ as a group of matrices, and correspondingly $\mathfrak{g} \subset \mathbb{K}(n)$.

5.2 Maurier–Cartan forms

Lemma 5.6 *Let ϕ^i be local coordinates on G. Write the generic group element as a matrix $g(\phi) \in GL(N, \mathbb{K})$. Then $\omega \in \Lambda^1(G) \otimes \mathfrak{g} \subset \Lambda^1(G) \otimes \mathbb{K}(n)$ is given by*

$$\omega = g^{-1} dg. \qquad (5.21)$$

Thinking of ω as a gauge field on G, it is "pure gauge."

Proof ω is invariant under $g(\phi) \to h\,g(\phi)$, so is a left–invariant form. We have to check its value at the identity e. Write $g(\phi) = \exp \phi^i T_i$. One has $g^{-1} dg|_{\phi=0} = d\phi^i X_i \equiv \mathrm{Id}_{\mathfrak{g}} \in \mathrm{End}(\mathfrak{g})$. $\square$

5.2.2 Maurier–Cartan equations

Since ω is "pure gauge" it has a vanishing "field–strength." This statement is known as the Maurier–Cartan equations

Proposition 5.7 (Maurier–Cartan) *One has*

$$d\omega + \frac{1}{2}[\omega, \omega] = 0, \qquad (5.22)$$

or in components

$$d\omega^i + \frac{1}{2} f_{kl}{}^i \omega^k \wedge \omega^l = 0. \qquad (5.23)$$

5.2.3 Haar invariant measure

Left Haar measure

Let $\{\omega^i\}_{i=1}^n$ be a basis of Maurier–Cartan forms as in Eq. (5.17). The n–form

$$\Omega = \omega^1 \wedge \omega^2 \wedge \cdots \wedge \omega^n \qquad (5.24)$$

is left–invariant. Any other left–invariant n–form may be written as $f\Omega$ with f a left–invariant *function*, hence a constant. Ω is a volume form on G.

Definition 5.8 *The unique (up to normalization) left–invariant volume form on G is called the* left Haar measure *on G. We write*

$$\int_G f(g)\,dg \qquad (5.25)$$

for the integral of the function $f: G \to \mathbb{C}$ *with respect to the Haar measure.*

The left Haar measure is characterized by the property

$$\int_G f(hg)\, dg = \int_G f(g)\, dg \qquad \forall\, h \in G. \tag{5.26}$$

Unimodular Lie groups

On a group manifold G we have an action of $G \times G$ given, respectively, by left and right multiplications

$$L_{h_L}, R_{h_R} : g \mapsto h_L\, g\, h_R. \tag{5.27}$$

The two actions *commute* (they act on "different indices"). By construction, Ω is invariant under left multiplication. Consider the n–form $R_g^* \Omega$. One has

$$L_h^* R_g^* \Omega = R_g^* L_h^* \Omega = R_g^* \Omega, \qquad \forall\, h, g \in G, \tag{5.28}$$

where we used the commutativity of the two actions. Hence $R_g^* \Omega$ is also a left–invariant n–form, and hence it should be a multiple of Ω,

$$R_g^* \Omega = \varrho(g)\, \Omega, \tag{5.29}$$

$$\text{with} \quad \varrho(gh) = \varrho(g)\, \varrho(h). \tag{5.30}$$

Definition 5.9 The quasicharacter $\varrho \colon G \to \mathbb{R}$ is called the *module* of the Lie group G. A Lie group with $\varrho(g) = 1$ for all $g \in G$ is called *unimodular*.

Corollary 5.10 *On a Lie group there is a unique (up to normalization) right–invariant measure (the* RIGHT HAAR MEASURE*), given by $\varrho(g)^{-1} dg$. For a* UNIMODULAR *group the left and right Haar measures coincide.*

Exercise 5.2.1 Show that a semi–simple Lie group is unimodular.

5.3 Invariant metrics on a compact group

5.3.1 Bi–invariant metrics

Proposition 5.11 *If G is compact, there exists a metric g_{ij} (unique up to normalization) which is both left– and right–invariant.*

Proof Let $(\cdot, \cdot)_h$ be any positive–definite smooth symmetric tensor. Given two vectors $X, Y \in T_h G$, define a new inner product

$$\langle X, Y \rangle_h = \int_G (L_g^* X, L_g^* Y)_{gh}\, dg. \tag{5.31}$$

5.3 Invariant metrics on a compact group

One has

$$\langle L_f^* X, L_f^* Y\rangle_{fh} = \int_G (L_{fg}^* X, L_{fg}^* Y)_{fgh} \, dg = \int_G (L_g^* X, L_g^* Y)_{gh} \, dg = \langle X, Y\rangle_h \quad (5.32)$$

so $\langle \cdot, \cdot \rangle_h$ is left–invariant. Then

$$g(X, Y)_h = \int_G \langle R_g^* X, R_g^* Y\rangle_{hg} \, \varrho(g)^{-1} \, dg \quad (5.33)$$

is right–invariant for construction and left–invariant because the two actions commute. □

In Section 5.1.4 we have seen that, for a *simple* G, a left–invariant metric – if it exists – should be proportional to the Cartan–Killing form on the left–invariant vector fields. Hence:

Corollary 5.12 *G is a semi–simple compact Lie group. Then the Cartan–Killing form $\mathfrak{g} \times \mathfrak{g} \to \mathbb{R}$*

$$\mathrm{Kil}(X, Y) := \mathrm{Tr}(\mathrm{adj}\, X \circ \mathrm{adj}\, Y) \quad (5.34)$$

is negative–definite.

If G is compact, we take as metric $g(X, Y) = -\mathrm{Kil}(X, Y)$. Being the unique left–invariant metric, $g(X, Y)$ should be also right–invariant, by Proposition 5.11. Hence it has the full isometry group $G \times G$, associated with the two commuting actions of G on itself (left and right). The corresponding Killing vectors are the X_i and the $\widetilde{X}_i$ (Eq. (5.7)),

$$[X_i, X_j] = f_{ij}{}^k X_k, \quad [X_i, \widetilde{X}_j] = 0, \quad [\widetilde{X}_i, \widetilde{X}_j] = -f_{ij}{}^k \widetilde{X}_k. \quad (5.35)$$

5.3.2 Explicit formula for the invariant metric

In practice, we have our matrix representation for the group elements $g(\phi) \in GL(n, \mathbb{K})$ (cf. Lemma 5.6). Then we write the bi–invariant metric in the obvious form

$$g_{ij}(\phi) \, d\phi^i \, d\phi^j = -\lambda \, \mathrm{tr}[(g^{-1} \partial_i g)(g^{-1} \partial_j g)] d\phi^i \, d\phi^j, \quad (5.36)$$

that is, as minus the trace of the square of the Maurier–Cartan "pure gauge field" $g^{-1} \partial_i g$. λ is a normalization constant, which depends on the particular representation of G we use to define the trace in the RHS.

Now *everything is very explicit*. To prove Eq. (5.36), one has only to check that it is invariant under left/right translations, which is obvious, and the rest follows from the above uniqueness and existence results. Notice that this metric is positive–definite if G is compact, but not in general.

We can rephrase the above result in a useful way. Take a basis of $\mathfrak{g}$ that is orthonormal with respect to the form $-\text{Kil}$, i.e., $-\text{Kil}(X_i X_j) = \delta_{ij}$. Then the coefficient 1–forms ω^i in $\omega \equiv \omega^i X_i$ form an *orthonormal coframe* in T^*G with respect to the invariant metric $(\cdot, \cdot)$

$$(\omega^i, \omega^j) = \delta^{ij}. \tag{5.37}$$

Therefore the ω^i can be identified with the (orthonormal) vielbein form $e^i \equiv e^i{}_i \, d\phi^i$ of the $(G \times G)$–invariant metric. Thus we can write

$$g^{-1} dg = e^i X_i \qquad (X_i \text{ an orthonormal basis in } \mathfrak{g}). \tag{5.38}$$

5.3.3 G is an Einstein space

From Section 5.1.3 the Riemann tensor of the invariant metric on G should be

$$R(X, Y) = -\frac{1}{4}[X, Y]. \tag{5.39}$$

Let us now compute the Ricci tensor:

$$\begin{aligned}
\text{Ric}(Y, Z) &:= \omega^i(R(X_i, Y)Z) = -\frac{1}{4}\omega^i([[X_i, Y], Z]) \\
&\equiv -\frac{1}{4}\omega^i\bigl(\text{adj}\, Z \circ \text{adj}\, Y(X_i)\bigr) = -\frac{1}{4}\text{Tr}\bigl(\text{adj}\, Z \circ \text{adj}\, Y\bigr) \\
&= -\frac{1}{4}\text{Kil}(Y, Z) = \frac{1}{4}g(Y, Z).
\end{aligned} \tag{5.40}$$

Hence *the Killing metric on a compact group G is Einstein with <u>positive</u> "cosmological constant"* $+1/4$ (in our normalization).

5.4 Chiral models

A σ–model with target space a compact Lie group G is called a *chiral model*. It has a symmetry group $G_L \times G_R$, which is suggestive of the flavor group in massless QCD, $SU(N)_L \times SU(N)_R$. In fact, it is used to model the low–energy theory of hadrons; to get the right physics one has to add to the basic σ–model Lagrangian

$$\begin{aligned}
\mathcal{L}_{\sigma \text{ model}} &= \frac{1}{2}\text{tr}[(g^{-1}\partial_\mu g)(g^{-1}\partial^\mu g)] \equiv -\frac{1}{2}\text{tr}[\partial_\mu g^{-1} \partial^\mu g] \\
&\equiv -\frac{1}{2}\text{tr}[(\partial_i g^{-1} \partial_j g]\, \partial_\mu \phi^i \, \partial^\mu \phi^j,
\end{aligned} \tag{5.41}$$

another coupling, i.e., the Wess–Zumino–Witten term [303, 304, 305].

Due to the high symmetry, the chiral model in 2D is solvable both classically [253] and quantum mechanically [135, 261]. Indeed, the two differential operators ($x^\pm \equiv x \pm t$)

$$\partial_{x^+} + \frac{\lambda_0}{\lambda - \lambda_0} g^{-1} \partial_{x^+} g, \tag{5.42}$$

$$\partial_{x^-} - \frac{\lambda_0}{\lambda + \lambda_0} g^{-1} \partial_{x^-} g, \tag{5.43}$$

commute for all values of the spectral parameter $\lambda \in \mathbb{C}$.

Exercise 5.4.1 Write the conserved $G_L \times G_R$ currents for Eq. (5.41).

5.5 Geometry of coset spaces G/H

In this section we discuss the aspects of the geometry of a coset space G/H, *not necessarily symmetric,* needed for writing explicitly the Lagrangian of an $N_S > 8$ SUGRA. A more detailed analysis will be given in Chapter 11. We assume G *simple* and H *compact*, these properties being automatic for G/H indecomposable symmetric. In particular, the action of H on the Lie algebra of G, $\mathfrak{g}$, is *reductive*, and we have an H–invariant orthogonal decomposition $\mathfrak{g} = \mathfrak{h} \oplus \mathfrak{m}$ where $\mathfrak{h}$ is the Lie algebra of H and $\mathfrak{m}$ is an H–module.

5.5.1 The gauge point of view

To study the geometry of the coset space G/H, we adopt the physicists' strategy based on the gauge idea, Section 4.5. Instead of using G/H as target space for our σ–model, we may use the full group G, provided we consider gauge–equivalent two field configurations $g_1, g_2 \colon \Sigma \to G$, if they differ by an arbitrary, space–time dependent, $h(x) \in H$,

$$g_1(x) = g_2(x) h(x). \tag{5.44}$$

Choose a (non–trivial) representation R of the simple group G, and let t_a ($a = 1, \ldots, \dim H$) be the matrices representing in R the compact Lie subalgebra $\mathfrak{h} \subset \mathfrak{g}$, normalized so that

$$\mathrm{tr}_R(t_a t_b) = -\delta_{ab}. \tag{5.45}$$

Consider the following Lagrangian

$$\mathcal{L} = \frac{\lambda}{2} \mathrm{tr}_R \big[(g^{-1} \partial^\mu g - A^{a\mu} t_a)(g^{-1} \partial_\mu g - A^b_\mu t_b) \big], \tag{5.46}$$

where the independent fields are $g(x) \in G$ and the H–valued connection $A_\mu \equiv A_\mu^a t_a$. This Lagrangian is obviously invariant under (global) *left translation* $g(x) \mapsto g'g(x)$, $g' \in G$, and it is also invariant with respect to *local* right translation by the subgroup H:

$$g \mapsto g h(x), \qquad A_\mu \mapsto h(x)^{-1} A_\mu h(x) + h(x)^{-1} \partial_\mu h(x). \tag{5.47}$$

Integrate away the auxiliary gauge fields A_μ^a using their equations of motion,

$$A_\mu^a = -\mathrm{tr}_R[t^a g^{-1} \partial_\mu g]. \tag{5.48}$$

The result is

$$\mathcal{L} = \frac{\lambda}{2} \mathrm{tr}_R[(g^{-1}\partial^\mu g)^\perp (g^{-1}\partial_\mu g)^\perp], \tag{5.49}$$

where $(\cdots)^\perp$ stands for the projection on $\mathfrak{h}^\perp \equiv \mathfrak{m} \subset \mathfrak{g}$ orthogonal to $\mathfrak{h}$ with respect to the Killing form. By construction,

$$g_{ij} := -\lambda \, \mathrm{tr}_R[(g^{-1}\partial_i g)^\perp (g^{-1}\partial_j g)^\perp], \tag{5.50}$$

is a metric on G/H invariant under left translation by G. It is positive–definite if and only if the form $-\lambda \, \mathrm{Kil}(\cdot,\cdot)$ is positive–definite on $\mathfrak{h}^\perp$. We have two possibilities: either $-\mathrm{Kil}(\cdot,\cdot)$ is positive–definite and we take the overall factor $\lambda > 0$, or it is negative–definite and we take $\lambda < 0$. Both cases lead to sensible physical theories.

Let m^α be an orthonormal basis of $\mathfrak{m} \equiv \mathfrak{h}^\perp \subset \mathfrak{g}$ with respect to the inner product $\pm\mathrm{Kil}(\cdot,\cdot)$ (whichever is positive). Then we can expand the $\mathfrak{g} = \mathfrak{h} \oplus \mathfrak{m}$ valued 1–form $g^{-1} dg$ in the basis $\{t_a, m_\alpha\}$

$$g^{-1} dg = e^\alpha m_\alpha + A^a t_a, \tag{5.51}$$

where the 1–forms $A^a \in \Lambda^1(G/H)$ are related to the previous gauge fields of the same name by[4] $A_\text{gauge}^a = \phi^* A^a$. By construction, the 1–forms e^α are an *orthonormal coframe* (a vielbein) in T^*G/H with respect to the metric in Eq. (5.50) (for $\lambda = \pm 1$).

The differentials of the 1–forms e^α, A^a satisfy a set of identities following from the Maurier–Cartan equations for G:

$$\begin{aligned} d(e^\alpha\, m_\alpha + A^a\, t_a) \\ = -\frac{1}{2} e^\alpha \wedge e^\beta [m_\alpha, m_\beta] - A^a \wedge e^\alpha [t_a, m_\alpha] - \frac{1}{2} A^a \wedge A^b [t_a, t_b]. \end{aligned} \tag{5.52}$$

---

[4] As always, $\phi \colon \Sigma \to \mathcal{M} \equiv G/H$ is the scalar field configuration map.

5.6 Symmetric spaces

To write more explicit formulae, let us recall the structure of the Lie algebras $\mathfrak{h} \subset \mathfrak{g}$ for a *reductive* Lie subgroup H of the simple[5] group G:

$$[t_a, t_b] = C_{ab}{}^c t_c, \qquad (\mathfrak{h} \text{ is a Lie subalgebra of } \mathfrak{g}) \tag{5.53}$$

$$[t_a, m_\alpha] = M_{a\alpha}{}^\beta m_\beta, \qquad (\mathfrak{m} \text{ is a representation of } \mathfrak{h}) \tag{5.54}$$

$$[m_\alpha, m_\beta] = M_{\alpha\beta}{}^a t_a + D_{\alpha\beta}{}^\gamma m_\gamma. \tag{5.55}$$

Then Eq. (5.52) becomes

$$de^\alpha = -\frac{1}{2} D_{\beta\gamma}{}^\alpha e^\beta \wedge e^\gamma - M_{a\beta}{}^\alpha A^a \wedge e^\beta, \tag{5.56}$$

$$dA^a = -\frac{1}{2} C_{bc}{}^a A^b \wedge A^c - \frac{1}{2} M_{\alpha\beta}{}^a e^\alpha \wedge e^\beta. \tag{5.57}$$

These formulae are interesting, but their geometrical meaning is much more pregnant if the subgroup $H \subset G$ is such that G/H is actually a *symmetric* space in the sense of Section 3.2. Thus, from now on we specialize in the symmetric case, which is the one relevant for $N_S > 8$ SUGRA; see Chapter 4.

Exercise 5.5.1 Compute the invariant metric for $Sp(2n, \mathbb{R})/U(n)$.

5.6 Symmetric spaces

5.6.1 Vielbeins, connections, and curvatures

In Section 3.2 we saw that a coset G/H is a Riemannian symmetric space if and only if $\mathfrak{g} = \mathfrak{h} \oplus \mathfrak{m}$ with

$$[\mathfrak{h}, \mathfrak{h}] \subset \mathfrak{h} \qquad [\mathfrak{h}, \mathfrak{m}] \subset \mathfrak{m} \qquad [\mathfrak{m}, \mathfrak{m}] \subset \mathfrak{h}. \tag{5.58}$$

Comparing with Eqs. (5.53)–(5.55), we see that G/H is symmetric iff

$$D_{\alpha\beta\gamma} = 0. \tag{5.59}$$

We can rephrase this condition by saying that the Lie algebra $\mathfrak{g}$ has an involutive automorphism σ,

$$\sigma: \mathfrak{g} \to \mathfrak{g}, \qquad \sigma^2 = \text{Id}_\mathfrak{g}, \tag{5.60}$$

defined by $\sigma|_\mathfrak{h} = +\text{Id}_\mathfrak{h}$ and $\sigma|_\mathfrak{m} = -\text{Id}_\mathfrak{m}$. Conversely, given a (real) Lie algebra $\mathfrak{g}$ with a non–trivial involutive automorphism σ, we can construct a pair (G, H) of

---

[5] For a semi–simple group the structure constants are totally antisymmetric.

Lie algebras so that G/H is symmetric by simply setting $\mathfrak{h}$ to be the eigenspace of σ corresponding to the eigenvalue $+1$, which is automatically a subalgebra since σ is an automorphism.

If the coset G/H is a *symmetric* space, the Maurier–Cartan identities (5.56), (5.57) become

$$de^\alpha + M_{a\beta}{}^\alpha A^a \wedge e^\beta = 0, \tag{5.61}$$

$$dA^a + \frac{1}{2} C_{bc}{}^a A^b \wedge A^c = -\frac{1}{2} M_{\alpha\beta}{}^a e^\alpha \wedge e^\beta, \tag{5.62}$$

which have a transparent differential–geometric meaning. Define the $\mathfrak{h}$–valued 1–form

$$\omega^\alpha{}_\beta := A^a M_{a\beta}{}^\alpha, \tag{5.63}$$

called the *connection form*. Since e^α is an orthonormal frame, Eq. (5.61) is nothing other than the first Cartan structure equation with vanishing torsion:

$$de^\alpha + \omega^\alpha{}_\beta e^\beta \equiv T^\alpha = 0, \tag{5.64}$$

so $(e^\alpha, \omega^\alpha{}_\beta)$ are the vielbein and the torsionless connection corresponding to the (unique up to scale) left–invariant metric g_{ij} we constructed in Eq. (5.50). Recall that there is a unique connection that is both metric and torsionless, namely the Levi–Civita one; so we are guaranteed that the connection $\omega^\alpha{}_\beta$ is equivalent to the standard one given, in "curved" indices, by the Christoffel symbols.

A^a is an $\mathfrak{h}$–valued gauge–field and the matrices $M_{a\beta}{}^\alpha$ are the matrices representing $\mathfrak{h}$ on $\mathfrak{m} \simeq T(G/H)$. Hence $\omega^\alpha{}_\beta$ is nothing other than the $\mathfrak{h}$–gauge–field A^a in the representation appropriate for $T^*(G/H)$. Thus, we may rewrite Eq. (5.62) as

$$R_\alpha{}^\beta := \left(d\omega + \frac{1}{2}\omega \wedge \omega\right)_\alpha{}^\beta = -\frac{1}{2} e^\gamma \wedge e^\delta M_{\gamma\delta}{}^a M_{a\alpha}{}^\beta, \tag{5.65}$$

which is the second Cartan structural equation defining the curvature. This shows, again, that $\mathrm{Hol}(G/H) \equiv H$. Note that the curvature is given by minus a commutator, as expected.

5.6.2 Explicit formulae for the Killing vectors

It will be useful to have explicit and compact formulae for the Killing vectors K_a corresponding to the (left) isometries of a symmetric space G/H. They may be written in many ways.

Matrix form

Identify G with a group of matrices *via* some representation R. From the left–action on G/H of the one parameter subgroup $\exp(tK) \subset G$,

$$(t, g) \longmapsto e^{tK} g, \tag{5.66}$$

we get

$$\pounds_{K_a} g = t_a g, \tag{5.67}$$

where t_a is a matrix representing in End(R) the generator of $\mathfrak{g}$ corresponding to the left–invariant vector K_a.

Gauge viewpoint

To get a more useful formula, we argue from the physical side, that is from the gauge point of view. A simple way to extract the Killing vector associated with a one–parameter subgroup of Iso($\mathcal{M}$) is to gauge that subgroup. From Eq. (1.15), we know that gauging an isometry amounts to the replacement $\partial_\mu \phi^i \to \partial_\mu \phi^i - A_\mu^a K_a^i$,

$$\mathcal{L}_{\text{gauged}} = -\frac{1}{2} g_{ij} (\partial_\mu \phi^i - A_\mu^a K_a^i)(\partial^\mu \phi^j - A^{b\,\mu} K_b^j). \tag{5.68}$$

On the other hand, consider the symmetric space G/H where G acts on the left. The Lagrangian reads[6]

$$\begin{aligned}
\mathcal{L}_{G/H} &= \pm \frac{1}{2} \text{Tr}\left[(g^{-1} \partial^\mu g - B^\mu)(g^{-1} \partial_\mu g - B_\mu)\right] \\
&= \pm \frac{1}{2} \text{Tr}\left[(g \partial^\mu g^{-1} + g B^\mu g^{-1})(g \partial^\mu g^{-1} + g B^\mu g^{-1})\right] \tag{5.69} \\
&\equiv \pm \frac{1}{2} \text{Tr}\left[(g D^\mu g^{-1})(g D^\mu g^{-1})\right],
\end{aligned}$$

where B_μ is the $\mathfrak{h}$–valued gauge field gauging the local symmetry H acting on the right. Let us now gauge a subgroup $F \subset G$ acting on g on the left. The F gauge symmetry reads

$$g(x) \mapsto f(x) g(x), \qquad f(x) \in F. \tag{5.70}$$

Under this transformation

$$g D_\mu g^{-1} \longmapsto f(g D_\mu g^{-1}) f^{-1} + f \partial_\mu f^{-1}, \tag{5.71}$$

---

[6] The sign $+\,(-)$ is appropriate if the Killing form on $\mathfrak{m} \equiv \mathfrak{h}^\perp$ is negative– (positive–)definite.

so, introducing an $\mathfrak{f}$-valued gauge connection $A_\mu \equiv A_\mu^a t_a$ transforming as

$$A_\mu \longmapsto f A_\mu f^{-1} + f \partial_\mu f^{-1}, \tag{5.72}$$

we can write the *F-gauge*-invariant Lagrangian

$$\mathcal{L} = \pm \frac{1}{2} \mathrm{Tr}\left[(g D^\mu g^{-1} - A_\mu)(g D^\mu g^{-1} - A^\mu)\right]. \tag{5.73}$$

Comparing with Eq. (5.68), we see that the coefficient of A_μ^a is $-g_{ij} K_a^i \partial^\mu \phi^j$ (the extra minus from $f \leftrightarrow f^{-1}$ in (5.72)). Thus from Eq. (5.73) we get

$$K_{i\,a} \equiv g_{ij} K_a^j = \pm \mathrm{Tr}[t_a (g D_i g^{-1})], \tag{5.74}$$

where t_a is the generator of $\mathfrak{f}$ corresponding to the given Killing vector. Since $g_{ij} = \mp \mathrm{Tr}[(g D_i g^{-1})(g D_j g^{-1})]$, Eq. (5.74) implies

$$K_a^j (g D_j g^{-1}) = -t_a, \tag{5.75}$$

or[7]

$$K_a^j D_j g = t_a g, \tag{5.76}$$

which corresponds to Eq. (5.67). This formula may be interpreted by saying that the action of G by multiplication on the left is accompanied by a compensating H–gauge transformation on the right given (at the infinitesimal level) by $K_a^i B_i \in \mathfrak{h}$.

If $\{t_a\}$ are generators of G normalized as $\mathrm{Tr}[t_a t_b] = \kappa_{ab}$, Eq. (5.74) gives

$$g D_i g^{-1} = \pm K_{i\,a} \kappa^{ab} t_b. \tag{5.77}$$

Exercise 5.6.1 Write the three Killing vectors of $SU(1,1)/U(1)$.

5.7 Classification of symmetric manifolds

5.7.1 Duality

We have shown in the previous section that, given a simple *compact* Lie group G and an involutive automorphism of its Lie algebra,

$$\sigma: \mathfrak{g} \to \mathfrak{g}, \qquad \sigma^2 = 1, \tag{5.78}$$

---

[7] $D_i g$ is defined by the rule $D_i g = -g(D_i g^{-1})g$.

5.7 Classification of symmetric manifolds

we can construct a symmetric space G/H ($H = \exp[\mathfrak{h}]$, $\mathfrak{h} = \mathrm{Fix}(\sigma)$) with a positive–definite left–invariant metric

$$-\mathrm{tr}[(g^{-1}\partial_i g)_\mathfrak{m}(g^{-1}\partial_j g)], \tag{5.79}$$

where $(\cdots)_\mathfrak{m}$ is the projection on the -1 eigenspace of σ.

Consider now the following subspace of the complexification $\mathfrak{g}_\mathbb{C} = \mathbb{C} \otimes \mathfrak{g}$,

$$\mathfrak{g}' = \mathfrak{h} \oplus i\,\mathfrak{m} \quad (\subset \mathfrak{g}_\mathbb{C}), \tag{5.80}$$

which corresponds to a different real form, G', of the complexified Lie group $G_\mathbb{C}$. $\mathfrak{g}'$ is again a Lie algebra with an involutive automorphism, and the Killing form $\mathrm{Kil}(\cdot,\cdot)$ restricted to $\mathfrak{h}^\perp \equiv i\,\mathfrak{m}$ is now *positive*–definite. Then we may define a new symmetric space G'/H with positive–definite metric:

$$\mathrm{tr}[(g^{-1}\partial_i g)_\mathfrak{m}(g^{-1}\partial_j g)]. \tag{5.81}$$

Definition 5.13 The spaces G/H and G'/H are called *dual symmetric* manifolds.

Thus (non–flat) Riemannian symmetric spaces come in dual pairs. The two spaces in a dual pair have the same holonomy group H and the same holonomy representation. As we already anticipated in Chapter 4, they are distinguished by the *sign* of their Ricci curvature. This is our next topic.

5.7.2 G/H is Einstein

In Section 5.3.3 we computed the Ricci curvature of a compact Lie group, showing that it is Einstein with "cosmological constant"[8] $+1/4$. Let us redo that computation in the case of a general symmetric manifold G/H. We write X_α for the vector fields dual to the 1–forms e^α. Let $X, Y \in \mathfrak{m} \simeq T_e G/H$. Then:

$$\mathrm{Ric}(Y, Z) = e^\alpha(R(X_\alpha, Y)Z) = -\frac{1}{4}e^i([[X_\alpha, Y], Z])$$

$$\equiv -\frac{1}{4}e^\alpha\big(\mathrm{adj}\,Z \circ \mathrm{adj}\,Y(X_\alpha)\big) = -\frac{1}{4}\mathrm{Tr}\big(\mathrm{adj}\,Z \circ \mathrm{adj}\,Y\big|_\mathfrak{m}\big). \tag{5.82}$$

Now we claim that

$$\mathrm{Tr}\big(\mathrm{adj}\,X_\alpha \circ \mathrm{adj}\,X_\beta\big|_\mathfrak{m}\big) = \frac{1}{2}\mathrm{Tr}\big(\mathrm{adj}\,X_\alpha \circ \mathrm{adj}\,X_\beta\big). \tag{5.83}$$

---

[8] In our normalization. In the textbooks one finds also other numbers (typically 1). But ours are the natural normalizations: see ref. [263].

This is due to the special form of the commutation relations for a Lie algebra with symmetry σ. In fact Eq. (5.83) is equivalent to

$$\text{Tr}\big(\text{adj}\, X_\alpha \circ \text{adj}\, X_\beta\big|_{\mathfrak{m}}\big) = \text{Tr}\big(\text{adj}\, X_\alpha \circ \text{adj}\, X_\beta\big|_{\mathfrak{h}}\big) \tag{5.84}$$

and this last equation can be rewritten in terms of the structure constants of $\mathfrak{g}$ in the form

$$f_{\alpha\gamma a} f_{\beta a\gamma} = f_{\alpha a\gamma} f_{\beta\gamma a}, \tag{5.85}$$

which is trivially true. Then:

Proposition 5.14 *A symmetric manifold G/H, equipped with the left G–invariant metric (normalized as above) is Einstein with cosmological constant $+1/8$ if G is compact and $-1/8$ for its dual G'.*

Hence the pairs of dual symmetric spaces G/H and G'/H are, respectively, positively and negatively curved. We recall a useful result:

Theorem 5.15 (see ref. [51]) *Let $\mathcal{M}$ be a homogeneous[9] Einstein manifold with cosmological constant λ. Then:*

(1) *if $\lambda > 0$, $\mathcal{M}$ is compact with finite π_1;*
(2) *if $\lambda = 0$, $\mathcal{M}$ is flat;*
(3) *if $\lambda < 0$, $\mathcal{M}$ is non–compact.*

(1) is Myers's theorem. (2) is a theorem by D.V. Alekseevskii and B. N. Kimelfeld. (3) is elementary: by Bochner's theorem, if $\mathcal{M}$ is compact and has negative–definite Ricci tensor it has no Killing vector, contrary to the assumption that it is homogeneous.

5.7.3 Classification

The irreducible symmetric spaces are usually classified into four types:

Type I spaces of the form G/H with G a real simple compact Lie group;
Type II spaces $(G \times G)/G \simeq G$, G a real simple compact Lie group;
Type III the non–compact duals of Type I, i.e., G'/H.
Type IV the non–compact duals of Type II, namely $H_{\mathbb{C}}/H$, where $H_{\mathbb{C}}$ is a complex simple simply connected Lie group and H a maximal compact connected subgroup, and σ the complex conjugation of $H_{\mathbb{C}}$ whose fixed set is H.

---

[9] That is, $\text{Iso}(\mathcal{M})$ is transitive.

5.8 Totally geodesic submanifolds. Rank

From the discussion there follows:

Corollary 5.16 *A non–compact irreducible symmetric space is a quotient G/H, where G is a real simple non–compact Lie group with trivial center and H is a maximal compact subgroup of G.*

We deduce the classification of the symmetric spaces from that of the real Lie groups. See the tables in refs. [51, 69] and [184].

Exercise 5.7.1 Describe the isometry group of the space dual to $SO(2n)/U(n)$.

5.8 Totally geodesic submanifolds. Rank

In Section 4.3 we saw that the "structure transporting maps" $\varrho_{D,d}$ (i.e., dimensional reduction and group disintegration) embed $\mathcal{M}_D$ into $\mathcal{M}_d$ as *a totally geodesic submanifold*. These submanifolds are described by the following theorem.

Theorem 5.17 *Let $\iota\colon \mathcal{W} \hookrightarrow G/H$ be a totally geodesic submanifold:*

(1) *$\mathcal{W}$ is also a symmetric manifold;*
(2) *the geodesic submanifolds of G/H passing through a fixed point ϕ are in one-to-one correspondence with the subspaces $\mathfrak{s} \subset \mathfrak{m} \subset \mathfrak{g}$ such that*

$$[\mathfrak{s},[\mathfrak{s},\mathfrak{s}]] \subseteq \mathfrak{s}. \tag{5.86}$$

Proof (1) is Theorem 4.4. (2) Let $\mathfrak{s} = T_\phi \mathcal{W} \subset T_\phi G/H \simeq \mathfrak{m}$. In view of the fact that the Riemann tensor of G/H is given by a double commutator, Eq. (5.86) is equivalent to the usual property of a totally geodesic submanifold $\mathcal{W}$, namely

$$R_\mathcal{W} T\mathcal{W} \subset \mathcal{W}. \tag{5.87}$$

So $\mathcal{W}$ is totally geodesic $\Rightarrow$ (5.86). Conversely, consider the submanifold of G/H defined by

$$\mathcal{W} := \exp_\phi \mathfrak{s}. \tag{5.88}$$

It is easy to see (for instance with the help of normal coordinates) that it is totally geodesic only if (5.86) holds. $\square$

From the explicit form of the curvature, Eq. (5.65), it is clear that a totally geodesic submanifold $\mathcal{M} \hookrightarrow G/H$ is *flat* if and only if the Lie subalgebra of $\mathfrak{g}$ generated by $\mathfrak{s}$ is Abelian. This motivates the following:

Definition 5.18 The dimension of the maximal Abelian subalgebra of $\mathfrak{m}$ is called the RANK of the symmetric space G/H.

The rank is also the dimension of the *maximal flat totally geodesic submanifolds* of G/H. Two such submanifolds are related by left–translation by G. They are tori for G/H compact and Euclidean spaces otherwise.

A particular interesting class of symmetric spaces are those with *rank* 1.

Proposition 5.19 *A symmetric space G/H has rank 1 if and only if the action of H on the unit sphere in $T_e G/H$ is transitive. (That is, if H is one of the Berger groups or $Spin(9)$ in dimension 16).*

Proof Cf. **Proposition 3.16**. □

5.9 Other techniques

As we mentioned in Chapter 3, a symmetric space G/H with holonomy group $H = U(1) \times \widetilde{H}$, is Kählerian. In this case much more powerful techniques exist to study their geometry. We will discuss them after developing the theory of complex and Kähler spaces to a sufficient level. In particular, the non–compact, Type III Hermitian symmetric manifolds correspond to bounded domains, and in particular to classical domains, in $\mathbb{C}^N$ [57, 71, 78, 264]. We can use complex function theory to study them. We already saw a basic example. Our "duality target space"

$$\frac{Sp(2V, \mathbb{R})}{U(1) \times SU(V)} \tag{5.89}$$

is the *Siegel upper–half space,* $\mathfrak{H}_V$, which is one of the most relevant domains of several–complex–variables theory. The Cayley transform maps it into the generalized disk $\mathfrak{D}_V$, opening the way to the methods of bounded domain theory, based on Hilbert space techniques [44, 259, 323]. The bounded domain viewpoint is the most convenient one for $\mathcal{N} = 1$ SUGRA in 4D.

A useful approach to symmetric (or just homogeneous) non–compact Kähler space is the *thermodynamical analogy,* Section 9.6.

5.10 An example: $E_{7(7)}/SU(8)$

As an illustration of the geometric techniques developed in this chapter, we work out in some detail the case of $E_{7(7)}/SU(8)$, that is, of $\mathcal{N} = 8$ 4D SUGRA. All four–dimensional supergravities with $\mathcal{N} \geq 5$ can be obtained from this one by truncation [98], that is, by restriction to the appropriate totally geodesic submanifold.

The reader may have a feeling that the exceptional Lie groups G_2, F_4, E_6, E_7, and E_8 are quite mysterious objects, difficult to visualize and to understand. In fact, in the literature [2, 167] there are a number of explicit constructions of these groups (we have already mentioned the one based on the octonionic projective spaces [28]). It turns out that *each* of these mathematical constructions is word–for–word equivalent to the physical realization given by a specific SUGRA model. The various constructions in [2] are related by morphisms that correspond to our $\varrho_{D,d}$ and to truncations to lower $\mathcal{N}$'s. Our physical discussion of E_7 below is, mathematically, nothing else than the *grading model* of the group (see Example 1 on page 180 of ref. [167], or Chapter 12 of ref. [2]). The analogous construction for maximal SUGRA in 5D, based on the coset $E_{6(6)}/Sp(8)$, corresponds to the explicit model of E_6 in Chapter 13 of ref. [2]. The same procedure applied to $\mathcal{N} = 16$, $\mathcal{N} = 12$, $\mathcal{N} = 10$, and $\mathcal{N} = 9$ in $D = 3$ gives the "standard" construction of (respectively) E_8, E_7, E_6, and F_4. For E_8 this is, essentially, the same construction one gets from the $d = 2$ current algebra, which leads, say, to the $E_8 \times E_8$ gauge symmetry in the heteoric string [170]. The "magical" $\mathcal{N} = 2$ SUGRA's in 5D [179] correspond to the "magical square" construction of the exceptional Lie groups – see §5.1.7 of ref. [167]; and, from another viewpoint, to the construction of E_6 on page 181 of ref. [167].

So, SUGRA is also a very powerful mathematical tool!

5.10.1 The exceptional group $E_{7(7)}$

From the physics of $\mathcal{N} = 8$ SUGRA we know a number of things about its scalar space $\mathcal{M}$ which are indeed sufficient to construct a *concrete model* of the exceptional, 133–dimensional, $\mathbb{R}$–split group $E_{7(7)}$ from scratch.

First of all, we know that $E_{7(7)}$ has a real symplectic **56** dimensional representation on the field–strengths $\mathcal{F}^+$'s. So we identify $E_{7(7)}$ with a group of **56** $\times$ **56** real symplectic matrices. We perform the Cayley transform in Section 1.5.1 to rewrite it, more conveniently, as a group of complex symplectic matrices E,

$$E^t \Omega E = \Omega, \tag{5.90}$$

satisfying the reality condition in Eq. (1.114):

$$E^* = \Sigma_1 E \Sigma_1. \tag{5.91}$$

Let $E = \exp L$ be an element of the group (so $L \in \mathfrak{e}_{7(7)}$). Write

$$L = \begin{pmatrix} A & B \\ C & D \end{pmatrix}, \tag{5.92}$$

where A, B, C, D are 28×28 matrices. The conditions (5.90) and (5.91) become, respectively,

$$A^t = -D, \qquad\qquad B^t = B, \qquad (5.93)$$
$$A^* = D, \qquad\qquad B^* = C, \qquad (5.94)$$

which imply $A^\dagger = -A$, and $B^\dagger = C \equiv C^t$. Then the subalgebra of the block–diagonal elements of $\mathfrak{e}_{7(7)}$,

$$\begin{pmatrix} A & 0 \\ 0 & A^* \end{pmatrix}, \qquad (5.95)$$

consists of compact generators, whereas the complementary set

$$\begin{pmatrix} 0 & B \\ B^\dagger & 0 \end{pmatrix} \qquad (5.96)$$

consists of non–compact ones. Hence the subalgebra (5.95) is identified with the maximal compact subalgebra of $\mathfrak{e}_{7(7)}$.

The second fact we know from SUGRA is that this maximal compact subalgebra is $\mathfrak{aut}_R(8)$, namely $\mathfrak{su}(8)$, here in the **28**–dimensional representation. Writing an index of the **28** as an antisymmetric pair $[ab]$ of fundamental representation indices $(a, b = 1, 2, \ldots, 8)$, one has

$$A^{[ab]}{}_{[cd]} = \Lambda^{[a}{}_{[c}\delta^{b]}{}_{d]}, \qquad (5.97)$$

where $\Lambda^a{}_b$ are matrices in the defining representation of $\mathfrak{su}(8)$; that is, they are anti–Hermitian traceless 8×8 matrices.

As to the third piece of information from SUGRA: writing $\mathfrak{e}_{7(7)} = \mathfrak{su}(8) \oplus \mathfrak{m}$, one has the isomorphisms

$$\mathfrak{m} \simeq T_\phi \mathcal{M} \simeq [[\wedge^4 \mathcal{S}_\phi]]; \qquad (5.98)$$

that is, the off–diagonal entries of L are in the totally antisymmetric four–indices $\mathfrak{su}(8)$ representation, subject to the reality condition that complex conjugation is equal to duality. Hence we write

$$B^{[ab][cd]} \equiv B^{abcd} = \epsilon^{abcdefgh} B_{dfgh}, \qquad (5.99)$$

where we adopt the convention that lowering/raising indices is complex conjugation. Thus a general element of the Lie algebra of $\mathfrak{e}_{7(7)}$ is given by

$$L = \begin{pmatrix} \Lambda^{[a}{}_{[e}\delta^{b]}{}_{f]} & B^{ab\,gh} \\ B_{cd\,ef} & \Lambda_{[c}{}^{[g}\delta_{d]}{}^{h]} \end{pmatrix}, \qquad (5.100)$$

with

$$\Lambda^a{}_b = -\Lambda_b{}^a, \quad \Lambda^a{}_a = 0, \tag{5.101}$$

and

$$B^{ab\,cd} = \epsilon^{abcdefgh} B_{df\,gh}. \tag{5.102}$$

In total we have $63 + 70 = 133$ generators, so we have constructed the full $\mathfrak{e}_{7(7)}$ Lie algebra.

To construct the compact version of $\mathfrak{e}_7$, we have only to use duality:

$$L_{\text{compact}} = \begin{pmatrix} \Lambda^{[a}{}_{[e}\delta^{b]}{}_{f]} & i\,B^{ab\,gh} \\ i\,B_{cd\,ef} & \Lambda_{[c}{}^{[g}\delta_{d]}{}^{h]} \end{pmatrix}. \tag{5.103}$$

5.10.2 Explicit $\mathcal{N} = 8$ couplings

We can insert the above explicit formulae in our construction of the SUGRA Lagrangians in Chapter 4. We write $\mathcal{E} = \exp L$ where L is as in Eq. (5.100), and compute the Maurier–Cartan form

$$(\partial_i \mathcal{E})\mathcal{E}^{-1} = \begin{pmatrix} Q_i^{[A}{}_{[E}\delta^{B]}{}_{F]} & P_i^{AB\,GH} \\ P_{i\,CD\,EF} & Q_{i\,[C}{}^{[G}\delta_{D]}{}^{H]} \end{pmatrix}. \tag{5.104}$$

(We write $(\partial_i \mathcal{E})\mathcal{E}^{-1}$ instead of $\mathcal{E}^{-1}\partial_i\mathcal{E}$ because in SUGRA we think of the global group G as acting on the *right* of $\mathcal{E}$ and not on the *left* as we do in mathematics).

The indices A, B, C, D, E, F, G, H in Eq. (5.104) are *local* $SU(8)$ indices. The target space 1–form with values in $\mathfrak{su}(8)_R$, $Q_i^A{}_B$, which appears on the RHS of Eq. (5.104), is by definition the $\mathfrak{su}(8)_R$ gauge–connection. Its curvature is computed by the Maurier–Cartan equations (in full agreement with the direct computations in Chapter 2).

The off–diagonal terms, P_i^{ABCD}, are the target vielbein; that is, the tensors which represent the isomorphism $T\mathcal{M} \simeq [\![\wedge^4 S]\!]$ with respect to standard basis. This vielbein defines the χs SUSY transformation, as well as the scalar kinetic terms

$$-P_i^{ABCD} P_{j\,ABCD}\, \partial_\mu \phi^i\, \partial^\mu \phi^j. \tag{5.105}$$

The same vielbein P_i^{ABCD} defines the couplings in Eq. (4.44). The 4–fermions couplings are given by the curvature tensor that we have already computed many times, up to some terms that originate either from the torsion part of the space–time connection (see the 3D case in Chapter 2) or from the supercovariantization of the supercurrent coupling. They are all described by the pull–back $\varrho_{4,3}^*$ and are rather straightforward.

There is only one piece lacking in the picture. We can still gauge some subgroup of isometries, $\mathcal{G} \subset \mathrm{Iso}(\mathcal{M})$, getting a more general theory with all kinds of possible couplings.[10] Gauging will be a major theme in the next three chapters.

Exercise 5.10.1 Construct E_6 using the maximal 5D SUGRA.

Exercise 5.10.2 Construct F_4 using 3D $\mathcal{N} = 9$ SUGRA.

Exercise 5.10.3 Give an alternative model of E_7 from 3D $\mathcal{N} = 12$.

5.11 Symmetric and Iwasawa gauges

In this chapter we have adopted the gauge point of view to describe the coset G/H: we added to the SUGRA model $\dim H$ unphysical scalars and as many auxiliary gauge vectors to "eat" them. For many applications it is useful to work in a *unitary gauge*. In such a gauge, all the scalar fields in $\mathcal{L}$ are physical, and no auxiliary gauge vector is present.

There are two natural unitary gauges, each one with its own merits. The first one is the *symmetric gauge*. One writes $G = \exp[\mathfrak{m}]$; that is, in the $E_{7(7)}$ example, say,

$$\mathcal{E} = \exp \begin{pmatrix} 0 & B^{ab\,gh} \\ B_{cd\,ef} & 0 \end{pmatrix}. \quad (5.106)$$

Now $\mathcal{E}$ is parametrized by $\dim G - \dim H$ physical scalars $B^{ab\,fg}$ (70, in the $E_{7(7)}$ example). To justify the choice, one has to show that any configuration is gauge equivalent to one and only one of the above form. This follows from Theorem 5.20.

Theorem 5.20 (Cartan decomposition [184]) *G semi–simple and G/H a symmetric space with $\mathfrak{g} = \mathfrak{h} \oplus \mathfrak{m}$. Then the map*

$$\begin{aligned} H \times \mathfrak{m} &\longrightarrow G \\ (h, p) &\longmapsto h \exp(p), \end{aligned} \quad (5.107)$$

is a diffeomorphism.

Proof Let σ be the Cartan involution. Embed $G \hookrightarrow Sp(2n, \mathbb{R})$. Then σ acts on $\mathcal{E}$ as $\sigma(\mathcal{E}) = (\mathcal{E}^t)^{-1}$. $\mathcal{P} = \mathcal{E}\mathcal{E}^t$ is a positive–definite symmetric matrix in $G \subset Sp(2n, \mathbb{R})$ which has a *unique* logarithm with real eigenvalues $2p \in \mathfrak{m}$. □

---

[10] For $D \leq 5$. In higher dimensions we may have couplings to higher form fields.

5.11 Symmetric and Iwasawa gauges

There is another gauge, less symmetric but which has the merit of working well under the maps $\varrho_{D,d}$, in the sense that it makes the successive embeddings of totally geodesic submanifolds much more transparent. Indeed, in such a gauge, most of the couplings are *polynomial* in the scalar fields. This is the *Iwasawa gauge* [98]. One could argue in abstract terms, using the general theorem about Iwasawa decomposition of a Lie algebra [69, 127]. However, to be concrete, we look at G as a group of $2n \times 2n$ real symplectic matrices, $G \subset Sp(2n, \mathbb{R})$.

Working in the standard real realization of $Sp(2n, \mathbb{R})$, we define the subgroup $\mathcal{B}_n \subset Sp(2n, \mathbb{R})$ consisting of the real matrices of the form

$$\mathcal{B}_n = \left\{ \begin{pmatrix} h & X(h^t)^{-1} \\ 0 & (h^t)^{-1} \end{pmatrix}, \begin{array}{l} h, X \in \mathbb{R}(n), \ X^t = X; \\ h \text{ upper } triangular \text{ with} \\ positive \text{ diagonal entries} \end{array} \right\}. \tag{5.108}$$

Notice that the change of basis

$$\begin{pmatrix} \delta_{a,b} & 0 \\ 0 & \delta_{n+1-a,b} \end{pmatrix} \tag{5.109}$$

maps $\mathcal{B}_n$ into a group of *upper triangular* matrices, which can be uniquely parametrized as $\exp(D)\exp(N)$, with $D = \text{diag}(d_i)$ diagonal and N *strictly* upper triangular. The group element $\exp(D)\exp(N) \in \mathcal{B}_n$ depends only polynomially on N, while the diagonal part is just $\exp(D) = \text{diag}(e^{d_i})$. The triangular subgroup $\mathcal{B}_n$ has a very convenient parametrization. Luckily, each class in $Sp(2n, \mathbb{R})/U(n)$ admits one and only one triangular representative, so this convenient parametrization of $\mathcal{B}_n$ is, in fact, a very convenient parametrization of the coset space $Sp(2n, \mathbb{R})/U(n)$.

Lemma 5.21 *An element $S \in Sp(2n, \mathbb{R})$ has a* UNIQUE *decomposition of the form*

$$S = b \cdot u, \quad b \in \mathcal{B}_n, \ u \in U(n). \tag{5.110}$$

Proof We have to show that all classes in $Sp(2n, \mathbb{R})/U(n)$ admit a representative of the form b. Equivalently, we have to show $\pi(\mathcal{B}_n) = \mathfrak{H}_n$, where $\pi : Sp(2n, \mathbb{R}) \to \mathfrak{H}_n$ is the map

$$\pi : \begin{pmatrix} A & B \\ C & D \end{pmatrix} \mapsto (iA + B)(iC + D)^{-1}. \tag{5.111}$$

Now

$$\begin{pmatrix} h & X(h^t)^{-1} \\ 0 & (h^t)^{-1} \end{pmatrix} \xrightarrow{\pi} ihh^t + X, \tag{5.112}$$

and the result follows from the well–known fact that any real symmetric positive–definite matrix can be written in a *unique* way in the form hh^t, with h, an upper triangular matrix with positive diagonal entries (cf. the Iwasawa decomposition for $GL(n, \mathbb{R})$; see ref. [165], proposition 1.2.6). The argument also implies the uniqueness of the Iwasawa decomposition. □

Let $G \subset Sp(2n, \mathbb{R})$ be a subgroup which is stable under the involutive automorphism of $Sp(2n, \mathbb{R})$, $g \mapsto (g^t)^{-1}$, and a totally geodesic symmetric submanifold. By Lemma 5.21, given an element $g \in G$ we have unique $b \in \mathcal{B}_n$ and $u \in U(n)$ such that $g = bu$. The general theorem of Iwasawa guarantees that $b \in G$ and then $u \in G \cap U(n) = H$. Thus, each point in G/H gets represented by an element of G of the form $\exp(D)\exp(N)$, where D is a diagonal matrix that belongs to an *Abelian* subalgebra $\mathfrak{a} \subset \mathfrak{g}$, with $\dim \mathfrak{a} = \text{rank}(G/H)$, and N is an upper triangular matrix belonging to a *nilpotent* subalgebra $\mathfrak{n} \subset \mathfrak{g}$, with $[\mathfrak{a}, \mathfrak{n}] \subset \mathfrak{n}$. If $D_i \equiv \text{diag}(d_{i,a})$, and $N_\alpha \equiv \{n_{\alpha,ab} \mid b > a\}$ are the basis of $\mathfrak{a}$ and $\mathfrak{n}$, respectively, we get[11]

$$\exp(s^\alpha N_\alpha)\exp(t^i D_i)$$
$$= \begin{pmatrix} 1 & & P_{ab}(S^\alpha) \\ & \ddots & \\ 0 & & 1 \end{pmatrix} \begin{pmatrix} \exp(t^i d_{i,1}) & & 0 \\ & \ddots & \\ 0 & & \exp(t^i d_{i,n}) \end{pmatrix} \quad (5.113)$$

with $P_{ab}(S^\alpha)$ polynomials. In this gauge the Maurier–Cartan equations take the form

$$g^{-1}dg = D_i \, dt^i + e^{-t^i D_i} N_\alpha e^{t^i D_i} F^\alpha{}_\beta(s) \, ds^\beta, \quad (5.114)$$

with $F^\alpha{}_\beta(s)$ polynomials in the s^γ.

Exercise 5.11.1 Write the metric and $U(1)$ connection of $SL(2, \mathbb{R})/U(1)$ in the Iwasawa parametrization.

---

[11] For convenience we have changed the parametrization of $\mathcal{B}_n$ a little by interchanging the order.

6
Killing spinors and rigid SUSY in curved spaces

The results of the previous chapters allow us to write down the Lagrangian of any supergravity theory in D dimension but for a few terms: the gauge couplings, the Yukawa ones, and the scalar potential. One goal of the present chapter is to give a *universal* formula for the potential $V(\phi)$, valid for all D and $\mathcal{N}$. This will solve the problem of the scalar potential once and for all. It is easy to prove the formula directly [88], but instead we shall make a long detour: what we learn along the route, i.e., the basics of rigid SUSY in curved spaces, justifies the effort. A basic theme of this book is that the geometrical structures on the scalar manifold $\mathcal{M}$ have the same flavor as the ones in the physical space–time Σ. A field configuration is seen as a map transporting structure back and forward $\Sigma \leftrightarrow \mathcal{M}$. In this spirit, we introduce the relevant geometric objects in physical space–time, and then apply them to $\mathcal{M}$. The chapter focuses on the concept of *Killing spinor*, its geometrical meaning and physical implications for SUSY on curved manifolds. To motivate its introduction, we start by reviewing its bosonic counterpart, the Killing *vector*, in General Relativity.

6.1 Review: space–time charges in General Relativity

In General Relativity one cannot, in general, define a conserved energy–momentum. The generators of space–time symmetries, such as momentum or angular momentum, can be defined only in space–times that have the corresponding *asymptotic* Killing vectors. Let us review how it works [1]. One starts with a metric $\bar{g}_{\mu\nu}$ that solves the Einstein equations of motion and is invariant under a group of isometries, $\mathrm{Iso}(\bar{g})$. The corresponding Lie algebra, $\mathrm{iso}(\bar{g})$, is generated by Killing vectors K^m_μ with brackets,

$$[K^m, K^n] = -f^{mn}{}_p K^p. \tag{6.1}$$

For simplicity, we consider a vacuum–like reference configuration, with constant scalars and vanishing vector fields, which solves the Einstein equations[1]

$$\bar{R}_{\mu\nu} - \frac{1}{2}\bar{g}_{\mu\nu}\bar{R} - \bar{\Lambda}\,\bar{g}_{\mu\nu} = 0. \tag{6.2}$$

The effective cosmological constant $\bar{\Lambda}$ receives contributions also from the "vacuum" energy of the given field configuration. One expands around this solution, setting $g_{\mu\nu} = \bar{g}_{\mu\nu} + h_{\mu\nu}$, and $\Phi = \bar{\Phi} + \phi$. The Einstein equations are rewritten in the form

$$R^L_{\mu\nu} - \frac{1}{2}\bar{g}_{\mu\nu}R^L - \bar{\Lambda}\,h_{\mu\nu} = (T_{\mu\nu} + t_{\mu\nu}) \equiv \Theta_{\mu\nu}, \tag{6.3}$$

where $R^L_{\mu\nu}$ is the Ricci tensor *linearized* around $\bar{g}_{\mu\nu}$, and $t_{\mu\nu}$ is minus the higher–order terms in the Einstein tensor $R_{\mu\nu} - \frac{1}{2}g_{\mu\nu}R$. The background version of the Bianchi identity,

$$\bar{D}^\mu \left(R^L_{\mu\nu} - \frac{1}{2}\bar{g}_{\mu\nu}R^L - \bar{\Lambda}\,h_{\mu\nu} \right) \equiv 0, \tag{6.4}$$

holds, provided the background satisfies the equations of motion (6.2). Then the symmetric tensor $\Theta_{\mu\nu}$ is covariantly conserved with respect to the background connection

$$\bar{D}^\mu \Theta_{\mu\nu} = 0. \tag{6.5}$$

We stress that this is an *exact* result. The K^m_μ are Killing vectors for the background metric $\bar{g}_{\mu\nu}$

$$\bar{D}_\mu K^m_\nu + \bar{D}_\nu K^m_\mu = 0. \tag{6.6}$$

Combining Eqs. (6.5) and (6.6) we get

$$\partial_\mu(\sqrt{-\bar{g}}\,\Theta^{\mu\nu}K^m_\nu) \equiv \bar{D}_\mu(\Theta^{\mu\nu}K^m_\nu) = \Theta^{\mu\nu}\bar{D}_\mu K^m_\nu = 0, \tag{6.7}$$

and the corresponding charge

$$M^m = \int_t d^3x\,\sqrt{-\bar{g}}\,\Theta^{0\nu}K^m_\nu \tag{6.8}$$

is conserved provided the fluctuations $h_{\mu\nu}$ and ϕ vanish sufficiently rapidly at spatial infinity to justify our formal manipulations.

---

[1] A bar over a symbol denotes a quantity evaluated on the reference (background) configuration. Φ stands for all fields in the theory distinct from the metric.

The M^m generate a Lie algebra isomorphic to $\mathfrak{iso}(\bar{g})$,

$$[M^m, M^n] = f^{mn}{}_p M^p, \tag{6.9}$$

as one checks by canonical manipulations. Using this procedure, if the space–time geometry is asymptotic to flat Minkowski space, we construct the generators of the Poincaré group, P_μ, and $M_{\mu\nu}$, while if Σ is asymptotic to Anti–de Sitter (AdS), the maximal symmetric solution with negative cosmological constant $\bar{\Lambda} < 0$, we get the generators M^{AB} of its isometry group,

$$\text{Iso}(AdS_d) \simeq SO(d-1, 2). \tag{6.10}$$

For positive $\bar{\Lambda}$ the maximally symmetric solution is de Sitter space, and the same method gives us the generators of $\text{Iso}(dS_d) \simeq SO(d, 1)$.

Most importantly, the space–time charges M^m can be expressed as flux integrals at infinity [1, 160, 192, 245, 246, 300] as one would expect for a gauge theory (Gauss's law). The explicit expression is given in Section 6.5 below. This implies that only the fields at infinity enter in the computation of the conserved quantity M^m; in particular, the Killing vector K^m is required to exist only asymptotically.

Exercise 6.1.1 Show that if the asymptotic Killing vector K^m has an extension to an everywhere–defined Killing vector, then $M^m = 0$.

6.2 AdS space

For convenience of the reader, we briefly review anti–de Sitter space. The D dimensional anti–de Sitter space, AdS_D, corresponds to the hyperboloid

$$X_0^2 + X_D^2 - \sum_{i=1}^{D-1} X_i^2 = R^2, \tag{6.11}$$

in a flat $\mathbb{R}^{2,D-1}$ space with metric

$$ds^2 = dX_0^2 + dX_D^2 - \sum_{i=1}^{D-1} dX_i^2. \tag{6.12}$$

By construction, AdS_D has an isometry group $SO(2, D-1)$ and is manifestly homogeneous and isotropic. Consider the point $(R, 0, \ldots, 0) \in AdS_D$. It is invariant under the subgroup $SO(1, D-1) \subset SO(2, D-1)$ acting on the last D coordinates. Thus

$$AdS_D = SO(2, D-1)/SO(1, D-1), \tag{6.13}$$

which should be compared with the usual D–sphere, $S^D = SO(D+1)/SO(D)$. Thus anti–de Sitter and de Sitter (= $SO(1,D)/SO(1,D-1)$) are, in a sense, analytic continuations of the sphere. Indeed, equation (6.11) may be solved as

$$X_0 = R\cosh\rho\,\cos\tau,$$
$$X_D = R\cosh\rho\,\sin\tau, \qquad (6.14)$$
$$X_i = R\sinh\rho\,\Omega_i, \quad i=1,2,\ldots,D-1, \quad \sum_i \Omega_i^2 = 1,$$

leading to the global metric

$$ds^2 = R^2(\cosh^2\rho\,d\tau^2 - d\rho^2 - \sinh^2\rho\,d\Omega^2), \qquad (6.15)$$

where $d\Omega^2$ is the usual round metric on the sphere S^{D-2}. Here $\rho \geq 0$ and $0 \leq \tau \leq 2\pi$, if we wish to cover the hyperboloid once. However, the S^1 parameterized by τ is a closed time–like geodesic, and this is not good for causality. To fix this, we *define* AdS_D to be the *universal cover* of the above hyperboloid, which just means that τ takes value in the full real line.

From Eq. (6.15) the relation $AdS_D \leftrightarrow S^D$ is manifest: taking ρ imaginary we get the usual metric on the sphere. We stress that the analytical continuation of the AdS_D metric that gives the sphere is *not* the Wick rotation which gives the Euclidean version of AdS_D. Rather, the Euclidean AdS_D is the non–compact dual to the sphere $SO(D,1)/SO(D)$: see Chapter 5.

6.2.1 Spinorial realization

For later convenience, we write the coset representative of a point in AdS_D as a $Spin(2,D-1)$ element

$$\mathcal{E} \equiv \exp[t_{AB}\,\gamma^{AB}/4], \qquad (6.16)$$

where γ^{AB} are the usual Dirac γ–matrices in signature $(2,D-1)$. Since $\gamma_0^2 = 1$, we have an involutive automorphism

$$\sigma: Spin(2,D-1) \to Spin(2,D-1)$$

given by

$$\sigma(\mathcal{E}) = \gamma_0\,\mathcal{E}\,\gamma_0. \qquad (6.17)$$

The elements of $Spin(2,D-1)$ which are left fixed by σ are precisely the elements of $Spin(1,D-1)$ (Exercise 6.2.1). The involution σ gives to

$$AdS_D \equiv Spin(2,D-1)/Spin(1,D-1) = Spin(2,D-1)/\mathrm{Fix}(\sigma) \qquad (6.18)$$

the structure of a (Minkowskian) *symmetric space* to which we can apply the machinery introduced in Chapter 5. We write ($a, b = 1, 2, \ldots, D$)

$$\mathcal{E}^{-1} d\mathcal{E} = \frac{1}{4} \omega^{ab} \gamma_{ab} + \frac{1}{2} e^a \gamma_0 \gamma_a, \qquad (6.19)$$

where e^a and ω_{ab} are, respectively, the AdS_D vielbeins and spin–connection. The $SO(2, D-1)$ Killing vectors are given by Eq. (5.74):

$$K_i^{AB} = \eta \operatorname{tr}\left[\gamma^{AB} \mathcal{E} D_i \mathcal{E}^{-1}\right], \qquad (6.20)$$

where η is a normalization coefficient. From Eq. (6.19) we obtain vielbeins and connections for other spaces by analytic continuation:

1. $\gamma_0, \to i \gamma_0, \gamma_D \to i \gamma_D$ yields S^D (sphere);
2. $\gamma_0, \to i \gamma_0$ yields dS_D (de Sitter);
3. $\gamma_D \to i \gamma_D$ yields H_D, (hyperbolic space, the non–compact dual of S^D).

Wick rotation relates AdS_D to H_D and dS_D to S^D.

Exercise 6.2.1 Show that $\operatorname{Fix}(\sigma) = Spin(1, D-1)$.

6.3 Killing spinors

In supergravity one would like to extended the previous construction from the Poincaré (or AdS) algebra to the full superPoincaré (or super–AdS) algebra. Indeed, in SUGRA space-time symmetries and supersymmetries ought to be unified. Our goal is to construct the supergenerators Q^A by mimicking the construction of the M^m in Section 6.1. The central ingredient of M^m was the (asymptotic) Killing vector. We need a fermionic counterpart to a Killing vector: such geometric objects exist and are called *Killing spinors*.

There are two different notions of what a Killing spinor is. One is the physical definition, which is more general (and deep). The second one is the formal mathematical definition, which is more convenient for doing geometry. The two notions coincide if the background has enough isometries.

6.3.1 The physical definition

A *supersymmetric background* in a supergravity theory is a bosonic field configuration in which some supersymmetry is unbroken. In particular, the Fermi fields vanish: $\psi_\mu^A = \chi^I = 0$. A SUSY background does not need to be a solution of the equations of motion; however, when the model is described in terms of the physical fields only, i.e., the auxiliary fields have been set *on–shell,* a SUSY background automatically satisfies *most* of the equations of motion.

Let the spinorial parameter $\epsilon_\alpha^A(x)$ correspond to an unbroken SUSY in the given background. It should leave invariant the Fermi fields

$$\delta\psi_\mu^A = \overline{\mathcal{D}}_\mu \epsilon^A = 0, \quad \delta\chi^I = \overline{\Xi}_A^I \epsilon^A = 0, \tag{6.21}$$

where in the RHS the bosonic fields are replaced by their background values (denoted by overbars). The variations of the bosonic fields vanish automatically, since the background is purely bosonic. As the notation suggests (and we know from Chapter 2), the linear operator $\overline{\mathcal{D}}_\mu$ in the RHS of the first Eq. (6.21) is a first–order differential operator[2]

$$\overline{\mathcal{D}}_\mu = \bar{D}_\mu + \begin{pmatrix} \overline{Q}_{i\ C}^A \partial_\mu \phi^i & \frac{i}{2}\overline{M}^{AD}\bar{\gamma}_\mu + \overline{\mathfrak{F}}_{\rho\sigma}^{AD}\bar{\gamma}^{\rho\sigma}\bar{\gamma}_\mu \\ \frac{i}{2}\overline{M}_{BC}\bar{\gamma}_\mu + \overline{\mathfrak{F}}_{BC\rho\sigma}\bar{\gamma}^{\rho\sigma}\bar{\gamma}_\mu & \overline{Q}_{iB}{}^D\partial_\mu\phi^i \end{pmatrix}, \tag{6.22}$$

where $\bar{D}_\mu$ is the (background) $Spin(1, D-1)$ covariant derivative. We leave the indices mostly implicit in the following.

In order to get nice and universal formulae, it is fundamental to write the Lagrangian in a canonical form: in particular, the kinetic terms should be block–diagonalized between the gravitini and the spin–1/2 fields χ^I; terms of the form $\bar{\chi}\sigma^{\mu\nu}D_\mu\psi_\nu$ should be eliminated from $\mathcal{L}$ by a suitable redefinition of the gravitino fields, $\psi_\mu^A \mapsto \psi_\mu^A + A_I^A(\phi)\gamma_\mu\chi^I$. Then the only term in $\mathcal{L}$ containing *derivatives* of ψ_μ^A is the Rarita–Schwinger (RS) one, $\bar{\psi}_\mu\gamma^{\mu\nu\rho}\mathcal{D}_\nu\psi_\rho$, and the operator $\mathcal{D}_\mu$ appearing in the gravitino kinetic term is the same one entering in the SUSY transformations $\delta\psi_\mu = \mathcal{D}_\mu\epsilon$. Hermiticity of the action gives the identity,

$$\overline{(\mathcal{D}_\nu\epsilon)}\gamma^{\mu\nu\rho}\psi_\rho + \bar{\epsilon}\gamma^{\mu\nu\rho}(\mathcal{D}_\nu\psi_\rho) = D_\nu(\bar{\epsilon}\gamma^{\mu\nu\rho}\psi_\rho). \tag{6.23}$$

A non–zero solution to Eqs. (6.21) is called a *Killing spinor*. The given background configuration is invariant under as many supersymmetries as there are linearly independent solutions to Eqs. (6.21).

We linearize the gravitino equations of motion into the form

$$\bar{\gamma}^{\mu\nu\rho}\overline{\mathcal{D}}_\nu\psi_\rho = J^\mu, \tag{6.24}$$

by moving onto the RHS all the terms but the one explicitly written on the LHS. This equation defines what we mean by the effective supercurrent J^μ much as the linearized Einstein equations in Section 6.1 defined the effective energy–momentum tensor $\Theta^{\mu\nu}$. Let ϵ be a Killing spinor. We construct the $U(\mathcal{N})_R$–singlet vector which is the SUSY analogue to $\Theta^{\mu\nu}K_\nu^m$ of the gravitational case (cf. Section 6.1)

$$(\bar{\epsilon}J^\mu) \equiv \bar{\epsilon}\,\bar{\gamma}^{\mu\nu\rho}\overline{\mathcal{D}}_\nu\psi_\rho = \bar{D}_\nu(\bar{\epsilon}\,\bar{\gamma}^{\mu\nu\rho}\psi_\rho), \tag{6.25}$$

---

[2] For definiteness, we write the following equation in the metric $(+, -, -, \ldots, -)$. Hence the γ–matrices satisfy $\gamma_\mu^\dagger = \gamma_0\gamma_\mu\gamma_0$. Recall that the upper/lower gravitino indices correspond to positive/negative chirality.

6.3 Killing spinors

where we have used Eq. (6.23). This current is automatically conserved (Exercise 6.3.4). The supercharge is defined by

$$Q[\epsilon] = \int_t \sqrt{-g}\,(\bar{\epsilon}J^0)d^3x = \oint_\infty d\sigma_i(\bar{\epsilon}\bar{\gamma}^{0i\rho}\psi_\rho), \qquad (6.26)$$

where in the last equality we used the Stokes theorem to rewrite it as a surface integral at spatial infinity. More covariantly, we write [245, 246]

$$Q[\epsilon] = \frac{1}{2}\oint_\infty (\bar{\epsilon}\,\bar{\gamma}^{\mu\nu\rho}\psi_\rho)\,d\sigma_{\mu\nu}, \qquad (6.27)$$

where $d\sigma_{\mu\nu} = \frac{1}{2}\epsilon_{\mu\nu\rho\sigma}dx^\rho \wedge dx^\sigma$ is the area element. Note that, in order for the supercharge $Q[\epsilon]$ in Eq. (6.26) to be *fermionic*, we must choose the Killing spinor ϵ to be *commuting*.

Thus the supercharges, as the (asymptotic) isometry generators M^m, are given by flux integrals at infinity. For this reason, to define the supercharge we need only to have *asymptotic* Killing spinors. In fact, two spinors ϵ_1 and ϵ_2 that are asymptotic to the same Killing spinor give the same supercharge by the effect of the SUSY Gauss's law. The simplest way to see this is to consider the canonical current associated with a SUSY variation of parameter ϵ (here Φ stands for all fields but the gravitino)

$$\begin{aligned}Q[\epsilon] &= \int \left[\delta_\epsilon \overline{\psi}_\nu \frac{\partial \mathcal{L}}{\partial(\partial_\mu \overline{\psi}_\nu)} + \delta_\epsilon \Phi \frac{\partial \mathcal{L}}{\partial(\partial_\mu \Phi)}\right] dv^\mu \\ &= \int \left[\overline{\mathcal{D}_\nu \epsilon} \frac{\partial \mathcal{L}}{\partial(\partial_\mu \overline{\psi}_\nu)} + \delta_\epsilon \Phi \frac{\partial \mathcal{L}}{\partial(\partial_\mu \Phi)}\right] dv^\mu \qquad (6.28) \\ &= \int \mathcal{D}_\nu \left(\bar{\epsilon} \frac{\partial \mathcal{L}}{\partial(\partial_\mu \overline{\psi}_\nu)}\right) dv^\mu + \int \bar{\epsilon}\left(-\mathcal{D}_\nu \frac{\partial \mathcal{L}}{\partial(\partial_\mu \overline{\psi}_\nu)} + J^\mu\right) dv^\mu,\end{aligned}$$

where the precise form of differential operator $\mathcal{D}_\nu$ in the last line is defined by the above integration by parts and J^μ is the supercurrent. In the last line the supercharge is written as the sum of a surface term and a "bulk" term, which vanishes by the gravitino equations of motion or, more precisely, by the SUSY Gauss's law. This is required by consistency of the ψ_μ equations of motion with SUSY. The surface term gives Eq. (6.27). Then any two smooth ϵs asymptotic to the same spinor at infinity give the same supercharge (in backgrounds satisfying Gauss's constraint).

In the special case that the asymptotic Killing spinor can be continued in the "bulk" to a *bona fide* Killing spinor, the background is supersymmetric, and the associate generator $Q[\epsilon]$ is not only well defined but also identically zero, since it leaves the state unchanged. This is most easily seen by writing $Q[\epsilon]$ as a volume integral, as in the second line of Eq. (6.28).

Equation (6.21) has an integrability condition, namely

$$[\overline{\mathcal{D}}_\mu, \overline{\mathcal{D}}_\nu]\epsilon^A = 0. \tag{6.29}$$

Computing the commutator, one gets an equation of the generic form

$$\overline{R}_{\mu\nu\,ab}\,\gamma^{ab}\,\epsilon^A + \cdots = 0, \tag{6.30}$$

that is, an *algebraic* condition on the Riemann tensor, and hence on the *holonomy group* of the supersymmetric background. This leads us back to the geometric structures studied in previous chapters.

The Killing spinor equation, Eq. (6.21), simplifies for backgrounds that are *vacuum*–like: the scalar fields ϕ^i have constant values, the vector fields A^x_μ vanish, and the metric $\bar{g}_{\mu\nu}$ is (typically) maximally symmetric. Configurations which do *not* look like vacua – and hence are interpreted as some kind of *object*, typically solitonic, in some vacuum – but do have some (few) non–trivial Killing spinors, are called BPS objects.

Let us look more closely at the Killing spinor equation in a vacuum–like background. The terms containing the $U(\mathcal{N})_R$ connection, being proportional to $\partial_\mu \phi^i$, drop out. So do the terms containing the $U(\mathcal{N})_R$–covariant field–strength $\mathfrak{F}_{\rho\sigma}$. Changing notations to Majorana fermions ψ^a_μ, ϵ^a with $a = 1, \ldots, \mathcal{N}$, (to write more compact formulae), we remain with

$$\delta\psi^a_\mu \equiv \left(\delta^{ab}\partial_\mu + \frac{1}{4}\delta^{ab}\omega^{mn}_\mu \gamma_{mn} - \frac{i}{2}\left(M^{ab} + i\tilde{M}^{ab}\gamma_5\right)\gamma_\mu\right)\epsilon^b = 0, \tag{6.31}$$

$$\delta\chi^i \equiv (h^i{}_a + i\tilde{h}^i{}_a \gamma_5)\epsilon^a = 0, \tag{6.32}$$

where $M^{ab} \pm i\tilde{M}^{ab}$ and $h^i{}_a \pm i\tilde{h}^i{}_a$ are constant matrices (depending on the constant values taken by the scalars in the given background). The matrix $M^{ab} + i\tilde{M}^{ab}$ is Hermitian;[3] by a chiral redefinition of the gravitini, we can diagonalize it. In the new basis the first equation takes the form

$$D_\mu \epsilon^a + \frac{i}{2}m_a \gamma_\mu \epsilon^a = 0 \qquad \text{not summed over } a \tag{6.33}$$

The second equation, (6.32), is then an algebraic equation, requiring that $\epsilon^a_\pm$ is a zero eigenvector of the matrix $(h^i{}_a \pm i\tilde{h}^i{}_a)$.

---

[3] The fact that, in the canonical form of $\mathcal{L}$, the operator $\mathcal{D}_\mu$ in the SUSY variations is the same one which appears in the RS term implies, in particular, that the gravitino "mass" term is proportional to

$$M^{ab}\bar{\psi}^a_\mu \gamma^{\mu\nu}\psi^b_\nu + i\tilde{M}^{ab}\bar{\psi}^a_\mu \gamma^{\mu\nu}\gamma_5 \psi^b_\nu,$$

with M^{ab}, $\tilde{M}^{ab}$ the *same* matrices appearing in the SUSY transformation. The claim in the text that $M^{ab} + i\tilde{M}^{ab}$ is Hermitian then follows from the fact that the gravitino "masses" should be Hermitian.

6.3 Killing spinors

If the asymptotical background has maximal symmetry, namely $\text{Iso}(\bar{g})$ is

- the Poincaré group for $\Lambda = 0$,
- $SO(D-1, 2)$ for $\Lambda < 0$,
- $SO(D, 1)$ for $\Lambda > 0$,

the conserved supercharges, if present, should organize themselves into spinorial representations of the (asymptotic) space–time symmetry group. Hence they form $\mathcal{N}_0$ copies of the basic spinorial representation appropriate for the given D and isometry group. These conserved supercharges generate an $\mathcal{N}_0$-extended SUSY algebra that is linearly realized on the physical states of the theory. As for ordinary bosonic symmetries, we say that SUSY is *unbroken* if $\mathcal{N}_0 = \mathcal{N}$, that is, if all supersymmetries of the theory are linearly realized. In the opposite case, $\mathcal{N}_0 = 0$, SUSY is *completely broken*. The gravitini corresponding to the broken supersymmetries become massive by the SUSY variant of the Higgs effect (SUPERHIGGS): each gravitino combines with the corresponding spin–1/2 goldstino into a massive spin–3/2 particle. The intermediate possibility, $0 < \mathcal{N}_0 < \mathcal{N}$ is called PARTIAL SUPERHIGGS. $\mathcal{N}_0$ gravitini remain massless, while the other $\mathcal{N} - \mathcal{N}_0$ get masses. The possibility of a *partial* breaking is peculiar to *local* supersymmetry: in the rigid case this cannot happen.[4]

From the integrability condition of Eq. (6.31) we have

$$0 = 4\left[D_\mu + \frac{i}{2}m_a\gamma_\mu, D_\nu + \frac{i}{2}m_a\gamma_\nu\right]\epsilon_a$$
$$= \left[R_{\mu\nu\alpha\beta}\gamma^{\alpha\beta} - 2m_a^2\gamma_{\mu\nu}\right]\epsilon_a \qquad (6.34)$$
$$= -\left[\frac{\Lambda}{D-1}(g_{\mu\alpha}g_{\nu\beta} - g_{\mu\beta}g_{\nu\alpha})\gamma^{\alpha\beta} + 2m_a^2\gamma_{\mu\nu}\right]\epsilon_a,$$

where in the last line we have inserted the expression of the Riemann tensor for a *maximally symmetric D–space* (since we are assuming our asymptotic background to be vacuum–like). Equation (6.34) yields

$$R_{\mu\nu\alpha\beta} = \frac{K}{D-1}(g_{\mu\alpha}g_{\nu\beta} - g_{\mu\beta}g_{\nu\alpha}), \qquad (6.35)$$

where

$$R_{\mu\nu} = Kg_{\mu\nu} \Rightarrow K = -\Lambda. \qquad (6.36)$$

---

[4] In the rigid case, SUSY is broken if and only if the energy of the vacuum is not zero. If the vacuum energy vanishes, *all* supercharges are not broken; if it is non–zero *all* are broken. However, there are situations, in which the rigid SUSY algebra is deformed by brane charges, which effectively mimic a partial breaking of extended supersymmetry. See ref. [132].

Therefore, $\epsilon_a \neq 0$ implies

$$m_a^2 = -\frac{\Lambda}{D-1}. \qquad (6.37)$$

Hence, in a unitary theory, we must have $\Lambda \leq 0$, that is either (asymptotic) *anti–de Sitter* or *flat* space, and the m_as equal (up to a phase) to $\sqrt{-\Lambda/(D-1)}$ for all gravitini $a = 1, 2, \ldots, \mathcal{N}_0$ corresponding to *unbroken* supersymmetries. If we give up unitarity, m_a may be taken as imaginary, and $\Lambda > 0$ becomes allowed.

6.3.2 The mathematicians' definition

The mathematicians take the special case in Eq. (6.31) as their definition of Killing spinors. To make the dictionary conform with the maths literature we must remember that they usually work in Euclidean signature, and that their gamma–matrices are *anti*–Hermitian.

Definition 6.1 A *Killing spinor* is a non–zero spinor such that

$$D_i \epsilon + \frac{m}{2} \gamma_i \epsilon = 0, \qquad (6.38)$$

for some $m \in \mathbb{C}$. One says that the Killing spinor is *real* (resp. *imaginary*) if $m \in \mathbb{R}^\times$ (resp. $im \in \mathbb{R}^\times$) and *parallel* if $m = 0$.

With this definition, the concept of Killing spinor is purely geometric, depending only on the Riemannian geometry of Σ. As already mentioned, the geometers are mostly interested in Killing spinors for manifolds having Euclidean signature, whereas we, in physics, are interested in a variety of space–time signatures, depending on the particular application we are pursuing. For instance, the physical space–time, Σ, may be a generic manifold of signature $(D-1, 1)$, and we look for the Killing spinors on this Minkowskian space. Or we may have a Kaluza–Klein set-up, where the space–time has the form $\mathbb{R}^{d-1,1} \times \mathcal{K}$, with $\mathcal{K}$ Euclidean and compact. In this case, we look for Killing spinors of the compact space $\mathcal{K}$, which has positive signature.

However, conceptually, the main difference between the two definitions (for maximally symmetric configurations) is that in physics it is fundamental that also the variation of the spin–1/2 fields, $\delta \chi^I$, vanishes. This requirement is not geometric, at least *a priori*.

6.3.3 First properties of the Killing spinors

A Killing spinor on Σ is automatically a solution to the Dirac equation

$$\left(\gamma^i D_i - \frac{n}{2} m\right)\epsilon = 0, \qquad n \equiv \dim \Sigma. \tag{6.39}$$

Relation with the Killing vectors: The Euclidean case

Let α, β be two *commuting* real Killing spinors (according to the mathematical definition). We wish to show that the vector–bilinear

$$K_\mu(\alpha, \beta) = \alpha^\dagger \gamma_\mu \beta, \tag{6.40}$$

if non-vanishing, is a Killing vector. Indeed, one has

$$D_\mu(\alpha^\dagger \gamma_\nu \beta) = -\frac{m}{2}\left(\alpha^\dagger \gamma_\mu^\dagger \gamma_\nu \beta + \alpha^\dagger \gamma_\nu \gamma_\mu \beta\right) = m(\alpha^\dagger \gamma_{\mu\nu}\beta) = -D_\nu(\alpha^\dagger \gamma_\mu \beta), \tag{6.41}$$

which is the Killing vector condition. Notice that the *pseudo*vector $\alpha^\dagger \gamma_\mu \gamma_5 \beta$ leads to the equation

$$D_\mu(\alpha^\dagger \gamma_\nu \gamma_5 \beta) + D_\nu(\alpha^\dagger \gamma_\mu \gamma_5 \beta) = -\frac{m}{2}\alpha^\dagger(\gamma_\mu \gamma_\nu + \gamma_\nu \gamma_\mu)\gamma_5 \beta = g_{\mu\nu}(m\alpha^\dagger \gamma_5 \beta), \tag{6.42}$$

which is the equation for a *conformal* Killing vector (cf. Section 6.4.3).

Minkowski signature

The above results hold for Euclidean spaces. In Minkowski signature with metric $(+, -, \ldots, -)$ we have an i in the differential operator $D_\mu + \frac{i}{2}m\gamma_\mu$, and the bilinear reads

$$K_\mu(\alpha, \beta) = \bar{\alpha}\gamma_\mu \beta \equiv \alpha^\dagger \gamma_0 \gamma_\mu \beta. \tag{6.43}$$

Now

$$D_\mu(\bar{\alpha}\gamma_\nu \beta) = -i\frac{m}{2}\alpha^\dagger(\gamma_\mu^\dagger \gamma_0 \gamma_\nu - \gamma_0 \gamma_\nu \gamma_\mu)\beta = im\bar{\alpha}\gamma_{\nu\mu}\beta = -D_\nu(\bar{\alpha}\gamma_\mu \beta), \tag{6.44}$$

which, again, is the Killing vector condition.

6.3.4 Killing spinors on AdS_D and S^m

We are interested, in particular, in the Killing spinors for the AdS_D space. AdS_D is an analytic continuation of the sphere S^D. We solve first the Killing spinor equation on the sphere S^D, and then continue the solution to AdS_D.

Killing spinors for S^{2n}

We work out the details for $D = 2n$, the odd–dimensional case being essentially similar and left to the reader.

Write S^{2n} as $Spin(2n+1)/Spin(2n)$ and identify $Spin(2n+1)$ with the matrix group given by the (unique) irreducible spinorial representation. The generators of $Spin(2n+1)$ are then[5] $\frac{1}{2}\gamma_{ab}$ and $\frac{1}{2}\gamma_a$, with $a,b = 1,2,\ldots,2n$.

Let $\mathcal{E} \in Spin(2n+1)$ be a coset representative. From Chapter 5 we know that the Maurier–Cartan form decomposes as

$$\mathcal{E}^{-1} d\mathcal{E} = \frac{1}{4}\omega^{ab}\gamma_{ab} + \frac{1}{2}e^a \gamma_a, \qquad (6.45)$$

where the 1–forms ω^{ab} and e^a are, respectively, the connection and the metric vielbein of S^{2n}. The covariant derivative $\mathcal{D}_\mu = D_\mu + \frac{1}{2}\gamma_\mu$ is simply

$$\mathcal{D} := dx^\mu \mathcal{D}_\mu = d + \frac{1}{4}\omega^{ab}\gamma_{ab} + \frac{1}{2}dx^\mu \gamma_\mu = d + \mathcal{E}^{-1} d\mathcal{E}, \qquad (6.46)$$

so the equation defining the Killing spinors, $\mathcal{D}\epsilon = 0$, has the general solution

$$\epsilon = \mathcal{E}^{-1}\epsilon_0, \qquad (6.47)$$

with ϵ_0 an arbitrary *constant* spinor. Hence the number of linearly independent Killing spinors on the sphere S^{2n} is equal to the dimension of the Dirac spinor in $D = 2n+1$, namely 2^n. There are also 2^n solutions to the equation with the opposite sign, $(D_\mu - \frac{1}{2}\gamma_\mu)\epsilon = 0$, given by $\gamma_5 \mathcal{E}^{-1}\epsilon_0$.

From the above Killing spinors we can construct the Killing vectors of $\text{Iso}(S^{2n})$. Equation (6.40) gives the general formula

$$\epsilon^\dagger \gamma_\mu \epsilon = \epsilon_0^\dagger (\mathcal{E}^{-1})^\dagger \gamma_\mu \mathcal{E}^{-1} \epsilon = -2\epsilon_0^\dagger (\mathcal{E} D_\mu \mathcal{E}^{-1})\epsilon_0$$

since, for the compact group $Spin(2n+1)$, the spinorial representation is unitary. Inserting Eq. (5.74) of Chapter 5 into the RHS, we get

$$\epsilon^\dagger \gamma_\mu \epsilon = \frac{1}{2^{n+1}} \epsilon_0^\dagger \gamma^{AB} \epsilon_0 \, K_{\mu\, AB}, \qquad A, B = 1, 2\ldots, 2n+1, \qquad (6.48)$$

where

$$\gamma^{AB} = -\gamma^{BA} := \begin{cases} \gamma^{AB} & A, B = 1, 2, \ldots, 2n, \\ -\gamma^A & B = 2n+1. \end{cases} \qquad (6.49)$$

and $K_{\mu\, AB}$ are the Killing vectors of the $Spin(2n+1)$ isometry of S^{2n}.

---

[5] As always when we work in Euclidean signature, we take the γ_a to be anti–Hermitian.

6.3 Killing spinors

The AdS$_D$ case

From Eq. (6.19) we have

$$\mathcal{E}^{-1}d\mathcal{E} = \frac{1}{4}\omega^{ab}\gamma_{ab} + \frac{1}{2}e^a \gamma_0\gamma_a, \quad (6.50)$$

where the conventions are such that γ_0, γ_D are Hermitian and the γ_i, $i = 1, 2, \ldots, D-1$ anti–Hermitian. The linear map[6]

$$\varpi : \mathbb{C}l(1, D-1) \to \mathbb{C}l^0(2, D-1) \quad (6.51)$$

given by $\gamma_a \mapsto -i\gamma_0\gamma_a$ is an isomorphism since

$$(-i\gamma_0\gamma_a)(-i\gamma_0\gamma_b) + (-i\gamma_0\gamma_b)(-i\gamma_0\gamma_a) = \gamma_a\gamma_b + \gamma_b\gamma_a = 2\eta_{ab}. \quad (6.52)$$

One has $\varpi(\gamma_{ab}) = (-i\gamma_0\gamma_{[a})(-i\gamma_0\gamma_{b]}) = \gamma_{ab}$. Then

$$\varpi(d + \mathcal{E}^{-1}d\mathcal{E}) = \mathcal{D} := d + \frac{1}{4}\omega^{ab}\gamma_{ab} + \frac{i}{2}e^a\gamma_a. \quad (6.53)$$

An AdS$_D$ Killing spinor is a (non–zero) solution to $\mathcal{D}\epsilon = 0$. The general solution is

$$\epsilon = \varpi^{-1}(\mathcal{E}^{-1})\epsilon_0 \quad (6.54)$$

with ϵ_0 any constant spinor.

The Killing vectors are given by the bilinears $\bar{\epsilon}\gamma_\mu\epsilon' \equiv \epsilon^\dagger \gamma_D\gamma_\mu\epsilon'$,

$$\begin{aligned}\epsilon^\dagger \gamma_D\gamma_\mu\epsilon' &= \epsilon_0^\dagger \varpi^{-1}\left((\mathcal{E}^{-1})^\dagger\right)\gamma_D\gamma_\mu\, \varpi^{-1}(\mathcal{E}^{-1})\epsilon_0' \\ &= -\epsilon_0^\dagger \varpi^{-1}\left((\mathcal{E}^{-1})^\dagger \gamma_0\gamma_D\gamma_0\gamma_\mu\mathcal{E}^{-1}\right)\epsilon_0' \\ &= -\epsilon_0^\dagger \varpi^{-1}\left(\gamma_0\gamma_D(\mathcal{E}\gamma_0\gamma_\mu\mathcal{E}^{-1})\right)\epsilon_0' \\ &= 2\epsilon_0^\dagger \varpi^{-1}\left(\gamma_0\gamma_D(\mathcal{E}D_\mu\mathcal{E}^{-1})\right)\epsilon_0' \\ &= 2\alpha\, \epsilon_0^\dagger \varpi^{-1}(\gamma_0\gamma_D\gamma_{AB})\epsilon_0'\, K_\mu^{AB} \\ &= 2\alpha\, \bar{\epsilon}_0\gamma_{AB}\epsilon_0'\, K_\mu^{AB},\end{aligned} \quad (6.55)$$

where we have used the explicit form of the $SO(2, D-1)$ Killing vectors, Eq. (6.54) and the identity

$$\gamma_D\gamma_0\mathcal{E}^{-1}\gamma_0\gamma_D = \mathcal{E}^\dagger. \quad (6.56)$$

---

[6] $\mathbb{C}l(p, q)$ denotes the Clifford algebra in signature (p, q); $\mathbb{C}l^0(p, q)$ its even subalgebra. The isomorphism ϖ is the generalization to signature (p, q) of the one described in Theorem 2.6.

α is a normalization constant. We take $\alpha = 1$ as a choice of normalization of the $SO(2, D-1)$ Killing vectors. Then we have the crucial identity

$$\bar{\epsilon}\gamma_\mu \epsilon' = (\bar{\epsilon}_0 \gamma_{AB} \epsilon'_0) K_\mu^{AB}. \tag{6.57}$$

de Sitter: no Killing spinors

In the de Sitter case, we have an extra i around, and the Killing condition is not consistent with ϵ having the natural reality condition. For example, in $D = 4$, ϵ cannot be both Killing and Majorana. The simplest way to see this is to note that $Spin(4, 1)$ has no Majorana representation. Hence de Sitter SUSY is inconsistent with unitarity. If we give up unitarity, SUSY may be realized even in de Sitter space.

Exercise 6.3.1 Show that the supercharge (6.26) is conserved.

Exercise 6.3.2 Work out the details of the Killing spinors for S^{2n+1}.

Exercise 6.3.3 Show that in the Lorentzian signature the existence of a Killing spinor implies the existence of a time/light–like Killing vector.

6.4 The geometry of Killing spinors

We recall that, mathematically, a Killing spinor ψ on a Riemannian (spin) manifold $\mathcal{M}$ is a solution to the equation[7]

$$(D_i - \alpha \gamma_i)\psi = 0, \tag{6.58}$$

with $\alpha \in \mathbb{C}$. If $\alpha = 0$, a Killing spinor is the same as a parallel spinor. In Theorem 3.23 we already solved the problem of characterizing the manifolds $\mathcal{M}$ that have such spinors; we summarize the result in Table 6.1.

To get the $\alpha \neq 0$ case, we begin by generalizing Lemma 3.24.

Proposition 6.2 *Assume that on the Riemannian n–fold $\mathcal{M}$ there is a non–vanishing Killing spinor ψ satisfying $(D_i - \alpha \gamma_i)\psi = 0$. Then $\mathcal{M}$ is Einstein with $R_{ij} = 4(n-1)\alpha^2 g_{ij}$.*

---

[7] When doing geometry, we adhere to the mathematicians' convention $\gamma_i \gamma_j + \gamma_j \gamma_i = -2\delta_{ij}$, so the γ matrices are anti–Hermitian.

Table 6.1 *Riemannian manifolds with (N_+, N_-) parallel spinors of chirality ± 1*

Manifold	Holonomy	Dimension	(N_+, N_-)
Flat	1	$2n$	$(2^{n-1}, 2^{n-1})$
Flat	1	$2n+1$	2^n
Calabi–Yau	$SU(2m)$	$4m$	$(2, 0)$
Calabi–Yau	$SU(2m+1)$	$4m+2$	$(1, 1)$
hyperKähler	$Sp(2m)$	$4m$	$(m+1, 0)$
G_2-manifold	G_2	7	1
$Spin(7)$-manifold	$Spin(7)$	8	$(0, 1)$

Proof Consider the integrability condition[8]

$$0 = 4[D_i - \alpha \gamma_i, D_j - \alpha \gamma_j]\psi = \left(-R_{ijkl}\gamma^{kl} + 8\alpha^2 \gamma_{ij}\right)\psi \tag{6.59}$$

and contract it with γ^j. Using the identities

$$\gamma^j \gamma^{kl} = \gamma^{jkl} - \delta^{jk}\gamma^l + \delta^{jl}\gamma^k, \qquad \gamma^j \gamma_{ij} = (n-1)\gamma_i \tag{6.60}$$

we get

$$\left[-R_{ijkl}\gamma^{jkl} - 2\left(R_{ij} - 4(n-1)\alpha^2 g_{ij}\right)\gamma^j\right]\psi = 0. \tag{6.61}$$

The first term in the bracket vanishes by the first Bianchi identity. So

$$\left(R_{ij} - 4(n-1)\alpha^2 g_{ij}\right)\gamma^j \psi = 0. \tag{6.62}$$

Let A_{ij} be a symmetric tensor. One has

$$(A_{ij}\gamma^j)(A_{ik}\gamma^k) = -(A_{ij}A_{ij})\mathbf{1} = -\text{tr}(A^\dagger A)\mathbf{1}, \tag{6.63}$$

so $A_{ij}\gamma^j\psi = 0$ and $\psi \neq 0$ implies $A_{ij} \equiv 0$. Apply this remark to Eq. (6.62). □

For the case of $\mathcal{M}$ an n–fold of Lorentz signature, see Exercise 6.4.3.

Corollary 6.3 *Killing spinors may exist only for α real (real Killing spinors) or α purely imaginary (imaginary Killing spinors).*

We are expecially interested in the real ones.

---

[8] The sign $-$ in front of the curvature arises because, in the present conventions, the generators of $Spin(n)$ are $-\frac{1}{4}[\gamma_i, \gamma_j]$.

6.4.1 Real Killing spinors

The situation for $\alpha = 0$ is presented in Table 6.1. It remains to discuss the case $\alpha \neq 0$. In fact, we can reduce the general case to the one we have already solved. Let us start with a definition:[9]

Definition 6.4 Let $\mathcal{M}$ be a Riemann n–fold with metric $g_{\alpha\beta}(y)dy^\alpha dy^\beta$. The metric (or Riemannian) cone over $\mathcal{M}$, denoted $\mathcal{C}_\lambda(\mathcal{M})$, is the Riemannian $(n+1)$–fold $\mathbb{R}_+ \times \mathcal{M}$ endowed with the metric

$$ds^2 = dr^2 + \lambda^2 r^2 \, g_{\alpha\beta}(y)dy^\alpha dy^\beta, \tag{6.64}$$

where $\lambda \in \mathbb{R}^\times$.

Theorem 6.5 (Bär [33]) $\epsilon(y)$ *is a real Killing spinor on* $\mathcal{M}$ *for some* $\alpha \in \mathbb{R}$ $\iff$ $\epsilon(y)$ *is a parallel spinor on the cone* $\mathcal{C}_{2\alpha}(\mathcal{M})$ *(for suitable choice of spin connection).*

Proof Let e^a and ω^{ab} be the vielbein and connection 1–forms for the manifold $\mathcal{M}$, satisfying the torsionless constraint $de^a + \omega^{ab} \wedge e^b = 0$. Let E^m, Ω^{mn} be the corresponding forms for the cone $\mathcal{C}_\lambda(\mathcal{M})$. We can take

$$E^0 = dr, \qquad E^a = \lambda r e^a, \tag{6.65}$$

$$\Omega^{a0} = -\Omega^{0a} = \lambda e^a, \qquad \Omega^{ab} = \omega^{ab}. \tag{6.66}$$

One checks that $dE^m + \Omega^{mn} \wedge E^n = 0$. Let

$$\nabla_\mu = \partial_\mu - \frac{1}{4} \Omega^{mn}_\mu \gamma_{mn} \tag{6.67}$$

be the spin connection on the cone $\mathcal{C}_\lambda(\mathcal{M})$. Using the connection (6.66):

$$\nabla_0 = \partial_r. \tag{6.68}$$

$$\nabla_\alpha = \partial_\alpha - \frac{1}{4} \omega^{ab}_\alpha \gamma_{ab} - \frac{\lambda}{2} e^a_\alpha \gamma_a \gamma_0. \tag{6.69}$$

As in Theorem 2.6, there is an algebra isomorphism

$$\varpi : \mathbb{C}l^0(n+1) \to \mathbb{C}l(n) \tag{6.70}$$

given by

$$\varpi(\gamma_{ab}) \mapsto \gamma_{ab}, \qquad \varpi(\gamma_a \gamma_0) \mapsto \gamma_a. \tag{6.71}$$

---

[9] With respect to the standard definition, we introduce a free parameter, λ, for later convenience.

Table 6.2 *Manifolds having (N_+, N_-) real Killing spinors and their Riemannian cones $\mathcal{C}(\mathcal{M})$*

Manifold $\mathcal{M}$	dim $\mathcal{M}$	$\mathcal{C}(\mathcal{M})$	$\mathrm{Hol}_0(\mathcal{C}(\mathcal{M}))$	(N_+, N_-)
General Riemannian	n	Riemannian cone	$SO(n+1)$	$(0,0)$
Round sphere S^n	n	Flat space $\mathbb{R}^{n+1}$	Id	$(2^{[n/2]}, 2^{[n/2]})$
Sasaki	$2n-1$	Kähler	$U(n)$	$(0,0)$
Sasaki–Einstein	$4m+1$	Calabi–Yau	$SU(2m+1)$	$(1,1)$
Sasaki–Einstein	$4m-1$	Calabi–Yau	$SU(2m)$	$(2,0)$
3-Sasaki	$4m-1$	hyperKähler	$Sp(2m)$	$(m+1, 0)$
Strict nearly–Kähler 6–manifold	6	G_2–manifold	G_2	$(1,1)$
Proper G_2–metric	7	$Spin(7)$–fold	$Spin(7)$	$(1,0)$

Then

$$\varpi(\nabla_\alpha) = D_\alpha - \frac{\lambda}{2}\gamma_\alpha. \tag{6.72}$$

From Eqs. (6.68) and (6.69) we see that a parallel spinor on the cone is a spinor $\psi(y)$, independent of the coordinate r, which, after the change of the realization of the γ–matrices given by the isomorphism ϖ, is a solution on $\mathcal{M}$ to the equation

$$\left(D_\alpha - \frac{\lambda}{2}\gamma_\alpha\right)\psi = 0, \tag{6.73}$$

that is, to the real Killing spinor equation for $\alpha = 2\lambda$. □

6.4.2 Cones: Sasaki manifolds and all that

Riemannian spaces $\mathcal{M}$ such that their metric cones $\mathcal{C}(\mathcal{M})$ have special holonomy groups have got names (and deep theories) in the math literature. Here we just mention the topic. A complete reference is [62]. Definitions and results are summarized in Table 6.2.

To complete the discussion, we state the following (cf. Exercise 6.4.3):

Proposition 6.6 *A (simply connected) Riemannian cone is a product of irreducible Riemannian cones. A* SYMMETRIC *Riemannian cone is flat.*

This proposition allows us to reduce the analysis to Riemannian cones whose holonomy is in the Berger list. Then we get the results in Table 6.2.

Kähler and Quaternionic–Kähler cones

From Table 6.2 we see that if $\mathcal{M}$ is strictly Sasaki, so that $\mathcal{C}(\mathcal{M})$ is Kähler, there is no Killing spinor according to the above definition. The obstruction is the $U(1)$ part of the holonomy of $\mathcal{C}(\mathcal{M})$. Then we can find on $\mathcal{M}$ a solution to the equation

$$(\nabla_\mu + A_\mu)\psi = 0, \tag{6.74}$$

where A_μ is a background $U(1)$ gauge field with holonomy equal to *minus* the $U(1)$ part of the spin connection. Likewise, if $\mathcal{C}(\mathcal{M})$ is Quaternionic–Kähler, we may introduce a background $SU(2)$ connection that cancels the corresponding part of the spin connection. These solutions ψ are Killing spinors (in the physical sense) whenever the operator $\mathcal{D}_\mu$ of the corresponding SUGRA model contains a $U(1)$ (resp. $SU(2)$) connection which may be tuned to minus the part of the spin connection that obstructs the existence of a math Killing spinor.

More generally, we may enlarge the space of manifolds admitting Killing spinors (that is, rigid supersymmetry) by replacing ∇_μ with a torsionful connection (see Part III).

6.4.3 Conformal Killing spinors

A *conformal Killing spinor* (CKS) on a (pseudo)Riemannian manifold $\mathcal{M}$ is a pair of spinors (ϵ, η) satisfying the equation

$$D_i \epsilon = \gamma_i \eta. \tag{6.75}$$

The Killing spinor in the sense of Eq. (6.38) is the special case $\eta = m\epsilon$. Contracting (6.75) with γ^i we get $\eta = -\frac{1}{n}\slashed{D}\epsilon$ and we may rewrite the above equation as

$$D_i \epsilon + \frac{1}{n}\gamma_i \slashed{D}\epsilon = 0. \tag{6.76}$$

The operator $P_i \equiv D_i + \frac{1}{n}\gamma_i \slashed{D}$, acting on spinors, is called the twistor (or Penrose) operator. If ϵ_1, ϵ_2 are two solutions of Eq. (6.76), then the vector $V_i \equiv \epsilon_1^\dagger \gamma_i \epsilon_2$ is a conformal Killing vector, i.e., it satisfies the equation

$$\mathcal{L}_V g_{ij} \equiv D_i V_j + D_j V_i = 2\Omega\, g_{ij}, \tag{6.77}$$

for some function Ω. Indeed,

$$D_i(\epsilon_1^\dagger \gamma_j \epsilon_2) = \frac{1}{n}(\slashed{D}\epsilon_1)^\dagger \gamma_i \gamma_j \epsilon_2 - \frac{1}{n}\epsilon_1^\dagger \gamma_j \gamma_i \slashed{D}\epsilon_2, \tag{6.78}$$

and Eq. (6.77) is satisfied with

$$\Omega = -\frac{1}{n}\left((\slashed{D}\epsilon_1)^\dagger \epsilon_2 - \epsilon_1^\dagger \slashed{D}\epsilon_2\right). \tag{6.79}$$

By definition, a conformal Killing vector generates a diffeomorphism which, while not being an isometry (in general), maps the metric into one of the same conformal class. Therefore, the existence of conformal Killing vectors is a property of the *conformal class* not of the single manifold. The same should be true for the conformal Killing spinors. Indeed:

Proposition 6.7 *Let* (ϵ, η) *be a CKS on a Riemannian n-fold* (M, g).

$$(\tilde{\epsilon}, \tilde{\eta}) \equiv \left(e^{\phi/2}\epsilon, \ e^{\phi/2}\left(\eta - \frac{1}{2}\gamma^i \partial_i \phi \, \epsilon\right)\right) \tag{6.80}$$

is a CKS for the conformally equivalent metric $\tilde{g} = e^{2\phi} g$.

This result follows from the identity

$$\frac{1}{4}\gamma_{ab}\tilde{\omega}_i^{ab} = \frac{1}{4}\gamma_{ab}\omega_i^{ab} - \frac{1}{2}\gamma_{ij}\partial^j\phi, \tag{6.81}$$

relating the spin connections $\tilde{\omega}_i^{ab}$ and ω_i^{ab} corresponding, respectively, to the metrics $\tilde{g}$ and g.

Taking the derivative of Eq. (6.76) we get $D_i D^i \epsilon + \frac{1}{n}\slashed{D}^2 \epsilon = 0$. Using the Lichnerowicz formula [51],

$$\slashed{D}^2 = -D^i D_i + \frac{1}{4}R, \tag{6.82}$$

yields

$$\slashed{D}^2 \epsilon = \frac{nR}{4(n-1)}\epsilon. \tag{6.83}$$

Using this identity one shows the identity (Exercise 6.4.3) [38]:

$$\left(\frac{n-2}{n}\right)D_i\slashed{D}\epsilon = -\frac{1}{2}R_{ij}\gamma^j\epsilon + \frac{R}{4(n-1)}\gamma_i\epsilon. \tag{6.84}$$

It is convenient to define the *Rho tensor*

$$K_{ij} := \frac{1}{n-2}\left(\frac{R g_{ij}}{2(n-1)} - R_{ij}\right) \tag{6.85}$$

and the *Cotton tensor*

$$C_{kij} = D_i K_{jk} - D_j K_{ik}. \tag{6.86}$$

Then equation (6.84) reads

$$\epsilon \text{ a CKS} \implies D_i \not{D}\epsilon = \frac{n}{2} K_{ij} \gamma^j \epsilon. \tag{6.87}$$

Local integrability conditions. The CKS equations (6.76) imply some obvious integrability conditions. Following ref. [38], we introduce a new covariant derivative acting on the doubled spinor bundle $\mathcal{S} \otimes \mathbb{C}^2$,

$$\mathcal{D}_i = \begin{pmatrix} D_i & \frac{1}{n}\gamma_i \\ -\frac{n}{2} K_{ij} \gamma^j & D_i \end{pmatrix}, \tag{6.88}$$

whose curvature 2–form is

$$\mathcal{R}_{kl} = \begin{pmatrix} \frac{1}{4} W_{klij} \gamma^{ij} & 0 \\ -\frac{n}{2} C_{ikl} \gamma^i & \frac{1}{4} W_{klij} \gamma^{ij} \end{pmatrix}, \tag{6.89}$$

where W_{klij} is the Weyl tensor

$$\begin{aligned} W_{ijkl} &:= R_{ijkl} - \frac{1}{n-2}(R_{jk}g_{il} - R_{jl}g_{ik} + g_{jk}R_{il} - g_{jl}R_{ik}) \\ &+ \frac{R}{(n-1)(n-2)}(g_{jk}g_{il} - g_{jl}g_{ik}). \end{aligned} \tag{6.90}$$

Equations (6.76) and (6.87) give the following lemma:

Lemma 6.8 *If ϵ is a CKS then $\mathcal{D}_i \begin{pmatrix} \epsilon \\ \not{D}\epsilon \end{pmatrix} = 0$. Conversely, if $\mathcal{D}_i \begin{pmatrix} \psi \\ \chi \end{pmatrix} = 0$, ψ is a CKS and $\chi = \not{D}\psi$.*

Therefore, the local integrability condition of the CKS equation (6.76) takes the standard form of an algebraic condition on the curvature:

$$\mathcal{R}_{kl} \begin{pmatrix} \epsilon \\ \not{D}\epsilon \end{pmatrix} = 0. \tag{6.91}$$

We recall that in dimension $n \geq 4$ a Riemannian manifold $\mathcal{M}$ is conformally flat iff the Weyl tensor vanishes, $W_{ijkl} = 0$. $n = 3$ is special, since in this case $W_{ijkl} \equiv 0$, and a Riemannian 3–fold is conformally flat iff the Cotton tensor $C_{ijk} \equiv 0$. Then:

Corollary 6.9 *The number of linearly independent CKS is $< d_n := 2^{[n/2]+1}$ and if a Riemannian n–fold has d_n linearly independent CKS, then it is conformally flat. Conversely, if M is conformally flat and simply connected it has exactly d_n linearly independent CKSs.*

6.5 Nester form of the space–time charges

Further details on the geometry of CKS may be found in ref. [38]. For the Lorentzian case, see the survey in ref. [34].

Exercise 6.4.1 State the version of Lemma 6.2 for Lorentzian signature.

Exercise 6.4.2 Prove Proposition 6.6.

Exercise 6.4.3 Show that a Riemannian n-fold with a conformal Killing vector V such that $g_{ij} = D_i V_j$, is isometric to a cone and $V \equiv r\frac{\partial}{\partial r}$.

Exercise 6.4.1 Prove identity (6.84).

Exercise 6.4.2 Prove Eq. (6.89).

Exercise 6.4.3 Find the CKSs for the warped product $\mathbb{R} \times_f \mathbb{R}^n$, i.e., the space with metric $ds^2 = d\tau^2 + f(\tau)^2 \sum_{i=1}^n dx_i^2$.

Exercise 6.4.4 Find the CKSs for the warped product $\mathbb{R} \times_f S^n$, i.e., the space with metric $ds^2 = dr^2 + f(r)^2 d\Omega^2$.

6.5 Nester form of the space–time charges

The space–time charges M^m associated with asymptotic isometries can be written as surface integrals at infinity in various ways (see ref. [1]). Here we introduce a particular form, due to Nester [245, 246], that is quite elegant and practical and also well suited for applications to SUGRA.

We start from scratch.[10] We specialize in the case of a geometry that is asymptotic anti–de Sitter (AAdS) with cosmological constant Λ. The asymptotically flat case (AF) is recovered by taking $\Lambda = 0$ in our formulae. For simplicity, we work in $D = 4$.

We recall two identities:

$$\gamma_\sigma \gamma_{\mu\nu} + \gamma_{\mu\nu} \gamma_\sigma = 2\gamma_{\sigma\mu\nu} = 2\epsilon_{\sigma\mu\nu\tau} \gamma^\tau \gamma_5, \tag{6.92}$$

---

[10] In this section we use Minkowski signature with metric $(+, -, \ldots, -)$.

and

$$-\frac{1}{4}\delta^\alpha_\mu \epsilon^{\mu\nu\rho\sigma} \epsilon_{\alpha\beta\gamma\delta} R_{\rho\sigma}{}^{\gamma\delta} = -\frac{1}{4}\delta^{\nu\rho\sigma}_{\beta\gamma\delta} R_{\rho\sigma}{}^{\gamma\delta}$$

$$\equiv -\frac{1}{4}[\delta^\nu_\beta \delta^\rho_\gamma \delta^\sigma_\delta + \delta^\rho_\beta \delta^\sigma_\gamma \delta^\nu_\delta + \delta^\sigma_\beta \delta^\nu_\gamma \delta^\rho_\delta - \delta^\nu_\beta \delta^\sigma_\gamma \delta^\rho_\delta - \delta^\rho_\beta \delta^\nu_\gamma \delta^\sigma_\delta - \delta^\sigma_\beta \delta^\rho_\gamma \delta^\nu_\delta] R_{\rho\sigma}{}^{\gamma\delta}$$

$$= R_\beta{}^\nu - \frac{1}{2}\delta_\beta^\nu R.$$

(6.93)

Consider the commutator (6.59), i.e.,

$$4\left[D_\mu + \frac{i}{2}m\gamma_\mu, D_\nu + \frac{i}{2}m\gamma_\nu\right] \equiv (R_{\mu\nu}{}^{\alpha\beta}\gamma_{\alpha\beta} - 2m^2 \gamma_{\mu\nu}),$$

(6.94)

where $m = \sqrt{-\Lambda/3}$. In view of the above identities, one has

$$\gamma_\sigma (R_{\mu\nu}{}^{\alpha\beta}\gamma_{\alpha\beta} - 2m^2 \gamma_{\mu\nu}) + (R_{\mu\nu}{}^{\alpha\beta}\gamma_{\alpha\beta} - 2m^2 \gamma_{\mu\nu})\gamma_\sigma$$
$$= 2(\epsilon_{\sigma\tau\alpha\beta} R_{\mu\nu}{}^{\alpha\beta} - 2m^2 \epsilon_{\mu\nu\sigma\tau})\gamma^\tau \gamma^5.$$

(6.95)

Multiply this expression by $\epsilon^{\sigma\rho\mu\nu}$ and use the identity Eq. (6.93) to get

$$2\epsilon^{\sigma\rho\mu\nu}\left(\gamma_\sigma\left[D_\mu + \frac{i}{2}m\gamma_\mu, D_\nu + \frac{i}{2}m\gamma_\nu\right] + \left[D_\mu + \frac{i}{2}m\gamma_\mu, D_\nu + \frac{i}{2}m\gamma_\nu\right]\gamma_\sigma\right)$$

$$= \left[-4\left(R^{\rho\tau} - \frac{1}{2}g^{\rho\tau} R\right) - 2\cdot 3 m^2 g^{\rho\tau}\right]\gamma_\tau \gamma_5$$

$$= -4\left(R^{\rho\tau} - \frac{1}{2}g^{\rho\tau} R - \Lambda g^{\rho\tau}\right)\gamma_\tau \gamma_5.$$

(6.96)

Now we write

$$\mathcal{D}_\mu := D_\mu + \frac{i}{2}m\gamma_\mu = \overline{\mathcal{D}}_\mu + \Omega_\mu + O(h^2),$$

(6.97)

where an overbar means the value on the AdS background and Ω_μ is the linear part in $h_{\mu\nu} = g_{\mu\nu} - \bar{g}_{\mu\nu}$. We assume Ω_μ to be of order $O(1/r^2)$, in order to have the right asymptotics at spatial infinity. The omitted terms are of order $O(1/r^3)$. We have

$$\Theta^{\rho\lambda}\gamma_\lambda = \left(R^{\rho\lambda} - \frac{1}{2}g^{\rho\lambda} R - \Lambda g^{\rho\lambda}\right)^{\text{Lin.}} \gamma_\lambda$$

$$= \epsilon^{\rho\sigma\mu\nu}(\gamma_\sigma \overline{\mathcal{D}}_\mu \Omega_\nu + \overline{\mathcal{D}}_\mu \Omega_\nu \gamma_\sigma)\gamma_5 = (\gamma^{\rho\mu\nu}\overline{\mathcal{D}}_\mu \Omega_\nu + \overline{\mathcal{D}}_\mu \Omega_\nu \gamma^{\rho\mu\nu})$$

$$= \overline{\mathcal{D}}_\mu(\gamma^{\rho\mu\nu}\Omega_\nu + \Omega_\nu \gamma^{\rho\mu\nu}).$$

(6.98)

Now let α, β be *commuting* spinors satisfying the equation

$$\overline{\mathcal{D}}_\mu \alpha = 0, \quad \Rightarrow \quad \Omega_\mu \alpha = \mathcal{D}_\mu \alpha + O(1/r^3), \qquad (6.99)$$

that is, α, β are Killing spinors for the Anti–de Sitter metric $\overline{g}_{\mu\nu}$. One has[11]

$$\begin{aligned}\Theta^{\rho\lambda}\,\bar\alpha \gamma_\lambda \beta &= \overline{\mathcal{D}}_\mu \big(\bar\alpha(\gamma^{\rho\mu\nu}\Omega_\nu + \Omega_\nu \gamma^{\rho\mu\nu})\beta\big) \\ &= \overline{\mathcal{D}}_\mu \big(\bar\alpha \gamma^{\rho\mu\nu}\mathcal{D}_\nu \beta - \overline{(\mathcal{D}_\nu \alpha)} \gamma^{\rho\mu\nu}\beta\big).\end{aligned} \qquad (6.100)$$

The LHS is the correct integrand for the space–time charge associated with the AdS Killing vector $\bar\alpha \gamma_\mu \beta$. On the other hand, as we saw in Section 6.3.4, *all* Spin(3,2) Killing vectors are of this form. More precisely, we have

$$\bar\alpha \gamma_\mu \beta = \frac{1}{2}(\bar\alpha_0\, \gamma_{AB}\, \beta_0)\, K^{AB}_\mu \qquad (6.101)$$

(here we have changed the normalization of the Killing vectors to adhere to the usual SUGRA conventions). Then

$$\begin{aligned}\frac{1}{2}(\bar\alpha_0\, \gamma_{AB}\, \beta_0)\, M^{AB} &= \frac{1}{2}(\bar\alpha_0\, \gamma_{AB}\, \beta_0) \int_S \Theta^{\mu\nu} K^{AB}_\nu d\Sigma_\mu = \int_S \Theta^{\mu\nu} (\bar\alpha \gamma_\nu \beta) d\Sigma_\mu \\ &= \int_S \overline{\mathcal{D}}_\mu \big(\bar\alpha \gamma^{\rho\mu\nu}\mathcal{D}_\nu \beta - \overline{(\mathcal{D}_\nu \alpha)}\gamma^{\rho\mu\nu}\beta\big) d\Sigma_\rho \\ &= \oint_{\partial S} \frac{1}{2}\big(\bar\alpha \gamma^{\rho\mu\nu}\mathcal{D}_\nu \beta - \overline{(\mathcal{D}_\nu \alpha)}\gamma^{\rho\mu\nu}\beta\big) d\sigma_{\rho\mu},\end{aligned} \qquad (6.102)$$

where S is a space-like hypersurface at infinity in Σ. The surface integral on the RHS is the Nester form of the space–time charges [245, 246].

6.6 The *AdS/Poincaré* SUSY algebra

Armed with the explicit expressions of Section 6.5 it is very easy to check the algebra generated by our supercharges [192]. We write the supercharge associated to a given (asymptotic) *commuting* Killing spinor α as

$$Q(\alpha) = \int_{\partial S} \bar\alpha\, \gamma^{\mu\nu\rho}\, \psi_\rho\, d\sigma_{\mu\nu}. \qquad (6.103)$$

---

[11] The overline outside the parentheses means *background value*, whereas the one inside the parentheses stands for *Dirac conjugate of the spinor*. It is hoped this is not too confusing.

$Q(\alpha)$ generates the SUSY transformation $2\,\delta_Q(\alpha)$ of parameter α. Then

$$\{Q(\alpha), Q(\beta)\} = 2\,\delta_Q(\alpha)Q(\beta) = 2\int_{\partial S} \bar{\beta}\,\gamma^{\mu\nu\rho}\,\delta_Q(\alpha)\psi_\rho\,d\sigma_{\mu\nu}$$
$$= 2\int_{\partial S} \bar{\beta}\,\gamma^{\mu\nu\rho}\,\mathcal{D}_\rho\alpha\,d\sigma_{\mu\nu}, \qquad (6.104)$$

where now $\mathcal{D}_\rho$ is the operator appearing in the gravitino transformation, Eq. (6.21). If $\mathcal{D}_\rho$ is simply $D_\rho + i\sqrt{-\Lambda/12}\,\gamma_\rho$, the RHS of Eq. (6.104) is as already computed above to be equal to

$$(\bar{\alpha}_0 \gamma_{AB} \beta_0)\,M^{AB}, \qquad (6.105)$$

where $M^{AB} = -M^{BA}$ are the $SO(2, D-1)$ generators. Therefore in AAdS we get a SUSY algebra

$$\{Q_\alpha, \bar{Q}_\beta\} = (\gamma_{AB})_{\alpha\beta}\,M^{AB}, \qquad (6.106)$$

which is the standard form of the *AdS* SUSY.

Taking $\Lambda \to 0$, the *AdS* isometry group, $SO(2, D-1)$, contracts to the Poincaré group, and the above expression reduces to $(\gamma_\mu)_{\alpha\beta}\,P^\mu$, where now P^μ is the global energy–momentum of the asymptotically flat configuration. Indeed, Eq. (6.102) is exactly Nester's original formula for P_μ [245]. At $\Lambda = 0$ Eq. (6.106) gets replaced by Poincaré SUSY.

However, in general $\mathcal{D}_\rho$ contains also other terms, and we get a richer algebra. We shall be rather sketchy, looking for the general structures rather than the details of the computations (which may be found in the references). Since we are not going to write detailed formulae, it is better to revert to Majorana notation for the gravitini. In $D = 4$ SUGRA we have[12]

$$(\mathcal{D}_\rho)^{ab} = (D_\rho)^{ab} + \frac{i}{2}m(\phi)^{ab}\gamma_\rho - \frac{1}{4}\mathfrak{F}^{ab}_{\mu\nu}\gamma^{\mu\nu}\gamma_\rho + \text{Fermi bilinears}, \qquad (6.107)$$

where $\mathfrak{F}^{ab}_{\mu\nu}$ is the local $U(\mathcal{N})_R$–covariant field–strength defined in Section 4.7.1.

In order to have a well–defined SUSY algebra, the asymptotic regime should be reached rapidly enough. One must require that $h_{\mu\nu} = O(1/r)$, such that the Ricci tensor deviates from the asymptotic value to order $O(1/r^3)$ and so does the matter energy–momentum tensor. This, in particular, requires $m(\phi)^{ab} =$

---

[12] Recall that the gravitino fields are redefined in such a way as to eliminate the cross–terms (spin–3/2)–(spin–1/2) from the kinetic (derivative) terms. This also eliminates terms in the SUSY transformations different from those we write, up to higher terms in the fermions.

$\sqrt{-\Lambda/3}\,\delta^{ab} + O(1/r^3)$ for $a,b = 1,2,\ldots,\mathcal{N}_0$; that is, for the indices a,b corresponding to the *unbroken* SUSY subalgebra. Thus, in these particular backgrounds, there is no additional contribution to the surface integral from the second term on the RHS of Eq. (6.107). Instead there may be contributions from the field–strength $\mathfrak{F}^{ab}_{\mu\nu}$. As we have already anticipated, for AF backgrounds the flux at infinity of the $U(\mathcal{N})_R$–covariant field–strength $\mathfrak{F}^{ab}_{\mu\nu}$ and its dual gives rise to the central charges Z^{ab}, which are linear combinations of the electric and magnetic charges of the basic vector fields with coefficients depending on the values of the scalars at infinitiy, as we discussed in Section 2.10.3. Thus, in the AF case, one gets a superalgebra with the general structure

$$\{Q^a, \bar{Q}^b\} = \delta^{ab} \gamma^\mu P_\mu + Z^{ab} + i\gamma_5 \tilde{Z}^{ab}. \tag{6.108}$$

The *AdS* case is more interesting. First of all, what do we expect to find? In five dimensions we have the following Lie group isomorphisms:[13]

$$Spin(5) \simeq Sp(4), \quad Spin(4,1) \simeq Sp(2,2), \quad Spin(3,2) \simeq Sp(4,\mathbb{R}). \tag{6.109}$$

So

$$\mathfrak{iso}(AdS_4) \simeq \mathfrak{sp}(4,\mathbb{R}), \tag{6.110}$$

which is, in particular, simple. The full bosonic symmetry realized on the supercharges should be $\mathfrak{sp}(4,\mathbb{R}) \oplus \widetilde{\mathfrak{aut}}_R$. The SUSY automorphism algebra, $\widetilde{\mathfrak{aut}}_R$, is, however, smaller than in the Poincaré case where it is $\mathfrak{u}(\mathcal{N}_0)$. In fact, the kinetic terms for the gravitini associated with the $\mathcal{N}_0$ unbroken supersymmetries is

$$-i\bar{\psi}^a_\mu \gamma^{\mu\nu\rho} \mathcal{D}_{\nu\,ab} \psi^b_\rho = -i\bar{\psi}^a_\mu \gamma^{\mu\nu\rho} D_\nu \psi^a_\rho - \sqrt{-\Lambda/3}\,\bar{\psi}^a_\mu \gamma^{\mu\nu} \psi^a_\nu + \cdots \tag{6.111}$$

In flat space, the second term on the RHS would be a mass for the spin–3/2 particle. In *AdS*, provided the coefficient is precisely $\sqrt{-\Lambda/3}$, this "mass term" combines with the curvature of the space in such a way that the net effect is that the gravitini propagate on the light–cone, that is, are physically massless (Exercise 6.8.2). However, the mass term breaks the chiral symmetry $U(\mathcal{N})$ down to the vector–like subgroup $SO(\mathcal{N}_0)$, just as it will do in flat space. Then we expect the bosonic part of the *AdS* SUSY algebra to be

$$SO(\mathcal{N}_0) \times Sp(4,\mathbb{R}). \tag{6.112}$$

This fits perfectly well into the classification theorem of superalgebras [211] which predicts the existence of a superalgebra, $Osp(\mathcal{N}_0 \mid 4)$, which has this bosonic subalgebra.

---

[13] Recall that, in our notation, $Sp(4,\mathbb{F})$ stands for the symplectic group having fundamental representation of dimension 4. It is often called $Sp(2,\mathbb{F})$, especially in the mathematics literature.

Let us check that this is precisely the result one obtains by inserting Eq. (6.107) into Eq. (6.104). Using the identity

$$\gamma^{\alpha\beta\sigma}\gamma^{\mu\nu}\gamma_\beta = 2i\gamma_5\,\epsilon^{\sigma\alpha\mu\nu} + 2g^{\sigma\mu}g^{\alpha\nu} - 2g^{\sigma\nu}g^{\alpha\mu} \qquad (6.113)$$

one finds [159]

$$-\bar\epsilon\gamma^{\mu\nu\rho}\mathcal{D}_\rho\epsilon = \cdots + \bar\epsilon\,(\mathfrak{F}^{\mu\nu} + i\gamma_5\widetilde{\mathfrak{F}}^{\mu\nu})\epsilon + (\cdots), \qquad (6.114)$$

where $\cdots$ stands for the terms (which we have already computed) that produce the expression $\bar\epsilon_0\gamma^{AB}\epsilon_0\,M_{AB}$ in the anticommutator of two Qs. Using the explicit formula for the (asymptotic) Killing spinors, Eq. (6.54), we get

$$\begin{aligned}
\bar\epsilon^a\epsilon^b &= (\epsilon^a)^\dagger\,\gamma_4\,\epsilon^b = (\epsilon_0^a)^\dagger\,\varpi^{-1}((\mathcal{E}^{-1})^t)\gamma_4\,\varpi^{-1}(\mathcal{E}^{-1})\epsilon_0^b \\
&= -i(\epsilon_0^a)^\dagger\,\varpi^{-1}((\mathcal{E}^{-1})^t\,\gamma_0\gamma_4\,\mathcal{E}^{-1})\epsilon_0^b = (\epsilon_0^a)^\dagger\,\varpi^{-1}(-i\gamma_0\gamma_4\,\mathcal{E}\mathcal{E}^{-1})\epsilon_0^b \\
&= (\epsilon_0^a)^\dagger\,\varpi^{-1}(-i\gamma_0\gamma_4\,\mathcal{E}\mathcal{E}^{-1})\epsilon_0^b = (\epsilon_0^a)^\dagger\,\gamma_4\,\epsilon_0^b \\
&= \bar\epsilon_0^a\epsilon_0^b.
\end{aligned} \qquad (6.115)$$

Thus the first contribution linear in the field–strength on the RHS of Eq. (6.114) reduces to $\mathfrak{F}^{\mu\nu}_{ab}\,(\bar\epsilon_0^a\epsilon_0^b)$ or, more elegantly,[14] to

$$\mathfrak{F}^{\mu\nu}_{ab}\,(\bar\epsilon_0 L^{ab}\epsilon_0), \qquad (6.116)$$

where $(L^{ab})_{cd} = \delta^{ab}_{cd}$ are the generators of $SO(\mathcal{N}_0)$ in the vector representation. Then this first term gives a contribution to the anticommutator of two Qs of the form

$$(\bar\epsilon\,L_{ab}\,\epsilon')\int_{\partial S}*\mathfrak{F}^{ab} = (\bar\epsilon\,L_{ab}\,\epsilon')\,J^{ab}, \qquad (6.117)$$

where the J^{ab} are *electric*–like charges which generate the global $SO(\mathcal{N}_0)$ symmetry. Very nice! However, the other term, $\bar\epsilon\,(*\mathfrak{F})\gamma_5\epsilon'$ (which is linear in the $\mathfrak{F}$–magnetic charges $m^{ab} = \int_{\partial S}\mathfrak{F}^{ab}$), is definitely *not* nice. The expression $\bar\epsilon^a\gamma_5\epsilon^b$ reduces to something like $\bar\epsilon_0^a(\mathcal{E}\gamma_5\mathcal{E}^{-1})\epsilon_0^b$, which is *not* a constant at infinity, nor an asymptotic Killing vector. So this contribution is ugly. It is not only ugly, it is also unexpected. The electric charges J^{ab} are enough to generate $SO(\mathcal{N}_0)$, and we do not expect any other bosonic generator of the global AdS SUSY algebra besides (M^{AB}, J^{ab}), since – by the arguments of [211] – we know it should correspond to $Osp(\mathcal{N}_0\,|\,4)$.

---

[14] Recall that the ϵ_0^a are *commuting* spinors!

The magnetic term, however, is there and it *should* be there. Why? Because in the limit $\Lambda \to 0$, we should recover the Poincaré algebra, and both the magnetic and electric charges are needed to get the full SUSY algebra (with all possible central charges). This is related to the fact that, as $\Lambda \to 0$ the automorphism group gets enhanced from $SO(\mathcal{N}_0)$ to $U(\mathcal{N}_0)$ and the $\mathfrak{F}$-fluxes should make a complete representation of this larger group. *How do we solve this paradox?*

Luckily, Hawking has already solved it for us [183]. From equation (6.15), we see that a massless photon moving radially satisfies the equation

$$d\tau = \frac{d\rho}{\cosh(\rho)} \equiv 2\frac{d\tanh(\rho/2)}{1 + \tanh^2(\rho/2)}, \qquad (6.118)$$

and the "boundary" $\rho = \infty$ is reached in a *finite* time $\tau = \pi$. Thus, AdS is not globally hyperbolic (i.e., there is no space–like Cauchy surface on which we can specify the initial data to get a unique time evolution). To get unique solutions, we need also to specify the boundary conditions at $\rho = \infty$ for the massless fields. For the gauge vectors, unitarity implies *reflective* boundary conditions. There are two types: either the electric fields $F_{0i} = O(1/r^2)$ and the magnetic fields $F_{ij} = O(1/r^3)$, or the other way around. Thus we can have either electric or magnetic fluxes at infinity, but not both. In the special case of $\mathfrak{F}_{\mu\nu}$, SUSY relates its boundary conditions to those of the graviton. The conclusion is that only the electric fluxes (charges) survive [183]. This is perfectly in agreement with our findings, and solves (quite elegantly) our little paradox.

Exercise 6.6.1 Show that a gravitino with a "mass" term $\sqrt{-\Lambda/3}$ propagates in AdS along the light cone.

6.7 Positive mass and BPS bounds

As we have already mentioned, the above machinery was introduced in order to prove a long–standing conjecture in General Relativity: namely, that the total mass of any asymptotically flat configuration is non–negative (and vanishes if and only if the space is everywhere flat). This theorem also implies the stability of Minkowski space as a solution to the equations of Einstein. In fact, R. Schoen and S.T. Yau did produce a proof of this crucial fact in refs. [268, 269, 270, 271], but their proof is long and rather technical, and also it depends on details of the minimal submanifolds which are not true for $D \geq 8$. Then Witten presented in ref. [300] a "simple" proof of the positivity theorem, which was further simplified and generalized by many authors [160, 192, 245, 246].

6.7.1 Positive energy in Minkowski space

The simplest (and most basic) result states that, in all *physically sound* theories describing gravity coupled to matter, the total mass M of any asymptotically Minkowskian solution to the Einstein equations is *non–negative* (and it vanishes if and only the space is globally Minkowski). The condition of being *physically sound* is reflected in the requirement that the matter energy–momentum tensor $T_{\mu\nu}$ satisfies the *dominant energy condition*, namely the matter energy density is required to be positive at all points and in all frames, or

$$T_{\mu\nu} U^\mu V^\nu \geq 0 \qquad \forall\, U^\mu,\, V^\nu \text{ time–like.} \tag{6.119}$$

To prove positivity, one starts from Eq. (6.102), written for an asymptotically Minkowski background, for $\alpha = \beta$,

$$\begin{aligned}\frac{1}{2}(\bar{\alpha}_0 \gamma^\mu \alpha_0) P_\mu &= \oint_{\partial S} \frac{1}{2} \left(\bar{\alpha}\gamma^{\rho\mu\nu} \mathcal{D}_\nu \alpha - (\overline{\mathcal{D}_\nu \alpha})\gamma^{\rho\mu\nu}\alpha \right) d\sigma_{\rho\mu} \\ &= \int_S \mathcal{D}_\mu \left(\bar{\alpha}\gamma^{\rho\mu\nu}\mathcal{D}_\nu\alpha - (\overline{\mathcal{D}_\nu\alpha})\gamma^{\rho\mu\nu}\alpha \right) dS_\rho,\end{aligned} \tag{6.120}$$

where $\mathcal{D}_\mu$ is the usual covariant derivative with respect to the (spin lift of the) Levi–Civita connection, and we used the divergence theorem to rewrite the second line as a volume integral over the space–like 3–surface S (we choose local coordinates so that S is the surface $x^0 = $ const.). Note that $\bar{\alpha}_0 \gamma_\mu \alpha_0$ is a *time–like* vector.

Now the fundamental idea (due to Witten) is that in evaluating the RHS of Eq. (6.120) we can use *any* (commuting) spinor α, as long as it has the correct asymptotics

$$\alpha(x) = \alpha_0 + O\left(\frac{1}{r}\right), \tag{6.121}$$

and the strategy is to make a smart choice of α which makes the RHS manifestly non–negative. Using the identities (6.92) and (6.93), and the very same manipulations we used in Section 6.5 to prove Nester's formula, we get for the RHS of Eq. (6.120):

$$\int_S \left\{ T^\rho{}_\mu \bar{\alpha}\gamma^\mu \alpha + (\mathcal{D}_\mu \bar{\alpha})(\gamma^\rho \gamma^{\mu\nu} + \gamma^{\mu\nu}\gamma^\rho)(\mathcal{D}_\nu \alpha) \right\} dS_\rho. \tag{6.122}$$

Since $\bar{\alpha}\gamma^\mu\alpha$ is non–space–like, the first term in the brace, $T^\rho{}_\mu \bar{\alpha}\gamma^\mu\alpha$ is non–negative provided the matter satisfies the dominant energy condition. The second term is, explicitly, $(i, j = 1, 2, 3)$

$$(\mathcal{D}_i\bar{\alpha})(\gamma^0\gamma^{ij} + \gamma^{ij}\gamma^0)(\mathcal{D}_j\alpha) = 2(\mathcal{D}_i\alpha)^\dagger \gamma^{ij}(\mathcal{D}_j\alpha)$$
$$= -2g^{ij}(\mathcal{D}_i\alpha)^\dagger (\mathcal{D}_j\alpha) + (\mathcal{D}_i\alpha^\dagger \gamma^i)(\gamma^j \mathcal{D}_j\alpha). \quad (6.123)$$

The first term is positive (our metric is $(+, -, -, -)$). Then the positivity theorem is proven provided we can choose our spinor α in such a way that $\gamma^i \mathcal{D}_i \alpha \equiv 0$. Spinors satisfying this equation and the boundary condition (6.121) are called *Witten spinors*. Witten spinors do exist, as one shows using standard linear analysis techniques. The proof of the positive–energy theorem in General Relativity is thus completed.

6.7.2 Stability of AdS space

In the proof of positive mass we have, on purpose, used for the covariant derivative the symbol $\mathcal{D}_\mu$ which in the rest of the chapter stands for the more complicated first-order differential operator appearing in the RS term (or in the gravitino SUSY transformations). The $\mathcal{D}_\mu$ used in the above proof corresponds to the SUSY transformation of *pure* $\mathcal{N} = 1$ supergravity with no cosmological constant. Taking $\mathcal{D}_\mu$ to be the SUSY transformation operator corresponding to various SUGRA theories, we get other positive–energy theorems. For instance,

$$\mathcal{D}_\mu = D_\mu + i\sqrt{-\tfrac{\Lambda}{12}}\,\gamma_\mu \quad (6.124)$$

leads to the *AdS* space positive–energy theorem.

6.7.3 Gravitational BPS bounds

In the same fashon, taking $\mathcal{D}_\mu$ to be the differential operator associated with an extended SUGRA

$$(\mathcal{D}_\mu \epsilon)^A = D_\mu \epsilon^A - \tfrac{1}{4}\mathfrak{F}^{AB}_{\rho\nu}\gamma^{\rho\nu}\epsilon_B \quad (6.125)$$

we get (in the Majorana notation)

$$\bar{\alpha}_0^A[\delta_{AB} P_\mu \gamma^\mu - (Z_{AB} + i\gamma_5 \widetilde{Z}_{AB})]\alpha_0^B$$
$$= \int_S [\hat{T}^\rho_\mu \bar{\alpha}\gamma^\mu\alpha - g^{ij}(\mathcal{D}_i\alpha)^\dagger (\mathcal{D}_j\alpha)]dS_\rho \geq 0, \quad (6.126)$$

where $\hat{T}_{\mu\nu}$ is the contribution to the energy–momentum from the matter *but* the gauge–vectors whose field–strengths enter in the differential operator (6.125). $\hat{T}_{\mu\nu}$ is assumed to satisfy the dominant energy condition.

Thus the matrix

$$\delta_{AB} P_\mu \gamma^\mu - (Z_{AB} + i\gamma_5 \widetilde{Z}_{AB}) \geq 0, \qquad (6.127)$$

which is equivalent to the BPS bound for the mass M

$$|M| \geq |Z + i\widetilde{Z}|^2. \qquad (6.128)$$

6.7.4 General facts about BPS objects

BPS objects enjoy some general properties. First of all, a static BPS object in a background which is time–translational invariant is automatically a solution to the equations of motion. Indeed, the corresponding field profile minimizes the action in the sector of given central charge.

Let α be a (non–zero) Killing spinor for the given BPS object. Then, in Lorentzian signature, $\bar{\alpha}\gamma_\mu\alpha$ is a non–zero Killing vector which is either time–like or light–like. In the first case the object is static, and the previous remark applies. In the second case, it is moving at the speed of light; one can show that, once we impose the generalized Gauss's constraints, the dynamical equations of motion are also satisfied.

BPS objects are protected by SUSY in the sense that they belong to shorter representations of the SUSY algebra, and hence under continuous deformations of the parameters, including time evolution, they should remain BPS. This, in particular, applies to BPS black holes. They cannot radiate, since this would imply a violation of the bound, and hence should have zero Hawking temperature (while having, in general, a non–trivial entropy). They are called *extremal* black holes [159, 236].

6.8 SUGRA Ward identities

6.8.1 SUSY Ward identities from positive mass

As we saw in Section 6.7, Nester–type expressions for the space–time global charges, associated with asymptotic Killing vectors, were invented to prove *positivity of mass* in general relativity for *any* reasonable matter content, whether supersymmetric or not. Of course, if the particular model *is* locally supersymmetric, we get stronger results: the SUSY algebra, the BPS bounds, etc. Now we apply this technology to a theory that *does have* $\mathcal{N}$–extended supersymmetry, but somehow forgetting this fact and treating it just as gravity coupled to some *cleverly chosen* matter. In this context, we are free to add spurious fields producing an additional contribution to the energy–momentum tensor, $\theta_{\mu\nu}$, which acts as an

6.8 SUGRA Ward identities

external source for gravity

$$R_{\mu\nu} - \frac{1}{2}g_{\mu\nu}R = T_{\mu\nu} + \theta_{\mu\nu}, \qquad (6.129)$$

where $T_{\mu\nu}$ is the energy–momentum tensor produced by the SUGRA fields. Nester-like formulae hold on–shell, which means that the Einstein equations (6.129) should be satisfied. In the manipulation below we shall use the analogue expressions for the supercharges, so we should require that the Rarita–Schwinger equations

$$R^{\mu} \equiv -\gamma^{\mu\nu\rho}D_{\nu}\psi_{\rho} + J^{\mu} = 0 \qquad (6.130)$$

are also satisfied. However, there is no need for the other fields to be on–shell, as long as the configuration is such that $D_{\mu}(T^{\mu\nu} + \theta^{\mu\nu}) = 0$, as required by the consistency of the Einstein equations (via the Bianchi identity). Thus, we can take as our background *any* constant value of the scalar fields $\phi^i = \phi_0^i =$ const. The scalar contribution to $T^{\mu\nu}$ then has the form $V(\phi_0) g^{\mu\nu}$, which is just another contribution to the cosmological constant, and hence it is consistent with the Bianchi identities for any constant value of $V(\phi_0)$. In our *gedanken* configuration the vector fields are taken to vanish. It is also convenient to choose the external sources to be $\theta_{\mu\nu} = -V(\phi_0) g_{\mu\nu}$, so that the effective cosmological constant is zero, and we can work in flat space–time and use Killing spinors (in the geometric sense!) which are strictly constant.[15] The Nester energy of such a configuration is zero,

$$0 = \int_{\partial S} d\sigma_{\mu\nu} \, \bar{\epsilon} \, \gamma^{\mu\nu\rho} D_{\rho}\epsilon = \int_{\partial S} d\sigma_{\mu\nu} \left(\bar{\epsilon} \, \gamma^{\mu\nu\rho} D_{\rho}\epsilon - \frac{i}{2} m\bar{\epsilon} \, \gamma^{\mu\nu}\epsilon \right)$$

$$= \delta_Q(\epsilon) \int_{\partial S} d\sigma_{\mu\nu} \, \bar{\epsilon} \, \gamma^{\mu\nu\rho} \, \psi_{\rho} - \frac{i}{2} \int_{\partial S} d\sigma_{\mu\nu} \, m\bar{\epsilon} \, \gamma^{\mu\nu} \epsilon, \qquad (6.131)$$

where the $SU(\mathcal{N})_R$ indices are left implicit and ϵ is a *commuting* spinor. We evaluate the surface integrals on the RHS using the divergence theorem. The last term is covariantly constant, and hence has vanishing divergence; it drops out of the

---

[15] Recall that we can choose the spinors in any way we like, provided they have the correct asymptotics.

computation. We are left with

$$0 = \delta_Q(\epsilon) \int_{\partial S} d\sigma_{\mu\nu} \, \bar{\epsilon} \, \gamma^{\mu\nu\rho} \, \psi_\rho$$

$$= \delta_Q(\epsilon) \int_S d\sigma_\nu \left(\overline{D_\mu \epsilon} \, \gamma^{\mu\nu\rho} \, \psi_\rho + \bar{\epsilon} \, \gamma^{\mu\nu\rho} \, D_\mu \psi_\rho \right) \tag{6.132}$$

$$= \delta_Q(\epsilon) \int_S d\sigma_\nu \left(\overline{D_\mu \epsilon} \, \gamma^{\mu\nu\rho} \, \psi_\rho + \bar{\epsilon} \, \gamma^{\mu\nu\rho} \, D_\mu \psi_\rho \right).$$

In the last line we have used a standard identity we have already used to deduce the global SUSY algebra (and which should hold if the underlying theory is supersymmetric [196]). Next we recall a couple of useful relations:

$$\gamma^{\mu\nu\rho} D_\nu \psi_\rho - J^\mu := R^\mu, \tag{6.133}$$

$$\delta_Q(\epsilon) \bar{\epsilon} R^\mu = -\frac{1}{2}\left(R^{\mu\nu} - \frac{1}{2}g^{\mu\nu} - T^{\mu\nu}\right)\bar{\epsilon}\gamma_\nu\epsilon \equiv -\frac{1}{2}\theta^{\mu\nu}\bar{\epsilon}\gamma_\nu\epsilon. \tag{6.134}$$

The second equation follows from the first since the gravitino equations of motion, $R^\mu = 0$, get transformed into the Einstein equations, which, *in the absence of external sources*, would read $\theta^{\mu\nu} = 0$; the overall coefficient, $-1/2$, is easily fixed by checking that the leading term in the curvatures,

$$\gamma^{\mu\nu\rho} D_\nu D_\rho \epsilon = \frac{1}{8}\gamma^{\mu\nu\rho}\gamma^{\alpha\beta} R_{\nu\rho\alpha\beta}, \tag{6.135}$$

is equal, for a maximally symmetric space, to $-\frac{1}{2}(R^{\mu\nu} - \frac{1}{2}g^{\mu\nu}R)\gamma_\nu$. Then, the RHS of Eq. (6.132) becomes

$$0 = \delta_Q(\epsilon) \int_S d\sigma_\nu \left(\overline{D_\mu \epsilon} \, \gamma^{\mu\nu\rho} \, \psi_\rho + \bar{\epsilon} \, \gamma^{\mu\nu\rho} \, D_\mu \psi_\rho \right)$$

$$= \int_S d\sigma_\nu (\overline{D_\mu \epsilon}) \gamma^{\mu\nu\rho}(D_\rho \epsilon) - \delta_Q(\epsilon) \int_S d\sigma_\mu \, \bar{\epsilon} \left(R^\mu - J^\mu \right) \tag{6.136}$$

$$= \int_S d\sigma_\nu (\overline{D_\mu \epsilon}) \gamma^{\mu\nu\rho}(D_\rho \epsilon) + \frac{1}{2} \int_S d\sigma_\mu \, \theta^{\mu\nu} \, \bar{\epsilon}\gamma_\nu\epsilon + \delta_Q(\epsilon) \int_S d\sigma_\mu \, \bar{\epsilon} \, J^\mu.$$

One has $\bar{\epsilon} J^\mu = \overline{(\delta_Q(\epsilon)\chi^I)} Z(\phi)_I{}^J \gamma^\mu \chi_J$, where $Z(\phi)_I{}^J$ is the positive–definite matrix which enters the kinetic terms for the physical spin–1/2 fields,

$$\frac{i}{2} e \, Z(\phi)_I{}^J \, \bar{\chi}_I \gamma^\mu D_\mu \chi_J. \tag{6.137}$$

Finally, Eq. (6.136) reduces to the identity

$$
0 = \int_S d\sigma_\nu (\overline{\mathcal{D}_\mu \epsilon}) \gamma^{\mu\nu\rho} (\mathcal{D}_\rho \epsilon) \frac{1}{2} \int_S d\sigma_\mu \, \theta^{\mu\nu} \, \bar\epsilon \gamma_\nu \epsilon
$$
$$
+ \int_S d\sigma_\mu \, \overline{\delta_Q(\epsilon) \chi^I} \, Z(\phi)_I{}^J \, \gamma^\mu \, \delta_Q(\epsilon) \chi^J. \tag{6.138}
$$

Defining the matrices $m(\phi)^{AB}$ and $C(\phi)^I{}_A$ by[16]

$$
\delta \psi^A_\mu = \mathcal{D}_\mu \epsilon^A + \frac{1}{2} m(\phi)^{AB} \gamma_\mu \epsilon_B, \tag{6.139}
$$

$$
\delta \chi^I = \frac{1}{2} C(\phi)^I{}_A \, \epsilon^A, \tag{6.140}
$$

and replacing them in the above identity we get the final formula

$$
0 = \left(-\frac{3}{2} m(\phi)_{AC} \, m(\phi)^{CB} - \frac{1}{2} V(\phi) \, \delta_A{}^B \right.
$$
$$
\left. + \frac{1}{4} C(\phi)_A{}^I Z(\phi)_I{}^J C(\phi)^B{}_J \right) \bar\epsilon^A \gamma_\mu \epsilon_B. \tag{6.141}
$$

Since it holds for any (constant) value of ϕ, we can interpret it as a general formula for the scalar potential $V(\phi)$:

General Lesson 6.10 *The scalar potential $V(\phi)$ of any $D = 4$ supersymmetric theory is given by the formula* [89]

$$
V(\phi) \delta_A{}^B = \frac{1}{2} C(\phi)_A{}^I Z(\phi)_I{}^J C(\phi)^B{}_J - 3 \, m(\phi)_{AC} \, m(\phi)^{CB}. \tag{6.142}
$$

This equation allows us to write the scalar potential $V(\phi)$ of any 4D SUSY theory in terms of the *fermionic shifts* $C(\phi)_A{}^I$ and $m(\phi)_{AB}$. From now on, we shall always compute $V(\phi)$ from this formula.

Comparing with the integrability condition for the AdS_D Killing spinors, Eq. (6.34), we easily infer the general formula valid for any D:

$$
V(\phi) \delta_A{}^B = \frac{1}{2} C(\phi)_A{}^I Z(\phi)_I{}^J C(\phi)^B{}_J - (D-1) \, m(\phi)_{AC} \, m(\phi)^{CB}. \tag{6.143}
$$

The first term on the RHS of Eq. (6.142) yields precisely the formula for $V(\phi)$ in *rigid* SUSY, which corresponds to the operator statement[17]

$$
H \delta^A{}_B = \{ Q^A, (Q^B)^\dagger \} \geq 0 \qquad \text{in the sector } P_i = 0. \tag{6.144}
$$

---

[16] We use the chiral convention; so ϵ^A has chirality $+1$ and ϵ_A chirality -1. Raising/lowering indices corresponds to complex conjugation.

[17] Assuming the Hilbert space has positive norm, as required by unitarity.

If only the first term were present, as happens in global SUSY, we would have concluded that the scalar potential is ≥ 0 (unitary requires the matrix $Z_I{}^J$ to be positive–definite); then $V(\phi)$ would vanish if and only if all $C_I^A = 0$, that is, only if supersymmetry is *unbroken*; moreover, since the LHS is proportional to $\delta^A{}_B$, if *one* supersymmetry is unbroken, *all* supersymmetries must be unbroken.[18] Instead, in local SUSY, we have, in addition, the last term in Eq. (6.142), which is the contribution from the gravitini. It is *negative* semi–definite. One can understand it as the contribution from the *negative–norm* "longitudinal" (helicity $\pm 1/2$) gravitini whose existence spoils naive arguments based on the positivity of the Hilbert space (just as happens in any gauge theory, where we cannot have – simultaneously – a covariant formalism and a positive–definite Hilbert space). This new term changes everything with respect to the rigid case. Now it is the matrix

$$V(\phi)\,\delta_A{}^B + 3\, m_{AC}\, m^{CA} \qquad (6.145)$$

that is positive semi–definite. If this matrix has a zero eigenvalue corresponding, say, to the eigenvector $\epsilon_A^{(0)}$, then the supersymmetry generated by $\epsilon_A^{(0)}$ is *automatically unbroken* in *AdS*/Minkowski space with $\Lambda \equiv V(\phi) \leq 0$. Indeed, if (6.145) has a zero eigenvector $\epsilon_A^{(0)}$:

(1) The "mass" m of the associated gravitino,[19] $\psi_\mu^{(0)}$, is related to the cosmological constant Λ by the "magical" relation

$$m = \sqrt{-\Lambda/3},$$

i.e., the gravitino is massless in the *AdS* sense (it propagates along *null* geodesics). This condition guarantees

$$\delta\psi_\mu^{(0)} = \overline{\mathcal{D}}_\mu \epsilon^{(0)} = 0. \qquad (6.146)$$

(2) From (6.142), $C_A{}^I Z_I{}^J C^B{}_J \epsilon_B^{(0)} = 0$, and we have automatically

$$\delta_\epsilon^{(0)} \chi^I = 0 \qquad (6.147)$$

for all spin $-1/2$ fermions, since $Z_I{}^J$ is positive–definite.

However, now the non–negative matrix (6.145) is no longer proportional to $\delta_A{}^B$, and thus it may have $\mathcal{N}_0 < \mathcal{N}$ zero eigenvalues and $\mathcal{N} - \mathcal{N}_0$ non–zero ones. In this case, we have the spontaneous breaking from $\mathcal{N}$–SUSY to $\mathcal{N}_0$–SUSY. This possibility is called *partial superHiggs* [89].

---

[18] Except for brane charges [132].

[19] "Mass" in quotation marks stands for the coefficient of the bilinear $\frac{1}{2}\bar{\psi}_\mu \gamma^{\mu\nu} \psi_\nu$ in the Lagrangian. It would coincide with the physical mass in the linearized theory around flat Minkowski space, but not otherwise.

6.8 SUGRA *Ward identities*

We may also have SUSY breaking, total or partial, at zero–vacuum energy, $V(\phi_{\min}) = 0$, if the negative gravitino contribution $-3m^2$ exactly cancels the positive spin–1/2 contribution $\frac{1}{2}C^\dagger ZC$. In some classes of model this cancellation is guaranteed (at the classical level!) by U–duality–like symmetries. The breaking scales of the diverse supersymmetries are, *a priori*, unrelated.

6.8.2 A simpler technique

We can obtain the above result by simpler techniques. We assume a bosonic background in which the scalar fields ϕ^i are constant and the vector ones vanish. Then we consider the obvious identity

$$\delta_\epsilon \mathcal{L}\big|_{\text{linear in } \psi_\mu} = 0. \tag{6.148}$$

Using the fact that (in our background)

$$\delta_\epsilon \left(e\, V(\phi) \right)\big|_{\text{linear in } \psi_\mu} = e\, V(\phi)\, \overline{\psi}^A_\mu \gamma^\mu \epsilon_A, \tag{6.149}$$

while the variation of the Noether term $e\, \overline{\psi}^A_\mu J^\mu_A$ gives

$$\delta_\epsilon \left(e\, \overline{\psi}^A_\mu J^\mu_A \right)\big|_{\text{linear in } \psi_\mu} = e\, \overline{\psi}^A_\mu \delta_\epsilon (J^\mu_A), \tag{6.150}$$

and inserting the definitions (6.139)(6.140), we recover Eq. (6.142) [88].

We preferred to deduce the formula for $V(\phi)$ starting from positive–mass–like theorems in order to stress that the additional term one gets in local SUSY is strongly related to the geometry of *AdS*, the Killing spinors, and their integrability conditions, and that its physical motivation can be really *understood* precisely in these geometrical terms.

Exercise 6.8.1 Fill in the details in the direct derivation of Eq. (6.142).

7
Parallel structures and isometries

The geometric results of Chapters 3–5 fully determine the Lagrangian of any *ungauged* supergravity. Our next task is to *gauge* our SUGRAS. In particular, we wish to understand which subgroups of their global symmetry group may be consistently gauged while preserving $\mathcal{N}$–extended SUSY. The global symmetry group acts on the scalars by isometries of the manifold $\mathcal{M}$. Our task is to understand the special properties of the isometry group of the manifolds $\mathcal{M}$ which are compatible with *extended* SUSY. The chapter is devoted to this purely geometric analysis. For definiteness, we adopt the language of SUSY in 3D, the geometric results being, of course, indepedent of the dimension of space–time. We start with the manifold relevant for *rigid* SUSY. The geometry of those corresponding to SUGRA is described in Sections 7.3–7.8.

7.1 Rigid SUSY: momentum maps
7.1.1 Canonical symplectic structures on $\mathcal{M}$

From the analysis of Chapter 2, we know that, in 3D, $\mathcal{N}$–extended rigid SUSY implies the existence of $\mathcal{N}(\mathcal{N}-1)/2$ canonical *parallel* 2–forms on $\mathcal{M}$,

$$(\Sigma^{AB})_{ij} = -(\Sigma^{BA})_{ij} := \begin{cases} g_{il}\,(f^B)^l{}_j, & \text{for } A = 1, B \neq 1, \\ g_{ik}\,(f^A)^k{}_l(f^B)^l{}_j, & \text{for } A, B \neq 1, A \neq B, \end{cases} \quad (7.1)$$

where $(f^a)^i{}_j$, $a = 2, 3, \ldots, \mathcal{N}$ are the parallel complex structures. The 2–forms Σ^{AB} transform in the adjoint of $Spin(\mathcal{N})_R$, and – being parallel – are, in particular, *closed*. Fix the indices A, B and consider $(\Sigma^{AB})_{ij}$ as a $\dim \mathcal{M} \times \dim \mathcal{M}$ (skew–symmetric) matrix in the indices i, j. Then

$$\det \Sigma^{AB} = \det(f^A)\,\det(g)\,\det(f^B) = \det g \neq 0. \quad (7.2)$$

7.1 Rigid SUSY: momentum maps

This means that *each* 2–form $\Sigma^{AB} = -\Sigma^{BA}$ satisfies[1]

$$d\Sigma^{AB} = 0, \quad \text{and} \quad (\Sigma^{AB})^n = n!\,\text{vol} \neq 0. \tag{7.3}$$

Equations (7.3) are the two conditions defining a symplectic 2–form. Thus

General Lesson 7.1 *Let $\mathcal{M}$ be the target manifold of a rigid $\mathcal{N}$–extended[2] 3D* SUSY. *On $\mathcal{M}$ there are $\mathcal{N}(\mathcal{N}-1)/2$ symplectic structures, Σ^{AB}, transforming in the adjoint representation of Spin($\mathcal{N}$)$_R$.*

The geometry of a symplectic manifold is equivalent to the Hamilton–Jacobi theory of classical mechanics.[3] In that case, the (single) symplectic structure is given by the closed 2–form $dp_i \wedge dq^i$ in phase space. The theorem of Darboux states that any symplectic manifold has *local* coordinates such that the symplectic 2–form has this canonical expression.

In rigid SUSY we have the same geometric structure but replicated $\mathcal{N}(\mathcal{N}-1)/2$ times, and we also have a nice group action rotating the $\mathcal{N}(\mathcal{N}-1)/2$ symplectic structures. The 2–forms Σ^{AB} are canonically identified with endomorphisms of $T^*\mathcal{M}$, that is with $(\Sigma^{AB})_i{}^j \equiv (\Sigma^{AB})_{ik}\, g^{kj}$. These automorphisms generate an algebra with multiplication table

$$\begin{aligned}\Sigma^{AB}\Sigma^{CD} = &(\delta^{BC}\delta^{AD} - \delta^{AC}\delta^{BD})\mathbf{1} \\ &+ \delta^{AC}\Sigma^{BD} - \delta^{AD}\Sigma^{BC} - \delta^{BC}\Sigma^{AD} + \delta^{BD}\Sigma^{AC} + \Sigma^{ABCD},\end{aligned} \tag{7.4}$$

corresponding to the Clifford multiplication in $\mathbb{C}\ell^0(\mathcal{N})$.

Our next job is to construct the Hamiltonian functions generating the flows corresponding to the multi–symplectic isometries of $\mathcal{M}$.

7.1.2 Symplectic momentum maps

As discussed in Chapter 2, we cannot gauge in a supersymmetric way *all* the isometries of $\mathcal{M}$. In rigid SUSY, the symmetry Aut$_R \equiv$ *Spin($\mathcal{N}$)$_R$* is necessarily rigid, and we can gauge only the isometries which *commute* with the *Spin($\mathcal{N}$)$_R$* group. These isometries have peculiar geometric properties. Let $K^i\partial_i$ be a Killing vector generating an isometry commuting with *Spin($\mathcal{N}$)$_R$*. It should leave invariant

---

[1] Here $2n \equiv \dim \mathcal{M}$ which is even, since we assume $\mathcal{N} \geq 2$ (otherwise there is no form Σ^{AB}), and vol stands for the canonically normalized volume $2n$–form associated with the metric g_{ij}.
[2] The number of supercharges is $N_S = 2\mathcal{N}$. Stated in terms of the absolute number of supercharges N_S, the result is true in *any* number of space–time dimensions D.
[3] General references for symplectic geometry and its relations to Hamiltonian mechanics are [19, 49, 76, 143, 177, 229].

the generators of the (global) $Spin(\mathcal{N})_R$ symmetry, namely the 2–forms Σ^{AB}. Thus

$$\pounds_K \Sigma^{AB} = 0. \tag{7.5}$$

A Killing vector K satisfying this equation for all A, B will be called a *multi–symplectic* Killing vector. Only multi–symplectic isometries can be gauged while preserving the full $\mathcal{N}$–SUSY. (Of course, one may envisage the case in which this equation holds for a subset of the 2-forms, in such a way that the gauging still preserves an $\mathcal{N}_0 < \mathcal{N}$ SUSY subalgebra.) Then, for the moment,[4] we limit ourselves to such symmetries. We denote by Iso_0 the subgroup of $\mathrm{Iso}(\mathcal{M})$ of multi–symplectic isometries. Explicitly, Eq. (7.5) reads

$$0 = \pounds_K \Sigma^{AB} = (i_K d + d i_K) \Sigma^{AB} = d\Big(i_K \Sigma^{AB}\Big). \tag{7.6}$$

By the Poincaré lemma, locally on $\mathcal{M}$ there exist functions $\mu^{AB}(K)$ such that

$$i_K \Sigma^{AB} = -d\mu^{AB}(K) \qquad \text{(Hamiltonian functions)}. \tag{7.7}$$

If $H^1(\mathcal{M}, \mathbb{C}) = 0$, the $\mu^{AB}(K)$ exist globally; for simplicity (by replacing $\mathcal{M}$ with its universal cover $\widetilde{\mathcal{M}}$, if necessary) we assume $\mathcal{M}$ to be simply connected. Then $\mu^{AB}(K)$ is well defined up to an additive constant. $\mu^{AB}(K)$ is clearly linear in K. We can interprete $\mu^{AB}(\,\cdot\,)$ as a linear map from the Lie algebra $\mathfrak{iso}_0$ to $\mathfrak{spin}(\mathcal{N})$, that is,

$$\mu \colon \mathcal{M} \to (\mathfrak{iso}_0)^{\vee} \otimes \mathfrak{spin}(\mathcal{N}). \tag{7.8}$$

The Hamiltonian functions, seen as a map $\mathcal{M} \to (\mathfrak{iso}_0)^{\vee} \otimes \mathfrak{spin}(\mathcal{N})$, are called the *momentum map*. Momentum maps have a rich algebraic structure. To be concrete, we fix a basis $\{K^{i\,m}\}$ ($m = 1, 2, \ldots, \dim \mathrm{Iso}_0$) of the multi–symplectic Killing vectors, and write $\mu^{AB\,m} \equiv \mu^{AB}(K^m)$. Then

$$\partial_i \mu^{AB\,m} = (\Sigma^{AB})_{ij} K^{j\,m}, \tag{7.9}$$

$$K^{i\,m} = -(\Sigma^{AB})^{ij} \partial_j \mu^{AB\,m} \qquad \textit{not}\text{ summed over } A, B. \tag{7.10}$$

From the Clifford algebra (7.4), one has

$$D_k((\Sigma^{AB})_i{}^k \mu^{CD\,m}) = (\Sigma^{AB})_i{}^k \partial_k \mu^{CD\,m} = (\Sigma^{AB}\Sigma^{CD})_{ik} K^{k\,m}$$
$$= -(\delta^{AC}\delta^{BD} - \delta^{AD}\delta^{BC})K_i^m + (\Sigma^{ABCD})_{ij} K^{j\,m}$$
$$+ \partial_i(\delta^{AC}\mu^{BD\,m} - \delta^{AD}\mu^{BC\,m} - \delta^{BC}\mu^{AD\,m} + \delta^{BD}\mu^{AC\,m}). \tag{7.11}$$

---

[4] Non–multi–symplectic isometries have deep physical consequences (being related to R–symmetries) and will be addressed in Section 7.5.

7.1.3 Action of $\mathfrak{iso}_0$ on the momentum maps

Let us compute $\pounds_{K^n}\mu^{ABm}$. We can evaluate it in two different ways. First

$$\pounds_{K^n}\mu^{ABm} = K^{in}\partial_i\mu^{AB,m} = (\Sigma^{AB})_{ij} K^{in} K^{jm}$$
$$= (\Sigma^{AB})^{ij} \partial_i\mu^{ABm} \partial_j\mu^{ABn} \quad \text{(Poisson bracket)} \tag{7.12}$$

or

$$\partial_i\left(\pounds_{K^n}\mu^{ABm}\right) = \pounds_{K^n}\partial_i\mu^{ABm} = (\Sigma^{AB})_{ij}\pounds_{K^n}K^{jm} = f^{mn}{}_p\partial_i(\mu^{ABp}), \tag{7.13}$$

that is

$$\pounds_{K^n}\mu^{ABm} = f^{mn}{}_p \mu^{ABp} + \text{const}. \tag{7.14}$$

We can reabsorb the constant by a redefinition[5] of the μ^{ABm}. Then the Poisson bracket of the momentum maps is equal to the Lie algebra commutator. Of course, this is well known from classical mechanics.

Therefore the momentum map $\mu^\bullet$ is a function

$$\mu^\bullet: \mathcal{M} \to \mathfrak{spin}(\mathcal{N}) \otimes \mathfrak{iso}_0^\vee \tag{7.15}$$

transforming according to the representation

$$\text{Adj}(\mathfrak{spin}(\mathcal{N})) \otimes \text{Coadj}(\mathfrak{iso}_0^\vee). \tag{7.16}$$

7.1.4 "Killing" maps

For $\mathcal{N} \geq 5$, the Clifford algebra of the *parallel* tensors contains also *symmetric* 2–tensors *not* proportional[6] to g_{ij}, e.g. Σ^{ABCD}. One may wonder what is the analogue of the momentum map for such *symmetric* parallel tensors. For any parallel 2–tensor T_{ij} one has[7]

$$\pounds_K T_{ij} \equiv K^k D_k T_{ij} + T_{kj} D_i K^k + T_{ik} D_j K^k$$
$$= D_i(K^k T_{kj}) + D_j(K^k T_{ik}), \tag{7.17}$$

if, in addition, T_{ij} is K–invariant, $\pounds_K T_{ij} = 0$, and symmetric we get

$$D_i(K^k T_{kj}) + D_j(K^k T_{ki}) = 0, \tag{7.18}$$

that is, $T_{ik}K^k$ is a *Killing vector*. Thus, say, $(\Sigma^{ABCD})_{ij}$ is a map $\mathfrak{iso}_0 \to \mathfrak{iso}$.

---

[5] This is *not* an elementary result. In general, there is a Lie algebra cocycle which may be an obstruction to a suitable redefinition of the momentum map. See the quoted references.
[6] Thus, by the corollary to Berger's theorem, for $\mathcal{N} \geq 5$ $\mathcal{M}$ is flat. We already know this result.
[7] Written in this way, the formula holds both for symmetric and antisymmetric tensors.

7.1.5 Properties of $\mathfrak{iso}_0$ Killing vectors

The $\mathfrak{iso}_0$ condition

$$\pounds_K \Sigma^{AB} = 0 \tag{7.19}$$

may be written as in Eq. (7.17), i.e.,

$$\begin{aligned} 0 &= D_i K^k (\Sigma^{AB})_k{}^j + (\Sigma^{AB})_i{}^k D^j K_k \\ &= D_i K^k (\Sigma^{AB})_k{}^j - (\Sigma^{AB})_i{}^k D_k K^j. \end{aligned} \tag{7.20}$$

Thus:

Proposition 7.2 *The Killing vector K^i belongs to $\mathfrak{iso}_0$ iff the skew–symmetric endomorphism*

$$(A_K)^j{}_i := -D_i K^j \in \mathrm{End}(T\mathcal{M}) \tag{7.21}$$

commutes *with all the (skew–symmetric) endomorphisms $(\Sigma^{AB})^i{}_j$.*

Under the isomorphism

$$T\mathcal{M} \otimes \mathbb{C} \simeq S \otimes U \oplus \widetilde{S} \otimes \widetilde{U}, \tag{7.22}$$

where $S, \widetilde{S}$ are irreducible $\mathbb{C}\ell^0(\mathcal{N}) \otimes \mathbb{C}$–modules[8] (see Section 2.4.1), the skew–symmetric endomorphism A_K is mapped to

$$A_K \mapsto 1 \otimes \kappa \oplus 1 \otimes \widetilde{\kappa}, \qquad \kappa \in \begin{cases} \mathfrak{so}(U) & S \text{ has } \mathbb{R} \text{ structure,} \\ \mathfrak{u}(U) & S \text{ has } \mathbb{C} \text{ structure,} \\ \mathfrak{sp}(U) & S \text{ has } \mathbb{H} \text{ structure,} \end{cases} \tag{7.23}$$

and analogously for $\widetilde{\kappa}, \widetilde{U}$.

Example 7.1.1 For $\mathcal{N} = 2$, with respect to the unique complex structure, the Killing vectors in $\mathfrak{iso}_0$ are holomorphic.

Example 7.1.2 For $\mathcal{N} = 3$, $\mathcal{M}$ is hyperKähler. Using the double index notation, Eq. (7.23) becomes $D_{\alpha a} K_{\beta b} = \epsilon_{\alpha\beta} \kappa_{ab}$. with $\kappa_{ab} \in \mathfrak{sp}(\dim \mathcal{M}/2)$. This is precisely Eq. (3.53) of Gaiotto–Witten [151].

---

[8] The summand $\widetilde{S} \otimes \widetilde{U}$ is present only for $\mathcal{N}$ even (in even dimensions we have two spinorial representations (chirality ± 1) while in odd dimension there is only one).

7.1 Rigid SUSY: momentum maps

Corollary 7.3 *For $\mathcal{N} \leq 4$ the image of the Killing map Σ^{ABCD} is contained in $\mathfrak{iso}_0$.*

Proof (For $\mathcal{N} \leq 3$ the statement is trivial.) Under the isomorphism, the Σ^{ABCD} map reads

$$1 \otimes \kappa \longmapsto \Sigma^{ABCD} \otimes \kappa \qquad (7.24)$$

and the image is in $\mathfrak{iso}_0$ iff Σ^{ABCD} acts as ± 1 on the irreducible Clifford–module. Notice that $\Sigma^{ABCD}(\mathfrak{iso}_0) \not\subset \mathfrak{iso}_0 \Rightarrow \mathcal{M}$ is flat. $\square$

More properties of the momentum maps

Consider the 1–form

$$dx^i (\Sigma^{AB})_i^{\ k} \partial_k \mu^{CD\,m} \in \mathfrak{spin}(\mathcal{N}) \otimes \mathfrak{spin}(\mathcal{N}) \otimes \mathfrak{iso}_0 \otimes \Lambda^1$$
$$\in \big(\odot^2 \mathfrak{spin}(\mathcal{N}) \otimes \mathfrak{iso}_0 \otimes \Lambda^1 \big) \oplus \big(\wedge^2 \mathfrak{spin}(\mathcal{N}) \otimes \mathfrak{iso}_0 \otimes \Lambda^1 \big). \qquad (7.25)$$

Comparing with Eq. (7.11) and Section 7.1.4, the symmetric component (= the projection on the first direct summand in the RHS) is a Killing vector,

$$(\delta^{AD}\delta^{BC} - \delta^{AC}\delta^{BD}) K_i^m + (\Sigma^{ABCD})_{ij} K^{jm}, \qquad (7.26)$$

while the skew–symmetric part (second summand) is an exact form

$$d(\delta^{AC} \mu^{BD\,m} - \delta^{AD} \mu^{BC\,m} - \delta^{BC} \mu^{AD\,m} + \delta^{BD} \mu^{AC\,m}). \qquad (7.27)$$

Therefore, one has

$$D_i[(\Sigma^{AB})_j^{\ k} \partial_k \mu^{CD\,m} \pm (\Sigma^{CD})_j^{\ k} \partial_k \mu^{AB\,m}]$$
$$\pm D_j[(\Sigma^{AB})_i^{\ k} \partial_k \mu^{CD\,m} \pm (\Sigma^{CD})_i^{\ k} \partial_k \mu^{AB\,m}] = 0 \qquad (7.28)$$

and so

$$(\Sigma^{AB})_i^{\ k} D_j \partial_k \mu^{CD\,m} = -(\Sigma^{CD})_j^{\ k} D_i \partial_k \mu^{AB\,m}. \qquad (7.29)$$

We write

$$(H[\mu^{AB\,m}])^i_{\ j} := D^i \partial_j \mu^{AB\,m} \qquad (7.30)$$

for the (covariant) Hessian of the momentum map (seen as an element of $\mathrm{End}(T\mathcal{M})$). Then

Lemma 7.4 *One has*

$$\Sigma^{AB} H[\mu^{CD\,m}] = H[\mu^{AB\,m}] \Sigma^{CD}. \qquad (7.31)$$

7.1.6 Σ^{AB}–harmonic functions

Definition 7.5 A function $F(\phi)$ on $\mathcal{M}$ is Σ^{AB}–harmonic if there exists a function $G(\phi)$ such that

$$\partial_i F = -(\Sigma^{AB})_{ij}\, \partial^j G. \tag{7.32}$$

If $F(\phi)$ is Σ^{AB}–harmonic, so is $G(\phi)$ (because $(\Sigma^{AB})^2 = -1$). $G(\phi)$ is called *the Σ^{AB}–conjugate function* to $F(\phi)$.

A Σ^{AB}–harmonic function $F(\phi)$ is, in particular, harmonic in the usual sense:

$$D^i \partial_i F = -(\Sigma^{AB})_{ij}\, D^i \partial^j G = 0, \tag{7.33}$$

since $(\Sigma^{AB})_{ij} = -(\Sigma^{AB})_{ji}$. The definition (7.32) is equivalent to

$$\left(D_i - i(\Sigma^{AB})_i{}^j D_j\right)(F + iG) = 0, \tag{7.34}$$

and we say that the complex function $F + iG$ is Σ^{AB}–holomorphic. A Σ^{AB}–harmonic function is the real part of a Σ^{AB}–holomorphic one. If the functions Φ_i are Σ^{AB}–holomorphic, any analytic function $F(\Phi_1, \Phi_2, \ldots, \Phi_k)$ is also Σ^{AB}–holomorphic.

The condition of being Σ^{AB}–harmonic can be written as the condition of integrability of Eq. (7.32):

$$0 = D_{[j}\left((\Sigma^{AB})_{i]}{}^k \partial_k F\right) = (\Sigma^{AB})_{[i}{}^k D_{j]}\partial_k F. \tag{7.35}$$

Defining the Hessian (symmetric) endomorphism $H[F]^i{}_j = D^i \partial_j F$, the above condition reads

$$\Sigma^{AB} H[F] = -H[F]\, \Sigma^{AB}, \tag{7.36}$$

that is, F is Σ^{AB}–harmonic if and only if its Hessian $H[F]$ *anticommutes* with Σ^{AB}.

Next, we ask under which condition a function F can be simultaneously harmonic (or holomorphic) with respect to a *set* $\{\Sigma^{A_1,B_1}, \Sigma^{A_2,B_2}, \ldots, \Sigma^{A_r,B_r}\}$ of parallel 2–forms (with square -1). In particular, we are interested in holomorphic functions with respect to two particular sets: the set of *all* $\mathcal{N}(\mathcal{N}-1)/2$ parallel 2–forms Σ^{AB}, and the set of the $(\mathcal{N}-1)$ parallel forms of the form Σ^{1a}, i.e., the set of the generators of $\mathbb{C}l(\mathcal{N}-1)$. Then:

Corollary 7.6 *A non–trivial harmonic/holomorphic function with respect to all Σ^{AB} exists only for $\mathcal{N} \leq 2$; a non–trivial harmonic/holomorphic function with respect to the $\Sigma^{1,a}$ ($a = 2, 3, \ldots, \mathcal{N}$) exists only for $\mathcal{N} \leq 3$.*

Non–trivial means that the Hessian $D_i \partial_j F$ is not identically zero. Indeed, the Hessian has to anticommute with all Σ^{AB}. But if it anticommutes with, say, Σ^{12} and Σ^{13}, it automatically *commutes* with $\Sigma^{23} = \Sigma^{12}\Sigma^{13}$.

Of course, we can fix one parallel 2–form, say Σ^{AB}, use it to define a complex structure, prove that it is integrable; that is, that there exist complex local coordinates x^i such that Σ^{AB} takes the special form:

$$(\Sigma^{AB})_{i\bar{j}} = ig_{i\bar{j}}, \quad (\Sigma^{AB})_{\bar{i}j} = -ig_{\bar{i}j}, \quad (\Sigma^{AB})_{ij} = (\Sigma^{AB})_{\bar{i}\bar{j}} = 0. \tag{7.37}$$

In these coordinates, a Σ^{AB}–harmonic function is just the real part of a holomorphic function of the x^i. Thus, for a *single* 2–form, our notion is just the standard one. We are, however, also interested in the interplay between different choices of complex structures Σ^{AB}.

7.1.7 Σ–harmonicity of momentum maps

From Eq. (7.11) one has

$$(\Sigma^{AB})_i^{\ j} \partial_j \mu^{CD}$$
$$= -2 \delta^{A[C} \delta^{D]B} K_i - 4 \delta^{[A[C} \partial_i \mu^{D]B]} + (\Sigma^{ABCD})_i^{\ j} K_j. \tag{7.38}$$

Taking, say, $D = A$ with A, B, and C all distinct:

$$(\Sigma^{AB})_i^{\ j} \partial_j \mu^{CA} = -\partial_i \mu^{BC}, \tag{7.39}$$

and therefore:

Lemma 7.7 *The following momentum maps are Σ^{AB}–harmonic:*

(1) $\mu^{AC} = -\mu^{CA}$ with $C \neq B$.
(2) $\mu^{CB} = -\mu^{BC}$ with $C \neq A$.

Moreover

$$\mu^{AC} + i\mu^{BC}, \quad C \neq A, B, \text{ is } \Sigma^{AB}\text{–holomorphic.} \tag{7.40}$$

Exercise 7.1.1 Write the momentum map for $\mathfrak{su}(n+1)$ on $P\mathbb{C}^n$.

7.2 T–tensors I

The momentum map $\mu^\bullet$ is a function from $\mathcal{M}$ to $\mathrm{spin}(\mathcal{N}) \otimes \mathfrak{iso}_0$ which is not $\mathfrak{iso}_0$–gauge–invariant; rather, it transforms in the adjoint representation. In order to construct a *gauge–invariant* theory, we need to work with gauge–invariant quantities.[9] Therefore we need to study the gauge–invariant functions on $\mathcal{M}$. The peculiar geometry of the isometries of a "supersymmetric" manifold $\mathcal{M}$ gives us a set of natural functions directly related to the symmetries we wish to gauge, namely the momentum maps $\mu^{AB\,m}$. Then it is natural to construct the gauge–invariant expressions starting from these basic, God–given functions.

Let $\mathfrak{g} \subset \mathfrak{iso}_0$ be the Lie subalgebra we wish to gauge. The simplest way to construct a $\mathfrak{g}$–invariant function out of the $\mu^{AB\,m}$ is to take an *invariant symmetric tensor* in $\odot^2 \mathfrak{g} \subset \odot^2 \mathfrak{iso}_0$, l_{mn}, and consider the following tensor belonging to $\odot^2 \mathfrak{spin}(\mathcal{N})$

$$T^{AB,CD} = \mu^{AB\,m}\, l_{mn}\, \mu^{CD\,n}. \tag{7.41}$$

$T^{AB,CD}$ is $\mathfrak{g}$–invariant by construction. In the SUGRA jargon, it is called the *T–tensor associated with the invariant tensor $l_{mn} \in \odot^2 \mathfrak{g}$*. $T^{AB,CD}$ is basically the invariant "square" of the momentum map. This object has a long and glorious history in both geometry and physics (in the particular case of just one symplectic structure, i.e., $\mathcal{N} = 2$). Atiyah and Bott [24, 25] showed that (for $\mathcal{G}$ simple) $T^{AB,AB}$ is a $\mathcal{G}$–*equivariant perfect Morse function*.[10] This should not be a surprise. Recall the relation between SUSY and Morse theory discussed in Chapter 2. An informal proof of the fact that the diagonal entries of the T–tensor are *perfect equivariant Morse functions* can be obtained by applying the logic of Chapter 2 to the gauged SUSY models we shall construct in the next chapter out of $T^{AB,CD}$. Stated differently, to gauge an isometry of $\mathcal{M}$, we need gauge–invariant functions on $\mathcal{M}$ which *do encode* the $\mathcal{G}$–covariant topology of $\mathcal{M}$ in the correct way since

$$\{\text{gauged SUSY vacua}\} \equiv \{\text{equivariant cohomology classes}\}. \tag{7.42}$$

The T–tensor has precisely the right properties.

Witten and others used these functions to get *exact quantum* expressions for the path–integral (the non–Abelian localization formula [53, 206, 312, 317]), generalizing the $U(1)$ result of Duistermaat and Heckman [22, 131]. Again, these nice quantum properties are expected on grounds of (extended) supersymmetry. Hence:[11]

General Lesson 7.8 *The gauging of a 3D supersymmetric theory is totally encoded in the corresponding T–tensor.*

---

[9] Think, for instance, the superspace approach. The pre/super–potentials should be gauge–invariant functions on $\mathcal{M}$.

[10] For a redable summary see Atiyah [22] Section 5. $\mathcal{G}$ stands for the gauge Lie group corresponding to the Lie algebra $\mathfrak{g}$.

[11] 3D is singled out as the dimension in which all bosonic (resp. fermionic) degrees of freedom live on $\mathcal{M}$ (resp. $T\mathcal{M}$) and the full physics of the gauging is encoded in the geometry of $\mathcal{M}$.

Therefore we expect that the $\mathcal{N}$–SUSY completion of the minimal gauge coupling (i.e., the induced Yukawa terms and scalar potential) in 3D takes a *universal* form in terms of $T^{AB,CD}$.

7.2.1 Properties of the T–tensor

Lemma 7.7 yields the following corollary.

Corollary 7.9 *Let $A \neq B$. Then the expressions*[12]

$$T^{AC,AD} - T^{BC,BD}, \qquad C \neq A, B, \; D \neq A, B, \tag{7.43}$$

$$T^{AC,BD} + T^{BC,AD}, \qquad C \neq A, B, \; D \neq A, B, \tag{7.44}$$

are Σ^{AB}–harmonic functions. In particular, the component

$$T^{AC,BC} \qquad C \neq A, B \tag{7.45}$$

is a Σ^{AB}–harmonic function. If A, B, C are all distinct,

$$(\Sigma^{AB})_i^{\;j} \partial_j (T^{AC,AC} - T^{BC,BC}) = 2\, \partial_i T^{AC,BC}. \tag{7.46}$$

The T–tensors satisfy a number of quite intricate (and interesting) differential identities. We write only the *very* simplest.

Lemma 7.10 *One has the identities:*

$$(\Sigma^{AB})_i^{\;k} \left\{ K_k^m \, l_{mn} \, K_j^n + D_j \partial_k \left(\frac{1}{2} T^{AB,AB} \right) \right\} + (i \leftrightarrow j) = 0 \tag{7.47}$$

not summed over A, B!

$$(\Sigma^{AB})_i^{\;k} (\Sigma^{AB})_j^{\;h} \left[K_k^m \, l_{mn} \, K_h^n + D_k \partial_h \left(\frac{1}{2} T^{AB,AB} \right) \right]$$
$$= K_i^m \, l_{mn} \, K_j^n + D_i \partial_j \left(\frac{1}{2} T^{AB,AB} \right), \tag{7.48}$$

not summed over A, B!

---

[12] In the statement Corollary 7.9 there is *no* sum over the repeated indices!

230   *Parallel structures and isometries*

Proof The second identity is an immediate consequence of the first one: just contract it with $(\Sigma^{AB})_l^j$. For the second one, consider

$$\begin{aligned}
(\Sigma^{AB})_i{}^k K_k^m l_{mn} K_j^n + (i \leftrightarrow j) &= (\partial_i \mu^{AB\,m}) l_{mn} K_j^n + (i \leftrightarrow j) \\
&= D_i \left(\mu^{AB\,m} l_{mn} K_j^n \right) + (i \leftrightarrow j) \\
&= -D_i \left(\mu^{AB\,m} l_{mn} (\Sigma^{AB} \Sigma^{AB})_j{}^k K_k^n \right) + (i \leftrightarrow j) \\
&= -(\Sigma^{AB})_j{}^h D_i \left(\mu^{AB\,m} l_{mn} \partial_h \mu^{AB\,n} \right) + (i \leftrightarrow j) \\
&= -(\Sigma^{AB})_j{}^h D_i \partial_h \left(\frac{1}{2} \mu^{AB\,m} l_{mn} \mu^{AB\,n} \right) + (i \leftrightarrow j) \\
&= -(\Sigma^{AB})_i{}^k D_j \partial_k \left(\frac{1}{2} T^{AB,AB} \right) + (i \leftrightarrow j). \quad (7.49)
\end{aligned}$$

□

7.3 Target space isometries in supergravity

In Sections 7.1 and 7.2 we exploited the symplectic structures of the target space in *rigid* SUSY to construct natural $\mathcal{G}$–invariant functions on $\mathcal{M}$ that encode all the relevant aspects of the isometry group to be gauged. These functions, the components of the T–tensor, are "perfect" geometric objects playing a crucial role in both the symplectic geometry and in the quantization procedure. This magic happened because $\mathcal{M}$ was a (multi–)symplectic manifold, while the symplectic forms Σ^{AB} have a direct relation to the action of extended SUSY on $T\mathcal{M}$. Our next task is to gauge subgroups of Iso($\mathcal{M}$) in the *local* SUGRA case. After what we learned above in rigid SUSY, the obvious idea is to try to generalize the momentum maps $\mu^{\bullet}$ and the T–tensors to the local situation. *However,* in SUGRA, $\mathcal{M}$ is certainly *not* symplectic! This means that we have to extend $\mu^{\bullet}$ outside the realm of symplectic geometry. In the literature there are a few generalizations of the momentum map (e.g., ref. [6]), but we are *not* looking for *some* generalization of $\mu^{\bullet}$: we are looking for a generalization which is intrinsically linked to the geometry and physics of SUGRA. Luckily, such a generalization exists. It turns out that we can define naturally $\mu^{\bullet}$ and $T^{\bullet,\bullet}$ precisely for two classes of manifold: (i) the (multi)symplectic ones (rigid SUSY), and (ii) the "SUGRA *manifolds,*" i.e., Riemannian manifolds with holonomy $Spin(\mathcal{N}) \times H$ and *covariantly constant* spin($\mathcal{N}$)–valued 2–forms Σ^{AB} such that the $Spin(\mathcal{N})$ part of the Riemann curvature is $\frac{1}{2}\Sigma^{AB}$. This remarkable fact is deeply rooted in the special parallel structures over the SUGRA manifolds. In order to understand the issue geometrically, we have to consider the relations of the holonomy and isometry groups.

7.4 Holonomy vs. isometries

In Section 7.1 we deduced the interplay between the holonomy of $\mathcal{M}$ (encoded in the system of parallel forms, according to Fundamental Principle 3.3) using well-known results in symplectic geometry. There exists a more general theory relating the holonomy and isometry groups of a Riemannian manifold, of which the symplectic case is just a special instance.

7.4.1 The derivation A_X

We begin by introducing[13] a new derivation acting on the tensor fields (in particular, forms) over a Riemannian manifold $\mathcal{M}$.

Let X be any vector field. Set

$$A_X = \pounds_X - D_X. \tag{7.50}$$

The operator A_X, being the difference of two derivations, is also a derivation. Let f be a smooth function: $A_X f = (\pounds_X - D_X)f = 0$, so A_X is a derivation that is trivial on functions, and hence is represented by a tensor (that is, it is an algebraic rather than a differential operator).

Proposition 7.11 (see [214, 215]) $\mathcal{M}$ is Riemannian. X, Y any vector fields on $\mathcal{M}$. Then

$$A_X Y = -D_Y X. \tag{7.51}$$

Let K be a Killing vector. Then, for any vector field X,

$$D_X(A_K) = R(K, X), \tag{7.52}$$

($R(\cdot, \cdot)$ is the curvature endomorphism). If K^1, K^2 are both Killing vectors,

$$A_{[K^1, K^2]} = [A_{K^1}, A_{K^2}] + R(K^1, K^2). \tag{7.53}$$

Proof (1) One has

$$\begin{aligned} A_X Y &\equiv \pounds_X Y - D_X Y && \text{[by definition of the torsion } T(X, Y)\text{]} \\ &= [X, Y] - D_X Y - (T(X, Y) - D_X Y + D_Y X + [X, Y]) \\ &= -D_Y X - T(X, Y) && \text{[since the Levi–Civita connection is torsionless]} \\ &= -D_Y X. && (7.54) \end{aligned}$$

---

[13] References for this subsection are: Chapter VI of ref. [215], expecially propositions 2.5 and 2.6 and Section 4; ref. [214], Chapter II; and ref. [263], Chapter 26 Section 4.

(2) By the definition of curvature:

$$R(K,X) = [D_K, D_X] - D_{[K,X]} = [\mathcal{L}_K - A_K, D_X] - D_{[K,X]}$$
$$= [\mathcal{L}_K, D_X] - D_{[K,X]} - [A_K, D_X]. \tag{7.55}$$

For a Killing vector, however, one has also[14]

$$[\mathcal{L}_K, D_X] = D_{[K,X]}. \tag{7.56}$$

Thus

$$R(K,X) = -[A_K, D_X] \equiv D_X(A_K). \tag{7.57}$$

For reference, we write this equation in coordinates,

$$D_i D_j K_k = -R_{jkil} K^l. \tag{7.58}$$

(3) Let K^1, K^2 be Killing fields.

$$0 = \mathcal{L}_{K^1}\mathcal{L}_{K^2} - \mathcal{L}_{K^2}\mathcal{L}_{K^1} - \mathcal{L}_{[K^1,K^2]}$$
$$= (D_{K^1} + A_{K^1})(D_{K^2} + A_{K^2}) - (D_{K^2} + A_{K^2})(D_{K^1} + A_{K^1}) - (D_{[K^1,K^2]} + A_{[K^1,K^2]})$$
$$= R(K^1, K^2) + D_{K^1}A_{K^2} - D_{K^2}A_{K^1} + [A_{K^1}, A_{K^2}] - A_{[K^1,K^2]}. \tag{7.59}$$

The identity (7.52) yields

$$D_{K^1}A_{K^2} = -R(K^1, K^2), \qquad D_{K^2}A_{K^1} = -R(K^2, K^1), \tag{7.60}$$

so (7.59) becomes

$$[A_{K^1}, A_{K^2}] = A_{[K^1,K^2]} + R(K^1, K^2). \tag{7.61}$$

$\square$

Example. In Proposition 7.2 we described the properties of the derivation A_K for a (pluri)symplectic manifold. One checks that the properties we describe in the present section hold, in particular, for the symplectic A_K.

Example. Let $\mathcal{M} = G$ be a group manifold. Take the left–invariant vectors X^a as a basis of $T\mathcal{M}$, Then from Proposition 5.2 we have

$$A_{X^a}X^b = -D_{X^b}X^a = -\frac{1}{2}[X^b, X^a] \equiv \frac{1}{2}[X^a, X^b]. \tag{7.62}$$

So (up to normalization) A_X gives the adjoint action of the Lie algebra $\mathfrak{g}$.

---

[14] The Levi–Civita connection D_X is constructed out of the metric, and hence should transform covariantly under isometries.

7.4.2 Kostant's theorem

Theorem 7.12 (Kostant [215, 219]) *Let $\mathfrak{hol}$ be the holonomy algebra of the Riemannian manifold $\mathcal{M}$ and $\mathfrak{n}(\mathfrak{hol})$ its normalizer in $\mathfrak{so}(\dim \mathcal{M})$. Let K be a Killing vector, and let $A_K \in \text{End}(T\mathcal{M})$ be the derivation $\pounds_K - D_K$. Let $\phi \in \mathcal{M}$ be any point. Then*[15]

$$A_K|_\phi \in \mathfrak{n}(\mathfrak{hol}|_\phi) \subset \text{End}(T_\phi \mathcal{M}). \tag{7.63}$$

Proof By equation (7.51), $(A_K)_{ij} \equiv -D_j K_i$ is a 2–form. Decompose the 2–form $(A_K)_{ij}|_\phi \in \mathfrak{so}(\dim \mathcal{M})$ as

$$A_K = B_K + E_K, \tag{7.64}$$

where $B_K \in \mathfrak{hol} \subset \wedge^2 T\mathcal{M}$ and $E_K \in \mathfrak{hol}^\perp \subset \wedge^2 T\mathcal{M}$. We claim that the 2–form E_K is *parallel*. Indeed, for any vector X

$$D_X A_K = R(K, X) \in \mathfrak{hol} \subset \wedge^2 T\mathcal{M}, \tag{7.65}$$

or, in coordinates,

$$D_k (A_K)_{ij} = D_k D_i K_j = -R_{ijkl} K^l \in \mathfrak{hol} \subset \wedge^2 T\mathcal{M}, \tag{7.66}$$

where we have used the the Ambrose–Singer theorem. Equation (7.65) implies

$$D_k (E_K)_{ij} = 0, \tag{7.67}$$

i.e., E_K is *parallel*. By Fundamental Principle 3.3, E_K commutes with the action of $\mathfrak{hol}$. Hence Eq. (7.63) is proven. □

The proof implies the stronger result

$$E_K \in \mathfrak{c}(\mathfrak{hol}) \setminus (\mathfrak{hol} \cap \mathfrak{c}(\mathfrak{hol})), \tag{7.68}$$

the centralizers $\mathfrak{c}(\mathfrak{hol})$ being described (for $\mathcal{M}$ simply connected irreducible) in Proposition 3.21. See Table 7.1. In particular, $E_K \equiv 0$ if $\mathcal{M}$ is compact or irreducible non–Ricci–flat [219]. Then Eq. (7.53) gives

$$[E_{K^1}, E_{K^2}] - E_{[K^1, K^2]} \in \mathfrak{hol} \cap \mathfrak{c}(\mathfrak{hol}) \equiv \text{Abelian summand of } \mathfrak{hol}. \tag{7.69}$$

Comparing with Table 7.1 gives us Corollary 7.13:

Corollary 7.13 *$\mathcal{M}$ simply connected and complete:*

$$[E_{K^1}, E_{K^2}] - E_{[K^1, K^2]} = 0. \tag{7.70}$$

---

[15] $\mathfrak{hol}|_\phi$ is the Lie algebra of the holonomy group viewed as the group of parallel transports along loops starting and ending at the point ϕ.

Table 7.1 $\mathfrak{c}(\mathfrak{hol}) \setminus (\mathfrak{hol} \cap \mathfrak{c}(\mathfrak{hol}))$ *for simply connected* $\mathcal{M}$

$\mathcal{M}$	$\mathfrak{c}(\mathfrak{hol}) \setminus (\mathfrak{hol} \cap \mathfrak{c}(\mathfrak{hol}))$
Flat	$\mathfrak{so}(\dim \mathcal{M})$
Irreducible Calabi–Yau	$\mathfrak{u}(1)$
Irreducible hyperKähler	$\mathfrak{sp}(2)$
All other irreducible	0

7.4.3 First consequences of Kostant's theorem

Theorem 7.12 is the fundamental link between $\mathfrak{hol}$ and $\mathfrak{iso}$. Let us see what it implies for the isometry groups of the SUGRA manifolds. Recall that, in a SUGRA manifold, $\mathfrak{hol} = \mathfrak{spin}(\mathcal{N}) \oplus \mathfrak{h}'$ so $E_K \equiv 0$ for $\mathcal{N} > 2$. Hence:

Corollary 7.14 $\mathcal{M}$ *a* SUGRA *n-fold for* $\mathcal{N} > 2$. *There exist* A_K^{AB} *such that*

$$-D_j K_i \equiv (A_K)_{ij} = \frac{1}{4} A_K^{AB} (\Sigma^{AB})_{ij} + (H_K)_{ij} \qquad (7.71)$$

with $H_K \in \mathfrak{h}'$. *In particular,*

$$[H_K, \Sigma^{AB}] = 0. \qquad (7.72)$$

Corollary 7.15 *There is a map* $\sigma : \mathfrak{iso}(\mathcal{M}) \to \mathfrak{spin}(\mathcal{N})$ *such that*[16]

$$\pounds_{K^m} \Sigma = \mathrm{adj}_{\sigma(K^m)} \Sigma \qquad (7.73)$$

or, explicitly,

$$\pounds_{K^m}(\Sigma^{AB})_{ij} = \sigma_{K^m}^{AC} (\Sigma^{CB})_{ij} - (\Sigma^{AC})_{ij} \sigma_{K^m}^{CB}, \qquad (7.74)$$

for certain $\sigma_{K^m}^{AB} = -\sigma_{K^m}^{BA} \equiv \sigma(K^m)$: *see Eq.* (7.77). *Moreover,*

$$\pounds_{K^m} Q_i^{AB} = \mathcal{D}_i \sigma_{K^m}^{AB}. \qquad (7.75)$$

---

[16] Compare with Eq. (7.82) for the RIGID case!

7.4 Holonomy vs. isometries

Proof From the previous Corollary 7.14, one has,[17]

$$\begin{aligned}\pounds_K(\Sigma^{AB})_i^j &= K^k D_k(\Sigma^{AB})_i^j + (D_i K^k)(\Sigma^{AB})_k^j + (\Sigma^{AB})_i^k (D_j K^k) \\ &= -(K^k Q_k)^{AC}(\Sigma^{CB})_i^j + (\Sigma^{AC})_i^j (K^k Q_k)^{CB} \\ &\quad + \frac{1}{4} A_K^{CD} (\Sigma^{CD} \Sigma^{AB} - \Sigma^{AB} \Sigma^{CD})_i^j \\ &= \sigma_K^{AC}(\Sigma^{CB})_i^j - (\Sigma^{AC})_i^j \sigma_K^{CB}, \end{aligned} \quad (7.76)$$

where

$$\sigma_K^{AB} = A_K^{AB} - K^i Q_i^{AB}. \quad (7.77)$$

To get (7.75), recall that $\frac{1}{2}\Sigma$ is the field–strength of the $Spin(\mathcal{N})$ connection Q_i^{AB}; then we recognize in Eq. (7.74) the gauge transformation of the $Spin(\mathcal{N})$ field–strength under an infinitesimal $Spin(\mathcal{N})$ gauge transformation of parameter σ_{Km}^{AB}. Equation (7.75) is the usual gauge transformation of a connection. □

Corollary 7.16 $\mathcal{M}$ *an* $\mathcal{N} \geq 3$ SUGRA *Riemannian manifold. Let* $P(\cdot)$ *be a homogeneous* invariant *polynomial on* spin($\mathcal{N}$) *of degree k (i.e., a generalized Casimir invariant),*

$$P(L) = P_{A_1 B_1, A_2 B_2 \cdots A_k B_k} L^{A_1 B_1} L^{A_2 B_2} \cdots L^{A_k B_k}. \quad (7.78)$$

Consider the parallel *2k–form* $P(\Sigma)$:

$$P(\Sigma) = P_{A_1 B_1 A_2 B_2 \cdots A_k B_k} \Sigma^{A_1 B_1} \wedge \Sigma^{A_2 B_2} \wedge \cdots \wedge \Sigma^{A_k B_k}. \quad (7.79)$$

Let K be ANY *Killing vector on* $\mathcal{M}$. *Then*

$$\pounds_K P(\Sigma) = 0. \quad (7.80)$$

If $\mathcal{M}$ is compact, this result already follows from the theorem in the footnote 13 of Chapter 2 (p. 57). Indeed, a parallel form is *a fortiori* a harmonic form.

Since $\frac{1}{2}\Sigma$ is the $Spin(\mathcal{N})$ curvature, the closed $2k$–form

$$\frac{1}{2^k} P(\Sigma) \quad (7.81)$$

represents a *characteristic class* [94, 215, 235] of the $Spin(\mathcal{N})$ bundle $M^{\oplus p} \oplus \widetilde{M}^{\oplus q}$ ($p + q = \dim \mathcal{M}/\mathbf{N}(\mathcal{N})$, see Chapter 2).

---

[17] Recall that Q_i^{AB} is the canonical $Spin(\mathcal{N})$ connection defined by SUGRA. Σ^{AB} is covariantly constant with respect to the sum of the Levi–Civita and $Spin(\mathcal{N})$ connections.

7.5 The rigid case revisited. Superconformal geometries

Let us reconsider the rigid SUSY manifolds in the light of Kostant's theorem. $\mathcal{M}$ is a (pluri)symplectic manifold with parallel forms Σ^{AB}. From the definition of E_K in terms of $\pounds_K - D_K$, Eq. (7.64), and the parallel property of the 2–forms Σ^{AB}, we get

$$\pounds_K \Sigma^{AB} = [A_K, \Sigma^{AB}] \equiv [E_K, \Sigma^{AB}]. \tag{7.82}$$

$K \in \mathfrak{iso}_0$ iff the LHS vanishes; the first equality then gives back Proposition 7.2. In particular,

$$E_K = 0 \implies K \in \mathfrak{iso}_0. \tag{7.83}$$

Now suppose, for $\mathcal{N} \le 4$, that $\mathcal{M}$ is *strict* in the sense that its holonomy group is the generic one for the given $\mathcal{N}$: that is, $\mathfrak{hol}$ is equal to the centralizer of $\mathfrak{spin}(\mathcal{N})$ in $\mathfrak{so}(\dim \mathcal{M})$. In this case, the *only* parallel tensors are those in the canonical Clifford algebra $\mathbb{C}l(\mathcal{N} - 1)$. Moreover, for $\mathcal{N} > 2$, $\mathfrak{hol}$ is semi–simple, and $\mathfrak{hol} \cap \mathfrak{c}(\mathfrak{hol}) = 0$. Hence, for $\mathcal{N} = 3, 4$, $E_K \in \mathfrak{spin}(\mathcal{N})$ and the converse to Eq. (7.83) holds

$$\mathcal{N} = 3, 4, \text{ strict} \qquad E_K = 0 \iff K \in \mathfrak{iso}_0. \tag{7.84}$$

Another interesting class of Killing vector is those with $B_K = 0$. Consider the following situation: we have an $\mathcal{N}$–SUSY manifold $\mathcal{M}$ with an isometry group isogeneous to $Spin(\mathcal{N})$ under which Σ^{AB} transforms in the adjoint representation. The corresponding physical symmetries will act on the supercharges Q^A according to the vector representation, and hence will be a group of R–symmetries. These isometries are generated by Killing vectors $K_i^{AB} = -K_i^{BA}$ such that

$$\pounds_{K^{AB}} \Sigma^{CD} = [\Sigma^{AB}, \Sigma^{CD}] = f^{ABCD}{}_{EF} \Sigma^{EF}. \tag{7.85}$$

For $\mathcal{N} > 2$ this equation implies

$$E_{K^{AB}} = \Sigma^{AB}. \tag{7.86}$$

Moreover, from Eq. (7.85) we see that

$$\pounds_{K^{AB}} \Sigma^{CD} + \pounds_{K^{CD}} \Sigma^{AB} = 0, \tag{7.87}$$

that is,

$$d\big(i_{K^{AB}} \Sigma^{CD} + i_{K^{CD}} \Sigma^{AB}\big) = 0. \tag{7.88}$$

7.5 The rigid case revisited. Superconformal geometries

Assuming $\mathcal{M}$ to be simply connected, this allows us to define a kind of "$\odot^2\mathfrak{spin}(\mathcal{N})$–valued momentum map" as the function $M^{AB\,CD}$ such that

$$i_{K^{AB}}\Sigma^{CD} + i_{K^{CD}}\Sigma^{AB} = dM^{AB\,CD}, \qquad M^{ABCD} \in \odot^2(\wedge^2\mathcal{N}). \tag{7.89}$$

We take $\mathcal{N} > 2$ so that the adjoint representation $\mathfrak{spin}(\mathcal{N})$ is non–trivial. Then we may decompose $M^{AB\,CD}$ into irreducible representations of $Spin(\mathcal{N})$. Suppose that only the singlet and the totally antisymmetric components are non–trivial (that is, non–constant). One has:

Proposition 7.17 *Assume $\mathcal{N} > 2$. $\mathcal{M}$ a (rigid) $\mathcal{N}$–SUSY manifold whose isometry group contains a subgroup (isogenous to) $Spin(\mathcal{N})$ under which the parallel symplectic forms transform in the adjoint representation and the decomposition of $M^{AB\,CD}$ in irreducible representations contains only the singlet and the totally antisymmetric one:*

$$M^{AB\,CD} = 2M(\delta^{AC}\delta^{BD} - \delta^{AD}\delta^{BC}) + N^{[ABCD]} \in \mathbb{R} \oplus \wedge^4\mathcal{N}. \tag{7.90}$$

Then $\mathcal{M}$ is (locally isometric to) a <u>metric cone</u>, i.e., there is a function V such that

$$g_{ij} = D_i\partial_j V, \qquad g^{ij}\partial_i V\partial_j V = 2V, \tag{7.91}$$

and explicitly

$$V = \frac{K^{ABk}K_k^{AB}}{2\mathcal{N}(\mathcal{N}-1)}. \tag{7.92}$$

Moreover,

$$D_i K_j^{AB} = \Sigma_{ij}^{AB}, \tag{7.93}$$

i.e., $B_{K^{AB}} = 0$. <u>Conversely:</u> if a rigid $\mathcal{N}$–SUSY manifold (for any $\mathcal{N}$) is a metric cone, then its isometry algebra has the form $\mathfrak{iso}_o \oplus \mathfrak{spin}(\mathcal{N})$ with the $\mathfrak{spin}(\mathcal{N})$ Killing vectors satisfying Eqs. (7.90)–(7.93).

Proof For $\mathcal{N} \geq 5$ the statement is trivially true since $\mathcal{M}$ is flat; hence its universal cover is $\mathbb{R}^n$, which is a cone over S^{n-1}. In the flat coordinates, one has $K_i^{AB} = x^j \Sigma_{ji}^{AB}$, and (7.93) holds. For $\mathcal{N} = 3, 4$ $\mathcal{M}$ is a (product of) hyperKähler manifolds and the statement reduces to proposition 1.6 of ref. [64] (see also [282]). We sketch the argument.

We start by showing the inverse implication: $\mathcal{M}$ a rigid $\mathcal{N}$–SUSY cone $\Rightarrow$ $\mathfrak{iso}(\mathcal{M}) = \mathfrak{iso}_o \oplus \mathfrak{spin}(\mathcal{N})$, where the summand $\mathfrak{spin}(\mathcal{N})$ is generated by Killing vectors K^{ABi} satisfying Eqs. (7.91)–(7.93). We recall that a vector field V_i such that $g_{ij} = D_i V_j$ is called *concurrent*. A Riemannian manifold is a metric cone if and only if it has a concurrent vector field; see theorem III.5.4 and theorem III.5.5 in ref. [325]. Then $V_i = D_i V$ for some function V, while V_i is a conformal Killing vector, $\pounds_V g_{ij} = 2 g_{ij}$, and V may be chosen so that $\pounds_V V = 2V$, which is Eq. (7.91). Then, given an $\mathcal{N}$–SUSY cone with concurrent vector V^i, we define the vector fields $K_j^{AB} = V^k (\Sigma^{AB})_{kj}$. One has

$$D_i K_j^{AB} = (D_i V^k)(\Sigma^{AB})_{kj} = (\Sigma^{AB})_{ij} \equiv -(\Sigma^{AB})_{ji}, \tag{7.94}$$

so the K_i^{AB} are $\mathfrak{spin}(\mathcal{N})$ Killing vectors with all the stated properties.

For the implication in the other direction (assuming $\mathcal{N} \geq 3$ and (7.90)) the only non–trivial cases are $\mathcal{N} = 3, 4$. The manifolds for $\mathcal{N} = 4$ are a product of two hyperKähler spaces, so we are reduced to the case $\mathcal{N} = 3$. We write $\Sigma^A = \frac{1}{2} \epsilon^{ABC} \Sigma^{BC}$ and $K^{Ai} = \frac{1}{2} \epsilon^{ABC} K^{BCi}$; Eq. (7.4) becomes

$$\Sigma^A \Sigma^B = -\delta^{AB} + \epsilon^{ABC} \Sigma^C. \tag{7.95}$$

Equations (7.89) and (7.90) imply

$$K^{Ai} \Sigma_{ij}^A = \partial_j M \quad \text{not summed over } A \tag{7.96}$$

$$\implies K_i^A = \Sigma_{ij}^A \partial^j M \tag{7.97}$$

and using Eq. (7.95),

$$\begin{aligned} \Sigma^{Ai}{}_j K^{Bj} &= -\delta^{AB} \partial^i M + \epsilon^{ABC} \Sigma^{Ci}{}_j \partial^j M \\ &= -\delta^{AB} \partial^i M + \epsilon^{ABC} K^{Ci}. \end{aligned} \tag{7.98}$$

From the last equation and (7.85), we infer

$$\begin{aligned} (B_{K^A})_{ij} + \Sigma_{ij}^A \equiv (A_{K^A})_{ij} = D_i K_j^A &= \frac{1}{4} \epsilon^{ABC} [A_{K^B}, \Sigma^C]_{ij} \\ &= \frac{1}{4} \epsilon^{ABC} [\Sigma^B, \Sigma^C]_{ij} \equiv \Sigma_{ij}^A, \end{aligned} \tag{7.99}$$

that is, $B_{K^A} = 0$. This proves Eq. (7.93). To show the rest, let

$$\begin{aligned} V_i &= \frac{1}{\mathcal{N}(\mathcal{N}-1)} (\Sigma^{AB})_i{}^k K_k^{AB} = \frac{1}{\mathcal{N}(\mathcal{N}-1)} D_i(K^{ABk}) K_k^{AB} \\ &= \partial_i \left(\frac{1}{2\mathcal{N}(\mathcal{N}-1)} K^{ABk} K_k^{AB} \right) \equiv \partial_i V \end{aligned} \tag{7.100}$$

(summation over repeated indices implied). From the Clifford algebra, one gets

$$g_{ij} = D_i V_j = D_i \partial_j V, \tag{7.101}$$

7.5 The rigid case revisited. Superconformal geometries

so V_i is a *concurrent* Killing vector, $\pounds_V g_{ij} = 2g_{ij} \Rightarrow \pounds_V V = 2V$, i.e., $g^{ij}\partial_i V \partial_j V = 2V$, and $\mathcal{M}$ is a cone. □

Corollary 7.18 *Let $\mathcal{M}$ be an $\mathcal{N}$–SUSY cone. The $\odot^2 \mathfrak{spin}(\mathcal{N})$ momentum map is given by*

$$M^{AB\,CD} = -K^{AB\,i} K_i^{CD}. \tag{7.102}$$

Proof From Eq. (7.93) one has

$$(i_{K^{AB}} \Sigma^{CD} + i_{K^{CD}} \Sigma^{AB})_j \equiv K^{AB\,i}(\Sigma^{CD})_{ij} + K^{CD\,i}(\Sigma^{AB})_{ij}$$
$$= -K^{AB\,i} D_j K_i^{CD} - K^{CD\,i} D_j K_i^{AB} = -\partial_j(K^{AB\,i} K_i^{CD}). \ \square$$

A similar formula holds for the usual momentum map

Proposition 7.19 *Let $\mathcal{M}$ be a conical rigid $\mathcal{N}$–SUSY manifold. Let K^m be a Killing vector belonging to $\mathfrak{iso}_0$. Its momentum map is given by*

$$\mu^{AB\,m} = \frac{1}{2} K^{AB\,j} K_j^m. \tag{7.103}$$

Proof

$$\partial_i(K^{AB\,j} K_j^m) = (\Sigma^{AB})_i^{\ j} K_j^m + K^{AB\,j} D_i K_j^m$$
$$= (\Sigma^{AB})_i^{\ j} K_j^m - K^{AB\,j} D_j K_i^m$$
$$= (\Sigma^{AB})_i^{\ j} K_j^m - (K^{AB\,j} D_i K_j^m - K^{m\,j} D_i K_j^{AB}) - K^{m\,j} D_i K_j^{AB}$$
$$= (\Sigma^{AB})_i^{\ j} K_j^m - K^{m\,j} D_i K_j^{AB}$$
$$= (\Sigma^{AB})_i^{\ j} K_j^m - K^{m\,j} (\Sigma^{AB})_{ij} = 2(\Sigma^{AB})_i^{\ j} K_j^m, \tag{7.104}$$

where we have used $\pounds_{K^{AB}} K^m = 0$. □

Physical meaning: superconformal invariance

It follows from Chapter 2 that if our 3D (ungauged) SUSY model has a $Spin(\mathcal{N})$ R–symmetry, under which the supercharges Q^A transform in the vector representation, the corresponding scalar manifold $\mathcal{M}$ has a $Spin(\mathcal{N})$ isometry group under which the parallel forms Σ^{AB} transform in the adjoint. The conserved R–currents are of the form $K_i^{AB} \partial_\mu \phi^i + \cdots$ so at weak coupling the two–current correlation function has a term proportional to $M^{AB\,CD}$. The requirement in Proposition 7.17 on the representation content of $M^{AB\,CD}$ may be interpreted physically in terms of usual current algebra properties of the R–currents. In particular, this is certainly the case whenever the rigid model is *superconformal*. In that case we have a full $Spin(\mathcal{N})$ R–symmetry group, under which the supercharges transform in the vector representation and the two–points of the R–currents are proportional to invariant tensors. Comparing with Proposition 7.17:

General Lesson 7.20 *In rigid $\mathcal{N} > 2$ 3D* SUSY *the theory is superconformal* $\Rightarrow$ $\mathcal{M}$ *is a metric cone. Conversely* [46], *if $\mathcal{M}$ is a cone the* SUSY *model is classically superconformal.*

In the case $\mathcal{N} \leq 2$ the fact that $\mathcal{M}$ is a cone does not guarantee superconformal invariance at the quantum level; in particular, for $\mathcal{N} = 2$ the condition of *quantum* superconformal invariance leads to a different (but still nice) geometric characterization of $\mathcal{M}$. For $\mathcal{N} \geq 3$ the above statement may be seen as a non-renormalization theorem which protects superconformal invariance at the quantum level of the 3D σ–models based on *conical* geometries. Note, however, that the conical geometries are everywhere smooth only when they are flat.

7.6 The Cartan–Kostant isomorphism

In ref. [120] the authors construct, by clever manipulations, a remarkable Lie algebra homomorphism, which we like to present in the original version due to E. Cartan (for the symmetric spaces) and B. Kostant (in general).

7.6.1 General theorems

Let $\mathcal{M}$ be *any connected* Riemannian manifold and $\phi \in \mathcal{M}$ an arbitrary point. $\wedge^2 T_\phi \mathcal{M}$, viewed as a subspace of $\text{End}(T_\phi \mathcal{M})$ via $\Lambda_{ij} \leftrightarrow \Lambda^i{}_j \equiv g^{ik}\Lambda_{kj}$, is a Lie algebra $\mathfrak{s} \simeq \mathfrak{so}(\dim \mathcal{M})$. Set

$$\mathfrak{G} = \mathfrak{s} \oplus T_\phi \mathcal{M}. \qquad (7.105)$$

On the vector space $\mathfrak{G}$ we introduce a skew–symmetric bilinear bracket $[\cdot, \cdot]$

$$[s_1, s_2] = s_1 s_2 - s_2 s_1 \in \mathfrak{s}, \qquad \text{for } s_1, s_2 \in \mathfrak{s}, \qquad (7.106)$$

$$[s, v] = -[v, s] = s(v) \in T_\phi \mathcal{M}, \qquad \text{for } s \in \mathfrak{s}, \ v \in T_\phi \mathcal{M}, \qquad (7.107)$$

$$[v_1, v_2] = R_\phi(v_2, v_1) \in \mathfrak{s}, \qquad \text{for } v_1, v_2 \in T_\phi \mathcal{M}, \qquad (7.108)$$

where $R_\phi(\cdot, \cdot)$ is the Riemann curvature at the given (arbitrary) point ϕ. In general, $\mathfrak{G}$ is *not* a Lie algebra since the Jacobi identity does not hold.

We define a map

$$\theta_\phi : \mathfrak{iso}(\mathcal{M}) \to \mathfrak{G} \qquad (7.109)$$

7.6 The Cartan–Kostant isomorphism

by setting

$$\theta_\phi(K) = A_K|_\phi + K|_\phi \in \mathfrak{s} \oplus T_\phi \mathcal{M}. \tag{7.110}$$

The image $\theta_\phi(\mathfrak{iso}(\mathcal{M})) \subset \mathfrak{S}$ *is* a Lie algebra! Indeed:

Theorem 7.21 (Kostant [219]) *Let $\mathfrak{g}_\phi = \theta_\phi(\mathfrak{iso}(\mathcal{M})) \subset \mathfrak{S}$. The map $\theta_\phi \colon \mathfrak{iso}(\mathcal{M}) \to \mathfrak{g}_\phi$ is a Lie algebra* ISOMORPHISM.

Proof Let K^1, K^2 be Killing vectors. From Eqs. (7.51) and (7.53) we have

$$[K^1, K^2] = D_{K^1} K^2 - D_{K^2} K^1 = A_{K^1} K^2 - A_{K^2} K^1, \tag{7.111}$$

$$[A_{K^1}, A_{K^2}] = A_{[K^1, K^2]} + R(K^1, K^2). \tag{7.112}$$

Evaluating both sides at the point ϕ, we get

$$[\theta_\phi(K^1), \theta_\phi(K^2)] = \theta_\phi([K^1, K^2]), \tag{7.113}$$

so θ_ϕ is a Lie algebra homomorphism. It remains to show that θ_ϕ is injective; that is, that a Killing vector field K is uniquely determined over all $\mathcal{M}$ by the restrictions $A_K|_\phi$ and $K|_\phi$ at any given point ϕ. Since $\mathcal{M}$ is connected, this follows from the fact that (K, A_K) satisfy the first–order linear equation

$$D_i \begin{pmatrix} K_j \\ (A_K)_{jk} \end{pmatrix} = \begin{pmatrix} 0 & \delta_{il} \delta_{jm} \\ R_{jki}{}^l & 0 \end{pmatrix} \begin{pmatrix} K_l \\ (A_K)_{lm} \end{pmatrix}. \tag{7.114}$$

$\square$

Let $\varrho_\phi \colon \mathfrak{iso}(\mathcal{M}) \to \mathfrak{s}$ be the map $K \mapsto A_K|_\phi$, corresponding to the first summand of θ_ϕ. The image $\varrho_\phi(\mathfrak{iso}) \subset \mathfrak{s}_\phi$ is a Lie subalgebra. Let us specialize in the case for which $\mathcal{M}$ is an irreducible symmetric space G/H, so that $\mathfrak{iso}(G/H) = \mathfrak{Lie}(G) \equiv \mathfrak{g}$. From Section 3.2 we know that $\mathfrak{g} = \mathfrak{h} \oplus \mathfrak{m}$ with $\mathfrak{m} \simeq T_{\phi_0}(G/M)$ and $\mathfrak{h} \simeq \mathfrak{hol}(G/H)$. Comparing the bracket (7.106)–(7.108) with the one defined in Eqs. (3.36)–(3.38), we see that the two are equal under the identification $\mathfrak{s}_\phi \equiv \mathfrak{hol}_\phi$. The identification is consistent since $A_K|_\phi \equiv B_K|_\phi \in \mathfrak{hol}(G/H)$; indeed, by Proposition 3.21,

$$A_K - B_k \equiv E_K \in \mathfrak{c}(\mathfrak{hol}(G/H)) = 0. \tag{7.115}$$

Then:

Proposition 7.22 (Cartan/Kostant [219]) *Let $\mathcal{M} = G/H$ be irreducible symmetric. We have the isomorphism*

$$\varrho_\phi(\mathfrak{iso}(\mathcal{M})) \simeq \mathfrak{hol}_\phi \simeq \mathfrak{h}. \tag{7.116}$$

7.6.2 Explicit formulae for G/H

In view of the application to SUGRAs with $N_S > 8$, we are interested in explicit expressions of θ_ϕ for the irreducible symmetric spaces G/H. We write $\{t^m\}$ for a chosen basis of $\mathfrak{g}$, and $\{K^m\}$ for the corresponding Killing vectors under an isomorphism $\mathfrak{iso} \simeq \mathfrak{g} \equiv \mathfrak{h} \oplus \mathfrak{m}$. Consider the map

$$\Theta: (G/H, \mathfrak{iso}) \to \mathfrak{g} \tag{7.117}$$

given by

$$(\phi, K) \mapsto \theta_\phi(K) \in \mathfrak{g}. \tag{7.118}$$

Let

$$\mathfrak{h}_\phi \equiv \{K \in \mathfrak{g} \mid K|_\phi = 0\} \subset \mathfrak{iso}. \tag{7.119}$$

$\mathfrak{h}_\phi$ is the Lie algebra of the isotropy subgroup at ϕ, $H_\phi \subset G$, and $K \in \mathfrak{h}_\phi \Leftrightarrow A_K|_\phi \equiv \theta_\phi(K)$. Let $g \in G$ be a representative of the point $\phi \in G/H$. $k \in G$ belongs to the subgroup H_ϕ if and only if

$$kg = gh \quad \text{for some } h \in H. \tag{7.120}$$

Passing to the Lie algebras, $\tau \in \mathfrak{h}_\phi$ if and only if $(g^{-1}\tau g) \in \mathfrak{h}$. Hence the Lie algebra isomorphism θ_ϕ is given simply by

$$\theta_\phi(K^m) = (g^{-1} t^m g)_\mathfrak{h} + (g^{-1} t^m g)_\mathfrak{m}. \tag{7.121}$$

Notice that it is unique up to "H–gauge transformations." Let h^I (resp. m^a) be a basis of $\mathfrak{h}$ (resp. $\mathfrak{m}$); so $\{t^m\} \equiv \{h^I, m^a\}$. We get the formulae

$$g^{-1} t^m g = L_I^m h^I + L_a^m m^a, \tag{7.122}$$

$$A_{K^m} = L_I^m h^I. \tag{7.123}$$

We specialize in the case of a SUGRA manifold, as in Corollary 7.14. Thus $\mathfrak{h} = \mathfrak{spin}(\mathcal{N}) \oplus \mathfrak{h}'$, and we split the basis $\{h^I\}$ as $\{\frac{1}{2}\gamma^{AB}\}$, $\{h^\alpha\}$. Comparing with that corollary, we have:

Corollary 7.23 $G/[\mathrm{Spin}(\mathcal{N}) \otimes H']$ a symmetric manifold. One has

$$g^{-1} t^m g = \frac{1}{4} A_{K^m}^{AB} \gamma^{AB} + H_{K^m}^\alpha h^\alpha + L_a^m m^a. \tag{7.124}$$

The Cartan–Kostant isomorphism then says that (up to normalization) the Killing vector K^m is

$$K_a^m = L_a^m \propto \mathrm{tr}[m^a g^{-1} t^m g]. \tag{7.125}$$

Is this consistent with the formula for the Killing vectors we got in Chapter 5? *Yes:* Eq. (7.125) gives the Killing vector in "flat" indices; to convert it to "curved" indices one uses the vielbein which is given by the m–projection of the Maurer–Cartan form. Thus, assuming (7.125), we have

$$K_i^m d\phi^i \propto (g^{-1}dg)_i \, \text{tr}[m^i \, g^{-1} t^m g] = \text{tr}[(g^{-1}dg)_m \, g^{-1} t^m g]$$
$$= -[(Dg^{-1})g \, g^{-1} t^m g] = -\text{tr}[t^m (gDg^{-1})] \qquad (7.126)$$

which is our formula in Chapter 5.

7.7 The covariant momentum map

In this section we define the *covariant momentum map* for a SUGRA manifold $\mathcal{M}$ in perfect analogy with the symplectic one for the rigid case, Section 7.1, just making everything $Spin(\mathcal{N})$–covariant. The following construction reduces in the $\mathcal{N} = 3, 4$ cases (i.e., for $\mathcal{M}$ Quaternionic–Kähler) to the one given by Galicki and Lawson [154, 155].

7.7.1 Definition and first properties

Let $\mathcal{Q}^\vee \subset \wedge^2 T^*\mathcal{M}$ be the subbundle spanned by the 2–forms Σ^{AB}. One would like to state a definition of the following sort:

Definition 7.24 Let $\mathcal{M}$ be a SUGRA manifold. The *covariant momentum map* $\mu^\bullet \in \Gamma(\mathcal{Q} \otimes \mathfrak{iso}^\vee)$ is the unique (smooth) section such that

$$\mathcal{D}\mu^{ABm} = -i_{K^m} \Sigma^{AB} \qquad (7.127)$$

where $\mathcal{D} = d + Q$ is the $Spin(\mathcal{N})$–covariant exterior derivative.

Is this definition well posed? That is, does Eq. (7.127) have a solution μ^{ABm} for all Killing vectors K^m? Is it unique? We claim that, for $\mathcal{N} \geq 3$, the answer is *yes* to both questions. The uniqueness of $\mu^\bullet$ has to be understood (of course) as a section of $\Gamma(\mathcal{Q} \otimes \mathfrak{iso}^\vee)$; that is, μ^{ABm} is unique *up to a $Spin(\mathcal{N})$ (local) gauge transformation* (i.e., up to changes of trivialization of the bundle $\mathcal{Q}$). In the case $\mathcal{N} = 2$, $Spin(2)$ is Abelian, and hence the connection is trivial in the adjoint representation; thus $\mathcal{D} = d$ acting on $\mu^\bullet$. Thus, for $\mathcal{N} = 2$, the covariant momentum map is just the ordinary (symplectic) momentum map. In this case uniqueness holds up to an additive constant. Note that for $\mathcal{N} \geq 3$, $\mu^\bullet$ exists for *all* Killing vectors K^m, while for $\mathcal{N} = 2$ it makes sense only for $K^m \in \mathfrak{iso}_0$.

We recall the identity

$$\mathcal{L}_K = i_K d + d i_K = i_K(\mathcal{D} - Q) + (\mathcal{D} - Q)i_K = i_K \mathcal{D} + \mathcal{D} i_K - (i_K Q), \qquad (7.128)$$

where (again) $\mathcal{D}$ is the $Spin(\mathcal{N})$–covariant exterior derivative and the 1–form $Q \equiv d\phi^i\, Q_i^{AB}$ is the $Spin(\mathcal{N})$–connection. Then, as claimed,

Proposition 7.25 $\mathcal{M}$ *is an* $\mathcal{N} \geq 3$ SUGRA *manifold,* K^m *a Killing vector of* $\mathcal{M}$. *There* EXISTS *a* UNIQUE[18] *function* $\mu^{AB\,m}$ *such that*

$$\mathcal{D}\mu^{AB\,m} = -i_{K^m}\Sigma^{AB}, \tag{7.129}$$

where $\mathcal{D}$ *is* $Spin(\mathcal{N})$*–covariant. Explicitly,*

$$\mu^{AB\,m} = 2\,A^{AB}_{K^m}, \tag{7.130}$$

where $A^{AB}_{K^m}$ *are the coefficients of* A_{K^M} *(cf. Corollary 7.14).*

Proof Act with $\mathcal{D}$ on both sides of Eq. (7.129). $\mathcal{D}^2$ is the $Spin(\mathcal{N})$ curvature that we computed in Chapter 2, i.e., $\frac{1}{2}\Sigma$. So,

$$\begin{aligned}-\frac{1}{2}\mathrm{adj}(\Sigma)\,\mu^{\bullet\,m} &= \mathcal{D}(i_{K^m}\Sigma) = \pounds_{K^m}\Sigma - i_{K^m}\mathcal{D}\Sigma + (i_{K^m}Q) \\ &= \pounds_{K^m}\Sigma + (i_{K^m}Q)\Sigma = -\mathrm{adj}(\Sigma)\big((i_{K^m}Q) + \sigma(K^m)\big),\end{aligned} \tag{7.131}$$

where we have used (7.128) and Corollary 7.15. Then

$$\mu^{AB\,m} = 2\left(K^{i\,m}\,Q_i^{AB} + \sigma^{AB}(K^m)\right) \equiv 2A^{AB}_{K^m}. \tag{7.132}$$

$\square$

Notice that $\mu^\bullet$ is (up to normalization) just the $\mathfrak{spin}(\mathcal{N})$–projection of the basic endomorphism A_K defined by [215, 219]

$$\pounds_K = D_K + A_K, \tag{7.133}$$

whose geometric meaning is explained in Section VI.4 of ref. [215]. Compare also with refs. [154, 155], or Eq. (12.4.5) of ref. [62]. Explicitly, one has

$$A_K = \frac{1}{8}\mu^{AB}\,\Sigma^{AB} + H_K, \tag{7.134}$$

i.e., $\frac{1}{2}\mu^\bullet$ is the $\mathfrak{spin}(\mathcal{N})$ projection of the Cartan–Kostant morphism

$$\theta_\phi \colon \mathfrak{iso} \to \mathfrak{spin}(\mathcal{N}) \oplus \mathfrak{h}' \oplus T_\phi\mathcal{M}. \tag{7.135}$$

It is geometrically natural (and physically useful – see, e.g., ref. [120]) – to generalize the momentum map to the full space on the RHS. Then:

---

[18] In the above sense.

7.7 The covariant momentum map

Definition 7.26 $\mathcal{M}$ is a SUGRA n–fold. The *generalized momentum map* is

$$\tilde{\mu}: \mathcal{M} \times \mathfrak{iso} \to \mathfrak{spin}(\mathcal{N}) \oplus \mathfrak{h}' \oplus T_\phi \mathcal{M}, \qquad (7.136)$$

$$(\phi, K^m) \mapsto \left(2\theta_\phi(K^m)\big|_{\mathfrak{spin}(\mathcal{N})},\ 2\theta_\phi(K^m)\big|_{\mathfrak{h}'},\ 2\theta_\phi(K^m)\big|_{T\mathcal{M}}\right) \qquad (7.137)$$

$$\mapsto \left(\tfrac{1}{4}\mu^{AB\,m}\Sigma^{AB},\ 2H_{K^m},\ 2K^{i\,m}\right).$$

In the case $\mathcal{N} = 3$, the construction in Proposition 7.25 reduces to the one given by Galicki and Lawson [154, 155] for Quaternionic–Kähler manifolds. The authors of ref. [120] write some of the above formulae for the covariant momentum map for general $\mathcal{N}$s (cf. equations (2.27)–(2.31) of [120]), although they arrive at them from a different route. Their S^{IJ} correspond to our σ^{AB}.

7.7.2 The map $\mu^\bullet$ and the isometry algebra $\mathfrak{iso}(\mathcal{M})$

Uniqueness, in particular, implies

$$\mu^\bullet(\varphi_* K) = \varphi^* \mu^\bullet(K) \quad \forall\, \varphi \in \mathrm{Iso}(\mathcal{M})\ \text{and}\ K \in \mathfrak{iso}(\mathcal{M}), \qquad (7.138)$$

where φ, in general, induces also a change of trivialization of $\mathcal{Q}$ ($\equiv$ a $Spin(\mathcal{N})$ gauge transformation). Let us check this at the infinitesimal level. Recall that Eq. (7.52) gives

$$[A_{K^m}, A_{K^n}] = A_{[K^m, K^n]} + R(K^m, K^n). \qquad (7.139)$$

Then

$$\pounds_{K^m}(\mu^{AB}(K^n))\,\Sigma^{AB}$$
$$= \pounds_{K^m}\!\left(\mu^{AB\,n}\Sigma^{AB}\right) - \mu^{AB\,n}\,\pounds_{K^m}\Sigma^{AB}$$
$$= 8\big\{(D_{K^m} + A_{K^m})A_{K^n} - A_{K^n}(D_{K^m} + A_{K^m})\big\}_{\mathfrak{spin}(\mathcal{N})} - \mu^{AB\,n}[\sigma_{K^m}, \Sigma]^{AB}$$
$$= 8\big\{-R(K^m, K^n) + [A_{K^m}, A_{K^n}]\big\}_{\mathfrak{spin}(\mathcal{N})} - [\sigma_{K^m}, \mu^{\bullet\,n}]^{AB}\Sigma^{AB} \qquad (7.140)$$
$$= 8\big\{-A_{[K^m,K^n]}\big\}_{\mathfrak{spin}(\mathcal{N})} - [\sigma_{K^m}, \mu^{\bullet\,n}]^{AB}\Sigma^{AB}$$
$$= -f^{mn}{}_p\, \mu^{AB\,p}\Sigma^{AB} - [\sigma_{K^m}, \mu^{\bullet\,n}]^{AB}\Sigma^{AB},$$

that is, the $\mu^\bullet$s transform according to the adjoint of $\mathfrak{iso}$ up to the usual $Spin(\mathcal{N})$ gauge transformation of parameter $(\sigma_{K^m})^{AB}$.

Projecting Eq. (7.139) on $\mathfrak{spin}(\mathcal{N})$, we get the nice identity

$$\left[\tfrac{1}{2}\mu(K^m), \tfrac{1}{2}\mu(K^n)\right]^{AB} = -f^{mn}{}_l\,\tfrac{1}{2}\mu^{AB\,l} - \tfrac{1}{2}K^{i\,m}K^{j\,n}\Sigma^{AB}_{ij}, \qquad (7.141)$$

where $\mu^{AB}(K^m)$ are seen as matrices in the $Spin(\mathcal{N})$ indices A, B.

Other properties of the covariant momentum map

Let $A \neq B$, and $A, B \neq C$. Then

$$[(\Sigma^{AB})_i^j + i\delta_i^j]\mathcal{D}_j(\mu^{ACm} + i\mu^{BCm}) = 0, \qquad (7.142)$$

which is the covariant counterpart to Σ^{AB}–holomorphicity. In general, all rigid case identities with only one derivative take the same form in the local case. Identities with two derivatives get new terms from the curvatures. These terms can be always re–expressed in terms of commutators of $\boldsymbol{\mu}^\bullet$, as we did above. We leave it to the reader to perform such mental gymnastics.

7.7.3 Symmetric spaces $G/[Spin(\mathcal{N}) \times H']$

The case of a symmetric space, the *covariant momentum map* is explicitly computed with Eq. (7.124). One has

$$\mu^{ABm} = -\eta \, \mathrm{tr}_R[\gamma^{AB} g^{-1} t^m g], \qquad (7.143)$$

where γ^{AB} are the matrices representing the subalgebra $\mathfrak{spin}(\mathcal{N}) \subset \mathfrak{g}$ in the representation R and η is the obvious normalization coefficient (depending on the representation R that we use to write the representative group elements as matrices).

We can also easily write down the *generalized momentum map*. The generators of $\mathfrak{g}$ can be decomposed as $\{t^m\} \equiv \{\gamma^{AB}, h^\alpha, m^i\}$, where $\{h^\alpha\}$ span $\mathfrak{h}' \equiv \mathfrak{hol} \ominus \mathfrak{spin}(\mathcal{N})$ and $\{m^i\}$ generate $\mathfrak{m} = \mathfrak{g} \ominus \mathfrak{hol}$. Then we can formally extend our definition of the $Spin(\mathcal{N})$–covariant map to a G–covariant *generalized momentum map*,

$$\mu^{ABm} = -\eta \, \mathrm{tr}_R[\gamma^{AB} g^{-1} t^m g], \qquad (7.144)$$

$$\mu^{\alpha m} = -\eta \, \mathrm{tr}_R[h^\alpha g^{-1} t^m g], \qquad (7.145)$$

$$\mu^{im} = -\eta \, \mathrm{tr}_R[m^i g^{-1} t^m g]. \qquad (7.146)$$

The properties of the momentum map follow from the Cartan–Kostant isomorphism. In particular, Eq. (7.146) corresponds to Eq. (7.125).

7.8 T–tensors II. Generalized $\mathcal{T}$ in SUGRA

$\mathcal{M}$ a SUGRA manifold. Let $\mathcal{G} \subset \mathrm{Iso}(\mathcal{M})$ be a subgroup, and $\mathfrak{L}$ its Lie algebra. Let $l_{mn} \in \odot^2 \mathfrak{L}^\vee$ be a symmetric $\mathrm{Adj}_\mathcal{G}$–invariant pairing $\mathfrak{L} \times \mathfrak{L} \to \mathbb{R}$. In perfect

7.8 T–tensors II. Generalized $\mathcal{T}$ in SUGRA

analogy to Section 7.2, we define the *covariant T–tensor* as

$$T^{AB,CD} \equiv \frac{1}{4} \mu^{AB\,m}\, l_{mn}\, \mu^{CD\,n}. \tag{7.147}$$

In view of the Cartan–Kostant isomorphism, it is convenient to extend this definition to a symmetric tensor $\mathcal{T} \in \odot^2 \mathfrak{g}_\phi \simeq \odot^2 \mathfrak{iso}(\mathcal{M})$, namely

$$\mathcal{T}(\phi) = \theta_\phi(K^m)\, l_{mn}\, \theta_\phi(K^n). \tag{7.148}$$

Notice that, say,

$$\mathcal{T}\big|_{\mathfrak{hol}^\perp \times \mathfrak{hol}^\perp} = K_i^m\, l_{mn}\, K_j^n, \tag{7.149}$$

and similar identifications hold for the other components.

General Lesson 7.8 has a local counterpart.

General Lesson 7.27 *In 3D* SUGRA *a gauging* $(\mathcal{G}, l_{mn})$ *is completely described by the corresponding covariant T–tensor. The covariant T–tensor can be generalized to an* $\mathrm{Iso}(\mathcal{M})$*–covariant object (the generalized momentum map) transforming in the representation* $\odot^2 \mathfrak{iso}$ *of* $\mathrm{Iso}(\mathcal{M})$.

Remark. To simplify the analysis we have fixed our attention on the 3D gaugings. However, the deep relations between the parallel geometric structures of SUSY/SUGRA and the isometry algebra are (obviously) universal, and in fact in the next chapter we shall see that the gauging in any $D > 3$ will require only small modifications (conceptually, at least).

8
Gauging and potential terms

In this chapter we study (geometrically) the problem of gauging, in an $\mathcal{N}$-supersymmetric way, the symmetries of the ungauged SUSY/SUGRA models we constructed above. This completes our program of constructing the most general $\mathcal{N}$-supersymmetric field theory in D space–time dimensions. In particular, we wish to know which gauge groups and which matter representations are compatible with $\mathcal{N}$-extended supersymmetry, and to understand the geometry and physics of these couplings. As in Chapter 2, we start with 3D. This is the "universal" dimension: any theory in $D \geq 3$ can be dimensionally reduced to 3D, while in 3D we have a larger span of possible values of $\mathcal{N}$, and hence more general geometrical structures. The conclusions will hold in any dimension (in which the theory makes sense) up to straightforward *mutatis mutandis*, which we discuss in detail.

8.1 Gaugings in rigid SUSY

We begin with rigid SUSY in 3D. We adopt the *diet* version of the theory; that is, we dualize all vectors à la de Wit, Herger, and Samtleben (dWIIS) [120] in such a way that the vector derivatives enter the Lagrangian $\mathcal{L}$ only through Chern–Simons terms

$$\frac{k^{mn}}{4\pi}\left(A_m \wedge dA_n + \frac{2}{3}A_m \wedge [A,A]_n\right). \tag{8.1}$$

Notice that gauge invariance requires k^{mn} to be a constant (i.e., independent of the scalars ϕ^i). The matrix k^{mn} may be assumed to be symmetric.

8.1.1 CS coupled to $\mathcal{N} = 1$ gauged σ–models

Pure Chern–Simons (CS) theory is a TFT (topological field theory) without any local degree of freedom. We can write an $\mathcal{N} = 1$ supersymmetric version of the

8.1 Gaugings in rigid SUSY

CS TFT, namely

$$\mathcal{L} = \frac{k^{mn}}{4\pi}\left(A_m \wedge dA_n + \frac{2}{3} A_m \wedge [A,A]_n - \lambda_m^\alpha \lambda_{\alpha\, n}\right). \tag{8.2}$$

where the Majorana fermion $\lambda_{\alpha\, m}$ is the gaugino associated to the vector $A_{\mu\, m}$ (conventions as in [151]). This TFT is invariant under the SUSY transformations

$$\delta A_{\mu\, m} = \epsilon^\alpha (\sigma_\mu)_{\alpha\beta} \lambda_m^\beta, \tag{8.3}$$

$$\delta \lambda_{\alpha\, m} = -(\sigma_\mu)_{\alpha\beta} \epsilon^\beta F_m^\mu \tag{8.4}$$

($F_m^\mu = \frac{1}{2}\epsilon^{\mu\nu\rho} F_{\nu\rho\, m}$ is the field–strength), as is easily checked since the variation of both terms is proportional to $k^{mn} F_m^\mu \epsilon^\alpha (\sigma_\mu)_{\alpha\beta} \lambda_n^\beta$.

We couple this supersymmetric TFT to SUSY matter. The vectors enter into the kinetic terms of scalars and fermions *via* the covariant derivatives

$$D_\mu \phi^i = \partial_\mu \phi^i - A_{\mu\, m} K^{i\, m}, \tag{8.5}$$

$$D_\mu \chi^i = \partial_\mu \chi^i + \partial_\mu \phi^j \Gamma^i_{jk} \chi^k - A_{\mu\, m} D_j K^{im} \chi^j, \tag{8.6}$$

while the gaugini λ_m may enter the Lagrangian only through the Yukawa couplings

$$Y_i^m(\phi) \lambda_m^\alpha \chi_\alpha^i. \tag{8.7}$$

$\mathcal{N} = 1$ supersymmetry requires this coupling to have a special form. The terms linear in λ in the variation of $\mathcal{L}$ are

$$\delta A_m^\mu J_\mu^m + Y_i^m \lambda_m^\alpha \delta \chi_\alpha^i, \tag{8.8}$$

where J_μ^m is the matter part of the gauge current,

$$J_\mu^m = -g_{ij} K^{i\, m} \partial_\mu \phi^j + \cdots, \tag{8.9}$$

and hence the Yukawa coupling is identified with the Killing vector K_i^m,

$$\lambda_m^\alpha \chi_\alpha^i K_i^m, \tag{8.10}$$

where K_m^i is normalized so that the scalar covariant derivative reads as in Eq. (8.5) (the coupling constant is absorbed in the normalization of K_i^m). Notice that the gaugini enter the full Lagrangian only algebraically, and thus they can be integrated away. Eliminating the auxiliary fermion λs, we get a physical Yukawa coupling

$$l_{mn} K_i^m K_j^n \chi^i \chi^j, \tag{8.11}$$

where l_{mn} is the inverse of the matrix $k^{mn}/4\pi$.

As we saw in Chapter 2, in the $\mathcal{N} = 1$ theory, in addition to the Yukawa coupling (8.11) induced by the gauge interactions, we may have Yukawa terms coming from a (real) superpotential $W(\phi)$. $W(\phi)$ should be a gauge–invariant function on $\mathcal{M}$, namely

$$\pounds_{K^m} W = 0, \tag{8.12}$$

where $\{K^m\}$ are Killing vectors generating the subgroup $G \subset \text{Iso}(\mathcal{M})$ to be gauged. The complete (rigid) $\mathcal{N} = 1$ Yukawa couplings are then given by

$$\left(l_{mn} K_i^m K_j^n + D_i \partial_j W \right) \chi^i \chi^j, \tag{8.13}$$

while the scalar potential $V(\phi)$ is fixed by the general Ward identity we discussed in Chapter 6 to be

$$V(\phi) = \frac{1}{2} G^{ij} \partial_i W \partial_j W. \tag{8.14}$$

The final Lagrangian and SUSY transformations for $\mathcal{N} = 1$ 3D CS–matter theory is (see, e.g., [46, 277])

$$\begin{aligned}\mathcal{L} = \frac{k^{mn}}{4\pi} * \left(A_m \wedge dA_n + \frac{2}{3} A_m \wedge [A, A]_n \right) - \frac{1}{2} G_{ij} D^\mu \phi^i D_\mu \phi^j \\ + \frac{i}{2} G_{ij} \chi^i \slashed{D} \chi^j + (l_{mn} K_i^m K_j^n + D_i \partial_j W) \chi^i \chi^j \\ - \frac{1}{2} G^{ij} \partial_i W \partial_j W - \frac{1}{24} R_{ijkl} \overline{\chi}^i \gamma^\mu \chi^j \overline{\chi}^k \gamma_\mu \chi^l, \end{aligned} \tag{8.15}$$

$$\delta \phi^i = \overline{\epsilon} \chi^i, \tag{8.16}$$

$$\delta \chi^i = D_\mu \phi^i \gamma^\mu \epsilon - G^{ij} \partial_j W \epsilon + \text{3-fermions}, \tag{8.17}$$

$$\delta A_{m\mu} = l_{mn} K_i^n \overline{\epsilon} \gamma_\mu \chi^i. \tag{8.18}$$

8.2 $\mathcal{N}$-extended (rigid) CS gauge theories

8.2.1 Consistency conditions

Let our 3D model (with vectors dualized in the CS form) have (rigid) $\mathcal{N}$–supersymmetry. Forgetting its invariance under $Q^2, Q^3, \ldots, Q^\mathcal{N}$, we may consider it simply as an $\mathcal{N} = 1$ model. Hence its Yukawa couplings and potential should have the form (8.13), (8.14) for some function $W(\phi)$. However, we may forget $\mathcal{N} - 1$ supercharges in various ways; the Lagrangian $\mathcal{L}$ should be independent of the choice we make. As we discussed in detail in Chapter 2, replacing Q^1 with Q^a ($a = 2, 3, \ldots, \mathcal{N}$) as the unforgotten supercharge, amounts to the replacement

$$\chi^i \longmapsto (f^a)^i{}_j \chi^j \equiv (\Sigma^{1a})^i{}_j \chi^j, \tag{8.19}$$

where $f^a \in \text{End}(T\mathcal{M})$ are the basic parallel complex structures. Therefore, in order for Q^a to generate a symmetry of the Lagrangian $\mathcal{L}$, there must be a superpotential $W^a(\phi)$ such that

$$(\Sigma^{1a})^t (K^m l_{mn} K^n + D\partial W^a) \Sigma^{1a} = (K^m l_{mn} K^n + D\partial W), \tag{8.20}$$

$$G^{ij} \partial_i W^a \partial_j W^a = G^{ij} \partial_i W \partial_j W, \tag{8.21}$$

(*not* summed over a!). The first (second) condition arises from equating the Yukawa couplings (resp. potential) of the two $\mathcal{N} = 1$ Lagrangians. If such a W^a exists, we are guaranteed that the supercharge Q^a generates a supersymmetry of $\mathcal{L}$. Thus $\mathcal{N}$–SUSY requires the existence of a complete set

$$W \equiv W^1, \ W^2, \ W^3, \ \cdots, \ W^{\mathcal{N}} \tag{8.22}$$

of (real) superpotentials satisfying the equations (8.12), (8.20), and (8.21).

Notice that $Spin(\mathcal{N})$ itself is not necessarily a symmetry; $\mathcal{L}$ is invariant under $Spin(\mathcal{N})$ only if the functional forms of the various $W^a(\phi)$ are related by suitable isometries $\varphi^a \colon \mathcal{M} \to \mathcal{M}$:

$$W^a(\phi) = W(\varphi^a(\phi)). \tag{8.23}$$

In general, $Spin(\mathcal{N})$ is only an automorphism of the formalism.

The equations (8.20) and (8.21) for $a = 2, 3, \ldots, \mathcal{N}$ have *no* solution in general. However, we are interested precisely in the *special* conditions under which a solution does exist: we wish to solve the following problem.

Problem *Given a scalar manifold $\mathcal{M}$, compatible with $\mathcal{N}$–SUSY, determine the subgroups $\mathcal{G} \subset \text{Iso}_0(\mathcal{M})$ which can be gauged in an $\mathcal{N}$–supersymmetric way as well as the associated gauge couplings l_{mn} which preserve $\mathcal{N}$–SUSY.*

In view of Eqs. (8.20) and (8.21), we can formalize our problem in the following form:

Problem *Define for which data $(\mathcal{M}, \mathcal{G}, l_{mn})$ there exist functions*

$$\{W^1(\phi) \equiv W(\phi), \ W^2(\phi), \ \cdots, \ W^{\mathcal{N}}(\phi)\} \tag{8.24}$$

that are gauge–invariant

$$\pounds_{K^m} W^a = 0, \tag{8.25}$$

and solve Eqs. (8.20),(8.21).

As stated, this is a geometrical problem about the isometries of the Riemannian manifolds $\mathcal{M}$ having a special holonomy, $\mathfrak{hol}(\mathcal{M}) \subseteq C(\mathfrak{spin}(\mathcal{N}))$, allowing for $\mathcal{N}(\mathcal{N}-1)/2$ parallel 2–forms Σ^{AB}. We discussed such isometry groups in the previous chapter. The relevant objects were the *momentum map* and the *T–tensor*. By General Lesson 7.8, the possible solutions to the above problem should correspond to conditions on these objects. Indeed, the supersymmetric gaugings take a universal form when written in terms of the T–tensor. Stated differently, the data $(\mathcal{M}, \mathcal{G}, l_{mn})$ are encoded in the T–tensor

$$T^{AB,CD} = \mu^{AB\,m} l_{mn} \mu^{CD\,n}, \tag{8.26}$$

and, as we shall see shortly, the equations for W^a depend on the data $(\mathcal{M}, \mathcal{G}, l_{mn})$ only through the T–tensor. Thus, at the end of the day, the solution to our problem boils down to a geometric condition on the T–tensor.

8.2.2 The consistency conditions for $\mathcal{N} \leq 3$

Let us see the implications of the above constraints for increasing $\mathcal{N}$s. For $\mathcal{N} = 1$ there is no condition. In the $\mathcal{N} = 2$ case we have a single complex structure $f = \Sigma \equiv \Sigma^{12}$. From Eq. (8.20) we have

$$(\Sigma^{12})_i^{\ k} (\Sigma^{12})_j^{\ h} \{K_k^m l_{mn} K_h^n + D_h \partial_k W\} = \{K_i^m l_{mn} K_j^n + D_i \partial_j W^2\}, \tag{8.27}$$

while from identity (7.48),

$$\begin{aligned}(\Sigma^{12})_i^{\ k} (\Sigma^{12})_j^{\ h} &\{K_k^m l_{mn} K_h^n + D_h \partial_k \left(\frac{1}{2} T^{12,12}\right)\} \\ &= \{K_i^m l_{mn} K_j^n + D_i \partial_j \left(\frac{1}{2} T^{12,12}\right)\}.\end{aligned} \tag{8.28}$$

Subtracting the last two equations, we get

$$(\Sigma)_i^{\ k} (\Sigma)_j^{\ h} D_h \partial_k \left(W - \frac{1}{2} T^{12,12}\right) = D_i \partial_j \left(W^2 - \frac{1}{2} T^{12,12}\right). \tag{8.29}$$

In view of Chapter 7, this equation is equivalent to

$$\left(W - \frac{1}{2} T^{12,12}\right) - \left(W^2 - \frac{1}{2} T^{12,12}\right) = 2\mathcal{F} + 2\overline{\mathcal{F}}, \tag{8.30}$$

$$\left(W - \frac{1}{2} T^{12,12}\right) + \left(W^2 - \frac{1}{2} T^{12,12}\right) = 2M, \tag{8.31}$$

where $\mathcal{F}$ is a Σ–holomorphic function and M is the Σ–momentum map of a (holomorphic) isometry. Of course, $\mathcal{F}$ is just the holomorphic superpotential of the $\mathcal{N} = 2$ superspace formalism, while M is a *twisted (or real) mass*, a special coupling of 3D $\mathcal{N} = 2$ SUSY (see, e.g., [204]). Real masses play a marginal role in our discussion, and from now on *we set them to zero* for simplicity; the reader may easily reintroduce them. At $M = 0$,

$$W = \frac{1}{2}T^{12,12} + \mathcal{F} + \overline{\mathcal{F}}, \tag{8.32}$$

$$W^2 = \frac{1}{2}T^{12,12} - \mathcal{F} - \overline{\mathcal{F}}. \tag{8.33}$$

The $\mathcal{N} = 2$ Chern–Simons vector superfield contains, beside A_μ and two auxiliary spin–1/2 fermions, also two auxiliary (real) scalars D and σ (see, e.g., [152]). The additional auxiliary fermion produces new Yukawa terms, while the elimination of the auxiliary scalars leads to a gauge contribution to the scalar potential of the form $g^{ij}(\partial_i T)(\partial_j T)$ where $T \equiv T^{12,12}$. On top of these gauge contributions there are the usual Yukawa and potential terms arising from the superpotential $\mathcal{F}$ (in the $\mathcal{N} = 2$ sense). Thus

$$V(\phi) = \begin{cases} \frac{1}{2}g^{ij}\partial_i W \, \partial_j W, & \mathcal{N} = 1 \text{ formalism,} \\ \frac{1}{8}g^{ij}\partial_i T \, \partial_j T + g^{i\bar{j}} \, \partial_i \mathcal{F} \, \partial_{\bar{j}}\overline{\mathcal{F}}, & \mathcal{N} = 2 \text{ formalism.} \end{cases} \tag{8.34}$$

The $\mathcal{N} = 1$ and $\mathcal{N} = 2$ expressions differ by cross–terms of the form

$$g^{ij} \, \partial_i T \, \partial_j \mathrm{Re}(\mathcal{F}), \tag{8.35}$$

but these vanish identically since

$$g^{ij}\partial_i T \, \partial_j \mathcal{F} = -2i \, \mu^m \, l_{mn} \, K^{n\,i} \, \partial_i \mathcal{F} = -2i \, \mu^m \, l_{mn} \, \pounds_{K^n} \mathcal{F} = 0 \tag{8.36}$$

by holomorphicity and gauge invariance of $\mathcal{F}$. Therefore the $\mathcal{N} = 1$ and $\mathcal{N} = 2$ formalisms lead to the same Lagrangian. We conclude that:

Lemma 8.1 *In an $\mathcal{N} \geq 2$ gauged* SUSY *model, we have the relation*

$$W(\phi) = \frac{1}{2} T^{1a,1a} + F^a \tag{8.37}$$

$(a = 2, 3, \ldots, \mathcal{N})$, *where F^a is the real part of a f^a–holomorphic function. Moreover,*

$$W^a = T^{1a,1a} - W. \tag{8.38}$$

We interpret Eq. (8.37) (or, equivalently, Eq. (8.29)) as a set of *linear* equations for the unknown function $W(\phi)$. The general solution, $W(\phi)$, is the sum of a particular solution, $W^{(0)}(\phi)$, and the general solution of the homogeneous equation (i.e., the equations with $T^{1a,1a}$ set to zero). But a solution $W(\phi)$ of the homogeneous equation is precisely a function which is f^a–harmonic for all complex structures f^a. We already know that for $\mathcal{N} \geq 4$ (three complex structures) such a function is trivial, that is $D_i \partial_j W = 0$, while for $\mathcal{N} = 3$ $D_i \partial_j W$ is at most a constant. Thus we learn the following General Lesson:

General Lesson 8.2 *For a given gauging, $(\mathcal{G}, l_{mn})$, the Yukawa couplings and scalar potential which complete the Lagrangian to an $\mathcal{N} \geq 3$ model – if they exist at all – are essentially[1] unique.*

$\mathcal{N} = 3$ CS models

For $\mathcal{N} = 3$, $W(\phi)$ is (essentially) uniquely determined from the gauging data $(\mathcal{G}, l_{mn})$. Let us show explicitly that, in this case, *there is* a solution for each $\mathcal{G} \subset \text{Iso}_0(\mathcal{M})$ and each symmetric invariant tensor l_{mn}. Indeed, the function

$$W(\phi) = \frac{1}{2}\left(T^{12,12} + T^{13,13} - T^{23,23}\right) \tag{8.39}$$

is a solution to Eq. (8.37) (while $\pounds_{K^m} W = 0$ holds by construction). To prove this, we have only to check that $T^{13,13} - T^{23,23}$ is a Σ^{12}-harmonic function, and correspondingly that $T^{12,12} - T^{23,23}$ is Σ^{13}-harmonic. Both facts are guaranteed by Lemma 7.9.

Remark. Notice that the gauged model is invariant under three supercharges, Q^1, Q^2, and Q^3, but not under the fourth one, which in the ungauged model is automatically conserved, namely the one associated with the complex structure $f^3 \equiv f^1 f^2$. Indeed, in general, the function

$$W - \frac{1}{2}T^{23,23} \equiv \frac{1}{2}(T^{12,12} + T^{13,13} - 2T^{23,23}) \tag{8.40}$$

is *not* Σ^{23}-harmonic.

We stress that *all gaugings $(\mathcal{G}, l_{mn})$ are admissible in $\mathcal{N} = 3$* SUSY.

A no–go theorem for $\mathcal{N} \geq 4$

This is not true for larger $\mathcal{N}$. Gaiotto and Witten [151, 212] give a physical argument why *generic* gaugings are not compatible with $\mathcal{N} \geq 4$ SUSY. Consider the

---

[1] The residual non–uniqueness is related, in particular, to the fact that the momentum maps themselves are defined up to an additive constant (say, for Abelian groups) and hence the T–tensors may be shifted by a quadratic function.

subclass[2] of $\mathcal{N}$–supersymmetric models that are obtained, via dWHS duality, from 3D models whose vectors have both F^2 canonical kinetic terms and Chern–Simons interactions. Then the gauge vectors are massive [126] and have (say) helicity $+1$. The $\mathcal{N}$–SUSY algebra has $\mathcal{N}$ helicity lowering operators, so a massive vector multiplet should contain states with helicity

$$\lambda = 1, \frac{1}{2}, 0, \cdots, 1 - \frac{\mathcal{N}}{2}. \tag{8.41}$$

In particular, for $\mathcal{N} \geq 4$, we have states with helicity -1 which are also massive vectors. Since (rigid) SUSY commutes with the gauge symmetry, all the above states transform in the same way under $\mathcal{G}$, that is, in the adjoint representation (which is the representation for the gauge vectors $\lambda = +1$). For $\mathcal{N} \geq 4$, we have also $\lambda = -1$ vectors, always in the adjoint. Thus the gauge vectors transform according to several copies of the adjoint representation, but this is forbidden in non–Abelian gauge theories.

$\mathcal{N} \geq 4$ gaugings *do* exist, but they are somehow exceptional, being possible only for very specific gauge groups, matter representations, and couplings. Simple examples (to which we shall return) are given by the dimensional reduction of 4D $\mathcal{N} = 2$ and $\mathcal{N} = 4$ gauge theories, which in 3D have, respectively, $\mathcal{N} = 4$ and 8.

Given that the $\mathcal{N} = 3$ gauged model is essentially unique (for given $(\mathcal{G}, l_{mn})$), and that any $\mathcal{N} \geq 4$ model is, in particular, an $\mathcal{N} = 3$ model, *all $\mathcal{N} \geq 4$ gauged models should be defined by the $\mathcal{N} = 3$ superpotential W in Eq.* (8.39), which – for suitable T–tensors – magically happens to satisfy the conditions (8.37) for all as. For a generic T–tensor this is impossible, as the previous "generic no–go" result implies. Our next task is to describe the circumstances which make the miracle happen.

8.2.3 General $\mathcal{N}$s: the main theorem

To solve the consistency conditions (8.37) it is natural to decompose the "source" $T^{AB,CD}$, which transforms according to the *reducible* $Spin(\mathcal{N})$ representation[3] $\odot^2 \text{Adj} \simeq (\wedge^2 V) \odot (\wedge^2 V)$, into *irreducible* representations. In terms of Young tableaux the decomposition of the T–tensor reads

---

[2] This class is somewhat "generic," that is "dense" in coupling constant space. Roughly speaking, the *no–go* theorem forbidding $\mathcal{N} \geq 4$ gaugings may be evaded only in a "zero–measure set" of the coupling space. Then we expect that the solutions to our problem for $\mathcal{N} \geq 4$ constitute, at most, a zero–measure subset of all gaugings.
[3] V stands for the $\mathcal{N}$–dimensional vector representation of $Spin(\mathcal{N})$.

$$\square\!\odot\!\square \simeq 1 \oplus \square\square \oplus \begin{array}{c}\square\square\\\square\square\end{array} \oplus \begin{array}{c}\square\\\square\\\square\end{array} \tag{8.42}$$

The main result of this section is the following theorem:

Theorem 8.3 *A solution W to the consistency condition (8.37) exists iff the $\boxplus$ component of the T–tensor vanishes.*

Proof Assume $T^{AB,CD}\big|_{\boxplus} = 0$. Then we have

$$T^{AB,CD} = \delta^{AC} T^{BD} - \delta^{AD} T^{BC} - \delta^{BC} T^{AD} + \delta^{BD} T^{AC} + T^{[ABCD]} \tag{8.43}$$

with $T^{AB} = T^{BA}$. Then, for $a \neq 1$,

$$T^{1a,1a} = T^{11} + T^{aa}. \tag{8.44}$$

Therefore the consistency conditions (8.37) become

$$W - \frac{1}{2}\left(T^{11} + T^{aa}\right) \text{ is } \Sigma^{1a} - \text{harmonic for } a = 2,\dots,\mathcal{N}. \tag{8.45}$$

Let $c, d \neq 1$ or a (thus $\mathcal{N} \geq 3$). One has

$$T^{1c,1d} - T^{ac,ad} = (T^{cd} + \delta^{cd} T^{11}) - (T^{cd} + \delta^{cd} T^{aa}) = \delta^{cd}(T^{11} - T^{aa}). \tag{8.46}$$

By Lemma 7.9 the LHS is Σ^{1a}-harmonic. So is the RHS. We conclude that $(T^{11} - T^{aa})$ is Σ^{1a}-harmonic. Take $W = T^{11}$. Then

$$\begin{aligned}W - \frac{1}{2}\left(T^{11} + T^{aa}\right) &= \frac{1}{2}\left(T^{11} - T^{aa}\right)\\ &\equiv \Sigma^{1a} - \text{harmonic } \forall\, a = 2,\dots,\mathcal{N}.\end{aligned} \tag{8.47}$$

So a solution to Eq. (8.37) exists. We already know that it is unique. On the other hand, assume $T|_{\boxplus} \neq 0$. This can happen only for $\mathcal{N} \geq 4$, since $T|_{\boxplus} \equiv 0$ for $\mathcal{N} \leq 3$. Then the model is, in particular, an $\mathcal{N} = 4$ theory. But, in this special case, we already notice that the consistency conditions have no solution. (See also Section 8.3 below.) $\square$

One advantage of the above formulation of the gauging problem is that the statement for local SUGRA will be *exactly* the same. Indeed, one may obtain the rigid theory as a particular limit of the SUGRA case [46].

Corollary 8.4 Set $W^1 = W \equiv T^{11}$. Then for $A = 1, 2, \ldots \mathcal{N}$ one has

$$W^A = T^{AA}. \tag{8.48}$$

Proof For $A = 1$ this is the definition. For $A = a$ from Eqs. (8.38) and (8.44)

$$W^a = T^{1a,1a} - W = T^{11} + T^{aa} - T^{11} = T^{aa}. \tag{8.49}$$

Notice that

$$W^A = \frac{1}{2}\left(T^{AB,AB} + T^{AC,AC} - T^{BC,BC}\right) \quad \forall B, C : A, B, C \text{ all distinct}. \tag{8.50}$$

□

The result encodes the gaugings for all space–time dimensions D. We shall discuss the $D > 3$ case at the end of this chapter in the context of SUGRA. The reader may fill in the detail for the rigid case.

8.2.4 Fermionic shifts and potentials

By *fermionic shifts* we mean the SUSY transformation of the χ^i evaluated in a constant bosonic background, namely the matrices $C^i_a(\phi)$ such that

$$\delta \chi^i = \cdots + C^i_a(\phi)\,\epsilon^a. \tag{8.51}$$

In the $\mathcal{N} = 1$ formalism the SUSY transformation of fermions reads

$$\delta \chi^i = D\phi^i \epsilon - (\delta_Q \phi^j)\,\Gamma^i_{jk}\chi^k - g^{ij}\,\partial_j W\,\epsilon. \tag{8.52}$$

If $\mathcal{N} > 1$, we have a similar formula for each supercharge Q^a

$$\delta(f^a \chi)^i = -g^{ij}\,\partial_j W^a\,\epsilon^a \quad \text{(not summed over } a\text{)} \tag{8.53}$$

so (with the convention $f^1 = -1$)

$$C^i_A = (f^A)^i_{\ k}\,g^{kj}\,\partial_j W^A. \tag{8.54}$$

From the basic Ward identity of Chapter 6 we know that

$$\frac{1}{2} g_{ij}\,C^i_A\,C^j_B = \delta_{AB}\,V(\phi). \tag{8.55}$$

Let us check this. From Eq. (8.50), for $A \neq B$ we can write $(C \neq A, B)$

$$W^A = \frac{1}{2}\left(T^{AB,AB} + T^{AC,AC} - T^{BC,BC}\right), \tag{8.56}$$

$$W^B = \frac{1}{2}\left(T^{AB,AB} - T^{AC,AC} + T^{BC,BC}\right). \tag{8.57}$$

Then (taking $A = a$ and $B = 1$)

$$C_1^i \equiv \partial^i W^1 = \mu^{1am} l_{mn} (f^a)^{ij} K_j^n - \frac{1}{2} \partial^i (T^{aC,aC} - T^{1C,1C}), \quad (8.58)$$

$$C_a^i \equiv (f^a)^{ij} \partial_j W^a = -\mu^{1am} l_{mn} K^{in} + \frac{1}{2} (f^a)^{ij} \partial_j (T^{aC,aC} - T^{1C,1C}). \quad (8.59)$$

In view of Eq. (7.46) we can rewrite this in the form

$$C_i^a = \mu^{a1m} l_{mn} K_i^n + \partial_i T^{aC,1C}, \quad (8.60)$$

where $C \neq 1, a$ (not summed over C!). From these we have

$$C_i^a g^{ij} C_j^a = \mu^{a1m} l_{mn} \mu^{a1n} + (\partial^i T^{aC,1C})(\partial_i T^{aC,1C})$$
$$= C_i^1 g^{ij} C_j^1 = 2 V(\phi), \quad (8.61)$$

$$C_i^a g^{ij} C_j^1 = 0, \quad (8.62)$$

and more generally that

$$C_i^A G^{ij} C_j^B = 2 \delta^{AB} V(\phi). \quad (8.63)$$

8.2.5 Covariant expressions

For future reference we define

$$C_i^{AB} = \mu^{ABm} l_{mn} K_i^n + \partial_i T^{AB}, \quad (8.64)$$

whose diagonal entries are just $\partial_i W^A$ while $C_i^{A1} = C_i^A$. By $Spin(\mathcal{N})$ covariance, we have

$$\delta(f^A \chi)^i = \cdots + g^{ij} C_j^{AB} \epsilon^B + \cdots . \quad (8.65)$$

8.3 Example: $\mathcal{N} = 4$ and the Gaiotto–Witten theorem

In ref. [151] Gaiotto and Witten gave a very elegant interpretation of the constraint on the T–tensor for 3D $\mathcal{N} = 4$ CS–matter theories. Let us recall our findings for this case: the group $Spin(4)$ decomposes into the product $Spin(3)_1 \times Spin(3)_2$; correspondingly, the scalar manifold $\mathcal{M}$ splits into the product of two hyperKähler spaces, $\mathcal{M}_1 \times \mathcal{M}_2$. Each factor manifold $\mathcal{M}_i$ ($i = 1, 2$) has three parallel 2–forms[4] ω_i^a ($a = 1, 2, 3$). The ω_1^a transform in the adjoint of $Spin(3)_1$ and are inert under $Spin(3)_2$, while the opposite holds for the ω_2^a. The spin(4)–momentum

---
[4] $\omega_1^a = \eta^{aAB} \Sigma^{AB}$ where η^{aAB} is the 't Hooft tensor.

8.3 Example: $\mathcal{N}=4$ and the Gaiotto–Witten theorem

map decomposes into two *independent* $\mathfrak{spin}(3)$–momentum maps on the factor manifolds, μ^{am} and $\widetilde{\mu}^{am}$, respectively in the representations $(\mathbf{3},\mathbf{1})$ and $(\mathbf{1},\mathbf{3})$ of $Spin(3)_1 \times Spin(3)_2$. Then the T–tensor belongs to the following representation:

$$T \in \odot^2\big((\mathbf{3},\mathbf{1}) \oplus (\mathbf{1},\mathbf{3})\big) \simeq \underbrace{(\mathbf{1},\mathbf{1})}_{\mathbf{1}} \oplus \underbrace{(\mathbf{1},\mathbf{1})}_{T^{[ABCD]}} \oplus \underbrace{(\mathbf{5},\mathbf{1}) \oplus (\mathbf{1},\mathbf{5})}_{\boxplus} \oplus \underbrace{(\mathbf{3},\mathbf{3})}_{\Box\Box}. \tag{8.66}$$

The $(\mathbf{3},\mathbf{3})$ component of the T–tensor couples the momentum map of $\mathcal{M}_1$ to the momentum map of $\mathcal{M}_2$,

$$T\big|_{(3,3)} = \mu^{am}\, l_{mn}\, \widetilde{\mu}^{bn}, \tag{8.67}$$

while the other components are bilinear in the momentum map of a single factor manifold,

$$T\big|_{(1,1)_1 \oplus (5,1)} = \mu^{am}\, l_{mn}\, \mu^{bn}, \qquad T\big|_{(1,1)_2 \oplus (1,5)} = \widetilde{\mu}^{am}\, l_{mn}\, \widetilde{\mu}^{bn}. \tag{8.68}$$

The consistency constraint requires the components $(\mathbf{5},\mathbf{1}) \oplus (\mathbf{1},\mathbf{5})$ to vanish. These are components of the T–tensor which pertain to a single space. The (gauge) coupling between the two hyperKähler manifolds is not constrained by $\mathcal{N}=4$ SUSY, and the condition splits into two independent conditions for the factor spaces. Thus we can assume that we have just one manifold, say, $\mathcal{M}_1$. Now one has

$$T^{ab\,cd} = \epsilon^{abe}\epsilon^{cdf}\, T^{ef} \tag{8.69}$$

(all indices taking the values $1, 2, 3$), and the constraint is

$$T^{ab} \equiv \mu^{am}\, l_{mn}\, \mu^{bn} = \delta^{ab}\, T. \tag{8.70}$$

Comparing with Eq. (8.40), we see that this condition guarantees $\mathcal{N}=4$ SUSY, since, in this case, $T^{12,12} + T^{13,13} - 2\,T^{23,23}$ vanishes.

If $\mathcal{M}_1$ is symmetric, the Cartan–Kostant isomorphism allows us to rewrite the condition (8.70) in terms of the algebra of $\mathfrak{iso}(\mathcal{M}_1)$. A symmetric hyperKähler is necessarily flat. Then $\widetilde{\mathcal{M}}_1$, seen as a vector space, has a quaternionic structure, so we write the coordinates in the form q^I_α ($\alpha = 1, 2$ and $I = 1, 2, \ldots, 2m$), subjected to the usual symplectic reality condition

$$(q^I_\alpha)^* \stackrel{\text{def}}{=} q^\alpha_I = \epsilon^{\alpha\beta}\, \Omega_{IJ}\, q^J_\beta. \tag{8.71}$$

In this notation, the momentum map reads

$$\mu^m_{\alpha\beta} \equiv (\sigma^a)_{\alpha\beta}\, \mu^{am} = \kappa^m_{IJ}\, q^I_\alpha\, q^J_\beta, \tag{8.72}$$

where the symmetric matrices κ_{IJ}^m generate the symplectic representation of $\mathfrak{iso}_0$ defined by

$$A_{K^m} \longleftrightarrow \mathbf{1} \otimes \kappa^m. \tag{8.73}$$

Equation (8.70) becomes

$$0 = l_{mn}\, \mu^m_{(\alpha\beta}\, \mu^n_{\gamma)\delta} = (l_{mn}\, \kappa_{IJ}^m \kappa_{KL}^n)\, q^I_{(\alpha} q^J_\beta q^K_\gamma q^L_{\delta)}, \tag{8.74}$$

or [151]

$$l_{mn}\, \kappa^m_{(IJ}\, \kappa^n_{K)L} = 0. \tag{8.75}$$

A gauging of the model is defined by the following data: a group $\mathcal{G}$, with a symplectic representation $\{\kappa_{IJ}^m\}$, and a CS matrix l_{mn}. Our goal is to characterize the gaugings $(\mathcal{G}, \kappa_{IJ}^m, l_{mn})$ which preserve $\mathcal{N} = 4$–SUSY.

Theorem 8.5 (Gaiotto–Witten [151]) *For $\mathcal{M}_1$ flat, the $\mathcal{N} = 4$ gaugings $(\mathcal{G}, \tau_{IJ}^m, l_{mn})$ are in one-to-one correspondence with the Lie superalgebras, whose fermionic generators form a quaternionic representation of the bosonic subalgebra, having an invariant non-degenerate quadratic form (l_{mn}, Ω_{IJ}). The Chern–Simons couplings are determined by the restriction of the quadratic form to the bosonic subalgebra.*

Proof Let M^m and λ_I be, respectively, the bosonic and fermionic generators of the Lie subalgebra ($m = 1, 2 \ldots, \dim\mathcal{G}$, $I = 1, 2, \ldots, \dim\mathcal{M}/2$)). Consider the bracket

$$[M^m, M^n] = f^{mn}{}_p\, M^p, \tag{8.76}$$

$$[M^m, \lambda_I] = \kappa_{IJ}^m\, \Omega^{JK}\, \lambda_K, \tag{8.77}$$

$$\{\lambda_I, \lambda_J\} = \kappa_{IJ}^m\, l_{mn}\, M^n. \tag{8.78}$$

This bracket defines a Lie superalgebra if and only if it satisfies the super–Jacobi identity. The only case which is not automatic is the Jacobi identity with three λs,

$$[\lambda_I, \{\lambda_J, \lambda_K\}] + [\lambda_J, \{\lambda_K, \lambda_I\}] + [\lambda_K, \{\lambda_I, \lambda_J\}] = 0, \tag{8.79}$$

which corresponds to Eq. (8.75). The form l_{mn} may be assumed to be non–degenerate without loss, since its radical corresponds to auxiliary vectors, not entering into the CS terms, which may be eliminated. □

The *simple* Lie supergroups are classified in [211]. See Table 8.1 for the list of those relevant for the $\mathcal{N} = 4$ gaugings of CS–matter theories. A general $\mathcal{N} = 4$ model has a target space $\mathcal{M}_1 \times \mathcal{M}_2$ with a "doubling" of this structure.

Table 8.1 *Simple Lie supergroups with a symplectic action of the bosonic subgroup on the fermionic generators having a non–degenerate quadratic form (cf.* [48]*).*

Lie superalgebra	Gauge group	max $\mathcal{N}$
$U(2\|2)$	$SU(2) \times SU(2) \times U(1)$	8
$U(m\|n)$	$SU(m) \times SU(n) \times U(1)$	6
$Osp(2\|n)$	$SO(2) \times Sp(2n)$	6
$Osp(m\|n)$	$SO(m) \times Sp(2n)$	5
$F(4)$	$SO(7) \times SU(2)$	5
$G(3)$	$G_2 \times SU(2)$	5
$D(2\|1;\alpha)$	$SO(4) \times Sp(2)$	5

A rigid $\mathcal{N} \geq 5$ model is, in particular, an $\mathcal{N} = 4$ model with $\mathcal{M}$ necessarily flat (Section 4.1). Each $\mathcal{N} \geq 5$ scalar supermultiplet consists of a pair of $\mathcal{N} = 4$ hypermultiplet and twisted hypermultiplet in the same representation of the gauge group, i.e., the scalar manifold of an $\mathcal{N} \geq 5$ model, seen as an $\mathcal{N} = 4$ one, has the form $\mathcal{M} \times \mathcal{M}^{\text{op}}$, where $\mathcal{M}^{\text{op}}$ is $\mathcal{M}$ with the opposite orientation. Then all $\mathcal{N} \geq 5$ gauged CS–matter models should be described by the Gaiotto–Witten theorem. In the table we have added a column "max $\mathcal{N}$," meaning the maximal supersymmetry we can construct with that Lie superalgebra. We see that a 3D rigid $\mathcal{N} > 8$ SUSY theory is free. For details see [3, 47, 48, 190, 189], where one finds many other interesting results about superconformal gaugings of CS–matter models in 3D.

8.4 A *puzzle* and its resolution

The above results may look puzzling at first. We found that, for $\mathcal{N} \geq 4$, only *certain* gauge groups are allowed. For instance, for $\mathcal{N} = 8$ Table 8.1 gives only $\mathcal{G} = SO(4)$. On the other hand, in 3D there must exist the $\mathcal{N} = 4$ and $\mathcal{N} = 8$ super–Yang–Mills (SYM) theories for *any* (reductive) Lie group G: indeed, we can construct them by dimensionally reducing the 4D $\mathcal{N} = 2, 4$ SYMs, which exist for all *compact* G. In 3D $\mathcal{N} = 4, 8$ SYM may be formulated as CS–matter theories, thanks to the dWHS duality. It seems we have a paradox. Are $\mathcal{N} = 4, 8$ SYM consistent with the Gaiotto–Witten theorem?

Yes, they are consistent, thanks to a few subtleties. First of all, the gauge group of the dual CS theory is *not* that of the original SYM. If G is the SYM gauge group, (which we assume to be *semi–simple*) one has

$$\mathcal{G} = G \ltimes A, \tag{8.80}$$

where A is an Abelian Lie algebra whose generators $P^{\hat{a}}$ transform in the adjoint of G

$$[t^a, t^b] = f^{ab}{}_c\, t^c \qquad [t^a, P^{\hat{b}}] = f^{ab}{}_c\, P^{\hat{c}} \qquad [P^{\hat{a}}, P^{\hat{b}}] = 0. \tag{8.81}$$

The invariant tensor l_{mn} for the SYM *à la* CS is associated with the Casimir invariant $t^a\, P^{\hat{a}}$ of $\mathcal{G}$. Note that the gauge group $\mathcal{G}$ is *not* reductive; for a 3D CS–matter theory this is still consistent with unitarity.

8.4.1 $\mathcal{N} = 4$ SYM

Under dimensional reduction and duality, the 4D $\mathcal{N} = 2$ vector multiplet becomes a 3D twisted hypermultiplet [91]. Thus $\mathcal{M}$ has a single flat factor and the Gaiotto–Witten theorem applies. Calling M^a the generators of G and P^a those of A, and adding fermionic generators Q_I ($I = 1, 2, \ldots, 2\dim G$), the Gaiotto–Witten superalgebra reads

$$[M^a, M^b] = f^{ab}{}_c\, M^c, \tag{8.82}$$

$$[M^a, P^b] = f^{ab}{}_c\, P^c, \tag{8.83}$$

$$[P^a, P^b] = 0, \tag{8.84}$$

$$[M^a, Q_I] = (\tau^a)_{IJ}\, \Omega^{JK}\, Q_K, \tag{8.85}$$

$$[P^a, Q_I] = 0, \tag{8.86}$$

$$\{Q_I, Q_J\} = (\tau^a)_{IJ}\, k_{ab}\, P^b, \tag{8.87}$$

where the matrices $(\tau^a)_{IJ}$, giving the symplectic representation of G, in the SYM case are

$$\tau^a = \begin{pmatrix} f^{ab}{}_c & 0 \\ 0 & -f^{ab}{}_c \end{pmatrix}, \tag{8.88}$$

and k_{ab} is (minus) the Killing form of G. Again, the only non–trivial Jacobi identity is the one with three odd generators,

$$[Q_I, \{Q_J, Q_K\}] + \text{cyclic permutations} = 0. \tag{8.89}$$

In the SYM case, this is trivially satisfied since $\{Q_I, Q_J\}$ is proportional to P^a which commutes with Q_K. Thus the Gaiotto–Witten theorem is explicitly verified. Note that taking $G = Spin(3)$, the superalgebra (8.82)–(8.87) is nothing other than the Euclidean Poincaré superalgebra in 3D.

Note that this system looks like a model with gauge vectors making *two copies* of (the reductive part of) the gauge group G, as naively expected from the heuristic argument around Eq. (8.41).

8.4.2 $\mathcal{N} = 8$ SYM

The target space $\mathcal{M}$ for $\mathcal{N} = 8$ is flat i.e., $\mathbb{R}^{8m}$, $m = \dim G$. The isometry group is $O(8m) \ltimes \mathbb{R}^{8m}$, while the subgroup leaving invariant the symplectic forms Σ^{AB} is $O(m) \ltimes \mathbb{R}^{8m}$. The semi–simple part of the gauge group, G, embeds in $SO(m)$, while A embeds in $\mathbb{R}^{8m}$.

In fact we know that the scalars transform[5] in the adjoint of G, i.e.,

$$X^{\alpha a} \mapsto \lambda_b f^{abc} X^{\alpha c} \qquad (\mathfrak{Lie}(G)) \tag{8.90}$$

$$X^{\alpha a} \mapsto X^{\alpha a} + \delta^{\alpha 8} \lambda^{\hat{a}} \qquad (\mathfrak{Lie}(A)), \tag{8.91}$$

breaking $Spin(8)_R$ symmetry (and hence the superconformal invariance). The momentum map of $\mathfrak{Lie}(A)$ is *linear* in the scalar fields $X^{\alpha a}$

$$\mu^{AB\,\hat{a}} = (\Sigma^{AB})_{8\,\alpha} X^{\alpha a}, \tag{8.92}$$

whereas that of $\mathfrak{Lie}(G)$ is *quadratic*

$$\mu^{AB\,a} = -\frac{1}{2}(\Sigma^{AB})_{\beta\gamma} f^{abc} X^{\beta b} X^{\gamma c}. \tag{8.93}$$

Since the invariant tensor $l_{mn} \in \mathfrak{Lie}(G) \otimes \mathfrak{Lie}(A)$, the T-tensor is cubic in the $X^{\alpha a}$:

$$T^{AB,CD} = g f_{abd} X^{\alpha a} X^{\beta b} X^{\gamma c} (\Sigma^{AB})_{\alpha\beta} (\Sigma^{CD})_{\gamma 8} + ([AB] \leftrightarrow [CD]). \tag{8.94}$$

This T-tensor *does* satisfy the consistency constraints. To see this, we notice that, for $\mathcal{N} = 8$, Eq. (8.42) reads (we denote the various irrepresentations by their Dynkin label)

$$\bigodot^2 [0,1,0,0] = [0,0,0,0] \oplus [2,0,0,0] \oplus [0,2,0,0]$$
$$\oplus \left([0,0,2,0] \oplus [0,0,0,2]\right). \tag{8.95}$$

In particular, the obstructive Young tableaux ⊞ corresponds to the representation $[0, 2, 0, 0]$ (the **300**). The scalar field $X^{\alpha a}$ is in one of the spinorial representations of $Spin(8)$, say the $[0, 0, 1, 0]$. From Eq. (8.94), since f_{abc} is totally antisymmetric, we see that[6]

$$T^{AB,CD} \in \left(\wedge^3 [0,0,1,0]\right) \bigotimes [0,0,1,0]$$
$$= [1,0,1,1] \oplus [2,0,0,0] \oplus [0,0,0,2] \oplus [0,1,0,0]. \tag{8.96}$$

---

[5] The 3D $\mathcal{N} = 8$ SYM has 7 scalars $X^{\alpha a}$ ($\alpha = 1, 2, \ldots, 7$) in the adjoint of the SYM gauge group G and transforming in the vector representation of $Spin(7)$. Dualizing the 3D vectors, we get an additional scalar in the adjoint of G, which we call X^{8a}. By construction, X^{8a} is the only scalar not invariant under the shift gauge symmetry A.

[6] The decompostion of the tensor product of representations was computed using the computer package LiE [288]; another convenient package is given in ref. [137].

Comparing with (8.95), we get

$$T^{AB,CD} \in [2,0,0,0] \oplus [0,0,0,2], \tag{8.97}$$

and $T^{AB,CD}|_{\boxplus} = 0$. Alternatively, we may consider the dual $\mathcal{N} = 8$ SYM as a special instance of an $\mathcal{N} = 4$ model. Since the 4D $\mathcal{N} = 4$ SYM supermultiplet decomposes into an $\mathcal{N} = 2$ SYM multiplet plus an adjoint hypermultiplet, its dualized 3D reduction produces a pair of twisted/untwisted adjoint hypermultiplets, so the scalar manifold is $\mathbb{H}_L^{\dim G} \times \mathbb{H}_R^{\dim G}$. $\mathbb{H}_L^{\dim G}$, coming from the 4D vector multiplet, is identified with the manifold for the $\mathcal{N} = 4$ SYM, Section 8.4.1. The T–tensor restricted to $\mathbb{H}_L^{\dim G}$ then satisfies the Gaiotto–Witten condition, while the restriction to $\mathbb{H}_R^{\dim G}$ vanishes. By the argument in Eqs. (8.66)–(8.68) the model has at least $\mathcal{N} = 4$ SUSY. Using the unbroken $SO(7)$ R–symmetry, we get full $\mathcal{N} = 8$ supersymmetry.

8.4.3 The gauge group $\mathcal{G}$ of a superconformal CS–matter theory

$\mathcal{N} = 4, 8$ SYM have a non–compact gauge group $G \ltimes A$ with a T–tensor which is cubic in the scalars. $T^{AB,CD}$ is an operator of (canonical) dimension 3/2, while conformal invariance requires dimension 2. Hence the "trick" of considering non–compact gauge groups to enlarge the spectrum of possible theories besides those listed in Table 8.1 cannot produce new *superconformal* theories. Therefore the list in ref. [48], which was focused on the superconformal case, is complete for that purpose. This follows from the following:

General Lesson 8.6 ([46])   *The gauge group of a unitary $\mathcal{N} \geq 3$ superconformal CS–matter 3D theory is compact. In particular, the Lagrangian cannot contain YM terms F^2.*

Proof   For $\mathcal{N} \geq 3$ superconformal, $\mathcal{M}$ is a Ricci–flat cone over B. Unitarity requires the metric of $\mathcal{M}$ to be positive–definite and B complete. For all cones $\mathcal{M}$, the Ricci curvatures of $\mathcal{M}$ and its base B are related as

$$R_{ij}\big|_{\mathcal{M}} = R_{ij}\big|_B - (\dim B - 1)g_{ij} \tag{8.98}$$

(i, j being directions tangent to the base B). Hence $\mathcal{M}$ Ricci–flat implies B Einstein with positive Ricci curvature. Since B is complete, Meyers theorem ([51] theorem 6.51 or [63]) implies that the base B is compact with diameter

$$\text{diam}(B) \leq \pi/\sqrt{(\dim B - 1)}. \tag{8.99}$$

This means that Iso(B) is also compact ([51] corollary 1.78). Now the flat space scalar kinetic term

$$-\frac{1}{2} G_{ab}\, \partial^\mu \phi^a\, \partial_\mu \phi^b, \tag{8.100}$$

when coupled to a curved background in a *conformal* invariant way in $D = p + 1$ dimensions, requires an improving term proportional to the space–time scalar curvature R,

$$-\frac{1}{2}\sqrt{-g}\left(g^{\mu\nu} G_{ab}\, \partial_\mu \phi^a\, \partial_\nu \phi^b + \frac{p-1}{2p} V R\right), \tag{8.101}$$

where $V \equiv r^2/2$ is the potential of the concurrent vector field of the cone $\mathcal{M}$ (equal to the hyperKähler potential for $\mathcal{N} = 3, 4$), $V_i = \partial_i V$. Since the Lagrangian must be gauge–invariant, we may gauge only the subgroup of Iso($\mathcal{M}$) which leaves invariant V, i.e., whose Killing vectors K^m satisfy

$$0 = \pounds_{K^m} V \equiv K^{m\,i} \partial_i V \equiv V^i K_i^m \equiv i_V K^m = r K_r^m. \tag{8.102}$$

This equation implies that $K^{m\,i}\partial_i$ is a Killing vector of B. Hence the conformally gaugeable group is the subgroup $\text{Iso}_0(B) \subset \text{Iso}(B)$ consisting of multisymplectic isometries. The gauge group, being a closed subgroup of the compact group Iso(B), is compact. □

8.5 World–volume theory of M2 branes: the ABJM model

The most important application of 3D CS–matter theories is the world–volume theory of a stack of N M2 branes of M–theory, which is a candidate *fundamental* theory of physics. When the N branes are on top of each other, we expect the world–volume theory to be a 3D superconformal theory with $\mathcal{N} = 8$ and gauge group containing $SU(N)$. Moreover, we expect that in some limit, after breaking the conformal invariance, it reduces to the world–volume theory of a stack of $D2$ branes which is (at low energy) just 3D $\mathcal{N} = 8$ SYM with $G = SU(N)$.

At first we are puzzled. Since the theory is $\mathcal{N} = 8$ superconformal, if unitary, it should have a compact gauge group. From Table 8.1 we see that the only possibility is $G \subseteq SU(2) \times SU(2) \times U(1)$. This theory (called the Bagger–Lambert model [29, 30]) corresponds to $N = 2$ M2 branes, but what about $N > 2$? We *know* that the theory exists !

A possibility [166] is to start from $\mathcal{N} = 8$ SYM and its non–compact gauge group $G \ltimes A$. One introduces a new field Y^α and modifies (8.90) and (8.91) as

$$\delta X^{\alpha\,a} = \lambda_b f^{abc} X^{\alpha\,c} + \lambda^{\hat{a}} Y^\alpha, \qquad \delta Y^\alpha = 0, \tag{8.103}$$

to get a quartic T–tensor and hence $\mathcal{N} = 8$ superconformal invariance. Unfortunately, this possibility is hard to reconcile with unitarity.

A deeper way out of this embarassing situation was pointed out by Aharony, Bergman, Jafferis, and Maldacena [3]. They consider $M2$ branes in the geometry $\mathbb{R}^{2,1} \times \mathbb{R}^8/\mathbb{Z}_k$, $k \in \mathbb{N}$. Since the *Spin*(8) R–symmetry corresponds to rotations of the transverse space, replacing $\mathbb{R}^8$ with the orbifold $\mathbb{R}^8/\mathbb{Z}_k$ breaks *Spin*(8) down to *Spin*(6), and we expect to get an $\mathcal{N} = 6$ superconformal theory. Looking to Table 8.1, we see that the $\mathcal{N} = 6$ model $U(N|N)$ has the right symmetries. Going through the above arguments, we see that the CS levels of the two $SU(N)$ groups must be opposite[7] $(k, -k)$, where the integer k is identified with the order of the orbifold group $\mathbb{Z}_k$ [3]. The 3D theory hence has a coupling constant $1/k$, whose classical limit corresponds to $k \to \infty$. A stack of M2 branes on the space $\mathbb{R}^{2,1} \times \mathbb{R}^8$ then corresponds to $k = 1$, that is, extremely strong coupling. In this regime non–perturbative effects are important. There is convincing evidence that these effects enhance the classical $\mathcal{N} = 6$ superconformal invariance to $\mathcal{N} = 8$ for $k = 1, 2$ [3].

8.6 Gauged supergravities

Passing from rigid SUSY to SUGRA, several issues arise. First of all, there are $\mathcal{N}$ gravitini, ψ_μ^A, in the vector representation of $Spin(\mathcal{N})$. Second, the $Spin(\mathcal{N})$ symmetry is gauged: the $Spin(\mathcal{N})$ connection is identified with a part of the Christoffel connection of $\mathcal{M}$, in view of the isomorphism

$$T\mathcal{M} \simeq S \otimes U. \tag{8.104}$$

As in rigid SUSY, the conditions on the possible gaugings stem from consistency of the Yukawa couplings. In SUGRA there are *three* such couplings

$$e g \left\{ \frac{1}{2} A_1^{AB} \overline{\psi}_\mu^A \gamma^{\mu\nu} \psi_\nu^B + A_{2i}^A \overline{\psi}_\mu^A \gamma^\mu \chi^i + \frac{1}{2} A_{3ij} \overline{\chi}^i \chi^j \right\}, \tag{8.105}$$

where $A_1^{AB} = A_1^{BA}$ and $A_{3ij} = A_{3ji}$. From the General Lessons of Chapter 7 we know that the Yukawa tensors $A_1, A_2,$ and A_3 should have a universal form in terms of the T–tensor. The conditions on the T–tensor then express the requirement that the induced Yukawas have the correct properties.

---

[7] A simple way to see this is to use the fact that the gauge algebra should admit a contraction to the SYM one $G \ltimes A$.

8.6.1 Gauging supergravity: the embedding tensor

The gauge group $\mathcal{G}$ is a subgroup of $\mathrm{Iso}(\mathcal{M})$. In SUGRA it is usual to write the embedding $\mathfrak{L} \equiv \mathfrak{Lie}(\mathcal{G}) \hookrightarrow \mathfrak{iso}(\mathcal{M})$ in terms of an *embedding tensor* [120, 122, 142, 248, 249]. In the present (*diet*) 3D situation it is written as an element of $\mathfrak{iso} \otimes \mathfrak{iso}$, which we write as l_{mn}. The (infinitesimal) gauge transformations then have the form

$$\delta \Phi = \Lambda(x)^m \, l_{mn} \, \pounds_{K^n} \, \Phi, \tag{8.106}$$

where the $\Lambda(x)^m$ are (space–time–dependent) parameters. If $\mathcal{G}$ is semi–simple, l_{mn} is a multiple of the Cartan–Killing form on each simple group factor, and hence it is symmetric, $l_{mn} = l_{nm}$. This property holds in general: $\mathfrak{iso}(\mathcal{M}) \simeq \ker l \oplus \mathrm{im}\, l \equiv \ker l \oplus \mathrm{im}\, \mathfrak{g}$ should be an orthogonal decomposition. The requirement that the image of l is a Lie subalgebra $\mathfrak{L} \subset \mathfrak{g}$ reads

$$l_{mp} \, l_{nq} f^{pq}{}_k = c_{mn}{}^h \, l_{hk}, \tag{8.107}$$

with $c_{mn}{}^h$ the structure constants of $\mathfrak{L}$. The scalar covariant derivative is

$$\mathcal{D}_\mu \phi^i = \partial_\mu \phi^i + g \, l_{mn} A_\mu^m K^{in}, \tag{8.108}$$

where we have inserted the coupling g as an order–counting device. This derivative transforms covariantly,

$$\mathcal{D}_\mu \phi^i \to (\delta^i{}_j + g \, l_{mn} \Lambda^m \partial_j K^{in}) \, \mathcal{D}_\mu \phi^j, \tag{8.109}$$

provided the gauge fields transform as

$$l_{mn} \delta A_\mu^n = l_{mn} \left(\partial_\mu \Lambda^n - g \, c_{pq}{}^n A_\mu^p \Lambda^q \right). \tag{8.110}$$

Consider the gauge variation of the gravitino ψ_μ^A. One has

$$\delta_K \psi_\mu^A = \pounds_K \psi_\mu^A = (A_K + D_K) \psi_\mu^A$$
$$= A_K^{AB} \psi_\mu^B + K^i Q_i^{AB} \psi_\mu^B, \tag{8.111}$$

and the new covariant derivative is

$$\mathcal{D}_\mu \psi_\nu^A = \nabla_\mu \psi_\nu^A + \partial_\mu \phi^i \, Q_i^{AB} \psi_\nu^B + g \, l_{mn} A_\mu^m \, \mathcal{A}^{ABn} \, \psi_\nu^B, \tag{8.112}$$

where ∇_μ is the covariant derivative with respect to the curved space–time connection, and we write $\mathcal{A}^{ABn} \equiv A_{K^n}^{AB}$ for the A_K–derivation ($=$ one–half the covariant momentum map). Covariance requires also

$$\mathcal{D}_\mu \epsilon^A = \nabla_\mu \epsilon^A + \partial_\mu \phi^i \, Q_i^{AB} \epsilon^B + g \, l_{mn} A_\mu^m \, \mathcal{A}^{ABn} \, \epsilon^B. \tag{8.113}$$

Now the SUSY variation of the Rarita–Schwinger term, $-\frac{i}{2}\varepsilon^{\mu\nu\rho}\overline{\psi}_\mu^A \mathcal{D}_\nu \psi_\rho^A$, has an additional contribution steming from the fact that

$$[\mathcal{D}_\mu, \mathcal{D}_\nu] = \cdots + g\, l_{mn} F_{\mu\nu}^m \pounds_{K^n}, \tag{8.114}$$

where $\cdots$ stands for terms already present in the ungauged theory (i.e., for $g = 0$). A similar term appears in the variation of the χ kinetic terms, since $(f^A)^i{}_j \delta\chi^j = \gamma^\mu \mathcal{D}_\mu \phi^i \epsilon^A + 2\text{–fermions}$. The SUSY variation of the covariantized original terms then reads

$$\delta\mathcal{L}_0 = \frac{i}{2} g\, l_{mn} \varepsilon^{\mu\nu\rho} F_{\mu\nu}^m \left(\mathcal{A}^{ABn} \overline{\psi}_\mu^A \epsilon^B + \frac{1}{2} K_i^n (f^A)^i{}_j \overline{\chi}^j \gamma_\mu \epsilon^A \right). \tag{8.115}$$

In order to cancel this variation, one introduces the Chern–Simons term

$$\mathcal{L}_{CS} = \frac{i}{4} g\, \varepsilon^{\mu\nu\rho} A_\mu^m\, l_{mn} \left(\partial_\nu A_\rho^n - \frac{1}{3} g\, c_{pq}{}^n A_\nu^p A_\rho^q \right), \tag{8.116}$$

and prescribes the SUSY variation of the vector fields

$$l_{mn} \delta A_\mu^m = l_{mn} \left[2 \mathcal{A}^{ABm} \overline{\psi}_\mu^A \epsilon^B + K_i^m (f^A)^i{}_j \overline{\chi}^j \gamma_\mu \epsilon^A \right]. \tag{8.117}$$

However, this produces additional SUSY variations at order $O(g)$:

$$\begin{aligned}&\delta(\mathcal{L}_0 + \mathcal{L}_{CS}) \\ &= e\, g\, l_{mn}(2\, \mathcal{A}^{ABm} \overline{\psi}_\mu^A \epsilon^B + K_i^m (f^A)^i{}_j \overline{\chi}^j \gamma_\mu \epsilon^A)\, K_k^n \mathcal{D}^\mu \phi^k + \cdots \end{aligned} \tag{8.118}$$

where we write only the terms linear in the fermions. This variation is cancelled by introducing the Yukawa couplings

$$e \left\{ \frac{1}{2} A_1^{AB} \overline{\psi}_\mu^A \gamma^{\mu\nu} \psi_\nu^B + A_{2i}^A \overline{\psi}_\mu^A \gamma^\mu \chi^i + \frac{1}{2} A_{3ij} \overline{\chi}^i \chi^j \right\}, \tag{8.119}$$

together with fermionic shifts in the SUSY transformations

$$\delta\psi_\mu^A = \cdots + g A_1^{AB} \gamma_\mu \epsilon^B, \qquad \delta\chi^i = \cdots - g A_2^{iA} \epsilon^A, \tag{8.120}$$

where $\cdots$ stands for $O(g^0)$ terms already present in the ungauged theory. Finally, to cancel the variations at order $O(g^2)$, we should add a scalar potential given by the 3D version of the Ward identity of Chapter 6.

8.6.2 Constraints on the covariant T–tensor

Let us look at the conditions under which the term in $\delta\mathcal{L}$ proportional to $g\, \psi_\mu^A \psi_\nu^B \psi_\rho^C$ vanishes. There are two sources of such terms: the variation of A_μ

inside the gravitino kinetic term, and the variation of e in front of the Yukawa bilinear in ψ_μ^A,

$$-\frac{i}{2}g\,\overline{\psi}_\mu^A \gamma^{\mu\nu\rho}\psi_\rho^B\, l_{mn}\,\mathcal{A}^{ABn}\,\delta A_\nu^M\Big|_\psi + \frac{g}{2}\delta e\, A_1^{AB}\,\overline{\psi}_\mu^A \gamma^{\mu\nu}\psi_\nu^B$$

$$= -\frac{1}{2}g\,\overline{\psi}_\mu^A \gamma^{\mu\nu\rho}\psi_\rho^B\, l_{mn}\,\mathcal{A}^{ABn}\,\mathcal{A}^{CDm}\,\overline{\psi}_\nu^C\epsilon^D + \frac{1}{4}g(\overline{\epsilon}^C \gamma^\rho \psi_\rho^C)A_1^{AB}\,\overline{\psi}_\mu^A \gamma^{\mu\nu}\psi_\nu^B$$

$$= -\frac{g}{2}T^{AB,CD}(\overline{\psi}_\mu^A \gamma^{\mu\nu\rho}\psi_\rho^B\,\overline{\psi}_\nu^C\epsilon^D) + \frac{g}{4}(A_1^{AB}\delta^{CD})(\overline{\psi}_\mu^A \gamma^{\mu\nu}\psi_\nu^B\,\overline{\psi}_\rho^C\gamma^\rho\epsilon^D)$$

(8.121)

(the formulae are meant to be just schematic). Notice that the totally antisymmetric part of the T–tensor, $T^{[AB,CD]}$, corresponding to the one–column 4–box tableaux, decouples from the first term in the last line. Indeed, since in 3D the Majorana fermions have only two real components, $\psi_{[\mu}^{[A}\psi_\nu^B\psi_{\rho]}^{C]} \equiv 0$, by Fermi statistics. Therefore, the condition $\delta\mathcal{L}|_{\psi\psi\psi} = 0$ yields:

General Lesson 8.7 [120] *In order for a given 3D* SUGRA *gauging to be allowed, the associated T–tensor should satisfy the algebraic condition*

$$T^{AB,CD} = \frac{1}{4}\left(\delta^{AC}A_1^{BD} - \delta^{BC}A_1^{AD} - \delta^{AD}A_1^{BC} + \delta^{BD}A_1^{AC}\right) + T^{[AB,CD]}, \quad (8.122)$$

that is,

$$T^{AB,CD}\Big|_{\boxplus} = 0. \tag{8.123}$$

As anticipated, this is precisely the *same* condition as in rigid SUSY, Eq. (8.42). Comparing with the definition of T^{AB}, Eq. (8.43), we get:[8]

Corollary 8.8 *For $\mathcal{N} \geq 3$ we have:*

$$A_1^{AB} = T^{AB}. \tag{8.124}$$

The cases $\mathcal{N} = 1, 2$ are special:

(1) *For $\mathcal{N} = 1$, both sides of Eq. (8.122) are identically zero. This corresponds to the fact that the Yukawa couplings contain terms coming from a real superpotential (as in the rigid case) and hence may be not zero even if <u>no</u> gauging is present. We can parametrize all Yukawa/potential couplings in terms of the gravitino mass, A_1 (which is essentially the real superpotential) together with*

---

[8] Recall that there is a factor $1/4$ in the definition of the covariant T–tensor with respect to the rigid one. Here we are rather cavalier with numerical constants. They are *not* really computed, rather they are fixed at the end by geometric arguments. The final formulae are certainly correct; the intermediate ones are meant to be just schematic.

the symmetric tensor $T_{ij} \equiv K_i^m \, l_{mn} \, K_j^n$ which fully encodes the gauging of the given $\mathcal{N} = 1$ model.

(2) For $\mathcal{N} = 2$, the LHS of Eq. (8.122) is a singlet of Spin(2). Hence only the singlet part of A_1^{AB} (that is its trace $\delta_{AB} A_1^{AB}$) is determined in terms of the T–tensor. The components of A_1^{AB} having Spin(2) charge ± 2,

$$\tfrac{1}{2}(A_1^{22} - A_1^{11}) \pm i A_1^{12}, \tag{8.125}$$

are *not* determined by the gauging. Again, as in the rigid case, this corresponds to the fact that we may add a non–trivial complex superpotential (a general solution to the homogeneous consistency condition). We can parametrize the Yukawas and potential in terms of the T–tensor plus the complex quantity $\tfrac{1}{2}(A_1^{22} - A_1^{11}) + i A_1^{12}$. (See Chapter 10 for details.)

We have still to determine the Yukawa tensors A_2, A_3, the fermionic shifts and potential, and check that everything works well. We do this in the next subsection, exploiting the Cartan–Kostant isomorphism.

8.6.3 Fermionic shifts, Yukawa tensors and potential

The direct computation of the Yukawas and fermionic shifts, for the general case, is quite involved (see ref. [120] for details). Thus, instead of computing, let us argue their form on general grounds. First, the fermionic shifts can be read directly from the SUSY current, namely from the terms in the Lagrangian $\mathcal{L}$ linear in the gravitino fields ψ_μ^A. Hence

$$\delta \psi_\mu^A = \cdots + g A_1^{AB} \gamma_\mu \epsilon^B, \tag{8.126}$$

$$\delta \chi^i = \cdots - g\, g^{ij} A_{2\,j}^A \epsilon^A, \tag{8.127}$$

so it is enough to determine the Yukawa tensors (the scalar potential is then predicted by the universal formula of Chapter 6). We already know that they have a universal expression in terms of the T–tensor.

The first tensor, A_1^{AB}, was computed in Eq. (8.122). To determine $A_{2\,i}^A$ and A_{3ij}, we return to the scaling argument we used in Chapter 2 to predict the 4–Fermi couplings. Rescaling the volume of $\mathcal{M}$ (or, more correctly, the Planck scale) we can "switch off" the supergravity couplings, ending up with a rigid $\mathcal{N}$–SUSY model for which we know both the fermionic shift $A_{2\,i}^A$ and the Yukawa matrix A_{3ij}. Moreover, we have a scaling property with respect to the gauge coupling g. Then the expression for SUGRA should be that for rigid SUSY (minimally covariantized, that is, with derivatives replaced by $Spin(\mathcal{N})$–covariant derivatives) *plus* corrections which vanish in the above rigid limit. As in Chapter 2, the possible corrections

8.6 Gauged supergravities

should have (at least) one more factor of g_{ij} than the ones present in the rigid limit. Thus $A^A_{2\,i}$ should not be corrected (except for covariantization), while

$$A_{3\,ij} = A_{3\,ij}^{\text{rigid}}\big|_{\text{covariantized}} + g_{ij} F(T^{AB,CD}), \tag{8.128}$$

where, by scaling in g, F is some *linear* function of the T–tensor. To determine F, we use a clever trick of ref. [120]. In order to make the action of $Spin(\mathcal{N})$ more manifest, one introduces an overcomplete system of spin–1/2 fermions, χ^{Ai}, $A = 1, 2, \ldots, \mathcal{N}$, $i = 1, 2, \ldots, \dim \mathcal{M}$ by

$$\chi^{Ai} = (f^A)^i{}_j \chi^j. \tag{8.129}$$

The fact that these fermions are not independent is written as the constraint

$$\chi^{Ai} = \mathbb{P}^{Ai}_{Bj} \chi^{Bj}, \tag{8.130}$$

where $\mathbb{P}^{Ai}_{Bj}$ is the projector

$$\mathbb{P}^{Ai}_{Bj} := \frac{1}{\mathcal{N}}\left(\delta^{AB}\delta^i_j - (\Sigma^{AB})^i{}_j\right). \tag{8.131}$$

Now one writes the fermionic shift in the form

$$\delta \chi^{Ai} = \cdots - g^{ij} A^{BA}_{2\ j} \epsilon^B \tag{8.132}$$

and the $\chi\chi$ Yukawa couplings as

$$\frac{1}{2} e\, g\, A^{AB}_{3\,ij}\, \overline{\chi}^{Ai} \chi^{Bj}. \tag{8.133}$$

These matrices should satisfy the algebraic conditions

$$A^{AC}_{2\ k} \mathbb{P}^{Ck}_{Bi} = A^{AB}_{2\ i}, \tag{8.134}$$

$$A^{AC}_{3\ ik} \mathbb{P}^{Ck}_{Bj} = A^{AB}_{3\ ij}, \tag{8.135}$$

$$A^{AB}_{3\ ij} = A^{BA}_{3\ ji}, \tag{8.136}$$

which, as we shall see momentarily, uniquely fix the unknown function F.

The χ–shift is read directly from Eq. (8.64),

$$A^{AB}_{2\ i} = \mu^{ABm} l_{mn} K^n_i + \mathcal{D}_i T^{AB}, \tag{8.137}$$

where we used the fact that the T–tensor satisfies the constraint $T|_{\boxplus} = 0$. In fact, as written, Eq. (8.137) is correct for $\mathcal{N} \geq 3$, for reasons explained in Corollary 8.8. The general expression, valid for all $\mathcal{N}$'s, is

$$A^{AB}_{2\ i} = \mu^{ABm} l_{mn} K^n_i + \mathcal{D}_i A^{AB}_1, \tag{8.138}$$

which has the correct rigid limit. Notice that the first term on the RHS is proportional to

$$\theta(K^m) \, l_{mn} \, \theta(K^n)\big|_{\mathrm{spin}(\mathcal{N}) \otimes \mathfrak{m}}, \qquad (8.139)$$

so it is again a component of the T–tensor as generalized by the Cartan–Kostant isomorphism, Section 7.8. For this reason we denote it as T_i^{AB}:

$$T_i^{AB} \equiv -T_i^{BA} := \mu^{AB\,m} \, l_{mn} \, K_i^n. \qquad (8.140)$$

Notice that

$$\mathcal{D}_{(i} T_{j)}^{AB} = (\Sigma^{AB})_{(i}{}^k K_k^m \, l_{mn} \, K_{j)}^n. \qquad (8.141)$$

One shows that the $A_{2\,i}^{AB}$ in Eq. (8.138) satisfies Eq. (8.134) as a consequence of the fact that $T|_{\boxplus} = 0$. Indeed, as long as we have expressions with only one (covariant) derivative, the identities of the rigid case apply also to SUGRA. Instead, expressions which contain $\mathcal{D}_i \mathcal{D}_j T^{AB,CD}$ are different in the two cases due to the presence, in SUGRA, of an antisymmetric term in $i \leftrightarrow j$ due to the non–trivial $Spin(\mathcal{N})$ curvature. $A_{2\,i}^{AB}$ contains only first derivatives, and then the algebraic proof of $Spin(\mathcal{N})$ invariance, under the condition $T|_{\boxplus} = 0$ in Section 8.2.3 applies word for word.

On the contrary, the Yukawa couplings contain the second derivative and hence corrections are necessary to maintain the correct symmetry properties. The relation between the curvature terms, arising from the commutation of covariant derivatives, and the symmetry condition forces a term proportional to g_{ij} to be present, as we shall see shortly. No other corrections are needed, in agrement with the general arguments.

In the rigid case one has for the diagonal entries[9] (no sum over A!)

$$-\frac{\mathcal{N}}{2} A_{3\,ij}^{AA} = \partial_i A_{2\,j}^{AA} + K_i^m \, l_{mn} \, K_j^n, \qquad (8.142)$$

which we can upgrade to a fully $Spin(\mathcal{N})$ covariant expression with the help of the projector $\mathbb{P}_{B\,j}^{A\,i}$. Therefore, in the local case we must have

$$(-1/2)\mathcal{N}^2 A_{3\,ij}^{AB} = \mathcal{N} \mathcal{D}_i A_{2\,j}^{AB} - K_i^m \, l_{mn} \, K_k^n \mathcal{N} \mathbb{P}_{B\,j}^{A\,k} + g_{ik} F^{AC} \mathcal{N} \mathbb{P}_{B\,j}^{C\,k}, \qquad (8.143)$$

where F^{AB} is the covariant counterpart to the function F in Eq. (8.128), which should be linear in the components of the T–tensor. The first term on the RHS

---

[9] The normalization is different from the one we used in the rigid case to get formulae similar to those one finds in the SUGRA literature. Nothing in the argument depends on the specific numeric coefficients.

8.6 Gauged supergravities

already satisfies the projection constraint, as a consequence of $T|_\boxplus = 0$. Now,

$$\begin{aligned}
\mathcal{D}_i A^{AB}_{2j} &= \mathcal{D}_{(i} T^{AB}_{j)} + \mathcal{D}_{[i} T^{AB}_{j]} + \mathcal{D}_{(i} \mathcal{D}_{j)} T^{AB} + \mathcal{D}_{[i} \mathcal{D}_{j]} T^{AB} \\
&= \mathcal{D}_{(i} \mathcal{D}_{j)} T^{AB} + \mathcal{D}_{[i} T^{AB}_{j]} \\
&\quad + (\Sigma^{AB})_{(i}{}^k K^m_k l_{mn} K^n_{j)} + \frac{1}{2} (\Sigma^{AC})_{ij} T^{CB} + \frac{1}{2} (\Sigma^{BC})_{ij} T^{AC},
\end{aligned} \qquad (8.144)$$

where we have used Eq. (8.141) and the explicit form of the $Spin(\mathcal{N})$ curvature. On the RHS, the first line has already the correct symmetry under the interchange $(A, i) \leftrightarrow (B, j)$; the other terms should give symmetric terms after adding the other two terms on the RHS of Eq. (8.143). One has

$$\begin{aligned}
\text{RHS of Eq. (8.143)} &= \mathcal{D}_{(i} \mathcal{D}_{j)} T^{AB} + \mathcal{D}_{[i} T^{AB}_{j]} \\
&\quad + (\Sigma^{AB})_{(i}{}^k K^m_k l_{mn} K^n_{j)} - \frac{1}{2} (\Sigma^{AC})_{ij} T^{CB} - \frac{1}{2} (\Sigma^{BC})_{ij} T^{AC} \\
&\quad - \delta^{AB} K^m_i l_{mn} K^n_j + K^m_i l_{mn} K^n_k (\Sigma^{AB})^k_j + g_{ij} F^{AB} - F^{AC} (\Sigma^{CB})_{ij} \\
&= \mathcal{D}_{(i} \mathcal{D}_{j)} T^{AB} + \mathcal{D}_{[i} T^{AB}_{j]} + \delta^{AB} K^m_i l_{mn} K^n_j + K^m_{[i} l_{mn} K^n_k (\Sigma^{AB})^k_{j]} \\
&\quad + g_{ij} F^{AB} + (\Sigma^{BC})_{ij} \left(\frac{1}{2} T^{AC} + F^{AC} \right) + \frac{1}{2} (\Sigma^{AC})_{ij} T^{CB}.
\end{aligned} \qquad (8.145)$$

The first line on the RHS has the required symmetry. The second line has the correct symmetry iff

$$F^{AB} = F^{BA}, \qquad (8.146)$$

$$T^{AC} + 2F^{AC} = -T^{AC}, \qquad (8.147)$$

that is,

$$F^{AB} = -T^{AB}. \qquad (8.148)$$

Thus the Yukawa couplings are completely determined in terms of the T–tensor. Once the fermionic shifts are determined, the scalar potential is also determined by the Ward identity. The validity of the Ward identity requires the fermionic shifts to satisfy a number of algebraic relations which we shall not verify here. Diligent readers may check them for themselves (or see [120]).

8.6.4 Identities for reference

The tensor $A^{AB}_{2\,i}$ satisfies two equations: (i) the decomposition

$$A^{AB}_{2\,i} = \underbrace{T^{AB}_i}_{\text{antisymm.}} + \underbrace{\mathcal{D}_i A^{AB}_1}_{\text{symm.}}, \qquad (8.149)$$

and (ii) the projection (8.134), which we write as

$$(\mathcal{N}-1)A^{AB}_{2\,i} = A^{AC}_{2\,k}(\Sigma^{CB})^k{}_i. \qquad (8.150)$$

Taking the symmetric and antisymmetric parts in A, B, we get

$$2(\mathcal{N}-1)\mathcal{D}_i A^{AB}_1 = \left(\mathcal{D}_k A^{AC}_1 (\Sigma^{CB})^k{}_i + (A \leftrightarrow B)\right) + 4\mathcal{D}_i T^{AC,CD}, \qquad (8.151)$$

$$2(\mathcal{N}-1)T^{AB}_i = (T^{AC\,k} + \mathcal{D}^k A^{AC}_1)(\Sigma^{CB})_{ki} - (A \leftrightarrow B)), \qquad (8.152)$$

where we have used the definition of the covariant momentum map and T-tensor,

$$T^{AC\,k}(\Sigma^{CB})_{ki} + (A \leftrightarrow B) = \mathcal{D}_i(\mu^{AC\,m} l_{mn} \mu^{CB\,n}) = 4\mathcal{D}_i T^{AC,CD}. \qquad (8.153)$$

8.7 Symmetric target spaces

The expressions simplify if the scalar manifold $\mathcal{M}$ is *symmetric* (or just homogeneous). This condition holds automatically for $\mathcal{N} \geq 5$, but even for $\mathcal{N} = 2, 3, 4$ there are many interesting Kähler (resp. Quaternionic–Kähler) manifolds that are symmetric. In the Kähler case they are the Hermitian symmetric manifolds; in the Quaternionic–Kähler case, we have the Wolf spaces, see [320] and tables below. We may assume the symmetric space $\mathcal{M}$ to be irreducible without loss of generality. Then, in this section, we write G/H for $\mathcal{M}$, with G simple and $H \equiv \mathrm{Spin}(\mathcal{N}) \times H'$ its maximal compact subgroup. From Chapters 3 and 5 we see that the Lie algebra $\mathfrak{g}$ of G decomposes into irreducible representations[10] of $\mathrm{Spin}(\mathcal{N})$ as

$$\mathfrak{g} = \mathfrak{spin}(\mathcal{N}) \oplus \mathbf{1}^{\oplus \dim \mathfrak{h}'} \oplus S^{\oplus m}, \qquad (8.154)$$

where S is an irreducible spin representation, and $m = \dim \mathcal{M}/\mathbf{N}(\mathcal{N})$.

Homogeneous spaces. Although for simplicity we work with *symmetric* spaces, we stress that, by the Kostant theorem [219], the following results hold even if G/H is *not* symmetric but just homogeneous, provided that it is either *compact* or *not Ricci–flat*. Since all $\mathcal{N} \geq 3$ SUGRA manifolds have negative Ricci–curvature, they

---
[10] This is true for $\mathcal{N} \neq 4$. For $\mathcal{N} = 4$ the adjoint representation is not irreducible. The modifications are obvious.

apply to all homogeneous $\mathcal{N} \geq 3$ manifolds. The homogeneous, non–symmetric, Quaternionic–Kähler spaces are known as the Alekseevskiĭ manifolds: there are three infinite families [5, 115]. The results of this section apply, *mutatis mutandis*, also to them.

8.7.1 Lifting the consistency condition [120]

We saw in Section 8.6.2 that the consistency requirements on the gauging data $(\mathcal{G}, l_{mn})$ reduce to the single equation $T|_{\boxplus} = 0$. The generalized T–tensor is easily computed with the help of Eq. (7.124), which we rewrite for convenience (as always, $g \in G$ stands for a representative of the coset G/H)

$$g^{-1} t^m g = \frac{1}{4} A^{AB}_{Km} \gamma^{AB} + H^{\alpha}_{Km} h^{\alpha} + L^m_a m^a. \qquad (8.155)$$

In particular,

$$T^{AB,CD}(g) = \eta \, \mathrm{tr}\!\left[\gamma^{AB} g^{-1} t^m g\right] l_{mn} \, \mathrm{tr}\!\left[\gamma^{CD} g^{-1} t^n g\right], \qquad (8.156)$$

where η is a suitable normalization constant. Now we must require

$$T^{AB,BC}(g)\big|_{\boxplus} = 0 \quad \forall\, g \in G. \qquad (8.157)$$

The generalized T–tensor

We consider the generalized T–tensor defined in Eqs. (7.144)–(7.146),

$$T^{k,h}(g) = \eta \, \mathrm{tr}\!\left[t^k g^{-1} t^m g\right] l_{mn} \, \mathrm{tr}\!\left[t^h g^{-1} t^n g\right], \qquad (8.158)$$

$$T^{k,h}(g) \in \odot^2 \mathfrak{g} \simeq \odot^2(\mathfrak{spin}(\mathcal{N}) \oplus \mathfrak{h}' \oplus \mathfrak{m}), \qquad (8.159)$$

or, more abstractly,

$$\mathcal{T}(g) := \sum_{m,n} l_{mn} \, (g^{-1} t^m g) \otimes (g^{-1} t^m g) \in \mathfrak{g} \odot \mathfrak{g}. \qquad (8.160)$$

$\mathfrak{g} \odot \mathfrak{g}$ is a G–module (a representation) with respect to the tensor product of the adjoint representation $\varrho \equiv \mathrm{Adj}_G \otimes \mathrm{Adj}_G$,

$$\varrho(h)(x \odot y) = hxh^{-1} \odot hyh^{-1}, \qquad h \in G,\ x, y \in \mathfrak{g}, \qquad (8.161)$$

and the dependence of $\mathcal{T}(g)$ on the scalar fields g takes the algebraic form

$$\varrho(g^{-1}) \mathcal{T}(e) \equiv \mathcal{T}(g). \qquad (8.162)$$

Let us decompose the G–module $\mathfrak{g} \odot \mathfrak{g}$ into *irreducible G–modules* R_i, where $R_0 \equiv \mathbf{1}$ stands for the trivial representation (corresponding to the Cartan–Killing form of the isometry group G),

$$\mathfrak{g} \odot \mathfrak{g} = \mathbf{1} \oplus \left[\bigoplus_{i \geq 1} R_i\right]. \tag{8.163}$$

Obviously,

$$\mathcal{T} = \lambda_{\mathcal{T}} K + \sum_{i \geq 1} \mathcal{T}|_{R_i}, \tag{8.164}$$

K being the Cartan–Killing form and $\lambda_{\mathcal{T}}$ some constant. Comparing with Eq. (8.162), we see that the full functional dependence of $\mathcal{T}(g)$ on the scalars is captured by the G–action ϱ, that is, by the decomposition of $\mathcal{T}(g)$ into G–irreducible components. Each irreducible G–representation R_i can be further decomposed into irreducible representations of the subgroup $Spin(\mathcal{N}) \subset G$,

$$R_i = \bigoplus_j R_{i,j}. \tag{8.165}$$

Now, in terms of $Spin(\mathcal{N})$ representations, (cf. (8.154); here $k = \dim \mathfrak{h}'$)

$$\begin{aligned}
\bigoplus_{i,j} R_{i,j} = \mathfrak{g} \odot \mathfrak{g} &\equiv \odot^2(\mathfrak{spin}(\mathcal{N}) \oplus \mathbf{1}^{\oplus k} \oplus S^{\oplus m}) \\
&= \odot^2 \mathfrak{spin}(\mathcal{N}) \oplus \mathbf{1}^{\oplus k(k+1)/2} \oplus \mathfrak{spin}(\mathcal{N})^{\oplus k} \\
&\quad \oplus (S \odot S)^{m(m+1)/2} \oplus (S \wedge S)^{m(m-1)/2} \\
&\quad \oplus S^{\oplus km} \oplus (\mathfrak{spin}(\mathcal{N}) \otimes S)^{\oplus m}.
\end{aligned} \tag{8.166}$$

In the last line we have only *spinorial* representations. In the third line only irrepresentations which are contained in $S \otimes S$; all such irrepresentations have the form $\wedge^l V$, where V is the vector representation of $Spin(\mathcal{N})$ [65]. In the second line we have our friend $\odot^2 \mathfrak{spin}(\mathcal{N})$ up to copies of $\mathbf{1}$ and $\mathfrak{spin}(\mathcal{N})$. The decomposition of $\odot^2 \mathfrak{spin}(\mathcal{N})$ is given in (8.42) in terms of $SO(\mathcal{N})$ Young tableaux. Thus Eq. (8.166) proves the following:

Lemma 8.9 *In the $Spin(\mathcal{N})$ decomposition $\odot^2 \mathfrak{g} \simeq \bigoplus_{i,j} R_{i,j}$ there is only one copy of the irreducible $Spin(\mathcal{N})$ representation* ⊞. *Therefore, in the G decompostion $\odot^2 \mathfrak{g} \simeq \bigoplus_i R_i$, there is one and only one irreducible G–representation, written $R_⊞$, such that the irreducible $Spin(\mathcal{N})$ representation* ⊞ *appears (with multiplicity 1) in its $Spin(\mathcal{N})$ decomposition*

$$R_⊞ = \bigoplus_j R_{⊞,j} = ⊞ \oplus \cdots \tag{8.167}$$

Table 8.2 *Symmetric spaces for $\mathcal{N} \geq 5$ supergravity in 3D. The representation $R_⊞$ of G is underlined in the decompostion of $\odot^2(\mathrm{Adj}_G)$. Ellipses ... represent zero weights. Taken from* [120].

$\mathcal{N}$	G/H	dim.	Adj_G	$\odot^2(\mathrm{Adj}_G)$
5	$\frac{Sp(4,2k)}{Sp(4)\times Sp(2k)}$	$8k$	$[2,0,\ldots]$	$[0,\ldots] \oplus [0,1,\ldots] \oplus [0,2,\ldots] \oplus \underline{[4,0,\ldots]}$
6	$\frac{SU(4,k)}{SU(4)\times U(k)}$	$8k$	$[1,0,\ldots,1]$	$[0,\ldots] \oplus [1,\ldots,1] \oplus [0,1,\ldots,1,0] \oplus \underline{[2,0,\ldots,2]}$
8	$\frac{SO(8,k)}{SO(8)\times SO(k)}$	$8k$	$[0,1,\ldots]$	$[0,\ldots] \oplus [0,0,0,1,\ldots] \oplus [2,\ldots] \oplus \underline{[0,2,\ldots]}$
9	$\frac{F_{4(-20)}}{SO(9)}$	16	52	$1 \oplus 324 \oplus \underline{1053}$
10	$\frac{E_{6(2)}}{SO(10)\times U(1)}$	32	78	$1 \oplus 650 \oplus \underline{2430}$
12	$\frac{E_{7(-5)}}{SO(12)\times Sp(2)}$	64	133	$1 \oplus 1539 \oplus \underline{7371}$
16	$\frac{E_{8(8)}}{SO(16)}$	128	248	$1 \oplus 3875 \oplus \underline{27000}$

Corollary 8.10 *The condition $T^{AB,CD}(g)|_⊞ = 0$, identically on G/H, IS EQUIVALENT to the condition*

$$\mathcal{T}|_{R_⊞} = 0, \tag{8.168}$$

where $\mathcal{T}$ is the G–representation in Eq. (8.162).

Thus the gauging condition reduces to a simple group–theoretic criterion.

Corollary 8.11 $\mathcal{N} \geq 5$: *the full isometry group G is always an admissible gauge group $\mathcal{G}$.*

Proof G is simple. The G–invariant tensor l_{mn} should be proportional to the Cartan–Killing form K of $\mathfrak{g}$, and hence $\mathcal{T} = \lambda K$, which is always a solution to Eq. (8.168). □

In Table 8.2 (taken from ref. [120]) we write the representation $R_⊞$ for all symmetric spaces arising in $\mathcal{N} \geq 5$ 3D SUGRA. The analogous results $\mathcal{N} = 4$ symmetric spaces (Wolf spaces) are presented in Table 8.3 (again taken from ref. [120]). We do not derive these tables here (they were obtained from computer calculations using the LiE package [288]). The reader may easily check all the results using the on–line version of LiE at the website

www-math.univ-poitiers.fr/ maavl/LiE/.

Table 8.3 *Symmetric spaces for $\mathcal{N} = 4$ supergravity in 3D. The representation $R_\boxplus$ of G is underlined in the decompostion of $\odot^2(\mathrm{Adj}_G)$. Ellipses ... represent zero weights. Taken from [120].*

G/H	dim.	Adj_G	$\odot^2(\mathrm{Adj}_G)$
$\frac{Sp(2m,2)}{Sp(2)\times Sp(2m)}$	$4m$	$[2,0,\ldots]$	$[0,\ldots]\oplus[0,1,\ldots]\oplus[0,2,\ldots]\oplus\underline{[4,0,\ldots]}$
$\frac{SU(m,2)}{SU(2)\times U(m)}$	$4m$	$[1,0,\ldots,1]$	$[0,\ldots]\oplus[1,\ldots,1]\oplus[0,1,\ldots,1,0]\oplus\underline{[2,0,\ldots,2]}$
$\frac{SO(4,m)}{SO(4)\times SO(m)}$	$4m$	$[0,1,\ldots]$	$[0,\ldots]\oplus[0,0,0,1,\ldots]\oplus[2,\ldots]\oplus\underline{[0,2,\ldots]}$
$\frac{G_{2(2)}}{SO(4)}$	8	14	$1\oplus 27\oplus\underline{77}$
$\frac{F_{4(4)}}{Sp(6)\times Sp(2))}$	28	52	$1\oplus 324\oplus\underline{1053}$
$\frac{E_{6(2)}}{SU(6)\times Sp(2)}$	40	78	$1\oplus 650\oplus\underline{2430}$
$\frac{E_{7(-5)}}{SO(12)\times Sp(2)}$	64	133	$1\oplus 1539\oplus\underline{7371}$
$\frac{E_{8(-24)}}{E_7\times Sp(2)}$	112	248	$1\oplus 3875\oplus\underline{27000}$

The check takes just a few seconds. The group theory tables of ref. [279] are also quite helpful.

Notice that the gauging criterion depends essentially only on the isometry group G. Two distinct SUGRAs with the same transitive group of isometries have the same gaugings: e.g. $\mathcal{N} = 12$ SUGRA and $\mathcal{N} = 4$ SUGRA with

$$\mathcal{M} = E_{7(-5)}/[SO(12)\times Sp(2)], \tag{8.169}$$

having the same isometry group, have the same possible gaugings $(\mathcal{G}, l_{mn})$, since in both cases the unwanted representation is the **7371** of $E_{7(-5)}$.

8.7.2 Summary of 3D and some general lessons

Before going to higher space–time dimensions D, let us pause a while to review what we have done in $D = 3$, restate it in a way suited for further generalization, and list the lessons we have learned.

1. The gauge group is embedded in the isometry group of the scalar manifold, $\mathcal{G} \hookrightarrow G \equiv \mathrm{Iso}(\mathcal{M})$. At the Lie algebra level this corresponds to a Lie algebra monomorphism

$$\mathfrak{Lie}(\mathcal{G}) \equiv \mathfrak{L} \to \mathfrak{g} \equiv \mathfrak{iso}(\mathcal{M}), \tag{8.170}$$

which is expressed by a tensor[11]

$$l \in \mathfrak{L} \otimes \mathfrak{g} \subset \mathfrak{g} \otimes \mathfrak{g} \qquad \text{("embedding tensor")}. \qquad (8.171)$$

Gauge invariance requires l to be invariant under the adjoint action of $\mathfrak{L}$.

2. We use the invariant tensor l to make the derivatives in the Lagrangian to be covariant with respect to the gauge group $\mathring{\mathcal{G}}$. The SUSY variation of $\mathcal{L}$ acquires new terms, proportional to the $\mathcal{G}$ field–strengths, from the commutator of the covariantized derivatives. In order to cancel them, in 3D, we must add a Chern–Simons term, whose structure requires the tensor l to be symmetric, namely

$$l \in \odot^2 \mathfrak{g}. \qquad (8.172)$$

3. To cancel the residual terms in the SUSY variation $\delta \mathcal{L}$, we must add to $\mathcal{L}$ the Yukawa terms (8.119),

$$e \left\{ \frac{1}{2} A_1^{AB} \overline{\psi}_\mu^A \gamma^{\mu\nu} \psi_\nu^B + A_{3\,i}^A \overline{\psi}_\mu^A \gamma^\mu \chi^i + \frac{1}{2} A_{3\,ij} \overline{\chi}^i \chi^j \right\}, \qquad (8.173)$$

and a potential V, while we modify the Fermi SUSY variations by some scalar shifts. These corrections are expressed in terms of the Yukawa tensors A_1^{AB}, $A_{2\,i}^A$ and $A_{3\,ij}$ via the universal (algebraic) Ward identities of SUGRA (cf. Chapter 6).

4. The interplay between the isometries and the parallel structures of $\mathcal{M}$ implies that the three Yukawa tensors are linear in the components of the generalized T–tensor

$$\mathcal{T}(\phi) := \theta_\phi(K^m)\, l_{mn}\, \theta_\phi(K^n), \qquad (8.174)$$

where

$$\theta_\phi : \mathfrak{iso}(\mathcal{M}) \longrightarrow \mathrm{End}(T_\phi \mathcal{M}) \oplus T_\phi \mathcal{M}, \qquad (8.175)$$

is the Cartan–Kostant monomorphism[12] (cf. Chapter 7).

5. The above linear map may be inverted; that is, we can write the T–tensor $\mathcal{T}$ as a linear combination of the Yukawa tensors A_1^{AB}, $A_{2\,i}^A$, and $A_{3\,ij}$.

6. Since $\mathfrak{hol}(\mathcal{M}) \simeq \mathfrak{spin}(\mathcal{N}) \oplus \mathfrak{h}'$, and the action of $\mathfrak{spin}(\mathcal{N}) \subset \mathfrak{hol}$ on $T\mathcal{M}$ is through a *spinorial* representation S, one has

$$\mathcal{T} \in \odot^2 \mathfrak{spin}(\mathcal{N}) \oplus \mathfrak{spin}(\mathcal{N})^{\oplus k_1} \oplus \mathbf{1}^{\oplus k_2} \oplus (S \otimes S)^{\oplus k_3} \oplus \text{(spinorial)}, \qquad (8.176)$$

in terms of $Spin(\mathcal{N})$–representations.

---

[11] For simplicity, we assume $\mathcal{G}$ and G to be semi–simple. In this case the adjoint representation coincides with its dual, and we omit the dual mark $\vee$ in our formulae. We also identify $\mathfrak{L}$ and its image in $\mathfrak{g}$. The modifications for the non–semi–simple case are obvious, see ref. [142].

[12] *I.e.* a linear map which is an isomorphism onto its image.

7. On the other hand, by item 5, the tensor $\mathcal{T}$ cannot contain any irreducible $Spin(\mathcal{N})$–representation which is not present in the tensors A_1^{AB}, $A_{2\,i}^A$ and $A_{3\,ij}$, namely

$$A_1^{AB} \longrightarrow V \odot V, \qquad (8.177)$$

$$A_{2\,i}^A \longrightarrow V \otimes S, \qquad (8.178)$$

$$A_{3\,ij} \longrightarrow S \odot S, \qquad (8.179)$$

where V is the vector representation $\mathcal{N}$. Since $\odot^2 S$ is the direct sum of totally antisymmetric representations $\wedge^r V$, we learn that $\mathcal{T}$ should satisfy the constraint

$$\mathcal{T}\big|_{\odot^2 \mathrm{spin}(\mathcal{N})} \in \odot^2 V \oplus \wedge^4 V. \qquad (8.180)$$

Of course, this is the same constraint on the T–tensor we obtained by direct computation, Eq. (8.123). Here we see that we could have deduced it from symmetry considerations alone.

8. If the isometry group G is transitive (and $\mathcal{M}$ is irreducible), the Cartan–Kostant *isomorphism*, together with the tt^*–structure of the Aut_R curvatures, allows us to identify the $\mathrm{Adj}^{\odot^2} G$–orbits of the "embedding tensor" l with the generalized T–tensor

$$\mathcal{T}(gH) = l(g) \equiv \mathrm{Adj}^{\odot^2}(g^{-1})(l), \qquad (8.181)$$

and the constraint (8.180) becomes

$$l(g) \in \oplus_k R_k \subset \odot^2 \mathfrak{g} \qquad \text{as } G\text{–representations!} \qquad (8.182)$$

where the irreducible representations R_k may contain only the following representations of the subgroup $Spin(\mathcal{N}) \subset G$:

$$\mathbf{1}, \ (\odot^2 V)_{\text{traceless}}, \ \wedge^2 V, \ \wedge^4 V,$$

as well as the spinorial irrepr.contained in $\qquad (8.183)$

$$(V \otimes S) \cap \odot^2 \mathfrak{g} \ \text{ and } \ S \cap \odot^2 \mathfrak{g}.$$

Again this statement is equivalent to the one given before.

In the next section we rephrase these items into a General Lesson valid for all space–time dimensions D.

8.8 *Gauged* supergravity in $D \geq 4$

Our last task is to extend the gaugings to different space–time dimensions D. Our analysis below holds for any D, although, in the presence of higher form–fields

$A_{\mu_1\mu_2\cdots\mu_k}$ ($k \geq 2$), there are some additional subtleties which are outside the scope of this book, see [111, 112, 123]. For $D \leq 5$ such fields may be dualized away, and we may formulate any SUGRA theory with a bosonic sector consisting just of scalars and vectors besides the graviton.

8.8.1 Review of the ungauged theory

There are a few physical differences between $D \geq 4$ and $D = 3$. First of all, in 3D (in the dWHS dual version) the vector fields do not propagate physical states; they are "auxiliary" fields, and we may add to the ungauged Lagrangian $\mathcal{L}_0$ as many of them as we wish or need. On the contrary, in $D \geq 4$, A_μ propagates physical states, which should belong to SUSY representations, so their *number* and *quantum numbers* are uniquely determined by the SUSY content of the ungauged theory. More concretely, on the target space $\mathcal{M}$ we have a flat vector bundle $\mathcal{V} \to \mathcal{M}$, such that the field–strength 2–forms F, and their $(D-2)$–form duals $G = *\frac{\partial \mathcal{L}_0}{\partial F}$, satisfy[13]

$$F \in \wedge^2 T^*\Sigma \otimes \Phi^*\mathcal{V}, \qquad G \in \wedge^{(D-2)} T^*\Sigma \otimes \Phi^*\mathcal{V}^\vee. \tag{8.184}$$

The situation in 4D is slightly different, since both F and G are 2–forms. As explained in Chapter 1, in this case, we should combine the Fs and the Gs into a double–size vector $\mathcal{F}$ with

$$\mathcal{F}_+ \in \wedge^2_+ T^*_\mathbb{C}\Sigma \otimes \Phi^*\mathcal{V}, \qquad \mathcal{F}_- \in \wedge^2_- T^*_\mathbb{C}\Sigma \otimes \Phi^*\overline{\mathcal{V}}. \tag{8.185}$$

On $\mathcal{M}$ we have two other natural vector bundles: (i) $\Psi \to \mathcal{M}$ is the vector bundle corresponding to the gravitino fields, ψ_μ^A; it has structure group Aut$_R$; (ii) $\Upsilon \to \mathcal{M}$ is the bundle associated with the spin–1/2 fields.

Υ decomposes into the Whitney sum of fiber bundles of the form $S_R \otimes U_\rho$, where S_R is the Aut$_R$–bundle associated with a suitable representation R, and U_ρ is the H'–bundle[14] associated with some representation ρ. In 4D, the relations between these fermionic bundles and $T\mathcal{M}$ is given in Table 2.6, where the representation pairs (R, ρ) are also listed. The subgroup $G \subset \text{Iso}(\mathcal{M})$, which is actually a symmetry of the ungauged theory, acts on the bundles $\mathcal{V}$, Ψ and Υ to guarantee invariance of the vector and fermion couplings (which have a non–trivial dependence on the scalars ϕ). In, say, 4D the action of G on $\mathcal{V}$ gives a group morphism

$$\mu^\sharp \colon G \to Sp(2n, \mathbb{R}), \tag{8.186}$$

---

[13] Recall that when we write (by abuse of notation) $F \in E$, with E a vector bundle, we really mean $F \in C^\infty(E)$, that is F is a smooth section of E.
[14] Recall that the Lie group H' is defined by the condition $\mathfrak{hol}(\mathcal{M}) \simeq \mathfrak{aut}_R \oplus \mathfrak{h}'$, where $\mathfrak{h}'$ is the Lie algebra of H'.

induced by the "suscetibility map"

$$\mu: \mathcal{M} \to Sp(2n, \mathbb{R}). \tag{8.187}$$

More generally, G acts on the field–strengths according to a linear representation ρ. On the other hand, the actions on Ψ and Υ are obtained by suitable projections of the Cartan–Kostant map θ, that is, by the generalized momentum map. In this context, our problem is to determine the possible gaugings consistent with $\mathcal{N}$-extended local SUSY in D dimensions.

8.8.2 General principles

What we have learned about supersymmetry and, in particular, the above discussion of 3D gaugings, suggests the following general statement:

General Lesson 8.12 *Let $\mathcal{V}$, Ψ and Υ be the target space bundles associated, respectively, with the vector, spin–3/2 and spin–1/2 fields. Let $\mathcal{G} \subset \mathrm{Iso}(\mathcal{M})$ be the symmetry of $\mathcal{L}_0$ to be gauged. Let*

$$l: \mathcal{V} \to \mathfrak{iso}(\mathcal{M}), \tag{8.188}$$

be a linear map whose image is a Lie subalgebra $\mathfrak{L} \subset \mathfrak{iso}(\mathcal{M})$ isomorphic to $\mathfrak{Lie}(\mathcal{G})$. Then the SUSY completion of the minimal gauge coupling[15]

$$D_\mu \Phi^a \to D_\mu \Phi^a - A_\mu^M l_{Mm} \pounds_{K^m} \Phi^a, \tag{8.189}$$

is completely encoded in the three Yukawa tensors

$$A_1 \in \widehat{\odot}^2 \Psi^\vee, \quad A_2 \in \Psi^\vee \otimes \Upsilon^\vee, \quad A_3 \in \widehat{\odot}^2 \Upsilon^\vee, \tag{8.190}$$

where the symbol $\widehat{\odot}$ stands for the non–symmetrized, symmetrized, or anti–symmetrized tensor product depending on the specific symmetry and reality properties of the spinors and their scalar bilinears in the given space–time dimension D (see Section 2.1.1 for the list). The three Yukawa tensors (8.190) can be linearly combined into a single tensor $\mathcal{T}(\phi)$

$$\mathcal{T}(\phi) := \rho(\widetilde{\mu}(\phi))^{-1} l \cdot \theta_\phi \tag{8.191}$$

where $\cdot$ stands for the Cartan–Killing inner product in $\mathfrak{g}$,

$$\theta_\phi \in (\mathrm{End}(T\mathcal{M}) \oplus T\mathcal{M})_\phi^\vee \otimes \mathfrak{g} \tag{8.192}$$

---

[15] Φ^a is a shorthand for all the fields in the theory *but* the vectors A_μ^M.

is the Cartan–Kostant morphism, and $\widetilde{\mu}$ is any lift of μ, making the following diagram commutative:

$$\begin{array}{ccc} \mathcal{M} & \xrightarrow{\widetilde{\mu}} & \begin{cases} Sp(2n, \mathbb{R}) & D = 4 \\ GL(n, \mathbb{R}) & D \geq 5 \end{cases} \\ \text{Id} \downarrow & & \downarrow \\ \mathcal{M} & \xrightarrow{\mu} & \begin{cases} Sp(2n, \mathbb{R})/U(n) & D = 4 \\ GL(n, \mathbb{R})/O(n) & D \geq 5 \end{cases} \end{array} \qquad (8.193)$$

with μ the "susceptibility" map defined in Section 1.4.2.[16]

A gauging l is allowed if and only if the decompostion of $\mathcal{T}(\phi)$ into irreducible representations of $\mathrm{Aut}_R \times H'$ contains only the representations associated with the three vector bundles in Eq. (8.190).

The $D = 4$ case is special, since the bundle $\mathcal{V}$ contains both the "electric" field strengths $F = dA$ (corresponding, say, to the upper block of the $2n$–vector $\mathcal{F}$) as well as their "magnetic" duals, $*\partial \mathcal{L}/\partial F$. In a local Lagrangian, A_μ^M and its magnetic counterpart $A_{\mu\,M}$ cannot be both present; hence, in a meaningful theory, only the "electric" vectors may be used to gauge the global symmetries. Of course, what we mean by "electric" depends on the duality frame we use. The invariant physical requirement is that the vectors entering into the Lagrangian are *mutually local fields*. In view of the Dirac quantization rule, this is equivalent to requiring the gauging l to satisfy [122]

$$l_{Mm}\, l_{Nn}\, \Omega^{MN} = 0, \qquad (8.194)$$

where Ω^{MN} is the $Sp(2n, \mathbb{R})$ symplectic matrix. This physical condition is *automatically satisfied for a gauging l whose T–tensor $\mathcal{T}(\phi)$ has the correct decomposition into* $\mathrm{Aut}_R \times H'$ *representations*.

If $\mathrm{Iso}(\mathcal{M}) = G$ is transitive,

$$\mathcal{T}_M{}^\alpha\, t_\alpha = \rho(g^{-1})_M{}^N\, l_{Nm}\, (g\, t^m\, g^{-1}) \qquad (8.195)$$

(see, e.g. [122, 121]), and the gauging constraints may be lifted to a problem about the representations of G as we did in Section 8.7.1 for the 3D case.

Let us argue that the above General Lesson is correct. Going through the same logical steps as in 3D, summarized in Section 8.7.2, we easily convince ourselves

---

[16] There we defined μ only for $D = 4$. Let us fill the gap. In $D \geq 5$ the vector kinetic terms have the form $f_{MN}(\phi)\, F_{\mu\nu}^M\, F^{N\,\mu\nu}$ for some real, symmetric, positive–definite matrix f_{MN}. The space of all the symmetric positive–definite metrics is identified with the coset $GL(n, \mathbb{R})/O(n)$ by the map $\mathcal{E} \mapsto \mathcal{E}^t \mathcal{E} \equiv f$. This identification defines the map $\mu \colon \phi \mapsto \mathcal{E}(\phi)$. The group $GL(n, \mathbb{R})$ acts on the vectors/scalars as $F \mapsto g F$, and $\mathcal{E} \mapsto \mathcal{E} g^{-1}$.

that the above conditions on the "embedding tensor" l_{Mm} are at least necessary in order to have local SUSY. On the other hand, we may think of relating a SUGRA model in D dimensions to one in 3D by "dimensional reduction." However, after the gauging (or the addition of a superpotential) the scalar potential (and hence the effective cosmological constant Λ) is no longer zero. Then, in general, we have no solution of the Einstein equations with a manifold of the form $T^{D-3} \times \Sigma_3$, and the dimensional reduction is not straightforward. However, $\Lambda = O(g^2)$, so the dimensional reduction argument still works at order $O(g)$ and we get the correct *linear* constraints on l_{Mm}. Moreover, the condition we get from SUSY at level $O(g^2)$ should be equivalent, on general grounds, to the Ward identity of Chapter 6, which has a universal form in terms of the tensors A_1^{AB} and A_{2i}^A, which, again, depends only on their representation content. Hence one does not expect any new independent constraint at the $O(g^2)$ level. In our view, the most convincing evidence for General Lesson 8.12 is that it is *very* geometric.

We present below a few checks of the above results. In 4D, for $\mathcal{N} \geq 3$ $\mathcal{M}$ is a symmetric space, and hence we need only to verify that the various representations match with the ones predicted. For $\mathcal{N} = 1, 2$ the situation is more "geometric" and we shall discuss the relevant topics in Part III. (See ref. [14] for a nice treatment of the general gauged $\mathcal{N} = 2$ supergravity in a language similar to the one used here.)

8.8.3 4D Gaugings and Peccei–Quinn symmetries

In 4D, the ungauged Lagrangian $\mathcal{L}_0$ is not invariant under the full U–duality group G which (for $\mathcal{N} \geq 3$) is identified with the isometry group $\mathrm{Iso}(\mathcal{M})$. Only the equations of motion are invariant. Gauging a continuous symmetry $\mathcal{G}$ of the ungauged theory which is not a symmetry of the ungauged Lagrangian is rather tricky. But it can be done, under certain circumstances, in a supersymmetric fashion [121].

We have already noted that the vectors A_μ^M gauging $\mathcal{G}$ should be *mutually local*. Without loss of generality,[17] we may assume that the first n–components of the field–strength $2n$–vector $\mathcal{F}_{\mu\nu}$ correspond to the curvatures of the mutually local connection fields A_μ^M. Then an element of $Sp(2n, \mathbb{R})$ acts as

$$\begin{pmatrix} F \\ G \end{pmatrix} \to \begin{pmatrix} A & C \\ B & D \end{pmatrix} \begin{pmatrix} F \\ G \end{pmatrix} = \begin{pmatrix} AF + CG \\ DG + BF \end{pmatrix}, \tag{8.196}$$

---

[17] In refs. [121, 122] it is discussed how different choices lead to *inequivalent* gauged theories. However, in our "abstract" geometric formulation, a different choice for the embedding of the curvatures of the local fields in the $2n$–vector $\mathcal{F}$ can be compensated by a corresponding modification of the "susceptibility" map μ, which need not be the canonical embedding $G/H \to Sp(2n, \mathbb{R})/U(n)$. However the map μ is still a totally geodesic embedding: by arguments presented in Chapter 5, μ encodes precisely the constant matrix $\mathsf{E} \in Sp(2n, \mathbb{R})$ of refs. [121, 122]. The inequivalent choices of E are labeled by the double coset $G \backslash Sp(2n, \mathbb{R})/GL(n, \mathbb{R})$.

and preserves mutual locality iff $C = 0$. Then the symplectic condition gives

$$D = (A^t)^{-1}, \qquad B = (A^t)^{-1} S \quad \text{with } S \text{ symmetric}. \tag{8.197}$$

Let $\mathcal{P}(2n) \subset Sp(2n, \mathbb{R})$ be the "parabolic" subgroup of matrices of the form

$$\begin{pmatrix} A & 0 \\ (A^t)^{-1} S & (A^t)^{-1} \end{pmatrix}. \tag{8.198}$$

From Lemma 5.21 we know that all coset elements $\xi \in Sp(2n, \mathbb{R})/U(n)$ have a representative in $\mathcal{P}(2n)$; in fact, we may even choose A to be lower–triangular with positive diagonal entries. Thus

$$\mathcal{P}(2n)/O(n) \simeq Sp(2n, \mathbb{R})/U(n). \tag{8.199}$$

Composing μ with isomorphism (8.199) yields a *reduced* susceptibility map

$$\mu^\flat \colon \mathcal{M} \to \mathcal{P}(2n)/O(n). \tag{8.200}$$

Now we may restate our General Lesson 8.12 with $\mathcal{G} \subseteq \mathcal{P}(2n)$ and μ replaced by $\mu^\flat$, and forget the locality condition which is now *automatic*.

Using the results of Chapter 1, we see that the ungauged Lagrangian changes under the transformation (8.198) by

$$\delta \mathcal{L}_0 = \frac{1}{4} S_{MN} F^M_{\mu\nu} \widetilde{F}^{N\,\mu\nu}, \tag{8.201}$$

which is a total derivative (and hence does not affect the equations of motion). A symmetry under which the Lagrangian transforms as in Eq. (8.201) is called *a Peccei–Quinn symmetry*. Gauging a generic subgroup $\mathcal{G} \subseteq \mathcal{P}(2n)$, we may end up gauging some Peccei–Quinn symmetry.[18] Now the group parameter, $S_{MN}(x)$, is not a constant, so (8.201) is not a total derivative any longer. Then we must cancel this variation by adding a Chern–Simons–like term (much as we did in 3D)

$$\mathcal{L}_{CS} \propto \varepsilon^{\mu\nu\rho\sigma}\, l_{M\,NP}\, A^M_\mu A^N_\nu \left(\partial_\rho A^P_\sigma - \frac{1}{2} f_{QR}{}^P A^Q_\rho A^R_\sigma \right), \tag{8.202}$$

where $l_{M\,NP}$ is the projection of l on[19]

$$\mathfrak{r}(\mathfrak{Lie}(\mathcal{G})) \otimes \mathcal{V}^\vee \subset \mathfrak{Lie}(\mathcal{G}) \otimes \mathcal{V}^\vee. \tag{8.203}$$

The resulting Lagrangian is supersymmetric. See ref. [121] for further details.

---

[18] By Cartan's criterion, the *gauged* Peccei–Quinn symmetries correspond to $\mathfrak{r}(\mathfrak{Lie}(\mathcal{G}))$, the *radical* of the Lie algebra of $\mathcal{G}$ (cf. ref. [61] Section I.5). Hence they cannot be present in *semi–simple* gaugings.

[19] See previous footnote for the definition of $\mathfrak{r}(\cdot)$.

8.9 An example: $\mathcal{N} = 3$ supergravity in 4D

To illustrate the above results, we discuss in some detail a simple example, namely gauged $\mathcal{N} = 3$ supergravity in 4D which is paradigmatic, as the title and abstract of ref. [79] imply.

As we saw in Chapter 4, the scalar manifold is

$$SU(3,k)/(SU(3) \times SU(k) \times U(1)), \tag{8.204}$$

where k is the number of matter gauge multiplets coupled to SUGRA. Again, we write $g \in SU(3,k)$ for a coset representative, with the global group $G \equiv SU(3,k)$ acting on the left, and the local one, $H = SU(3) \times SU(k) \times U(1)$, acting on the right. To be explicit, let

$$J = \begin{pmatrix} \mathbf{1}_{3\times 3} & 0 \\ 0 & -\mathbf{1}_{k\times k} \end{pmatrix}. \tag{8.205}$$

The elements $g \in SU(3,k)$ are identified with the $(3+k) \times (3+k)$ unimodular complex matrices such that

$$g^\dagger J g = J. \tag{8.206}$$

The Lagrangian itself is invariant only under the subgroup $SO(3,k) \subset G$, which does not mix "electric" and "magnetic" fields.

We have $(3+k)$ vector fields, so we can gauge at most a subgroup of G of dimension $(3+k)$. Following ref. [81], we shall limit ourselves to semi–simple subgroups $K \subset SO(3,k)$ (so no Peccei–Quinn gauging). The subgroup K to be gauged should fulfill the following requirements. (i) the fundamental representation D of $SU(3,k)$ must split as

$$D \xrightarrow[K]{} \text{adj} \oplus \text{adj}, \tag{8.207}$$

one summand for the "electric" fields and one for the "magnetic" ones. (ii) K must preserve the metric J, which therefore can be identified with the Cartan–Killing metric of K. Then the structure constants $f_{ab}{}^c$ of K become fully antisymmetric by lowering the upper index with the metric J. The Lie algebra of K, $\mathfrak{K}$, must be a real subalgebra of $\mathfrak{so}(3,k)$ of dimension $(3+k)$ and with a bilinear invariant of signature $(3,k)$. Thus we have at most three non–compact generators, and the only possibilities are [81]

$$K = SU(3) \times K_k, \tag{8.208}$$

$$K = SO(3,1) \times K_{k-3}, \tag{8.209}$$

where K_n stands for any compact Lie group of dimension n (recall that the adjoint representation embeds any compact Lie algebra $\mathfrak{L}$ of dimension n into the Lie algebra $\mathfrak{so}(n)$). In both cases, using our Eq. (8.195), the $\mathcal{T}$ tensor is easily computed. It corresponds to the "boosted structure constants" of the gauged subgroup K [79, 81],

$$\mathcal{T}^L{}_{MN}(g) = (g^{-1})^L{}_R f^R{}_{PQ} (g)^P{}_M (g)^Q{}_N. \tag{8.210}$$

Our General Lesson 8.12 predicts that the fermionic shifts (hence the Yukawa couplings) are linear in this $\mathcal{T}$ tensor. Explicitly, one finds [79, 81]

$$\delta\psi_\mu^A = \cdots - \frac{i}{8} g \big(\mathcal{T}^A_{PQ} \epsilon^{BPQ} + (A \leftrightarrow B)\big) \gamma_\mu \epsilon_B \quad \text{(gravitino)}, \tag{8.211}$$

$$\delta\chi = \cdots - \frac{1}{4} g \, \mathcal{T}^B_{BA} \, \epsilon^A \quad \text{(dilatino)}, \tag{8.212}$$

$$\delta\lambda^i = \cdots - \frac{1}{2} g \, \mathcal{T}^i_{BC} \, \epsilon^{ABC} \epsilon_A \quad \text{(gluino singlets)}; \tag{8.213}$$

$$\delta\lambda_{Ai} = \cdots - \left(\mathcal{T}^B{}_{iA} - \frac{1}{2}\mathcal{T}^B{}_{iB}\delta^B_A\right)\epsilon_B \quad \text{(gluino triplets)}. \tag{8.214}$$

in full agreement with the General Lesson. One checks [79, 81] that these fermionic shifts do satisfy the algebraic conditions following from the universal Ward identity relating the scalar potential to the fermionic shifts.

The important lesson is that the fermionic shifts and Yukawa tensors are precisely the projections of $\mathcal{T}$ on the representations of the $\text{Aut}_R \times H'$ local symmetry appropriated for the given Fermi bilinear. The numerical coefficients of these projection are universal, as we saw in 3D.

8.10 Gauging maximal supergravity in D dimensions

We apply the previous General Lesson to the case of *maximal* supergravity in D space–time dimensions, that is, SUGRA with $N_S = 32$ supercharges. All SUGRAs with $N_S \geq 18$ are truncations of these maximal theories. The scalar manifold is symmetric; the cosets G/H corresponding to maximal SUGRA in $D \geq 3$ are listed in Table 4.2; in Table 8.4 the reader finds the decompostion of the G–representation $\mathcal{V} \otimes \text{Adj}_G \ni \mathcal{T}$ into irreducible representations (following [121, 122]). The bottom line is the 3D case that we studied in detail. The table is easily checked using the LiE package or the Tables [279].

Table 8.4 *Decompostion of $\mathcal{V} \otimes \mathrm{Adj}_G$ into irreducible G–representations.*

D	G	H	Decomposition of the G–module $\mathcal{V} \otimes \mathrm{Adj}_G{}^a$
7	$SL(5)$	$Sp(4)$	$\mathbf{10} \otimes \mathbf{24} = \mathbf{10} \oplus \boxed{\mathbf{15}} \oplus \mathbf{40} \oplus \mathbf{175}$
6	$SO(5,5)$	$Sp(4) \times Sp(4)$	$\mathbf{16} \otimes \mathbf{45} = \mathbf{16} \oplus \boxed{\mathbf{144}} \oplus \mathbf{560}$
5	$E_{6(6)}$	$Sp(8)$	$\mathbf{27} \otimes \mathbf{78} = \mathbf{27} \oplus \boxed{\mathbf{351}} \oplus \mathbf{1728}$
4	$E_{7(7)}$	$SU(8)$	$\mathbf{56} \otimes \mathbf{133} = \mathbf{56} \oplus \boxed{\mathbf{912}} \oplus \mathbf{6480}$
3	$E_{8(8)}$	$SO(16)$	$\mathbf{248} \odot \mathbf{248} = \boxed{\mathbf{1}} \oplus \boxed{\mathbf{3875}} \oplus \mathbf{27000}$

[a] The (extended) $\mathcal{T}$–tensor belongs to $\mathcal{V} \otimes \mathrm{Adj}_G$. SUSY requires that only its components in the *boxed* irreducible representations do not vanish.

The basic criterion for a SUSY–preserving gauging, l_{Mm}, is that the associated $\mathcal{T}$–tensor,

$$\mathcal{T}(g) = \varrho_{\mathcal{V}\otimes\mathrm{Adj}}(g)\, l, \tag{8.215}$$

has non–vanishing components *only* in those irreducible G–representations $\subset \mathcal{V} \otimes \mathrm{Adj}_G$ which, when decomposed into representations of the holonomy subgroup H, contain only the H–irrepresentations appearing in the Yukawa tensors $A_1, A_2,$ and A_3. The H–representations of the Yukawa couplings are listed in Table 8.5, (again following [121, 122]). Note that in 7D, the mass–matrices A_1 and A_3 should be *anti*symmetric in the fermionic indices as the scalar bilinear in the 7D pseudoMajorana spinors.

8.10.1 Example: D = 4

Consider, for instance, $\mathcal{N} = 8$ supergravity in four dimensions. One has $G = E_{7(7)}$ and $H = SU(8)$. The three $E_{7(7)}$ representations appearing in $\mathcal{V} \otimes \mathrm{Adj}$ decompose under $SU(8)$ as follows:

$$\mathbf{56} \xrightarrow{SU(8)} \mathbf{28} \oplus \overline{\mathbf{28}}, \tag{8.216}$$

$$\mathbf{912} \xrightarrow{SU(8)} \mathbf{36} \oplus \overline{\mathbf{36}} \oplus \mathbf{420} \oplus \overline{\mathbf{420}}, \tag{8.217}$$

$$\mathbf{6480} \xrightarrow{SU(8)} \mathbf{28} \oplus \mathbf{420} \oplus \mathbf{1280} \oplus \mathbf{1512} \oplus \text{(conjugate reprs.)} \tag{8.218}$$

Comparing with Table 8.5, we see that the representation **6480** cannot be present. Also the **28** cannot be present; the reason is that $A_{2A}{}^{BCD}$ satisfies the less obvious identity $A_{2A}{}^{ACD} = 0$, and hence it is a section of a *proper* subbundle of $\Psi^\vee \otimes \Upsilon^\vee$,

Table 8.5 *Possible Yukawa couplings in various dimensions for maximal* SUGRA *and the corresponding* Aut_R–*representations.*

D	$H \equiv \text{Aut}_R$	$\widehat{\odot}^2 \Psi^\vee$	$\Psi^\vee \otimes \Upsilon^\vee$	$\widehat{\odot}^2 \Upsilon^\vee$
7	$Sp(4)$	$\mathbf{1} \oplus \mathbf{5}$	$\mathbf{5} \oplus \mathbf{10} \oplus \mathbf{15} \oplus \mathbf{35}$	$\mathbf{1} \oplus \mathbf{5} \oplus \mathbf{14} \oplus$ $\oplus \mathbf{30} \oplus \mathbf{35} \oplus \mathbf{35'}$
6	$Sp(4) \times Sp(4)$	$(\mathbf{4},\mathbf{4})$	$(\mathbf{4},\mathbf{4}) \oplus (\mathbf{4},\mathbf{4}) \oplus$ $\oplus (\mathbf{4},\mathbf{16}) \oplus (\mathbf{16},\mathbf{4})$	$(\mathbf{4},\mathbf{4}) \oplus (\mathbf{4},\mathbf{16}) \oplus$ $\oplus (\mathbf{16},\mathbf{4}) \oplus (\mathbf{16},\mathbf{16})$
5	$Sp(8)$	$\mathbf{36}$	$\mathbf{27} \oplus \mathbf{42} \oplus \mathbf{315}$	$\mathbf{1} \oplus \mathbf{27} \oplus \mathbf{36} \oplus \mathbf{308} \oplus$ $\oplus \mathbf{315} \oplus \mathbf{792} \oplus \mathbf{825}$
4	$SU(8)$	$\mathbf{36} \oplus \overline{\mathbf{36}}$	$\mathbf{28} \oplus \overline{\mathbf{28}} \oplus$ $\oplus \mathbf{420} \oplus \overline{\mathbf{420}}$	$\mathbf{420} \oplus \overline{\mathbf{420}} \oplus$ $\oplus \mathbf{1176} \oplus \overline{\mathbf{1176}}$
3	$SO(16)$	$\mathbf{1} \oplus \mathbf{135}$	$\mathbf{128} \oplus \overline{\mathbf{1920}}$	$\mathbf{1} \oplus \mathbf{1820} \oplus \overline{\mathbf{6435}}$

corresponding to the representation $\mathbf{420} \oplus \overline{\mathbf{420}}$ of $SU(8)$. Indeed, from our General Lesson, valid in any dimension, we know there exists an equation of the form[20]

$$c_1 A_{2A}{}^{BCD} + c_2 A_1^{B[C} \delta_A^{D]} = \mathcal{T}^{BCD}{}_A \equiv (g^{-1})^{BC}{}_{EF} l^{EF}{}_m (\boldsymbol{\mu}^m)^B{}_A, \qquad (8.219)$$

where the 8×8 matrix $(\boldsymbol{\mu}^\bullet)^B{}_A$ is the component of the covariant momentum map in the Lie algebra $\mathfrak{su}(8)$, which is traceless by definition. Taking the trace of both sides of Eq. (8.219), we get $A_{2A}{}^{ACD} = 0$, i.e., $A_2 \in \mathbf{420} \oplus \overline{\mathbf{420}}$.

Since the T–tensor is an $E_{7(7)}$–covariant object, it is enough to impose the representation constraint at one point of the G orbit, say at the origin. Then we get a condition in terms of the embedding tensor l only,

$$\mathbb{P}_{\mathbf{912}}\, l = l, \qquad (8.220)$$

where $\mathbb{P}_{\mathbf{912}}$ is the projector on the $\mathbf{912}$ representation of $E_{7(7)}$. The interested reader may find in ref. [121] the explicit form of this projector, as well as the projector for the other relevant representations for maximal SUGRA in 4D and 5D.

At this point, the complete classification of all possible SUSY–preserving gaugings of $\mathcal{N} = 8$ SUGRA is reduced to a problem in group theory, albeit *not* a trivial one. The full list of gaugings with gauge group $\mathcal{G} \subset SL(8, \mathbb{R})$ (no Peccei–Quinn gauge symmetries or other fancy mechanisms) can be found in refs. [97, 193, 194, 195, 197]. There exist other, trickier gaugings; see refs. [16, 17, 121, 122, 198] for examples.

---

[20] c_1 and c_2 are numerical constants whose precise value is immaterial for the argument.

Assuming $\mathcal{G} \subset SL(8, \mathbb{R})$, one gets all gauge groups of the form

$$SO(8-p, p), \qquad p = 0, 1, \ldots 4, \tag{8.221}$$

as well as some non–semi–simple groups – called $CO(p, q, 8-p-q)$ – whose Lie algebra is the semi–direct sum of $\mathfrak{so}(p, q)$ with a solvable Lie algebra. $CO(p, q, 8-p-q)$ is obtained from the group $SO(8-q, q)$ by a suitable contraction.

For the corresponding analysis of gauged maximal SUGRA in $D > 4$ dimensions, see, e.g., refs. [15, 180, 256].

Part III

Special geometries

9
Kähler and Hodge manifolds

The results of Part II allow us to write the Lagrangian of any SUSY/SUGRA theory for all Ds and $\mathcal{N}$s. However, the geometry was totally explicit only for *symmetric* scalar manifolds (automatically true for $N_S \geq 9$). To be more explicit for $N_S \leq 8$, we should describe the geometry of the corresponding manifolds. This is the main goal of Part III. In the present chapter we give a sketchy description of the Riemannian manifolds with $\mathrm{Hol}(g) \subseteq U(n)$.

9.1 Complex manifolds

9.1.1 Almost complex and almost Hermitian manifolds

Definition 9.1 By an *almost complex n–fold* we mean a smooth $2n$–fold M equipped with a $GL(n, \mathbb{C})$–structure. By Proposition A.2 this is equivalent to giving the pair (M, I), where I is a (smooth) section of $\mathrm{End}(TM) \equiv TM \otimes T^*M$ with $I^2 = -1$. I is called the *almost complex structure*.

According to Section 2.1.1, the complexified tangent space decomposes into *type components* $\mathbb{C} \otimes TM \simeq T_+ \oplus T_-$ with $I|_{T_\pm} = \pm i$ and $T_- = (T_+)^*$. A section of T_+ (resp. T_-) is said to be *a vector field of type* $(1, 0)$ (resp. $(0, 1)$).

Definition 9.2 By an *almost Hermitian n–fold* we mean a smooth $2n$–fold M equipped with a $U(n)$–structure.

Since $U(n) = GL(n, \mathbb{C}) \cap SO(2n)$, an almost Hermitian manifold is both almost complex and Riemannian. Let $g \in \odot^2 T^*M$ be its Riemannian metric and I its almost complex structure. The $U(n)$–structure implies the consistency condition for the pair (g, I),

$$g(Iv, Iw) = g(v, w) \quad \forall v, w \in TM, \tag{9.1}$$

i.e., $g_{kh} I^k_i I^h_j = g_{ij}$. Then g induces a Hermitian metric on T_+ such that

$$g(v, w) = 2 \operatorname{Re} g(v^*_+, w_+), \qquad (9.2)$$

where v_+ denotes the $(1, 0)$ component of the real vector v.

From Eq. (9.1) we get $g(Iv, w) = -g(Iv, I^2 w) = -g(v, Iw) = -g(Iw, v)$. Then $\omega(v, w) = g(Iv, w)$ is a 2–form (a section of $\wedge^2 T^*M$). One has

$$\omega(v, w) = 2 \operatorname{Im} g(v^*_+, w_+) \qquad (9.3)$$

and g and ω are respectively (twice) the real and imaginary parts of the Hermitian form $g(\bar{\cdot}, \cdot)$ on T_+.

Definition 9.3 *The 2–form of the $U(n)$–structure is $\omega = I^k_i g_{kj} dx^i \wedge dx^j$. Note that ω is real.*

9.1.2 Integrable complex structures. Nijenhuis tensor

An almost complex structure I is said to be a *complex* structure if the corresponding $GL(n, \mathbb{C})$–structure is *integrable*; that is, if in a neighborhood of each point of M there exist n smooth functions f_a such that the differentials df_a are linearly independent eigenvectors of I^t with eigenvalue i, namely we have $I^i_j \partial_i f_a = i \partial_j f_a$ and $df_1 \wedge \cdots \wedge df_n \neq 0$.

Definition 9.4 *A complex manifold M is a smooth manifold $M_\mathbb{R}$ together with an integrable complex structure. An almost Hermitian manifold whose complex structure is integrable is called Hermitian.*

This definition agrees with the usual one in terms of an equivalence class of atlases of holomorphic coordinates. In fact the equation $I^i_j \partial_i f_a = i \partial_j f_a$ is just the Cauchy–Riemann condition, and I integrable $\Leftrightarrow$ *there are enough local holomorphic functions to make a complex (local) coordinate system*.

Not all almost complex structures are integrable. There are obstructions to integrability. One such obstruction is obvious: in general, if a G–structure is integrable then its intrinsic torsion vanishes (i.e., it admits a torsionless connection). Thus the intrinsic torsion of the $GL(n, \mathbb{C})$–structure is an obstruction to the integrability of I. Luckily, there is no other:

Proposition 9.5 *A $GL(n, \mathbb{C})$–structure is integrable iff it is torsionless.*

Proof Let θ^i be the $GL(n, \mathbb{C})$ frame and $\omega^i{}_j$ a $\mathfrak{gl}(n, \mathbb{C})$–connection. The condition of being torsionless reads

$$T^i \equiv d\theta^i + \omega^i{}_j \wedge \theta^j = 0. \qquad (9.4)$$

9.1 Complex manifolds

A connection $\omega^i{}_j$ satisfying this equation exists iff $d\theta^i$ belongs to the differential ideal generated by the θ^j. Comparing with the Frobenius theorem in terms of differential forms [67], we see that the $GL(n, \mathbb{C})$-structure has zero intrinsic torsion iff the distribution $T_+ \subset \mathbb{C} \otimes TM$ is integrable, i.e., iff there exist (local) complex coordinates. $\square$

By the Frobenius theorem, T_+ is integrable iff it is *involutive*. We write the Cauchy–Riemann condition in the form $P_j f_a = 0$ where $P_j = I^k_j \partial_k - i\partial_j$. The vector fields P_j span T_+ which is involutive precisely iff $[P_i, P_j]$ is a linear combination of the P_j (i.e., the Cauchy–Riemann equations are integrable). One has

$$[P_i, P_j] = [(I^k_i \partial_k I^l_j - I^k_j \partial_k I^l_i) - (I^k_i \partial_j I^l_k - I^k_j \partial_i I^l_k)]\partial_l \quad (9.5)$$
$$+ (\partial_i I^l_j - \partial_j I^l_i) P_l.$$

The tensor

$$N^l{}_{ij} = (I^k_i \partial_k I^l_j - I^k_j \partial_k I^l_i) - (I^k_i \partial_j I^l_k - I^k_j \partial_i I^l_k) \quad (9.6)$$

is the obstruction to the integrability of the almost complex structure. It is called the *Nijenhuis tensor*.

Corollary 9.6 *The almost complex structure I is integrable if and only if its Nijenhuis tensor vanishes.*

$N^l{}_{ij}$ is a *bona fide* tensor in the Riemannian sense. In fact, assume that the almost complex manifold M is equipped with a Riemannian metric and let ∇ be the corresponding Levi–Civita connection. One has

$$N^l{}_{ij} = I^k_i(\nabla_k I^l_j - \nabla_j I^l_k) - I^k_j(\nabla_k I^l_i - \nabla_i I^l_k), \quad (9.7)$$

indeed, given the symmetry of the Christoffel symbols and $I^k_i I^m_k = -\delta^m_i$,

$$\text{RHS}(9.7) - \text{R. H. S.}(9.6) = (I^k_i \Gamma^l_{km} I^m_j - I^k_i \Gamma^l_{jm} I^m_k) - (I^k_j \Gamma^l_{km} I^m_i - I^k_j \Gamma^l_{im} I^m_k)$$
$$= -I^k_i \Gamma^l_{jm} I^m_k + I^k_j \Gamma^l_{im} I^m_k = \Gamma^l_{ji} - \Gamma^l_{ij} \equiv 0. \quad (9.8)$$

Corollary 9.7 *M Riemannian with Levi–Civita connection ∇. Let $I^i{}_j$ be a parallel almost complex structure, $\nabla_i I = 0$. Then I is integrable. In particular, if M is almost Hermitian and $\nabla_i \omega = 0$, M is Hermitian.*

Proof From Eq. (9.7) the Nijenhuis tensor is manifestly zero. $\square$

9.1.3 Holomorphic bundles on a complex manifold

Definition 9.8 Given two complex manifolds $\mathcal{M}_1$ and $\mathcal{M}_2$, a smooth map $\mathcal{M}_1 \xrightarrow{h} \mathcal{M}_2$ is *holomorphic* iff it is analytic in local complex coordinates. A *holomorphic bundle* is a smooth bundle $E \xrightarrow{\pi} B$, where E, B are complex manifolds and the projection π is holomorphic.

A rank m holomorphic vector bundle $V \to \mathcal{M}$ may be described through an open cover $\{U_\alpha\}$ of $\mathcal{M}$ such that $V|_{U_\alpha} \simeq \mathbb{C}^m \times U_\alpha$; the transition functions $\phi_{\alpha\beta}: U_\alpha \cap U_\beta \to GL(m, \mathbb{C})$ are required to be holomorphic. If the rank of V is 1 we say that V is a holomorphic *line* bundle.

Holomorphic tangent bundle. Given a complex manifold, we define the *holomorphic tangent bundle* to be the $(1, 0)$ part of the complexified tangent bundle. If $\mathcal{M}$ is complex, we reserve the notation $T\mathcal{M}$ for the holomorphic tangent bundle and use the notation $T\mathcal{M}_\mathbb{R}$ for the tangent bundle of the underlying smooth manifold.

$$\mathbb{C} \otimes T\mathcal{M}_\mathbb{R} = T\mathcal{M} \oplus \overline{T}\mathcal{M}, \qquad \overline{T}\mathcal{M} = (T\mathcal{M})^*. \tag{9.9}$$

If z^i are local holomorphic coordinates in the patch U, $T\mathcal{M}$ is locally spanned by the holomorphic section $\partial/\partial z^j$. Let w^j be local holomorphic coordinates in the patch V. The basis $\{\partial_{z^i}\}$ and $\{\partial_{w^j}\}$ are related in $U \cap V$ by the *holomorphic* transition functions $\partial z^i/\partial w^j$; hence $T\mathcal{M}$ is a holomorphic vector bundle. A smooth (resp. holomorphic) section of $T\mathcal{M}$ is called a smooth (resp. holomorphic) vector field of type $(1, 0)$; sections of $\overline{T}\mathcal{M}$ are said to have type $(0, 1)$.

Differential forms of type (p, q). Let $T^*\mathcal{M} := (T\mathcal{M})^\vee$ be the dual bundle. $T^*\mathcal{M}$ is called the *holomorphic cotangent bundle,* and its sections *differential forms of type $(1, 0)$*. The dual of Eq. (9.9) is $\mathbb{C} \otimes T^*\mathcal{M}_\mathbb{R} \simeq T^*\mathcal{M} \oplus \overline{T}^*\mathcal{M}$, with $T^*\mathcal{M}_\mathbb{R}$ the cotangent bundle in the C^∞ sense. The (complex-valued) differential forms of degree k on $\mathcal{M}$ are then sections of

$$\mathbb{C} \otimes \wedge^k T^*\mathcal{M}_\mathbb{R} \simeq \wedge^k(T^*\mathcal{M} \oplus \overline{T}^*\mathcal{M}) = \bigoplus_{p+q=k} \wedge^p T^*\mathcal{M} \otimes \wedge^q \overline{T}^*\mathcal{M}. \tag{9.10}$$

The sections of the bundle $\wedge^{p,q} := \wedge^p T^*\mathcal{M} \otimes \wedge^q \overline{T}^*\mathcal{M}$ are called *differential forms of type (p, q)*. Explicitly a (p, q)–form is written as

$$\sum \varphi_{i_1 i_2 \cdots i_p \bar{j}_1 \bar{j}_2 \cdots \bar{j}_q} \, dz^{i_1} \wedge \cdots \wedge dz^{i_p} \wedge d\bar{z}^{j_1} \wedge \cdots \wedge d\bar{z}^{j_q} \tag{9.11}$$

for some C^∞ functions $\varphi_{i_1 i_2 \cdots i_p \bar{j}_1 \bar{j}_2 \cdots \bar{j}_q}$. We conclude that, on a complex manifold $\mathcal{M}$, the forms of given degree k decompose into the sum of forms of *definite* type (p, q) with $p + q = k$.

9.1 Complex manifolds

Canonical line bundle. In particular, $\Lambda^{n,0}$ is a holomorphic line bundle. It is called the *canonical line bundle* of the manifold $\mathcal{M}$.

Dolbeault differentials $\bar{\partial}, \partial$. In complex coordinates the exterior derivative reads $d \equiv dz^i \, \partial_{z^i} + d\bar{z}^i \, \partial_{\bar{z}^i}$. Decomposing d into components of definite type we get the Dolbeault differentials: $\partial := dz^i \, \partial_{z^i}$ and $\bar{\partial} := d\bar{z}^i \, \partial_{\bar{z}^i}$. Hence

$$d = \partial + \bar{\partial}. \tag{9.12}$$

Let α be a differential form of type (p, q). $\partial \alpha$ (resp. $\bar{\partial}\alpha$) is a form of type $(p+1, q)$ (resp. $(p, q+1)$). Decomposing the relation $0 = d^2 = (\partial + \bar{\partial})^2$ into components of definite type we get the identities

$$\partial^2 = \bar{\partial}^2 = \bar{\partial}\partial + \partial\bar{\partial} = 0. \tag{9.13}$$

A form α is called $\bar{\partial}$–*closed* (resp. $\bar{\partial}$–*exact*) if $\bar{\partial}\alpha = 0$ (resp. if $\alpha = \bar{\partial}\beta$ for some form β). Since $\bar{\partial}^2 = 0$, exact forms are automatically closed. This motivates the following definition.

Definition 9.9 Let $\mathcal{Z}^{p,q} := \{\alpha \in \Lambda^{p,q} \mid \bar{\partial}\alpha = 0\}$ be the space of closed (p, q) forms. The *Dolbeault cohomology group* $H_{\bar{\partial}}^{p,q}(\mathcal{M})$ is defined as the space of the $\bar{\partial}$–closed (p, q) forms modulo the $\bar{\partial}$–exact ones

$$H_{\bar{\partial}}^{p,q}(\mathcal{M}) := \mathcal{Z}^{p,q} / \bar{\partial}\Lambda^{p,q-1}. \tag{9.14}$$

For $q > 0$, the cohomology is locally trivial; indeed one has:

Theorem 9.10 (Poincaré–Dolbeault lemma) *If a C^∞ (p,q)–form ϕ, $q \geq 1$, defined on the polydisk $\Delta = \{z \in \mathbb{C}^n \mid |z^i| < R, i = 1, \ldots, n\}$, is $\bar{\partial}$–closed, there is a C^∞ $(p, q-1)$–form ψ on Δ such that $\phi = \bar{\partial}\psi$, that is*

$$H_{\bar{\partial}}^{p,q}(\Delta) = 0 \quad \text{for } q \geq 1. \tag{9.15}$$

For a proof see ref. [217] theorem 3.3, or ref. [171] pages 25–27 or ref. [168] Chapter II.

This result implies that the groups $H_{\bar{\partial}}^{p,q}(\mathcal{M})$ may be understood in terms of sheaf cohomology. Let Ω^p be the sheaf of germs of local *holomorphic* sections of $\wedge^p T^*\mathcal{M}$. Then:

Theorem 9.11 (Dolbeault) *One has $H_{\bar{\partial}}^{p,q}(\mathcal{M}) = H^q(\mathcal{M}, \Omega^p)$.*

See ref. [171] pages 45–47, or [217] theorem 3.12, or ref. [186] theorem 15.4.1, or ref. [94], or ref. [181] Chapter VI, or ref. [168] Chapter II, or ref. [149] Secttion VI.2.

Definition 9.12 The non–negative integers $h^{p,q} := \dim H^{p,q}_{\bar\partial}(\mathcal{M})$ are called *Hodge numbers*.

Hodge theory implies that $h^{p,q}$ are finite if $\mathcal{M}$ is compact (see below). In general $h^{p,q} \neq h^{q,p}$, but we have equality in Kähler manifolds (Section 9.4). However, it is always true that $h^{p,q} = h^{n-p,n-q}$, since the pairing $\int_{\mathcal{M}} \alpha \wedge \beta$ is non–degenerate (Serre duality).

Lemma 9.13 ($\partial\bar\partial$–Poincaré Lemma; Exercise 9.1.3) *Let φ be a (k,k)–form ($k \geq 1$), defined in some polydisk $\Delta \subset \mathbb{C}^n$, which is d–closed, $d\varphi = 0$. Then there is in Δ a $(k-1, k-1)$–form β such that*

$$\varphi = \partial\bar\partial\beta. \tag{9.16}$$

Connections on holomorphic vector bundles

Let $\mathcal{M} = \cup_\alpha U_\alpha$ be a complex manifold and $V \to \mathcal{M}$ a holomorphic vector bundle. If t^i_α is a trivialization in U_α of a local section t, in $U_\alpha \cap U_\beta$ one has $t^i_\alpha = (\phi_{\alpha\beta})^i{}_j\, t^j_\beta$ with $\phi_{\alpha\beta}$ a rank $V \times$ rank V *holomorphic* invertible matrix.

Definition 9.14 A *Hermitian fiber metric* for V is a skew–linear map $\langle\cdot,\cdot\rangle : \bar V \otimes V \to \mathbb{C} \otimes \mathcal{M}$ which is fiberwise *positive–definite*.

Fix a local (holomorphic) trivialization for V. A Hermitian fiber metric is given, in U_α, by a (rank $V \times$ rank V) matrix $H^{(\alpha)}_{\bar k i}$,

$$\langle t, s\rangle\big|_{U_\alpha} = (t^k_{(\alpha)})^* H^{(\alpha)}_{\bar k j} s^j_{(\alpha)}. \tag{9.17}$$

Agreement on $U_\alpha \cap U_\beta$, requires

$$H^{(\alpha)} = (\phi^t_{\alpha\beta})^* H^{(\beta)} \phi_{\alpha\beta}, \tag{9.18}$$

namely $H^{(\alpha)}$ is a (smooth) section of $\bar V^\vee \otimes V^\vee$.

A connection $\nabla = d + A$ on V is *holomorphic* if $\nabla_{\bar k} t = \partial_{\bar k} t$, i.e., if its *anti–holomorphic* component vanishes, $A_{\bar k} = 0$.

Proposition 9.15 *Let V be a holomorphic vector bundle with a Hermitian fiber metric H. There is a unique connection on V which is both holomorphic and metric. It is called the* **Chern connection**.

Proof A connection ∇ is metric iff $d\langle t, s\rangle = \langle\nabla t, s\rangle + \langle t, \nabla s\rangle$. If ∇ is both metric and holomorphic, we have

$$\partial_i(\bar t H s) = \overline{\partial_{\bar i} t} H s + \bar t H \nabla_i s, \tag{9.19}$$

9.1 Complex manifolds

so $H\nabla_i s = \partial_i(H s)$, i.e., $\nabla_i s = H^{-1}\partial_i(H s)$ or

$$\nabla_i = \partial_i + H^{-1}(\partial_i H). \tag{9.20}$$

$\square$

Note that the connection in the holomorphic directions, $A_i = H^{-1}\partial_i H$, has the "pure–gauge" form. As a consequence, the $(2,0)$ part of the curvature vanishes, $F_{ij} = [\nabla_i, \nabla_j] = 0$, as does the $(0,2)$ part $F_{\bar{i}\bar{j}} = [\partial_{\bar{i}}, \partial_{\bar{j}}] = 0$. Then:

Proposition 9.16 *The curvature of a holomorphic Hermitian connection ∇ has pure type $(1,1)$. Explicitly*

$$F_{\bar{i}j} = [\partial_{\bar{i}}, \nabla_j] = \partial_{\bar{i}}(H^{-1}\partial_j H), \tag{9.21}$$

or, in form terms,

$$F = \bar{\partial}(H^{-1}\partial H) \in \mathfrak{Lie}(\mathrm{Aut}(V)) \otimes \Lambda^{1,1}. \tag{9.22}$$

The case of $T^*\mathcal{M}$: torsion

The above results apply, in particular, to $T^*\mathcal{M}$. Let $H^{\bar{i}j} = \langle dz^i, dz^j \rangle$ be the fiber metric, and E^a_j a $n \times n$ matrix such that $\bar{E}^t H E^t = \mathbf{1}$. The $(1,0)$ forms $\theta^a = E^a_j dz^j$ give a unitary frame in $T^*\mathcal{M}$, $\langle \theta^a, \theta^b \rangle = \delta^{\bar{a}b}$; that is, H defines a $U(n)$–structure. In analogy with the real case, we introduce a *unitary connection*: in the θ^a frame it is represented by a matrix of 1–forms, ω^a_b, such that $(\omega^t)^* + \omega = 0$. The first Cartan structure equation reads

$$d\theta^a + \omega^a_b \wedge \theta^b = \tau^a, \qquad d\bar{\theta}^{\bar{a}} + \bar{\omega}^{\bar{a}}_{\bar{b}} \wedge \bar{\theta}^{\bar{b}} = \bar{\tau}^{\bar{a}}, \tag{9.23}$$

where τ^a is the *torsion* 2–form.

Proposition 9.17 *There is a unique unitary connection ω^a_b on $T^*\mathcal{M}$ such that the torsion τ^a is a pure $(2,0)$–form; it corresponds to the unique Chern connection given by Eq. (9.20).*

Proof One has $\tau|_{(1,1)} = \bar{\partial}\theta + \omega|_{(0,1)} \wedge \theta$. Hence, if $\tau|_{(1,1)} = 0$, we must have $0 = (\bar{\partial}E^a_j)dz^j + \omega^a_b|_{(1,0)}E^b_j dz^j$, that is $\omega|_{(0,1)} = -(\bar{\partial}E)E^{-1}$. On the other hand, $\omega|_{(1,0)} = -(\omega^t|_{(0,1)})^* = -(\partial\bar{E}^{-1})\bar{E}$; hence the connection form ω is uniquely specified in terms of E. Since $\bar{E}^{-1} = EH^t$, ω can be rewritten

$$\begin{aligned}\omega|_{(1,0)} &= E(H^t\partial(H^t)^{-1})E^{-1} + E\partial E^{-1} \equiv E\nabla E^{-1},\\ \omega|_{(0,1)} &= E\bar{\partial}E^{-1} \equiv E\bar{\nabla}E^{-1},\end{aligned} \tag{9.24}$$

which is manifestly gauge–equivalent to the (dual) of the connection in Eq. (9.20).

$\square$

Thus for the Chern connection one has $\tau = \partial\theta + \omega|_{(1,0)} \wedge \theta$, or explicitly

$$\tau^a = E^a_l H^{\bar{k}l} \partial_i H_{j\bar{k}} \, dz^i \wedge dz^j. \tag{9.25}$$

The Chern class of a line bundle. We specialize the above discussion to rank 1, that is, to a *line bundle* $\mathcal{L}$. Let h_α be a fiber metric for $\mathcal{L}$ over U_α. Locally $|s|^2|_{U_\alpha} = h_\alpha (s_\alpha)^* s_\alpha$ for any section s. On the overlaps $U_\alpha \cap U_\beta$, $h_\alpha = h_\beta |\phi_{\alpha\beta}|^2$ for some holomorphic function $\phi_{\alpha\beta}$. The curvature is given by Eq. (9.22),

$$\Theta = \bar{\partial}\partial \log h_\alpha. \tag{9.26}$$

This formula has an important interpretation. First of all, note that Θ is a global d–closed form on $\mathcal{M}$. Indeed $\bar{\partial}\partial(\log h_\alpha - \log h_\beta) = -2\operatorname{Re} \partial\bar{\partial} \log \phi_{\alpha\beta} = 0$. Then one can consider the de Rham class

$$\left[\frac{i}{2\pi}\Theta\right] \in H^2(\mathcal{M}). \tag{9.27}$$

It turns out that the cohomology class $[i\Theta/2\pi]$ is always *integral*, i.e., it represents an element of $H^2(\mathcal{M}, \mathbb{Z})$. It is a characteristic class called *the first Chern class $c_1(\mathcal{L})$ of the line bundle $\mathcal{L}$*. $c_1(\mathcal{L})$ is independent of the particular metric used to compute it, and completely characterizes the line bundle $\mathcal{L}$ up to *topological* (or even differential) *equivalence*. Chern classes are important in physical applications since they correspond to the Dirac quantization rule for electric and magnetic charges.

One shows [171] that *every integral class in $H^2(\mathcal{M}, \mathbb{Z})$ which has a de Rham representative of type $(1, 1)$ is the Chern class of some holomorphic line bundle $\mathcal{L} \to \mathcal{M}$*.

Exercise 9.1.1 Show that a complex manifold is always oriented and that all holomorphic maps $M \xrightarrow{h} N$ preserve orientation.

Exercise 9.1.2 Prove the $\partial\bar{\partial}$–Poincaré lemma.

Exercise 9.1.3 Let g be a Hermitian metric on a n–fold, and ω the associated 2–form. Show that the volume $2n$–form induced by g is ω^n/n.

9.2 Kähler metrics and manifolds

Let $\mathcal{M}$ be a complex manifold with local holomorphic coordinates z^i and Hermitian metric g. One has

$$ds^2 = g_{i\bar{j}} \, dz^i \, d\bar{z}^j \quad \text{and} \quad \omega = \frac{i}{2} g_{i\bar{j}} \, dz^i \wedge d\bar{z}^j. \tag{9.28}$$

9.2 Kähler metrics and manifolds

Definition 9.18 A *Kähler metric* is a Hermitian metric g on the complex manifold $\mathcal{M}$ such that the associated $(1,1)$ form ω is closed, $d\omega = 0$. ω is called the *Kähler form*. A *Kähler manifold* is a complex manifold $\mathcal{M}$ which admits one Kähler metric.

The $\partial\bar{\partial}$–Poincaré lemma implies:

Proposition 9.19 Let ω be a Kähler form on $\mathcal{M}$ and $U \subset \mathcal{M}$ a sufficiently small open set. There is a real C^∞ function ϕ such that

$$\omega = i\partial\bar{\partial}\phi \quad \text{in } U. \tag{9.29}$$

ϕ is called the (local) Kähler potential. *It is unique up $\phi \to \phi + f + f^*$ with f a holomorphic function (in U).*

In the physics literature, the freedom to add the real part of a holomorphic function f to the potential ϕ is known as *Kähler gauge invariance*. The function ϕ does not exist globally, in general. In particular it cannot exist if $\mathcal{M}$ is compact (cf. Exercise 9.1.3).

Decomposing into type, $d\omega = 0$ implies $\partial\omega = \bar{\partial}\omega = 0$. In coordinate terms these conditions are[1]

$$\partial_i g_{j\bar{k}} = \partial_j g_{i\bar{k}}, \qquad \partial_{\bar{k}} g_{i\bar{h}} = \partial_{\bar{h}} g_{i\bar{k}}, \tag{9.30}$$

whose solution (locally) is

$$g_{i\bar{k}} = \partial_i \partial_{\bar{k}} \phi. \tag{9.31}$$

Equation (9.31) is often taken as the definition of a Kähler metric. Other equivalent definitions will be given below.

Let us explain why Kähler metrics are so nice and natural. We know that on a Riemannian manifold $(\mathcal{M}, g)$ there is a unique connection which is both metric and torsion–free, namely the Levi–Civita one. On the other hand, if $\mathcal{M}$ is complex and g Hermitian, we know that on the holomorphic tangent bundle $T\mathcal{M}$ there is a unique connection which is both metric and holomorphic, the Chern one. What is the relation between these two *unique* connections on the tangent bundle?

Proposition 9.20 *The Chern connection on $T\mathcal{M}$ is torsionless iff g is Kähler. For a Kähler g the Chern and Levi–Civita connections coincide.*

Hence, for a Kähler manifold the Kähler geometry has the virtues of both the holomorphic and Riemannian structures.

---

[1] We use the shorthand $\partial_i = \frac{\partial}{\partial z^i}$ and $\partial_{\bar{k}} = \frac{\partial}{\partial \bar{z}^k}$.

Proof We start from the Chern connection whose torsion has pure $(2,0)$ type (Proposition 9.17). From Eq. (9.25), we see that this torsion is proportional to $\partial_i g_{j\bar{k}} - \partial_j g_{i\bar{k}}$ and hence it vanishes precisely for g Kähler. □

The above proposition can be used as the definition of a Kähler metric. Another useful (equivalent) definition can be phrased in the language of the "equivalence principle." In General Relativity the equivalence principle states that, at each point x of space–time, we can find coordinates such that $(g_{ij})_x = \delta_{ij}$ and $(\Gamma^i_{jk})_x = 0$. Kähler manifolds are precisely the Hermitian manifolds in which the equivalence principle holds *in the holomorphic sense*.

Proposition 9.21 *At each point* $x \in \mathcal{M}$ *one can choose* holomorphic *coordinates such that* $(g_{i\bar{j}})_x = \delta_{i\bar{j}}$ *and* $(\Gamma^i_{jk})_x = 0$ *if and only if g is Kähler.*

Proof Let g be Kähler. We choose coordinates centered at x with $(g_{i\bar{j}})_x = \delta_{i\bar{j}}$. The local Kähler potential has an expansion;

$$\phi = \delta_{i\bar{j}} z^i \bar{z}^j + A_{\bar{i}jk} \bar{z}^i z^j z^k + A_{i\bar{j}\bar{k}} z^i \bar{z}^j \bar{z}^k + (A_{ijk} z^i z^j z^k + A_{\bar{i}\bar{j}\bar{k}} \bar{z}^i \bar{z}^j \bar{z}^k) + O(z^4).$$

A Kähler transformation cancels the term in the parentheses. Let $w^i = z^i + A_{\bar{i}jk} z^j z^k + O(z^3)$. In the coordinates w^i, $\phi = \delta_{i\bar{j}} w^i \bar{w}^j + O(w^4)$, and hence $(\Gamma^k_{ij})_0 = 0$. On the other hand a simple computation gives

$$2\bar{\partial}\omega = -i g_{\bar{k}l} \Gamma^l_{ij} dz^i \wedge d\bar{z}^i \wedge d\bar{z}^k, \tag{9.32}$$

so if $(\bar{\partial}\omega)_x \neq 0$ there are no complex coordinates in which all the Christoffel coefficients vanish at x. □

We state the equivalence principle in the form: *Any covariant identity that involves the metric together with its first derivatives and which is valid on $\mathbb{C}^n$ with the flat metric, is also valid on any Kähler manifold.*

Kähler spaces as symplectic manifolds. Isometries. The Kähler form ω is a closed, non–degenerate[2] 2–form: that is, a symplectic structure. Then a Kähler manifold is automatically a symplectic space. Let V be a holomorphic Killing vector, i.e., a holomorphic motion which is an infinitesimal isometry for the given Kähler metric. Since V preserves both the metric and the complex structure, it preserves the Kähler form, i.e., $\pounds_V \omega = 0$. Then

$$\pounds_V \omega = (d i_V + i_V d)\omega = d(i_V \omega), \tag{9.33}$$

---

[2] A 2–form in a $2n$–dimensional real space is said to be *non–degenerate* if $\omega^n \neq 0$ at each point.

9.2 Kähler metrics and manifolds

i.e., V is a holomorphic isometry iff the 1–form $i_V\omega$ is closed. Therefore locally there exists a function h_V such that $i_V\omega = dh_V$, called the *Hamiltonian*. The symplectic geometry was studied in more detail in Chapter 7.

9.2.1 Curvatures

Proposition 9.20 implies that the Riemann tensor for a Kähler metric is just the curvature of the holomorphic connection given by Eq. (9.21). Then

$$R^i{}_{j\bar{k}h} = \partial_{\bar{k}}\Gamma^i_{jh} = \partial_{\bar{k}}(g^{i\bar{l}}\partial_h g_{\bar{l}j}) = g^{i\bar{l}}\partial_{\bar{k}}\partial_h g_{\bar{l}j} - g^{i\bar{l}}\partial_{\bar{k}}g_{\bar{l}m}g^{m\bar{n}}\partial_h g_{\bar{n}j} \qquad (9.34)$$

and this is the only non–vanishing component (up to index permutations). In particular Eq. (9.22) gives

$$R^i{}_j \in \mathfrak{Lie}(U(n)) \otimes \Lambda^{1,1}, \qquad (9.35)$$

and hence $\mathrm{Hol}(g) \subseteq U(n)$. The first Bianchi identity reduces to

$$R_{i\bar{j}k\bar{h}} = -R_{\bar{j}ki\bar{h}} - R_{ki\bar{j}\bar{h}} = R_{k\bar{j}k\bar{h}} = R_{i\bar{h}k\bar{j}}, \qquad (9.36)$$

i.e. *the Riemann tensor is invariant under the interchange of the two unbarred (resp. barred) indices.*

The Ricci tensor is defined as $R_{j\bar{k}} = -R^i{}_{j\bar{k}i} \equiv -R^i{}_{ikj}$. Thus,

$$R_{j\bar{k}} = -\partial_{\bar{k}}(g^{i\bar{l}}\partial_j g_{\bar{l}i}) = -\partial_{\bar{k}}\partial_j \log \det g. \qquad (9.37)$$

Out of the Ricci tensor we can construct a real $(1,1)$ form, known as the Ricci form: $\rho := iR_{i\bar{j}}dz^i \wedge d\bar{z}^j$. $\rho = iR^i{}_{i l\bar{k}}dz^l \wedge d\bar{z}^k$ is the trace of the curvature $(1,1)$ form $\mathcal{R}^i{}_j = iR^i{}_{jl\bar{k}}dz^l \wedge d\bar{z}^k$ and hence it corresponds to the trace part of the holonomy, that is, to the curvature of the $U(1)$ factor of the holonomy group $U(1) \times SU(n)$. From Eq. (9.37) we see that *the Ricci form is closed* (this is simply the Bianchi identity for an Abelian connection). In fact

$$\rho = i\bar{\partial}\partial \log \det g, \qquad (9.38)$$

and ρ has the same structure as the Kähler form, with $(\log \det g)$ playing the role of potential. Note that ρ is well defined. Indeed in the overlap of two coordinate patches, $U_\alpha \cap U_\beta$, one has $g_\alpha = \overline{\phi_{\alpha\beta}} \, g_\beta \, \phi_{\alpha\beta}$, where the matrix $(\phi_{\alpha\beta})^i{}_j = \partial z^i_\beta/\partial z^j_\alpha$ is the *holomorphic* Jacobian. Then

$$\log \det g_\alpha = \log \det g_\beta + \log \det [\partial z_\beta/\partial z_\alpha] + (\log \det [\partial z_\beta/\partial z_\alpha])^* \qquad (9.39)$$

and the two potentials differ by the real part of a holomorphic function i.e., by an irrelevant Kähler gauge transformation.

Ricci Form and the canonical bundle

Let $\mathcal{M}$ be a complex manifold of dimension n. Recall that the holomorphic line bundle $K := \wedge^n T^*\mathcal{M}$ is called the canonical bundle. Locally on U_α, a section of K has the form $f_\alpha \, dz_\alpha^1 \wedge dz_\alpha^2 \wedge \cdots \wedge dz_\alpha^n$ some function f_α. On the overlaps $U_\alpha \cap U_\beta$, $f_\beta = f_\alpha \det[\partial z_\alpha/\partial z_\beta]$. Let h_α be a fiber metric for K. On the overlaps $h_\alpha = h_\beta \,|\det(\partial z_\beta/\partial z_\alpha)|^2$. Then Eq. (9.39) says that $\det g_\alpha$ is a fiber metric for the canonical bundle. Recalling the discussion in Section 9.1.3, we have shown:

Proposition 9.22 *Let ρ be the Ricci form of any Kähler metric g on the complex manifold $\mathcal{M}$. One has*

$$c_1(K) = \frac{i}{2\pi}[\rho]. \tag{9.40}$$

9.3 $U(n)$ manifolds

Theorem 9.23 *A Riemannian $2n$–manifold (M, g) has holonomy $U(n)$ if and only if it is Kähler.*

Proof The implication M Kähler $\Rightarrow$ M has holonomy $U(n)$ is obvious: in Section 9.2.1 we saw that, if g is Kähler, the non–vanishing components of the Riemann tensor have the form $R_{i\bar{j}k\bar{l}}$ and hence the curvature belongs to $\odot^2 \mathfrak{u}(n)$. The opposite implication follows from Fundamental Principle 3.3. The subgroup $U(n) \subset SO(2n)$ leaves invariant a matrix I with $I^2 = -1$, namely the complex structure of $\mathbb{C}^n \simeq \mathbb{R}^{2n}$. Hence,

$$\operatorname{Hol}(M) \subseteq U(n) \Rightarrow \begin{cases} \text{on } M \text{ there is a } parallel \text{ tensor } I^i{}_k, \\ \text{with } I^i{}_k I^k{}_j = -\delta^i{}_j \text{ and } \nabla_k I^i{}_j = 0. \end{cases} \tag{9.41}$$

Comparing with Eq. (9.7), we get $N^i_{jk} = 0$, that is, *the complex structure I is integrable and M is a complex manifold*. Let ω_{ij} be the 2–form associated with g. One has

$$\nabla_k \omega_{ij} = \nabla_k(g_{jl} I^l{}_i) = g_{jl} \nabla_k I^l{}_i = 0. \tag{9.42}$$

In particular, $d\omega = 0$ and (M, g) is Kähler. $\square$

The theorem gives us another definition of a Kähler manifold: *it is an almost Hermitian manifold such that $\nabla_i I^j{}_k = 0$*. More generally:

Corollary 9.24 *Let (M, I) be a complex manifold and g a Hermitian metric on M. The following three conditions are equivalent:*
(i) $d\omega = 0$, (ii) $\nabla_i \omega = 0$, and (iii) $\nabla_i I = 0$.

Corollary 9.25 *A Kähler metric is equivalent to a torsionless U(n)–structure.*

Indeed, by definition on M there is a connection which is metric, torsionless, and with holonomy contained in $U(n)$.

Calabi–Yau spaces

Obviously, $\mathfrak{u}(n) = \mathfrak{u}(1) \oplus \mathfrak{su}(n)$. The trace part of the holonomy is generated by $R_{\bar{i}jkl}\,g^{\bar{k}l} \equiv R_{\bar{k}\bar{j}l}\,g^{\bar{k}l} = -R_{\bar{j}i}$, the Ricci tensor. Hence the above proof also shows the following theorem.

Theorem 9.26 *A Riemannian 2n–manifold has holonomy $\subseteq SU(n)$ if and only if it is Kähler and Ricci-flat.*

From Berger's theorem, if $\mathrm{Hol}(g) \subseteq SU(n)$ and irreducible, we have three possibilities: (1) $\mathrm{Hol}(g) = SU(n)$ (Calabi–Yau); (2) $\mathrm{Hol}(g) = Sp(n/2)$ (hyperKähler); (3) M is locally symmetric. By Theorem 5.15 a Ricci–flat symmetric manifold is flat, so in the third case M is necessarily flat.

9.4 Hodge theory in Kähler spaces

In this section $\mathcal{M}$ is a *compact* complex manifold of dimension n.

Using the Hodge dual $*$, we introduce a Hermitian inner product on the space of smooth (p, q)–forms,

$$\langle \psi, \eta \rangle := \int_{\mathcal{M}} *\psi^* \wedge \eta, \qquad \eta, \psi \in \Lambda^{p,q}(\mathcal{M}), \tag{9.43}$$

making it a (pre)Hilbert space. We ask the question: *Given a $\bar{\partial}$–closed form η, can we find a representative in its Dolbeault class $\{\eta + \bar{\partial}\xi\}$ of smallest norm?*

To answer, we consider the adjoint of the operator $\bar{\partial}\colon \Lambda^{p,q}(\mathcal{M}) \to \Lambda^{p,q+1}(\mathcal{M})$ with respect to the product (9.43), $\eth := \bar{\partial}^\dagger$. Explicitly,

$$\eth = -*\partial*. \tag{9.44}$$

Lemma 9.27 *A $\bar{\partial}$–closed form η has minimal norm in its class iff $\eth\eta = 0$.*

Proof Let $\eth\eta = 0$. Then

$$\|\eta + \bar{\partial}\xi\|^2 = \|\eta\|^2 + |\bar{\partial}\xi\|^2 + 2\mathrm{Re}\,\langle \eta, \bar{\partial}\xi \rangle$$
$$= \|\eta\|^2 + |\bar{\partial}\xi\|^2 + 2\mathrm{Re}\,\langle \eth\eta, \xi \rangle = \|\eta\|^2 + |\bar{\partial}\xi\|^2 \geq \|\eta\|^2 \tag{9.45}$$

with equality iff $\bar{\partial}\xi = 0$. Conversely, if η has minimal norm,

$$\left.\frac{\partial}{\partial t} \|\eta + t\bar{\partial}\xi\|^2 \right|_{t=0} = 2\mathrm{Re}\,\langle \eth\eta, \xi \rangle = 0 \tag{9.46}$$

for all ξ, which again implies $\partial \eta = 0$. □

Thus, formally, the Dolbeault cohomology group is represented exactly by the space of solutions of the two first–order equations $\bar{\partial}\eta = \partial\eta = 0$. We can replace them by the single second–order equation

$$\Delta_{\bar{\partial}}\eta = (\bar{\partial}\partial + \partial\bar{\partial})\eta = 0. \tag{9.47}$$

Indeed, $\bar{\partial}\eta = \partial\eta = 0 \Rightarrow \Delta_{\bar{\partial}}\eta = 0$ and, conversely,

$$\langle \eta, \Delta_{\bar{\partial}}\eta \rangle = \langle \eta, \bar{\partial}\partial\eta \rangle + \langle \eta, \partial\bar{\partial}\eta \rangle = \|\partial\eta\|^2 + \|\bar{\partial}\eta\|^2, \tag{9.48}$$

which implies

$$\Delta_{\bar{\partial}}\eta = 0 \Rightarrow \bar{\partial}\eta = \partial\eta = 0. \tag{9.49}$$

The operator $\Delta_{\bar{\partial}}$ is called the $\bar{\partial}$–*Laplacian*. The solutions to the equation $\Delta_{\bar{\partial}}\eta = 0$ are called $\bar{\partial}$–harmonic forms. We write $\mathcal{H}^{p,q}_{\bar{\partial}}(\mathcal{M})$ for the vector space of $\bar{\partial}$–harmonic forms of type (p,q). Hilbert space arguments give the following.

Theorem 9.28 (Hodge) $\mathcal{M}$ *is compact Kähler.* $\dim \mathcal{H}^{p,q}_{\bar{\partial}}(\mathcal{M}) < \infty$, *and the orthogonal projector* $P_{\mathcal{H}} : \Lambda^{pq} \to \mathcal{H}^{p,q}_{\bar{\partial}}$ *is well defined. There exists a unique operator, the Green's operator,* $G_{\bar{\partial}} : \Lambda^{p,q} \to \Lambda^{p,q}$, *with* $G_{\bar{\partial}} P_{\mathcal{H}} = 0$ *and* $\bar{\partial} G_{\bar{\partial}} = G_{\bar{\partial}}\bar{\partial}$, $G_{\bar{\partial}}\partial = \partial G_{\bar{\partial}}$, *such that*

$$P_{\mathcal{H}} + \Delta_{\bar{\partial}} G_{\bar{\partial}} = 1. \tag{9.50}$$

Equation (9.50) gives

$$\psi = P_{\mathcal{H}}\psi + \bar{\partial}(\partial G_{\bar{\partial}}\psi) + \partial(\bar{\partial} G_{\bar{\partial}}\psi). \tag{9.51}$$

This equality is called *the Hodge decomposition*, since it implies the orthogonal direct–sum decomposition

$$\Lambda^{p,q} = \mathcal{H}^{p,q}_{\bar{\partial}} \oplus \bar{\partial}\Lambda^{p,q-1} \oplus \partial\Lambda^{p,q+1}. \tag{9.52}$$

On a complex manifold $\mathcal{M}$ we have three natural external derivatives, namely d, $\bar{\partial}$, and ∂. We may consider their respective adjoint operators δ, $\bar{\partial}$, and $\bar{\partial}$, and the corresponding Laplacians

$$\Delta_d = d\delta + \delta d \quad \text{(usual real Laplacian)}, \tag{9.53}$$

$$\Delta_{\bar{\partial}} = \bar{\partial}\partial + \partial\bar{\partial}, \tag{9.54}$$

$$\Delta_\partial = \partial\bar{\partial} + \bar{\partial}\partial. \tag{9.55}$$

9.4 Hodge theory in Kähler spaces

The above Hodge theorem applies to all three Laplacians, with different Green's functions and projectors. In particular, we have three different notions of harmonic forms according to which Laplacian we use. From the Hodge decomposition (9.51) we have the isomorphisms

$$\mathcal{H}^k_d(\mathcal{M}) \simeq H^k_d(\mathcal{M}), \tag{9.56}$$

$$\mathcal{H}^{p,q}_{\bar{\partial}}(\mathcal{M}) \simeq H^{p,q}_{\bar{\partial}}(\mathcal{M}), \tag{9.57}$$

$$\mathcal{H}^{p,q}_{\partial}(\mathcal{M}) \simeq H^{p,q}_{\partial}(\mathcal{M}). \tag{9.58}$$

The cohomology groups on the RHS are, in general, non–isomorphic. In the following subsection we shall see that the situation improves a lot if $\mathcal{M}$ is supposed to be Kähler.

Corollary 9.29 (Kodaira–Serre duality) $\quad H^n(\mathcal{M}, \Omega^n) \xrightarrow{\sim} \mathbb{C}$, while

$$H^q(\mathcal{M}, \Omega^p) \otimes H^{n-q}(\mathcal{M}, \Omega^{n-p}) \to H^n(\mathcal{M}, \Omega^n) \tag{9.59}$$

is a non–degenerate pairing.

Proof Since $\partial = \bar{\partial}^\dagger$, we have $\Delta_{\bar{\partial}} = \Delta^\dagger_{\partial}$. Then,

$$\psi \in \mathcal{H}^{p,q}_{\bar{\partial}}(\mathcal{M}) \implies \psi^\dagger \equiv *\psi^* \in \mathcal{H}^{n-p,n-q}_{\bar{\partial}}(\mathcal{M}). \tag{9.60}$$

$\square$

Hodge identities. We assume the compact complex manifold $\mathcal{M}$ to be Kähler; we write ω for the associated $(1, 1)$ closed form. On the space of smooth (p, q) forms, $\Lambda^{p,q}(\mathcal{M})$, we define an operator, denoted L, by

$$L \colon \Lambda^{p,q}(\mathcal{M}) \to \Lambda^{p+1,q+1}(\mathcal{M}), \tag{9.61}$$

$$\phi \mapsto \omega \wedge \phi, \tag{9.62}$$

and its adjoint, $\Lambda = (-1)^{p+q} * L *$. Λ maps (p, q) forms into $(p-1, q-1)$ forms. We compute the commutators of the various operators we introduced. We already know

$$\partial^2 = \bar{\partial}^2 = \bar{\partial}\partial + \partial\bar{\partial} = \mathfrak{d}^2 = \bar{\mathfrak{d}}^2 = \bar{\mathfrak{d}}\mathfrak{d} + \mathfrak{d}\bar{\mathfrak{d}} = 0, \tag{9.63}$$

as a consequence of the well–known identities $d^2 = 0$ and $\delta^2 = 0$.

Proposition 9.30 (Hodge identities) *On a Kähler manifold:*

$$[\partial, L] = [\bar{\partial}, L] = [\Lambda, \partial] = [\Lambda, \bar{\partial}] = 0, \tag{9.64}$$

$$[\Lambda, \bar{\partial}] = -i\partial, \quad [\Lambda, \partial] = i\bar{\partial}, \tag{9.65}$$
$$[\partial, L] = i\bar{\partial}, \quad [\bar{\partial}, L] = -i\partial,$$

acting on $\Lambda^{p,q}$, $[L, \Lambda] = p + q - n$. \hfill (9.66)

Proof The equalities to be shown are covariant equations involving only the metric and its first derivatives; by the *holomorphic equivalence principle* it is enough to show them in flat space. See Exercise 9.4.1. □

Corollary 9.31 *On a Kähler manifold:*

$$\partial\bar{\partial} + \bar{\partial}\partial = 0 \quad \text{and} \quad \bar{\partial}\bar{\partial} + \bar{\partial}\bar{\partial} = 0, \tag{9.67}$$

$$\Delta_{\bar{\partial}} := \bar{\partial}\partial + \partial\bar{\partial} = \partial\bar{\partial} + \bar{\partial}\partial =: \Delta_{\partial}, \tag{9.68}$$

$$\Delta_d := d\delta + \delta d = 2\Delta_{\bar{\partial}} = 2\Delta_{\partial}, \tag{9.69}$$

$$[\Delta, \Lambda] = 0 \quad \text{and} \quad [\Delta, L] = 0. \tag{9.70}$$

Proof Simple computation using the Hodge identities: Exercise 9.4.1 □

Hodge identities for holomorphic vector bundles. The above identities may be generalized by twisting $\Lambda^{p,q}(\mathcal{M})$ by a holomorphic vector bundle $\mathcal{V}$, which we assume to be equipped with a Hermitian fiber metric h. One considers differential forms with coefficients in $\mathcal{V}$, that is, elements of $\Lambda^{p,q}(\mathcal{M}) \otimes \mathcal{V}$. The connection is given by the exterior differentials $\bar{\partial}$ and $D = \partial + h^{-1}\partial h$ (see Section 9.1.3). Taking the adjoints, $D^\dagger = \bar{\partial} \equiv i[\Lambda, \bar{\partial}]$, whereas $\bar{\partial}^\dagger = h^{-1}\partial h =: \partial_h$.

Proposition 9.32 *One has*

$$[D, L] = [\bar{\partial}, L] = 0, \quad [\bar{\partial}, \Lambda] = [\partial_h, \Lambda] = 0, \tag{9.71}$$

$$[L, \bar{\partial}] = i\partial, \quad [L, \partial] = -iD, \tag{9.72}$$

$$[\Lambda, D] = i\partial_h, \quad [\Lambda, \bar{\partial}] = -i\partial. \tag{9.73}$$

However, Eq. (9.63) is replaced by

$$\bar{\partial}^2 = D^2 = 0, \quad \bar{\partial}D + D\bar{\partial} = F, \tag{9.74}$$

where F denotes the operation of exterior multiplication by the curvature $(1, 1)$ form (see Section 9.1.3). Therefore,

$$\partial_h^2 = \bar{\partial}^2 = 0, \quad \partial_h\bar{\partial} + \bar{\partial}\partial_h = -*^{-1}F*. \tag{9.75}$$

Again, we have a Laplacian acting on sections of $\Lambda^{p,q} \otimes \mathcal{V}$, namely

$$\Delta_{\bar{\partial}} = \bar{\partial}\,\eth_h + \eth_h\,\bar{\partial}, \tag{9.76}$$

to which the Hodge theorem applies, as does the Hodge orthogonal decomposition, Eq. (9.51). Let $\mathcal{H}^{p,q}(\mathcal{M}, \mathcal{V})$ be the space of the *harmonic* sections of $\Lambda^{p,q} \otimes \mathcal{V}$; that is, the space of sections η such that $\Delta_{\bar{\partial}}\eta = 0$. One has

$$\mathcal{H}^{p,q}(\mathcal{M}, \mathcal{V}) \simeq H^{p,q}_{\bar{\partial}}(\mathcal{M}, \mathcal{V}) \simeq H^q(\mathcal{M}, \Omega^p \otimes \mathcal{V}). \tag{9.77}$$

The Serre duality generalizes to this setting (Exercise 9.4.1)

Theorem 9.33 (Serre–Kodaira duality)

$$H^q(\mathcal{M}, \Omega^p \otimes \mathcal{V}) \simeq H^{n-q}(\mathcal{M}, \Omega^{n-p} \otimes \mathcal{V}^\vee). \tag{9.78}$$

9.4.1 Hodge decomposition

The Hodge identities have nice implications for the cohomology of a compact Kähler space. Corollary 9.31 shows that, on a compact Kähler manifold $\mathcal{M}$, a (p, q) form ψ which is harmonic in the $\bar{\partial}$ sense, i.e., $\bar{\partial}\psi = \eth\psi = 0$, is also harmonic in the ∂ sense: that is, $\partial\psi = \bar{\eth}\psi = 0$. Thus,

$$\mathcal{H}^{p,q}_{\bar{\partial}}(\mathcal{M}) = \mathcal{H}^{p,q}_{\partial}(\mathcal{M}) \equiv (\mathcal{H}^{q,p}_{\bar{\partial}}(\mathcal{M}))^*. \tag{9.79}$$

Hence the Hodge numbers[3] $h^{p,q} := \dim H^{p,q}(\mathcal{M})$ have the symmetry

$$h^{p,q} = h^{q,p} = h^{n-p,n-q} = h^{n-q,n-p}. \tag{9.80}$$

On a non–Kähler manifold $h^{p,q} \neq h^{q,p}$, in general (Exercise 9.4.1). To emphasize their symmetry, for a Kähler manifold, the integers $h^{p,q}$ are usually written in a diamond–shaped table called *the Hodge diamond*.

Corollary 9.31 also implies that the forms ψ which are harmonic in the usual smooth sense, namely the zero modes of Δ_d, automatically satisfy $\Delta_{\bar{\partial}}\psi = \Delta_\partial \psi = 0$, and hence are harmonic in the $\bar{\partial}$ and ∂ senses. Then the harmonic forms of degree k decompose into harmonic forms of definite type (p, q). Hence we have:

Theorem 9.34 (Hodge decompostion) *One has*

$$H^k(\mathcal{M}, \mathbb{C}) \simeq \bigoplus_{p+q=k} H^{p,q}(\mathcal{M}) \simeq \bigoplus_{p+q=k} H^q(\mathcal{M}, \Omega^p). \tag{9.81}$$

---

[3] We omit labels $\bar{\partial}$, ∂ from now on.

This result is quite remarkable: the vector space $H^k(\mathcal{M}, \mathbb{C})$, a topological invariant isomorphic to the holomology group $H_{2n-k}(\mathcal{M}, \mathbb{C})$, has an orthogonal decomposition into subspaces $H^{p,q}(\mathcal{M})$ with $p + q = k$ which depend on the complex structure of the manifold $\mathcal{M}$. The Hodge decomposition is at the root of the *period map,* which leads to a deep geometrical understanding of the target spaces of SUGRA; see Chapters 11 and 12.

From Theorem 9.34 we get strong results for the Betti numbers, B_k, of a compact Kähler manifold:

Proposition 9.35 *In a compact Kähler manifold: (i) the even Betti numbers B_{2k} are positive; (ii) the odd Betti numbers B_{2k+1} are even integers.*

Indeed, (1) $\omega^k = L^k \cdot 1$ is a non–vanishing $2k$ form which is harmonic since the constant 1 is, and $[\Delta, L^k] = 0$. (2) $B_{2k+1} = \sum_{p+q=k} h^{p,q} = 2 \sum_{p=0}^{k} h^{p,2k+1-p}$, by symmetry $h^{p,q} \leftrightarrow h^{q,p}$.

This result can be used to verify that certain compact complex manifolds are not Kähler and hence do not admit any metric of holonomy $U(n)$.

Proposition 9.36 *On a compact Kähler manifold, the holomorphic $(p, 0)$ forms are harmonic (and hence cannot be $\bar{\partial}$–exact).*

Proof Let φ be a holomorphic $(p, 0)$ form ($p = 0, 1, \ldots, n$). By definition, $\bar{\partial}\varphi = 0$. It remains to show that $\partial\varphi = 0$. But $\partial: \Lambda^{p,q} \to \Lambda^{p,q-1}$ and so ∂, acting on $(p, 0)$–forms, is the zero operator. □

Lefshetz's SU(2)

The Hodge diamond has subtler symmetries. Let us define a new operator acting on differential form, written H. Acting on a form ψ of type (p, q), H is simply multiplication by $(p + q - n)$:

$$H\psi = (p + q - n)\psi \qquad \psi \in \Lambda^{p,q}. \tag{9.82}$$

Since $L: \Lambda^{p,q} \to \Lambda^{p+1,q+1}$,

$$[H, L] = 2L \qquad [H, \Lambda] = -2\Lambda \qquad [L, \Lambda] = H, \tag{9.83}$$

where the second equation is the adjoint of the first one, and the third one is Eq. (9.66). Equations (9.83) are the commutation relations of the (complexified) Lie algebra $\mathfrak{su}(2)$. L, H, and Λ commute with the Laplacian Δ, and hence map harmonic forms to harmonic forms. Hence:

9.4 Hodge theory in Kähler spaces

Proposition 9.37 *M a compact Kähler manifold. The vector space*

$$\mathcal{H} \equiv \bigoplus_{p,q} \mathcal{H}^{p,q}(M) \simeq \bigoplus_{p,q} H^{p,q}(M) \tag{9.84}$$

is a finite–dimensional representation of $SU(2)$.

As a consequence, the space $\mathcal{H}$ of all harmonic forms may be decomposed into *irreducible* representations of $SU(2)$. An irreducible representation is constructed by acting with the raising operator L on a lowest–weight state, i.e., on a form annihilated by the lowering operator Λ. In the present context, a lowest–weight state is called a *primitive cohomology class*. A class is primitive if the corresponding harmonic form, η, satisfies $\Lambda \eta = 0$. We write $P^{p,q}(M)$ for the vector space of *primitive* type (p,q) cohomology classes.

Corollary 9.38 (Lefshetz decomposition)   *M compact Kähler,*

$$H^{p,q}(M) = \bigoplus_{k=0}^{\min(p,q)} L^k \, P^{p-k,q-k}(M). \tag{9.85}$$

Since $L^{n+1} = 0$, the maximal "spin" of a representation appearing in the Lefshetz decomposition is $\frac{n}{2}$. In fact there is only one representation with this spin corresponding to the $(n+1)$ harmonic forms ω^k, $k = 0, 1, \ldots, n$.

Following Witten [141, 315], we can understand the origin of the Lefshetz $SU(2)$ in physical terms. A 4D SUSY σ–model should have a Kählerian target space. The model has also an $SO(3,1)$ symmetry under the Lorentz group. We reduce the theory to $D = 1$ by simply dropping the dependence of the fields on the "extra" coordinates x^1, x^2, and x^3. The "internal" part of the Lorentz group $SO(3) \simeq SU(2)$ is then a global symmetry of the $D = 1$ model. One checks that it is equivalent to the Lefshetz one.

The same argument, applied to the reduction of a 6D SUSY σ–model, predicts that the cohomology of a compact hyperKähler manifold carries a representation of the Lie group $SO(6-1) \simeq Sp(4)$.

Exercise 9.4.1   Construct a simple example of a compact complex manifold with $h^{p,q} \neq h^{q,p}$.

Exercise 9.4.2   Prove the Hodge identities, Proposition 9.30.

Exercise 9.4.3   Prove the identities (9.67)–(9.70).

Exercise 9.4.4 Check Eqs. (9.71)–(9.75).

Exercise 9.4.5 Prove Theorem 9.33.

Exercise 9.4.6 Show geometrically that the cohomology of a compact hyperKähler $4n$–fold is an $Sp(4)$–module.

Exercise 9.4.7 Let M be a *strict* Calabi–Yau m–fold (i.e., Hol$(M) \equiv SU(m)$). Show that the Hodge numbers satisfy $h^{p,0} = \delta_{p\,0} + \delta_{p\,m}$.

9.5 Hodge manifolds

Let $\mathcal{M}$ be a Kähler manifold. The Kähler form ω is a *closed* $(1, 1)$ form which is real and *positive*, $\omega > 0$. By this we mean that $\omega \equiv ig_{\alpha\bar{\beta}}(z)\,dz^\alpha \wedge d\bar{z}^{\bar{\beta}}$ with $g_{\alpha\bar{\beta}}(z)$ a *positive–definite Hermitian form* at all points z. ω, being closed, defines a cohomology class $[\omega] \in H^2(\mathcal{M}, \mathbb{R})$ which is possibly trivial, but certainly non–trivial for $\mathcal{M}$ compact, since ω^n/n is the volume form. Inside the vector space $H^2(\mathcal{M}, \mathbb{R})$ sits the lattice $H^2(\mathcal{M}, \mathbb{Z})/$(torsion). Physically, we may see $H^2(\mathcal{M}, \mathbb{R})$ as the space of the $\mathfrak{u}(1)$ field–strength 2–forms F solving the vacuum (Euclidean) Maxwell equations $dF = \delta F = 0$. Then $H^2(\mathcal{M}, \mathbb{Z})/$(torsion) corresponds to the subset of solutions whose fluxes obey Dirac's quantization, that is, such that F is the curvature of a connection on a $U(1)$ vector bundle (whose sections are the charged particle wave–functions). The Kähler manifolds whose 2–form ω satisfies Dirac's quantization have special geometric properties.

Definition 9.39 A *Hodge manifold* is a Kähler manifold whose Kähler form represents a class $[\omega] \in H^2(\mathcal{M}, \mathbb{Z})$.

In a Hodge manifold, $\int_S \omega^k \in \mathbb{Z}$ for all $2k$–cycles S. If the cycle S may be represented by a complex submanifold $\int_S \omega^k$ is a *strictly positive* integer, since it is $k!$ times the volume of S with respect to the induced metric.

We have already mentioned that any class

$$[\phi] \in H^2(\mathcal{M}, \mathbb{Z}) \cap H^{1,1}(\mathcal{M}) \tag{9.86}$$

is the Chern class of some line bundle $\mathcal{L} \to \mathcal{M}$ [171]. Since the Kähler class is automatically in $H^{1,1}(\mathcal{M})$, we can equivalently say:

9.5 Hodge manifolds

Definition 9.40 A *Hodge manifold* is Kähler manifold such that

$$[\omega] = \left[\frac{i}{2\pi}\Theta\right] \equiv c_1(\mathcal{L}) \tag{9.87}$$

with Θ the curvature of a line bundle $\mathcal{L} \to \mathcal{M}$.

The equality at the level of cohomology classes can be uplifted *to an equality of forms*. Indeed, by the $\partial\bar{\partial}$–Poincaré lemma, Eq. (9.87) implies the existence of a global smooth real function Φ such that

$$\omega = \frac{i}{2\pi}\Theta + i\partial\bar{\partial}\Phi. \tag{9.88}$$

Changing the fiber metric on $\mathcal{L}$, $H_\alpha \to H'_\alpha \equiv e^{-\Phi}H_\alpha$ we get a new curvature form $\Theta' = -\partial\bar{\partial}\log H'_\alpha$ such that

$$\omega = \frac{i}{2\pi}\Theta'. \tag{9.89}$$

That is, *given a real closed* $(1,1)$ *form* ω *with* $[\omega] = c_1(\mathcal{L})$ *there exists a metric connection on* $\mathcal{L}$ *with curvature* $\frac{2\pi}{i}\omega$.

However, not all real closed integral $(1,1)$–forms may be Kähler forms. To be a Kähler form ω should be pointwise positive–definite. A generic line bundle $\mathcal{L}$ does not admit any positive–definite curvature form (indeed, there may be topological obstructions: say $\int_S c_1(\mathcal{L}) \not> 0$ for some analytic cycle S). The line bundles whose curvature is such that $\frac{i}{2\pi}\Theta$ is positive are special: we shall call them *positive line bundles*.

Example. The projective space $\mathbb{C}P^k$ admits a Hodge metric. Indeed, $\mathbb{C}P^k \equiv SU(k+1)/U(k)$ is a symmetric space whose holonomy is $U(k)$ and hence is Kähler. $H^2(\mathbb{C}P^k, \mathbb{Z}) = \mathbb{Z} \cdot \omega_0$, and $H^2(\mathbb{C}P^k, \mathbb{R}) = \mathbb{R} \cdot \omega_0$. Then the symmetric Kähler metric has $\omega = \lambda\omega_0$, for some $\lambda \neq 0$ (since $\mathbb{C}P^k$ is compact). Then rescale the metric by $|\lambda|^{-1}$ to get a Hodge metric. The dual non–compact symmetric space $\mathbb{H}^n \equiv SU(n,1)/U(n)$ is also Kähler and contractible; hence $[\omega] = 0$ is trivially integral, and then $\mathbb{H}^n$ is also (trivially) Hodge.

9.5.1 Kodaira embedding theorem

As we mentioned in Section 2.7, one reason why Hodge manifolds are special is Kodaira's intrinsic characterization of *compact* complex manifolds which are (analytically) isomorphic to *algebraic manifolds*; that is, manifolds which are the zero set of a finite collection of homogeneous polynomials in the homogeneous coordinates of some $\mathbb{C}P^k$. The first result is [171]:

Theorem 9.41 (Chow) *All complex submanifolds of $\mathbb{C}P^k$ are algebraic.*

Thus it is enough to give an intrinsic characterization of the complex submanifolds of $\mathbb{C}P^k$. It is easy to see that they are necessarily compact Hodge manifolds. Indeed, being closed subsets of compact spaces, they are compact. Let $\mathcal{M} \xrightarrow{\iota} \mathbb{C}P^k$ be such a submanifold (ι a holomorphic embedding). $\mathbb{C}P^k$ has a Hodge Kähler form ω. Then $\iota^*\omega$ is a Hodge Kähler form for $\mathcal{M}$. The Kodaira theorem states that this necessary condition is also sufficient:

Theorem 9.42 (Kodaira embedding) *A compact complex manifold is algebraic iff it is Hodge.*

We do not prove the theorem but give the general flavor of the argument, which will be useful to us. If $\mathcal{M}$ is Hodge, ω represents the Chern class of a positive line bundle $\mathcal{L}$. Let $m \gg 1$ be an integer and consider the (finite–dimensional) vector space of all the holomorphic sections of $\mathcal{L}^m$; let $\psi_s(z)$ ($s = 0, 1, \ldots, k$) be a basis in some local trivialization. Consider the vector

$$(\psi_0(z), \psi_1(z), \cdots, \psi_k(z)). \tag{9.90}$$

While it is not well defined as an element of $\mathbb{C}^{k+1}$, since it is trivialization–independent, it is well defined as an element of $\mathbb{C}P^k$. Hence we get a holomorphic map $\mathcal{M} \to \mathbb{C}P^k$. This map is an embedding if it is regular and injective. In fact, it has some mild singularity which may be repaired by blow–ups. The resulting regular map is the Kodaira embedding.

Exercise 9.5.1 Show that a compact stricly Calabi–Yau manifold Y with $\dim_\mathbb{C} Y \geq 3$ is algebraic.

Exercise 9.5.2 Give two examples of Calabi–Yau manifolds which are <u>not</u> algebraic.

9.6 Symmetric and homogeneous Kähler manifolds

We conclude this chapter by discussing the manifolds which are both symmetric (or just homogeneous) and Kähler; they are relevant for physical applications. Recall from Chapter 5 that we have *four* types of symmetric spaces. We start by showing that a Kählerian symmetric space of Type II or IV is necessarily *flat*. Indeed, a non–flat irreducible Riemannian manifold M is Kähler iff $\{e\} \neq \mathrm{Hol}_0(M) \subseteq U(n)$; by this we actually mean that the *holonomy representation* is induced by the fundamental of $U(n)$, that is, $TM_\mathbb{R} \otimes \mathbb{C} = \boldsymbol{n} \oplus \boldsymbol{\bar{n}}$. A non–flat irreducible symmetric space of type II or IV has holonomy group a simple compact

Table 9.1 *Compact irreducible Hermitian symmetric spaces G/H.*

G	H	$dim_{\mathbb{C}}$	Rank	Name
$SU(n+m)$	$U(1) \times SU(n) \times SU(m)$	nm	$\min\{n,m\}$	Grassmanian
$SO(2n)$	$U(n)$	$n(n-1)/2$	$[n/2]$	
$Sp(2n)$	$U(n)$	$n(n+1)/2$	n	Y_n
$SO(n+2)$	$SO(2) \times SO(n)$	n	2	Quadric
E_6	$SO(2) \times SO(10)$	16	2	
E_7	$SO(2) \times E_6$	27	3	

group H acting in the adjoint representation. Since the adjoint representation of a simple Lie group is never reducible (over $\mathbb{C}$), it cannot be induced by $\boldsymbol{n} \oplus \boldsymbol{\bar{n}}$.

9.6.1 Type I Hermitian symmetric manifolds

The (irreducible) compact Hermitian symmetric manifolds are precisely the Type **I** Cartan symmetric spaces G/H with $H = U(1) \times H'$. By comparing with the tables in refs. [51] and [184] we get Table 9.1 (see also Exercise 9.6.2). All the compact symmetric Hermitian spaces are Hodge, hence projective. Indeed, they are *flag manifolds*,[4] $G/H \simeq G_{\mathbb{C}}/P$, with $P = \exp(\mathfrak{h}_{\mathbb{C}} + \mathfrak{g}^-)$ where $\mathfrak{g}^-$ is the negative root subspace of $\mathfrak{g}_{\mathbb{C}}$. They are also *rational* varieties [4].

9.6.2 Type III. Explicit invariant metrics

From Section 4.9.1 it follows that the symmetric Hermitian spaces most relevant for the physical applications are the non–compact Type III ones dual to those listed in Table 9.1. They are all isomorphic to bounded domains in $\mathbb{C}^m$ [184]. We are interested in writing explicit invariant metrics.

Mathematical preliminaries. We begin with some definitions [161, 292].

Definition 9.43 A *convex cone* $V \subset \mathbb{R}^n$ is an *open* subset of $\mathbb{R}^n$ such that

$$x \in V \quad \Rightarrow \quad \lambda x \in V \qquad \text{for all } \lambda \in \mathbb{R}_+ \tag{9.91}$$

$$x, y \in V \quad \Rightarrow \quad \alpha x + (1-\alpha) y \in V \quad \text{for all } 0 \leq \alpha \leq 1. \tag{9.92}$$

For our purposes, it is convenient to assume that V is *strict*, i.e., does not contain any full straight line (not necessarily passing through the origin).

---
[4] Flag manifolds are discussed in Chapter 11.

Definition 9.44 The *automorphism group* of the convex cone V, $A(V)$, is

$$A(V) = \{g \in GL(n, \mathbb{R}) \mid gV = V\}. \tag{9.93}$$

V is said to be *homogeneous* if $A(V)$ is *transitive*.

Definition 9.45 Let $V \subset \mathbb{R}^n$ be a convex cone. The *dual convex cone*, $V^\vee$, is the cone

$$V^\vee := \{x \in (\mathbb{R}^n)^\vee \mid \langle x, y \rangle > 0 \ \forall y \in \overline{V},\ y \neq 0\}. \tag{9.94}$$

One has $V^{\vee\vee} = V$. The contragradient map $g \mapsto (g^{-1})^t$ gives an isomorphism $A(V) \simeq A(V^\vee)$.

Definition 9.46 A convex cone V is called *self–dual* if $V = V^\vee$. It is called *symmetric* if it is both homogeneous and self–dual.

Definition 9.47 The *Siegel domain of first kind* associated with the convex cone V, $\mathfrak{S}(V) \in \mathbb{C}^n$, is the domain

$$\mathfrak{S}(V) := \{Z \in \mathbb{C}^n,\ \operatorname{Im} Z \in V\}. \tag{9.95}$$

$\mathfrak{S}(V)$ is said to be homogeneous if V is; in this case the group of affine automorphisms of $\mathfrak{S}(V)$ is manifestly transitive.

Example: Siegel's space $Sp(2n, \mathbb{R})/U(n.)$ Let $V = S_n$ be the space of real symmetric matrices with positive eigenvalues. Then $\mathfrak{S}(S_n)$ is the symmetric space $Sp(2n, \mathbb{R})/U(n)$, see Section 1.5.

Definition 9.48 Let $V \subset \mathbb{R}^n$ be a convex cone. A *V–Hermitian function* F is a map

$$F: \mathbb{C}^m \times \mathbb{C}^m \to \mathbb{C}^n \tag{9.96}$$

such that:

(1) $F(\lambda u + \mu v, z) = \lambda F(u, z) + \mu F(v, z)$. for all $\lambda, \mu \in \mathbb{C}$;
(2) $F(u, v) = \overline{F(v, u)}$;
(3) $F(u, u) \in \overline{V}$;
(4) $F(u, u) = 0 \Leftrightarrow u = 0$.

Definition 9.49 Let V be a convex cone and F a V–Hermitian function. The domain

$$\mathfrak{S}(V, F) = \{(Z, W) \in \mathbb{C}^{n+m} \mid \operatorname{Im} Z - F(W, W) \in V\} \tag{9.97}$$

is called the *Siegel domain of the second kind* associated with V and F.

9.6 Symmetric and homogeneous Kähler manifolds

Example: $H\mathbb{C}^k \equiv SU(k,1)/U(k)$, i.e., the non–compact dual of $P\mathbb{C}^k$. $V = \mathbb{R}_+$ and $F(\cdot, \cdot) \equiv \langle \cdot, \cdot \rangle_{k-1}$, the standard Hermitian product in $\mathbb{C}^{k-1}$. Then

$$H\mathbb{C}^k = \mathfrak{S}(\mathbb{R}_+, \langle \cdot, \cdot \rangle_{k-1}). \tag{9.98}$$

Theorem 9.50 ([161]) *Every bounded homogeneous domain $\mathcal{D} \subset \mathbb{C}^n$ is equivalent to a homogeneous Siegel domain of the first or second kind.*

Since all non–compact simply connected Hermitian symmetric spaces are biholomorphically equivalent to bounded domains, Theorem 9.50 produces all of them. We do not prove the theorem; we just specify the Vs and Fs corresponding to the Type III Hermitian manifolds. The list of symmetric Siegel first kind domains is presented in Table 9.2. It corresponds to the known list of *symmetric* convex cones. It is easy to check the entries using either the Bergman kernel [71] or the approach discussed below. It remains to describe as Siegel domains of the second kind the symmetric Hermitian spaces:

$$\mathfrak{I}_{n,n+q} := \frac{SU(n, n+q)}{SU(n) \times U(n+q)} \qquad q \geq 1, \tag{9.99}$$

$$\mathfrak{D}_{2n+1} := \frac{SO^*(4n+2)}{U(2n+1)} \tag{9.100}$$

$$\mathfrak{E}_6 := \frac{E_{6(-14)}}{SO(10) \times SO(2)} \qquad \begin{pmatrix} \text{the } \mathcal{M} \text{ for } \mathcal{N} = 10 \\ \text{SUGRA in 3D} \end{pmatrix}. \tag{9.101}$$

As coordinates in $\mathfrak{I}_{n,n+q}$ we choose (Z, W), where Z is a $n \times n$ matrix and W is a $n \times q$ one. As V we take the cone PDH_n of positive–definite Hermitian $n \times n$ matrices, while[5]

$$F: \mathbb{C}(n, q) \times \mathbb{C}(n, q) \to \mathbb{C}(n), \tag{9.102}$$

$$F(W_1, W_2) = W_1 W_2^\dagger. \tag{9.103}$$

Then

$$\mathfrak{S}(PDH_n, F) = \frac{SU(n, n+q)}{SU(n) \times U(n+q)}. \tag{9.104}$$

The cone for $\mathfrak{D}_{2n+1}$ is the same for $\mathfrak{D}_{2n} \equiv SO^*(4n)/U(2n)$, while for $\mathfrak{E}_6$ it is the same for $SO(2, 8)/SO(2) \times SO(8)$. See ref. [323] for further details.

To extend the construction from the *symmetric* to the *homogenous* domains, one has to classify all homogenous (strict) convex cones. This is done very explicitly with the help of the T–algebras; see refs. [84, 292].

---

[5] For $\mathbb{A}$ a ring we denote $\mathbb{A}(n, q)$ the $n \times q$ matrices with entries in $\mathbb{A}$. As always $\mathbb{A}(n) \equiv \mathbb{A}(n, n)$.

Table 9.2 *Non–compact symmetric Hermitian spaces that are Siegel domains of the first kind.*

Space	Cone
$Sp(2n,\mathbb{R})/U(n)$	Positive–definite *symmetric* $n \times n$ matrices
$SU(n,n)/[U(n) \times SU(n)]$	Positive–definite *Hermitian* $n \times n$ matrices
$SO^*(4n)/U(2n)$	Positive–definite *quaternionic–Hermitian* $n \times n$ matrices
$E_{7(-25)}/[E_6 \times SO(2)]$	"Positive–definite" *octonionic–Hermitian* 3×3 matrices[a]
$SO(2,n)/[SO(2) \times SO(n)]$	The spherical cone $x_1 > \sqrt{x_2^2 + x_3^2 + \cdots + x_n^2}$

[a] The octonions are not associative but still *alternating*. Hence Hermitian matrices with octonionic entries are well defined for sizes up to 3.

Invariant Kähler metrics from 0D QFT

We now construct an invariant Kähler metric on any homogeneous bounded domain. The idea is that the Kähler potential may be seen as the quantum effective action of a 0D QFT. We interpret the coordinates y^i ($i = 1, 2, \ldots, n$) of $(\mathbb{R}^n)^\vee$ as "quantum fields," while the dual coordinates x_i in $\mathbb{R}^n$ are seen as "sources." The effective action $\Gamma_V[x]$ is

$$e^{-\Gamma_V[x]} = \int_{V^\vee} e^{-\sum_i x_i y^i} d^n y, \qquad (9.105)$$

which is convergent by the definition of $V^\vee$. The effective action defines a map $V \to V^\vee$, known as the Legendre transform, given by

$$\langle y^i \rangle_x = \frac{\delta \Gamma_V[x]}{\delta x_i} \equiv \frac{\int_{V^\vee} e^{-\sum_i x_i y^i} y_i\, d^n y}{\int_{V^\vee} e^{-\sum_i x_i y^i} d^n y} \in V^\vee, \qquad (9.106)$$

where $\langle \cdot \rangle_x$ is the VEV, in the QFT sense, in the presence of the source $x \in V$. Let $g \in A(V) \subset GL(n)$. One has

$$\Gamma_V[gx] = \Gamma_V[x] + \log \det[g] \qquad (9.107)$$

and

$$\langle y \rangle_{gx} = (g^{-1})^t \langle y \rangle_x, \qquad (9.108)$$

9.6 Symmetric and homogeneous Kähler manifolds

so the Legendre map $x \mapsto \langle y \rangle$ commutes with the action of $A(V)$. Positivity of the Hilbert space implies that

$$\langle y^i y^j \rangle_{\text{connected}} \equiv -\partial_{x_i} \partial_{x_j} \Gamma_V[x] \quad \text{is positive–definite}, \tag{9.109}$$

so it is a natural metric tensor on V (the 0D Zamolodchikov metric). This metric is $A(V)$–invariant by Eq. (9.107), and, if V is homogeneous, it must be the unique one (up to normalization) having this invariance. In a physically sensible theory, the Zamolodchikov metric is geodesically complete. One easily checks that V with the above metric is indeed complete.

Theorem 9.51 *Let $\mathfrak{S}(V, F) \subset \mathbb{C}^{n+m}$ be a Siegel domain of the second kind (first kind being the case $m = 0$) defined by the convex cone $V \subset \mathbb{R}^n$ and V–Hermitian function F. The Kähler metric*

$$g_{i\bar{j}} \equiv -\partial_i \partial_{\bar{j}} \, \Gamma_V\big[\operatorname{Im} Z - F(W, W)\big] \tag{9.110}$$

is invariant under the full automorphism group of $\mathfrak{S}(V, F)$. In particular, if (V, F) corresponds to a symmetric Hermitian space of Type III, Eq. (9.110) gives the unique symmetric Kähler metric (canonically normalized). Finally, the manifold $\mathfrak{S}(V, F)$ equipped with this Kähler metric, is homogeneous if and only if V is homogeneous.

In simple terms, the Kähler potential is minus the 0D effective action.

Proof We have to show that the metric is positive–definite. Let us observe that, for any V and F, the holomorphic transformation

$$\begin{cases} W & \to W + C \qquad C \in \mathbb{C}^m \\ Z & \to Z + i F(W, C) + \frac{i}{2} F(C, C) \end{cases} \tag{9.111}$$

leaves invariant the Kähler potential and hence, *a fortiori*, is a holomorphic isometry of the above Kähler metric. Using this isometry, we can map each point of $\mathfrak{S}(V, F)$ to a point of the submanifold $W = 0$. Thus it is enough to show positivity on this submanifold. In coordinates, $F(W, W)$ takes the form $f_{i\alpha\bar{\beta}} W^\alpha \overline{W}^{\bar{\beta}}$. By definition, the coefficients $f_{i\alpha\bar{\beta}}$ have the following property:

$$\text{for all } y \in V^\vee \text{ the Hermitian matrix } y^i f_{i\alpha\bar{\beta}} \text{ is positive–definite}. \tag{9.112}$$

At $W = 0$ the Kähler metric reads

$$\langle y^i y^j \rangle^{\text{connected}} dZ_i \, d\overline{Z}_j + \langle y^i \rangle f_{i\alpha\bar{\beta}} \, dW^\alpha \, d\overline{W}^{\bar{\beta}}. \tag{9.113}$$

Since $\langle y \rangle \in V^\vee$, this metric is positive–definite on the submanifold $W = 0$ and hence everywhere. The proof also shows that $\mathfrak{S}(V, F)$ is homogeneous if the associated domain of the first kind, $\mathfrak{S}(V)$ is. The transitive group of symmetries is given by $A(V)$, arbitrary translations of the real part of Z, and the symmetries (9.111). □

Exercise 9.6.1 Show (or check) that the compact symmetric Hermitian spaces are in 1–to–1 correspondence with the pairs (Γ, α_0), where Γ is a Dynkin graph and $\alpha_0 \in \Gamma$ is a node with Coxeter label 1.

10
$\mathcal{N} = 1$ supergravity in 4D

In this chapter we describe $\mathcal{N} = 1$ supergravity in four dimensions [100, 102, 125, 147]. It is the most interesting SUGRA for "phenomenology."

10.1 $\mathcal{N} = 2$ supergravity in 3D

Dimensionally reducing $D = 4$ $\mathcal{N} = 1$ supergravity to $d = 3$ we get $\mathcal{N} = 2$ SUGRA. Thus, as a warm–up, we review this case first.

The scalar manifold $\mathcal{M}$. According to Chapter 2, on the scalar manifold $\mathcal{M}$ we have a (smooth) real 2–form Σ such that:

(a) $\Sigma_{ik} g^{kl} \Sigma_{lj} = -g_{ij}$.
(b) Σ, being in the trivial representation of *Spin*(2), is parallel.
(c) The curvature of the *Spin*(2) $\simeq U(1)$ gravitino bundle is given by $-\frac{i}{2}\Sigma$.

(a) says that $I^i{}_j := g^{ik}\Sigma_{kj}$ is an *almost–complex structure*. In view of Corollary 9.7, (b) implies that I is integrable, and hence $\mathcal{M}$ is a *complex manifold*. Then (a) says that the scalar metric g_{ij} is Hermitian, with Kähler form Σ. Since Σ is closed, $\mathcal{M}$ is Kähler. Then (c) says that $-i\Sigma$ is (up to normalization) the curvature of a holomorphic[1] line bundle $\mathcal{L}$. Thus, as already anticipated in Section 2.7:

General Lesson 10.1 *In 3D* $\mathcal{N} = 2$ *SUGRA (and 4D* $\mathcal{N} = 1$ *SUGRA) the scalar manifold* $\mathcal{M}$ *is a* Hodge *manifold.*

Corollary 10.2 *If* $\mathcal{M}$ *is compact, it is a projective variety.*

---

[1] Why *holomorphic?* The curvature $-\frac{i}{2}\Sigma$ is a (1,1) form. So the (0,2) part of the curvature vanishes; let D'' be the (0,1)–part of the connection. One has $(D'')^2 = 0$ and locally we can write $D'' = e^{-\phi}\bar{\partial}e^{\phi}$ for some complex function ϕ. By the change of trivialization $1 \mapsto e^{-\phi}$, we get $D'' = \bar{\partial}$, which is the definition of a holomorphic bundle.

To state the next corollary, we reintroduce the physical dimensions in our formulae. Thus the bosonic terms of the Lagrangian are written as

$$-\frac{1}{2\kappa^2} eR - ef_\pi^2\, g_{i\bar{j}}(\phi,\bar{\phi})\, \partial_\mu \phi^i\, \partial^\mu \overline{\phi^j} + \cdots, \tag{10.1}$$

where the fields ϕ^i are dimensionless, while the overall *dimensionfull* coupling f_π – the analogue of the pion decaying constant in current algebra – has dimension [mass]$^{(D-2)/2}$ in D dimensions.

Corollary 10.3 (Witten–Bagger [318]) *Assume $\mathcal{M}$ contains a compact complex submanifold $S \subset \mathcal{M}$ (of positive dimension). Then all (regular) Kähler–Hodge metrics correspond to non–trivial elements of $H^2(\mathcal{M},\mathbb{Z})$. Normalize the reference Kähler metric $g_{i\bar{j}}$ in Eq. (10.1) in such a way that the Kähler class $[\omega] \in H^2(\mathcal{M},\mathbb{Z})$, while $[\omega]/k \notin H^2(\mathcal{M},\mathbb{Z})$ for $k = 2, 3, \ldots$. Then Newton's constant κ is quantized:*

$$\kappa^2 = \frac{n}{f_\pi^2}, \qquad n \in \mathbb{N}. \tag{10.2}$$

From now on, the indices i,j label *holomorphic* coordinates ϕ^i on $\mathcal{M}$, while the anti–holomorphic ones $\bar{\phi}^i$ are labeled by barred indices $\bar{i},\bar{j}$. Thus

$$\Sigma_i{}^j = i\delta_i{}^j, \qquad \Sigma_{\bar{i}}{}^{\bar{j}} = -i\delta_{\bar{i}}{}^{\bar{j}}, \qquad \Sigma_{i\bar{j}} = ig_{i\bar{j}}, \qquad \Sigma_{\bar{i}j} = -ig_{\bar{i}j}. \tag{10.3}$$

Let $K(\phi^i,\bar{\phi}^{\bar{j}})$ be the (local) Kähler potential. The metric is given by $g_{i\bar{j}} = \partial_i \partial_{\bar{j}} K$. Since it is proportional to the curvature of the $\mathfrak{u}(1)$ connection Q, we have (up to a $\mathfrak{u}(1)$ gauge transformation)

$$Q_i = -\tfrac{i}{4}\partial_i K, \qquad Q_{\bar{i}} = \tfrac{i}{4}\partial_{\bar{i}} K. \tag{10.4}$$

Gaugings and superpotential interactions. We are interested in the most general $\mathcal{N} = 2$ model; then we must allow the gauging of some subgroup $\mathcal{G}$ of $\mathrm{Iso}(\mathcal{M})$ as well as a superpotential term. From the analysis in Chapter 8, we know that the only consistency condition is that $\mathcal{G}$ should be generated by (the real part of) holomorphic Killing vectors. Since $\mathrm{Spin}(2)$ is Abelian, the covariant momentum map coincides with the symplectic one μ^m. From Eq. (10.3), *locally* we have

$$\partial_i \mu^m = -ig_{i\bar{j}} K^{\bar{j}\,m} = -i\partial_i(\partial_{\bar{j}} K) K^{\bar{j}\,m} = -i\partial_i(\partial_{\bar{j}} K\, K^{\bar{j}\,m}), \tag{10.5}$$

$$\partial_{\bar{i}} \mu^m = ig_{\bar{i}j} K^{j\,m} = \partial_{\bar{i}}(i\partial_j K\, K^{j\,m}), \tag{10.6}$$

where we have used that $K^{\bar{j}\,m}$ is anti–holomorphic. Thus,

$$\mu^m - iK^{j\,m}\partial_j K = \Phi \equiv \text{holomorphic}. \tag{10.7}$$

10.1 $\mathcal{N}=2$ supergravity in 3D

We consider the extended T–tensor, which we normalize as

$$T = \tfrac{1}{2}\,\mu^m\, l_{mn}\, \mu^n, \tag{10.8}$$

$$T_i = \tfrac{1}{2}\,\mu^m\, l_{mn}\, K_i^n, \tag{10.9}$$

$$T_{ij} = K_i^m\, l_{mn}\, K_j^n. \tag{10.10}$$

Then

$$\partial_i T = \mu^m l_{mn} \partial_i \mu^n = -\mu^m l_{mn} \Sigma_i^{\,j} K_j^n = -2i\, T_i, \tag{10.11}$$

$$D_i \partial_j T = -2i\, D_i T_j = -T_{ij}, \tag{10.12}$$

where in the last equality we used the identity $D_i K_j^n \equiv g_{j\bar{k}}\, D_i K^{n\bar{k}} = 0$, true for all anti–holomorphic vectors. Further identities may be obtained from Eqs. (8.151). Using the $\mathcal{N}=2$ identity $T^{AC,CB} \propto \delta^{AB}\, T$, Eqs. (8.151) yield

$$\mathcal{D}_i(A_1^{22} - A_1^{11}) = 2i\, \mathcal{D}_i A_1^{21}, \qquad \mathcal{D}_{\bar{i}}(A_1^{22} - A_1^{11}) = -2i\, \mathcal{D}_{\bar{i}} A_1^{21}, \tag{10.13}$$

$$\partial_i(A_1^{11} + A_1^{22} + 2T) = \partial_{\bar{i}}(A_1^{11} + A_1^{22} + 2T) = 0. \tag{10.14}$$

The second line gives $\delta_{AB} A_1^{AB} = -2T$. The first says that the quantity $\tfrac{1}{2}(A_1^{22} - A_1^{11}) + i A_1^{12}$ is covariantly holomorphic:

$$\mathcal{D}_{\bar{i}}\left(\tfrac{1}{2}(A_1^{22} - A_1^{11}) + i A_1^{12}\right) = 0. \tag{10.15}$$

The expression in parentheses has charge $+2$ with respect to $\mathfrak{u}(1) \simeq \mathfrak{spin}(2)$; thus the operator $\mathcal{D}_{\bar{i}}$ appearing in Eq. (10.15) is explicitly

$$\partial_{\bar{i}} + 2i\, Q_{\bar{i}}^{12} \equiv \partial_{\bar{i}} - \tfrac{1}{2}(\partial_{\bar{i}} K) = e^{K/2} \partial_{\bar{i}}\, e^{-K/2}, \tag{10.16}$$

where K is the Kähler potential (cf. Eq. (10.4)). Thus the general solution of the constraint (10.15) is simply

$$\tfrac{1}{2}(A_1^{22} - A_1^{11}) + i A_1^{12} = e^{K/2}\, W, \qquad W \text{ holomorphic.} \tag{10.17}$$

The holomorphic object W is called the *superpotential*. From the last equation we see that the combination

$$|e^{K/2} W|^2 = e^K\, W\, \overline{W} \tag{10.18}$$

is a $Spin(2)$–gauge–invariant scalar, and hence globally defined on $\mathcal{M}$. This expression has a simple meaning: recall from Chapter 9 that e^K is (locally) the fiber metric on the line bundle $\mathcal{L}^{-1}$, where $\mathcal{L}$ is a *positive* line bundle such that $[\omega] = c_1(\mathcal{L})$. Thus:

General Lesson 10.4 *The superpotential W of an* $\mathcal{N}=2$, *3D supergravity is a holomorphic section of the (negative) line bundle* $\mathcal{L}^{-1}$.

Corollary 10.5 *If* $\mathcal{M}$ *is compact, and W is regular,* $W \equiv 0$.

Proof Consider the smooth global function $\|W\|^2 \equiv e^K |W|^2$ and let $p \in \mathcal{M}$ be a point where it attains a maximum. If $W \not\equiv 0$, the value of $\|W\|^2$ at p is positive, and the function

$$\log \|W\|^2 = K + \text{harmonic} \tag{10.19}$$

is smooth in some neighborhood U of p and has a maximum in p. In U,

$$\partial_i \partial_{\bar{j}} \log \|W\|^2 = \partial_i \partial_{\bar{j}} K = g_{i\bar{j}} \quad \text{positive–definite} \tag{10.20}$$

but, since p is a maximum, $\partial_i \partial_{\bar{j}} \log \|W\|^2|_p$ is *negative*–definite. The only way out of the contradiction is $W \equiv 0$. □

Example: $\mathcal{H}/SL(2,\mathbb{Z})$. Consider a prototypical example (cf. Section 4.9.1): the fundamental domain $\mathcal{F}$ of the modular group $SL(2,\mathbb{Z})$ in the upper half–plane

$$\mathcal{H} = \{z = x + iy \in \mathbb{C} \mid y > 0\}. \tag{10.21}$$

As Kähler metric we take the $SL(2,\mathbb{R})$–invariant one

$$g_{z\bar{z}} = \frac{k}{y^2} \tag{10.22}$$

with a normalization constant k in front which we take to be an *integer* (the Witten–Bagger quantization condition). One has $K = -k \log y$, and $e^K = y^{-k}$. The meromorphic sections of $\mathcal{L}$ correspond to the weakly modular functions of weight $-k$ (twisted by unitary characters) [68, 276]. As an example, consider the Dedekind η–function:

$$\eta(z) = q^{\frac{1}{24}} \prod_{n=1}^{\infty} (1 - q^n), \qquad q = e^{2\pi i z}. \tag{10.23}$$

The following expression is globally defined in $\mathcal{F} = \mathcal{H}/PSL(2,\mathbb{Z})$,

$$\frac{e^{K(z,\bar{z})}}{|\eta(z)|^{4k}}, \tag{10.24}$$

and $W(z) = A\, \eta(z)^{-2k}$ is an allowed superpotential.

Other couplings. All other couplings may be written in terms of T and W by the formulae of Chapter 8 and the Ward identity of Chapter 6. We quote them without further discussion [120]:

$$A^{11}_{2i} = -i A^{12}_{2i} = -\tfrac{1}{2}(\partial_i T + e^{K/2}\mathcal{D}_i W), \tag{10.25}$$

$$A^{22}_{2i} = -i A^{21}_{2i} = -\tfrac{1}{2}(\partial_i T - e^{K/2}\mathcal{D}_i W), \tag{10.26}$$

$$A^{11}_{3ij} = \tfrac{1}{4} e^{K/2} \mathcal{D}_i \mathcal{D}_j W, \tag{10.27}$$

$$A^{11}_{2i\bar{j}} = \tfrac{1}{2} T_{i\bar{j}} - \tfrac{1}{4} g_{i\bar{j}} T + \tfrac{1}{4} \partial_i \partial_{\bar{j}} T, \tag{10.28}$$

$$V = -g^2 \big(4T^2 - 4 g^{i\bar{j}} \partial_i T \partial_{\bar{j}} T + 4e^K |W|^2 - g^{i\bar{j}} e^K \mathcal{D}_i W \mathcal{D}_{\bar{j}}\overline{W}\big). \tag{10.29}$$

10.2 $\mathcal{N} = 1$ D = 4 ungauged supergravity

We start by considering the general 4D $\mathcal{N} = 1$ supergravity without vector fields; that is, supergravity coupled to *chiral* supermultiplets.

10.2.1 Coupling $\mathcal{N} = 1$ SUGRA to a σ–model

The ungauged Lagrangian with *vanishing* superpotential, $W = 0$, and hence with *no* Yukawa or potential couplings, is fully determined by the arguments in Chapter 4. Again, the target space $\mathcal{M}$ should be a Kähler–Hodge manifold, and the consequences discussed above in the 3D $\mathcal{N} = 2$ case hold also in four dimensions [318].

In the 4–component spinor notation, the $W = 0$ Lagrangian reads

$$\begin{aligned}
e^{-1}\mathcal{L} = & -\frac{1}{2}R - \frac{1}{2e}\epsilon^{\mu\nu\rho\sigma}\overline{\psi}_\mu\gamma_5\gamma_\nu D_\rho\psi_\sigma \\
& - g_{i\bar{j}}\,\partial_\mu\phi^i\,\partial^\mu\overline{\phi^{\bar{j}}} - g_{i\bar{j}}\,\overline{\chi}^{\bar{i}}\gamma^\mu D_\mu \chi^j \\
& + \Big(g_{i\bar{j}}\,\partial_\nu\overline{\phi^{\bar{j}}}\,\overline{\chi}^{\bar{i}}\tfrac{(1+\gamma_5)}{2}\gamma^\mu\gamma^\nu\psi_\mu + \text{H.c.}\Big) \\
& + \tfrac{1}{8} g_{i\bar{j}}\,\overline{\chi}^{\bar{i}}(1+\gamma_5)\gamma_\sigma \chi^j \,[\epsilon^{\mu\nu\rho\sigma}\overline{\psi}_\mu\gamma_\nu\psi_\rho - \overline{\psi}_\rho\gamma_5\gamma^\sigma\psi^\rho] \\
& - \tfrac{1}{16}[g_{i\bar{j}}\,g_{k\bar{l}} - 2R_{i\bar{j}k\bar{l}}]\,\overline{\chi}^{\bar{i}}(1+\gamma_5)\gamma_\mu\chi^j\,\overline{\chi}^{\bar{k}}(1+\gamma_5)\gamma^\mu\chi^l.
\end{aligned} \tag{10.30}$$

In the last line we see the peculiar 4–χ couplings already discussed several times; the third line is the canonical Noether coupling, while the fourth line contains the covariantizing four Fermi couplings. The covariant derivatives appearing in $\mathcal{L}$ are (here $\chi_R^i = \frac{(1+\gamma_5)}{2}\chi^i$)

$$\mathcal{D}_\mu \chi_R^i = \left(\partial_\mu + \tfrac{1}{4}\omega_{\mu\alpha\beta}\gamma^{\alpha\beta}\right)\chi_R^i + \Gamma^i_{jk}\partial_\mu\phi^j \chi_R^k - \frac{i}{2}\mathrm{Im}(\partial_j K\, \partial_\mu\phi^j)\chi_R^i, \qquad (10.31)$$

$$\mathcal{D}_\mu \psi_\nu = \left(\partial_\mu + \tfrac{1}{4}\omega_{\mu\alpha\beta}\gamma^{\alpha\beta}\right)\psi_\nu + \frac{i}{2}\mathrm{Im}(\partial_j K\, \partial_\mu\phi^j)\gamma_5 \psi_\nu, \qquad (10.32)$$

where K is the Kähler potential of $g_{i\bar{j}}$. The last term in Eq. (10.32) is the $\mathfrak{u}(1)_R$ connection, corresponding to the $\mathfrak{spin}(2)_R$ connection of the 3D case. Again, it is the consistency of this coupling that forces $\mathcal{M}$ to be a Hodge manifold. Equation (10.31) reflects the fact that the chiralini χ_R^i have opposite $\mathfrak{u}(1)_R$ charge with respect to the right–handed gravitini and gaugini λ_R^m.

10.2.2 Adding W

The Yukawa tensor A_1 (the complex gravitino mass)

$$e\tfrac{1}{2}A_1 \overline{\psi}_\mu \gamma^{\mu\nu} \psi_{\nu R} + \mathrm{H.c} \qquad (10.33)$$

has $\mathfrak{u}(1)_R$ charge -2. A_1 corresponds to the traceless part of the $\mathfrak{spin}(2)_R$ Yukawa tensor A_1^{AB} of 3D. Again, we require that it contains only $\mathfrak{u}(1)_R$ representations which are allowed for the Yukawa couplings. In Chapter 8 we did not write explicitly the 4D differential constraints on the Yukawa tensors corresponding to Eqs. (8.151) for the 3D case. However, the general arguments we used to get those constraints remain valid in 4D: (i) the scaling behavior of the couplings in the rigid limit; (ii) the $\mathfrak{u}(1)_R$ representation content; (iii) consistency with dimensional reduction. Consider the couplings

$$e A_{2i}\overline{\psi}_\mu \gamma^\mu \chi^i - e A_{3ij}\overline{\chi}^i \chi^j + \mathrm{H.c.} \qquad (10.34)$$

(i) and (ii) together imply that A_{2i} and A_{3ij} should be linearly related to the covariant derivatives of A_1 and $\overline{A}_1$. χ^i with the holomorphic index i has $\mathfrak{u}(1)_R$ charge -1, as we read in the covariant derivative (10.31). This means that $A_{2i} = \mathcal{D}_i \overline{A}_1$, while $\mathcal{D}_i A_1 = 0$, since no Yukawa coupling has these quantum numbers. This is the same constraint we got in 3D, and has the same solution.

General Lesson 10.6 *In $\mathcal{N} = 1$ 4D* SUGRA,

$$A_1 = e^{K/2}\,\overline{W} \qquad (10.35)$$

with W a holomorphic section of $\mathcal{L}^{-1}$.

A priori, in the statement we should say $\mathcal{N} = 1$ 4D *ungauged* SUGRA, since this is the class of theories we are discussing. However, our arguments imply that Eq. (10.35) will *not* be modified by gauging. This follows from dimensional reduction, since in 3D the gauging affects only the singlet part of A_1^{AB}, while the components of charge ± 2 do not talk to the gauge couplings. More geometrically, in $\mathcal{N} = 1$ SUGRA the momentum map of any (holomorphic) isometry is a $\mathfrak{u}(1)_R$ singlet, and hence may appear only in *singlet* Yukawa couplings which should contain a gaugino λ^m.

Corollary 10.7 *From the explicit form of the covariant derivatives,*

$$A_{2\,i} = e^{K/2}\, \mathcal{D}_i W \equiv e^{K/2}(\partial_i W + \partial_i K\, W), \tag{10.36}$$

$$A_{3\,ij} = e^{K/2}\, \mathcal{D}_i \mathcal{D}_j W \equiv e^{K/2}(\partial_i \partial_j W - \Gamma^l_{ij} D_l W + \partial_i K\, D_j W), \tag{10.37}$$

$$V(\phi, \overline{\phi}) = e^K \big[\, g^{i\bar{j}}\, \mathcal{D}_i W\, \mathcal{D}_{\bar{j}} \overline{W} - 3|W|^2 \big], \tag{10.38}$$

where in the last line we have used the Ward identity for the scalar potential (recall that A_1 and $A_{2\,i}$ are also the fermionic shifts).

The Lagrangian, being geometric in nature, is invariant under a change of holomorphic trivialization of the line bundle $\mathcal{L}$ (also known as *a Kähler gauge transformation* in the SUGRA literature) which acts *locally* as

$$K \mapsto K + \Phi + \overline{\Phi}, \tag{10.39}$$

$$W \mapsto e^{-\Phi}\, W, \tag{10.40}$$

with Φ any (local) holomorphic function. Then:

Corollary 10.8 *Away from the support of the divisor of $W \in \Gamma(\mathcal{L}^{-1})$, we may set $W = 1$. Then the Lagrangian depends on the single function $\mathcal{G}$ [100]:*

$$\mathcal{G}(\phi, \overline{\phi}) := -\log \|W\|^2 \equiv -\log |A_1|^2 \equiv -K - \log |W|^2. \tag{10.41}$$

This completely determines the general *ungauged* $\mathcal{N} = 1$ 4D SUGRA, see [100, 102, 146] for details. We pause a little to discuss its "phenomenology."

10.3 SuperHiggs. Flat potentials

Let us study the "physics" of an (ungauged) $\mathcal{N} = 1$ model defined by given K and W. According to Chapter 6, a symmetric configuration (fermionic fields zero, scalar fields equal to a constant ϕ_0, $g_{\mu\nu}$ maximally symmetric) preserves SUSY iff the spin-$1/2$ shifts vanish:

$$e^{K/2}\, \mathcal{D}_i W \big|_{\phi_0} = 0. \tag{10.42}$$

Such a supersymmetric configuration corresponds to AdS–space with a cosmological constant

$$\Lambda = -3\, e^K |W|^2 \big|_{\phi_0} \equiv -3\, e^{-\mathcal{G}(\phi_0)}. \tag{10.43}$$

In particular, if at ϕ_0 we have also $W(\phi_0) = 0$, the cosmological constant Λ vanishes, and we have unbroken *Poincaré* SUSY. However, in many "phenomenological" analyses, one takes the divisor (W) to be trivial (i.e., W is set globally[2] to 1) and, in this case, a Poincaré supersymmetric vacuum is ruled out.

Since the potential (10.38) is not positive–definite, we can envisage the situation in which SUSY is broken but still $\Lambda = 0$. This is the superHiggs effect [100, 102]. In this case, the mass of the gravitino is related to order parameter of SUSY breaking, $\|\mathcal{D}_i W\|$, by the universal formula

$$m_{3/2} = \frac{1}{\sqrt{3}} \|\mathcal{D}_i W(\phi_0)\|. \tag{10.44}$$

In fact we can do better. We may have $V \equiv 0$, and SUSY broken for all (constant) values of the scalars. Models with V identically vanishing are called *flat potential* (or *no–scale*) supergravities [101, 223]. Indeed, rewrite the potential (10.38) in terms of the Kähler–invariant function $\mathcal{G}$:

$$V = (g^{i\bar{j}}\, \partial_i \mathcal{G}\, \partial_{\bar{j}} \mathcal{G} - 3)\, e^{-\mathcal{G}}, \tag{10.45}$$

where $g^{i\bar{j}}$ is the inverse of the Kähler metric $-\partial_i \partial_{\bar{j}} \mathcal{G}$. We can recast this expression in the form

$$V = e^{-(\dim \mathcal{M}+3)\mathcal{G}/3} \left(\frac{3}{\dim \mathcal{M}}\right)^2 g^{i\bar{j}}\, \partial_i \partial_{\bar{j}}\, e^{\frac{\dim \mathcal{M}}{3}\mathcal{G}}, \tag{10.46}$$

so

$$V \equiv 0 \iff \Delta\, e^{\frac{\dim \mathcal{M}}{3}\mathcal{G}} = 0, \tag{10.47}$$

where $\Delta \equiv g^{i\bar{j}}\, \partial_i \partial_{\bar{j}}$ is the scalar Laplacian in the Kähler space $\mathcal{M}$. In particular, if $\dim \mathcal{M} = 1$, the scalar Laplacian is conformal–invariant, and the condition (10.47) is independent of the metric

$$\partial \bar{\partial}\, e^{\mathcal{G}/3} = 0 \iff e^{\mathcal{G}/3} = \text{harmonic}, \tag{10.48}$$

so, up to coordinate redefinitions $z \mapsto f(z)$, there is a unique (local) solution

$$e^{\mathcal{G}/3} = i\bar{z} - iz, \tag{10.49}$$

---

[2] Then, in this class of models, $-\mathcal{G}$ is a *global* Kähler potential and $\mathcal{M}$ is *non–compact*.

10.3 SuperHiggs. Flat potentials

and the Kähler potential of the flat model is

$$K = -3 \log(i\bar{z} - iz). \tag{10.50}$$

Thus we recover the basic example in Section 10, namely the upper half–plane with its invariant Poincaré metric (10.22). The isometry group is $SL(2, \mathbb{R}) \simeq SU(1, 1)$. Notice that the flat V condition fixes the integer in Eq. (10.22) to $k = 3$. The $SL(2, \mathbb{R})$ symmetry is broken by the Yukawa couplings to the invariances of $\mathcal{G}$, i.e., to the parabolic subgroup $\mathbb{R}$ of unipotent upper triangular matrices, or

$$z \to z + \text{real constant}. \tag{10.51}$$

For $\dim \mathcal{M} \geq 2$ it is more convenient to start from the identity

$$\det\left[-\partial_i \partial_{\bar{j}} e^{\mathcal{G}/3}\right] \equiv \left(\frac{e^{\mathcal{G}/3}}{3}\right)^{\dim \mathcal{M}} \det(g_{i\bar{j}}) \left[1 - \frac{1}{3} g^{i\bar{j}} \partial_i \mathcal{G} \partial_{\bar{j}} \mathcal{G}\right] \tag{10.52}$$

and rewrite the condition of vanishing potential as a *complex Monge–Ampere equation* (a class of differential equations crucial in the theory of Calabi–Yau spaces [326]). In the differential–form notation,

$$V \equiv 0 \quad \Longleftrightarrow \quad \left(\partial\bar{\partial} e^{\mathcal{G}/3}\right)^{\dim \mathcal{M}} = 0. \tag{10.53}$$

In the $\dim \mathcal{M} = 1$ case the flat condition led us to a (locally) *symmetric* space. For $d \equiv \dim \mathcal{M} \geq 2$ we take the symmetry of $\mathcal{M}$ as an *ansatz*. From the previous chapter we know that a non–compact symmetric Hermitian manifold must be a Siegel domain. So we assume $\mathcal{M}$ to be a *Siegel domain* $\mathfrak{S}(V, F) \subset \mathbb{C}^{n+m}$, specified by the convex cone $V \subset \mathbb{C}^n$ and V–Hermitian function $F: \mathbb{C}^m \times \mathbb{C}^m \to \mathbb{C}^n$ (the Siegel domains of the *first* kind being the special case $m = 0$). The class of Siegel domains $\mathfrak{S}(V, F)$ contains *all* the negatively curved symmetric Kähler manifolds, but also many other homogeneous or *non–homogeneous* manifolds. The canonical Kähler metric on $\mathcal{M}$ defined by (cf. Eq. (9.110))

$$g_{i\bar{j}} = -\partial_i \partial_{\bar{j}} G(2 \operatorname{Im} z_i - F_i(w, w)), \tag{10.54}$$

$$e^{-G(y_i)} = \int_{V^\vee} d^n x \, e^{-y_i x^i}. \tag{10.55}$$

This canonical metric is invariant under all complex automorphisms of $\mathfrak{S}(V, F)$, and hence if $\mathfrak{S}(V, F)$ is *symmetric*, it coincides (up to an overall scale) with the *unique* symmetric metric. From Eq. (10.55) we see that the function $e^{G(y_i)}$ is homogeneous in the real variables $y_i \equiv 2 \operatorname{Im} z_i - F_i(w, w)$ of degree n. Set

$$e^{\mathcal{G}(z_i, \bar{z}_j, w_\alpha, \bar{w}_\beta)/3} = \exp\left[G(i\bar{z}_i - iz_i - F_i(w, w))/n\right], \tag{10.56}$$

that is, take as Kähler potential the canonical one re–scaled by the overall factor $\frac{3}{n}$ (and superpotential $W = 1$). The resulting Kähler metric is still positive and invariant under all automorphisms of $\mathfrak{S}(V, F)$. We claim that function (10.56) is a solution to the Monge–Ampere equation (10.53). Indeed, (10.53) is invariant under holomorphic redefinitions of the coordinates. Since by the holomorphic symmetry of $\mathcal{G}$

$$\begin{cases} w_\alpha \to w_\alpha - w_\alpha^{(0)} \\ z_i \to z_i + iF_i(w, w^{(0)}) + \frac{i}{2}F(w^{(0)}, w^{(0)}) \end{cases} \tag{10.57}$$

the point $(z, w^{(0)})$ is mapped to the point $(z', 0)$, it is enough to show that

$$\left(\partial\bar\partial\, e^{\mathcal{G}/3}\right)^{n+m}\bigg|_{w=0} = 0. \tag{10.58}$$

Now,

$$\partial\bar\partial\, e^{\mathcal{G}/3}\bigg|_{w=0} = \frac{\partial^2 e^{G(y)/n}}{\partial y_i\, \partial y_j}\, dz^i \wedge d\bar z^j - \frac{\partial e^{G(y)/n}}{\partial y_i}\, f_{i\alpha\bar\beta}\, dw^\alpha \wedge d\overline{w}^\beta, \tag{10.59}$$

where $y_i = 2\,\mathrm{Im}\, z_i$ and $F_i(w, w) \equiv f_{i\alpha\bar\beta}\, w^\alpha \overline{w}^\beta$. $e^{-K(y)/n}$ is homogeneous of degree 1 in the variables y_i, and hence, by Euler's theorem,

$$y_j\, \frac{\partial^2 e^{\mathcal{G}/3}}{\partial y_i\, \partial y_j} = 0, \tag{10.60}$$

and thus the vector $(y_i, 0)^t$ is a zero eigenvector of the Hermitian matrix $\partial\bar\partial\, e^{\mathcal{G}/3}|_{w=0}$. Then its determinant vanishes: that is, Eq. (10.53) holds.

General Lesson 10.9 *Let $\mathfrak{S}(V, F) \subset \mathbb{C}^{n+m}$ be a Siegel domain, and g its canonical Kähler metric g. The $\mathcal{N} = 1$ supergravity model defined by the Kähler metric $\frac{3}{n}g$ and superpotential $W = 1$ has flat potential.*

Corollary 10.10 *There is a flat potential* SUGRA *for any non–compact symmetric Kähler manifold. The corresponding $\mathcal{G}$ function is given by (10.56).*

Given a homogeneous domain, by the same technique we may construct many other flat potentials associated with the several homogeneous Kähler metrics on the domain. It is easy to construct continuous families depending on as many parameters as the rank of the associated T–algebra [83].

Examples. Consider the following symmetric spaces:

$$\mathfrak{I}_{n,n} \equiv \frac{SU(n,n)}{U(n) \times SU(n)}, \quad \mathfrak{H}_n \equiv \frac{Sp(2n, \mathbb{R})}{U(n)}, \quad \mathfrak{O}_n = \frac{SO^*(2n)}{U(n)}. \tag{10.61}$$

An element of $\mathfrak{I}_{n,n}$ is written as a complex $n \times n$ matrix Z with $i(Z^\dagger - Z)$ positive–definite. The restriction to symmetric (resp. antisymmetric) matrices then gives $\mathfrak{H}_n$ (resp. $\mathfrak{O}_n$). The flat potential $\mathcal{N} = 1$ SUGRA is then defined by the function

$$\exp[\mathcal{G}(Z,\bar{Z})/3] = \left(\det[\,iZ^\dagger - iZ\,]\right)^{1/n}. \tag{10.62}$$

10.4 Gauged $\mathcal{N} = 1$ 4D supergravity

Adding vector fields to our $\mathcal{N} = 1$ supergravity we switch on two kinds of coupling: (i) gauge couplings (together with their SUSY completion) and (ii) "magnetic susceptibility" couplings related to a non–trivial function $\mathcal{N}_{mn}(\phi)$ in the vector's kinetic terms,

$$-e\,\frac{1}{4}\mathcal{N}_{mn}(\phi)\,F_{\mu\nu}^{m+}\,F_{\rho\sigma}^{n+}\,g^{\mu\rho}\,g^{\nu\sigma} + \text{H. c.}\;(+\,\text{SUGRA completion}). \tag{10.63}$$

$\mathcal{N}_{mn}$ is best understood as a holomorphic map $\mu\colon \mathcal{M} \to \mathfrak{H}_n$.

It is convenient to discuss the two classes of coupling separately. We begin with the minimal gauge couplings. Thus for the moment we assume

$$\mathcal{N}_{mn} = \delta_{mn} \tag{10.64}$$

and we use this invariant tensor to freely raise/lower gauge indices. In this section g denotes the gauge coupling.

10.4.1 Gauge couplings

From Chapter 8 we know that the SUSY completion of the minimal gauge couplings is completely described by the Cartan–Kostant map. Since a Hodge manifold is in particular symplectic, the couplings should have a universal expression when written in terms of K_i^m, $K_{\bar{\imath}}^m$ and μ^m. To fix all the terms in the SUSY completion, it is enough to determine the $O(g)$ contributions to the Yukawa tensors A_1, A_2, and A_3: the gauge contribution to the scalar potential is then fixed by the universal Ward identity, while there are no induced interactions of higher dimension by g–scaling arguments. In the discussion following Eq. (10.35) we already saw that there is no $O(g)$ contribution to the gravitino mass A_1.

We use the following notation for the spin–$\tfrac{1}{2}$ fields: we denote by $\chi^i = \gamma_5 \chi^i$ the *chiralini*, i.e., the fermionic partners of the scalars ϕ^i which live in $T\mathcal{M}$, while the *gaugini* (the fermions of the vector multiplets) will be denoted as $\lambda^m = \gamma_5 \lambda^m$. χ^i and λ^m have opposite $u(1)_R$ charges. We shall use the overbar to denote Majorana conjugation.

The $O(g)$ terms in the Yukawa couplings A_2, A_3 have the general structure

$$A_{2\,m}\,\bar{\psi}_\mu\gamma^\mu\lambda^m + A_{3\,\bar{\imath}\,m}\,\bar{\chi}^{\bar{\imath}}\lambda^m + A_{3\,mn}\,\bar{\lambda}^m\lambda^n + \text{H.c.} \tag{10.65}$$

plus, *a priori*, $O(g)$ corrections to the old $\bar{\psi}_\mu \gamma^\mu \chi^i$ and $\bar{\chi}^i \chi^j$ terms. However, $O(g)$ scaling predicts $A_{2\,i}$ and $A_{3\,ij}$ to be at most linear in $K^{m\,i}$, μ^m and their covariant derivatives, and no covariant object in this class has the right $\mathfrak{u}(1)_R$ charge to contribute to these couplings. So only the Yukawas in (10.65) get $O(g)$ contributions.

$A_{2\,m}$ transforms in the adjoint of the gauge group and has $\mathfrak{u}(1)_R$ charge zero. The only object linear in g with these properties is μ_m. Thus,

$$A_{2\,m} = c_1 \mu_m \tag{10.66}$$

for a universal coefficient c_1 which, in the usual normalization, is equal to $-i/2$ (this is well known from the rigid limit: $A_{2\,m}$ is the gaugino fermionic shift, which is $-\frac{i}{2} D_m$ where D_m is the auxiliary D-field; more or less by definition, on–shell the D_m and μ_m are equal).

For $A_{3\,\bar{i}\,m}$ we have two candidates $K_{\bar{i}\,m}$ and $\partial_{\bar{i}} \mu_m$. Luckily, they just differ by an overall $-i$. Then

$$A_{3\,\bar{i}\,m} = c_2 K_{\bar{i}\,m}. \tag{10.67}$$

In some standard convention, $c_2 = -2$. Finally, $A_{3\,mn} = 0$ since no invariant with the right g scaling exists.

The scalar potential is deduced from the fermionic shifts to be

$$V = V^{(0)} + \frac{1}{2} \mu^m \mu_m, \tag{10.68}$$

where $V^{(0)}$ is given by Eq. (10.38).

There is a last class of couplings which needs to be determined: the Pauli ones. If $\mathcal{N}_{mn}$ is trivial, Pauli interactions are ruled out by the usual symmetry arguments, except for the term

$$\frac{1}{4} F^m_{\mu\nu} \bar{\psi}_\rho \gamma^{\mu\nu} \gamma^\rho \lambda_m + \text{H. c}, \tag{10.69}$$

which is a Noether term fully determined by the rigid limit supercurrent.

10.4.2 The general case

We know that the vector–scalar couplings in the ungauged theory are described by the μ map

$$\mu: \mathcal{M} \to \frac{Sp(2V, \mathbb{R})}{U(V)} \simeq \mathfrak{H}_V \quad \begin{pmatrix} \text{a Siegel domain} \\ \text{of the first kind} \end{pmatrix}. \tag{10.70}$$

10.4 Gauged $\mathcal{N} = 1$ 4D supergravity

In the case of $\mathcal{N} = 1$ SUSY/SUGRA the map μ should be *holomorphic*. Conceptually, the simplest way to construct the Lagrangian is to promote the map μ in Eq. (10.70) to a superfield map. As an aside, we review the subject.

Duality in superspace

In superspace the self–dual part of the Abelian field–strengths, $F_{\mu\nu}^{m\,+}$, are promoted to Weyl-spinor chiral superfields [297]

$$W_\alpha^m, \qquad \overline{\mathcal{D}}_{\dot\alpha} W_\alpha^m = 0. \qquad (10.71)$$

The Lagrangian is a scalar chiral density, $\mathcal{L} = \int d^2\theta\, L(W_\alpha^m, \Phi)$. The superfield Bianchi identities and the equations of motion read

$$\mathcal{D}^\alpha W_\alpha^m - \overline{\mathcal{D}}^{\dot\alpha} \overline{W}_{\dot\alpha}^m = 0 \qquad \text{(Bianchi identity)}, \qquad (10.72)$$

$$\mathcal{D}^\alpha \frac{\partial L}{\partial W^{\alpha\,m}} + \overline{\mathcal{D}}^{\dot\alpha} \frac{\partial L}{\partial \overline{W}^{\dot\alpha\,m}} = 0 \qquad \text{(equations of motion)}, \qquad (10.73)$$

the only difference between rigid SUSY and SUGRA being that in the latter case the spinorial covariant derivative $\mathcal{D}_\alpha$ is also gravitationally covariant.

From Equations (10.72) and (10.73) we see that the "magnetic" dual field-strengths

$$Y_{\alpha\,m} := i \epsilon_{\alpha\beta} \frac{\partial L}{\partial W_\beta^m} \qquad (10.74)$$

satisfy the same equations as the original field–strengths W_α^m. Equations (10.72) and (10.73) are invariant under any linear transformation of $(W_\alpha^m, Y_{\alpha\,m})$ with *real* coefficients. The invariance of Eq. (10.74) restricts, again, to the subgroup $Sp(2n, \mathbb{R}) \subset GL(2n, \mathbb{R})$. Just as in the bosonic case, we can rewrite the condition as

$$(1 - i\Omega)\mathcal{E}\begin{pmatrix} W_\alpha \\ Y_\alpha \end{pmatrix}, \qquad (10.75)$$

where now the "vielbein" $\mathcal{E}$,

$$\mathcal{E} = \begin{pmatrix} A & C \\ B & D \end{pmatrix}, \qquad (10.76)$$

is a $2n \times 2n$ matrix whose entries are superfields, such that the complex superfield $n \times n$ matrix

$$f_{mn} := [(C + iD)^{-1}(A - iB)]_{mn} \qquad (10.77)$$

is *chiral*,

$$\overline{\mathcal{D}}_{\dot\alpha} f_{mn} = 0. \tag{10.78}$$

Indeed,

$$Y_{\alpha\,m} = f_{mn}\, W_\alpha^n \quad \text{and} \quad \mathcal{L} = -\frac{i}{2} f_{mn}\, W_\alpha^m\, W^{m\alpha}. \tag{10.79}$$

f_{mn} terms in $\mathcal{L}$

The above couplings induce five changes in the Lagrangian:

1. The invariant "metric" δ_{mn} on the gauge Lie group is replaced by the field-dependent one $-i f_{mn}$, cf. Eq. (10.79).
2. The momentum map μ^m (that is, the auxiliary field D^m) gets replaced by

$$l_{mn}\, \mu^n \to f_{mn}\, \mu^m + 2 D_i f_{mn}\, \overline{\chi}^i \lambda^n. \tag{10.80}$$

3. There is a new Pauli coupling

$$2 D_i f_{mn}\, \overline{\chi}^i \gamma^{\mu\nu} \lambda^n\, F_{\mu\nu} + \text{H. c.} \tag{10.81}$$

4. The χ fermionic shifts (or the F_i auxiliary fields) are modified:

$$A_{2\,i} \to A_{2\,i} + D_i f_{mn}\, \overline{\lambda}^m \lambda^n. \tag{10.82}$$

5. The replacements in μ_m and A_{2i} when inserted in the Yukawa terms and in the scalar potential induce a bunch of new 4-Fermi terms. In addition, there is a coupling of the form

$$\frac{1}{4}(\partial_i \partial_j f_{mn})\, \overline{\chi}^i \chi^j\, \overline{\lambda}^m \lambda^n. \tag{10.83}$$

Equations (10.80)–(10.83) are determined, up to few numerical coefficients, by the requirements of covariance, g scaling, and the known form for f_{mn} constant. The numerical coefficients may be fixed (in principle) by comparing with 3D.

Further details may be found in the original papers [100, 102] and in a recent book [146].

11
Flag manifolds. Variations of Hodge structures

In Chapter 4 we saw that in 4D supergravities with $N_S \geq 9$ the geometric structure associated with the central charge (Section 4.7) is encoded in a map ζ from the scalar manifold $\mathcal{M}$ to a Grassmanian $\mathbb{G}r(k, V_\mathbb{F})$. Grassmanians are examples of *flag manifolds*: they are the flag manifolds that are also symmetric spaces. In a general 4D SUGRA the central charge structure defines a map from $\mathcal{M}$ to a polarized flag manifold whose kind depends on $\mathcal{N}$ and the matter supermultiplets coupled to SUGRA. In this chapter we study the differential geometry of these spaces, following ref. [172], as a preparation for the construction of 4D $\mathcal{N} = 2$ supergravity. The polarized flag manifolds coincide with the *Griffiths period domains*, already mentioned in Section 1.5.2, which parametrize the Hodge decomposition of the cohomology spaces of compact Kähler manifolds. In string theory this correspondence has a deep physical meaning, which we review in Chapter 12. For this reason, in describing these spaces we shall from the very start use the notations adapted to Hodge theory. General references for this chapter are [77, 173, 293, 294].

11.1 Hodge structures and Griffiths domains

Let $V_\mathbb{R}$ be a real vector space of dimension n, and $V_\mathbb{C} \equiv V_\mathbb{R} \otimes \mathbb{C}$ its complexification; equivalently, $V_\mathbb{C}$ is a $\mathbb{C}$–space equipped with a real structure[1] R such that $V_\mathbb{R}$ is its subspace of real vectors. We denote complex conjugation with respect to the real structure R by an overbar $v \mapsto \bar{v}$.

By a *polarization of weight* $w \in \mathbb{N}$ of the vector space $V_\mathbb{R}$ we mean the datum of a *non–degenerate* bilinear form, $Q\colon V_\mathbb{R} \otimes V_\mathbb{R} \to \mathbb{R}$, which is *symmetric* for w even and *skew–symmetric* for w odd:

$$Q(v, u) = (-1)^w Q(u, v), \qquad u, v \in V_\mathbb{R} \text{ or } V_\mathbb{C}. \tag{11.1}$$

---
[1] Cf. Definition 2.1.

11.1.1 Hodge structures

Definition 11.1 A polarized <u>real</u> Hodge structure of weight w of $(V_\mathbb{R}, Q)$ is a vector–space decomposition

$$V_\mathbb{C} = \bigoplus_{p+q=w} H^{p,q} \qquad (11.2)$$

such that

$$H^{q,p} = \overline{H^{p,q}}, \qquad (11.3)$$

and the following properties hold (the *"Riemann bilinear relations"*)

(i) $\quad Q(H^{p,q}, H^{p',q'}) = 0 \quad$ if $(q', p') \neq (p, q)$ $\qquad (11.4)$

(ii) $\quad h(u, v) \equiv Q(Cu, \bar{v})$ is a positive–definite Hermitian form $\qquad (11.5)$

where C, called the *Weil operator*, acts on $H^{p,q}$ as multiplication by i^{p-q}.

Remark. A *Hodge structure* is a real Hodge structure together with the datum of a lattice $V_\mathbb{Z}$ such that $V_\mathbb{R} = V_\mathbb{Z} \otimes \mathbb{R}$ and $Q(V_\mathbb{Z}, V_\mathbb{Z}) \subset \mathbb{Z}$. The lattice datum will be important for the physical applications.

A Hodge structure may equivalently be specified as a Hodge *filtration* (also called a *flag*)

$$F^w \subset \cdots \subset F^2 \subset F^1 \subset F^0 \equiv V_\mathbb{C}. \qquad (11.6)$$

instead of the decompostion (11.2). The dictionary decomposition $\leftrightarrow$ filtration is given by

$$F^p = \bigoplus_{k \geq p} H^{k, w-k}, \quad \text{and} \quad H^{p,q} = F^p \cap \overline{F^q}. \qquad (11.7)$$

Note that

$$V_\mathbb{C} = F^p \oplus \overline{F^{w-p+1}}, \qquad (11.8)$$

while the first Riemann bilinear relation (11.4) becomes[2]

$$(F^p)^\perp = F^{w-p+1}. \qquad (11.9)$$

Filtrations (11.6) satisfying Eq. (11.9) are called *isotropic* (or *self–dual*).

---

[2] $(F)^\perp$ means the subspace orthogonal to F with respect to the bilinear form Q.

Example. The canonical example of Hodge structure is given by the Hodge theory of a compact Kähler n–fold K with Kähler form ω. Let $V_\mathbb{R} = H^w(K, \mathbb{R})_{\text{pr}}$ be the Lefshetz *primitive* cohomology of K in degree w. A class $[\eta] \in V_\mathbb{R} = H^w(K, \mathbb{R})_{\text{pr}}$ is represented by a real harmonic w–form η with $\Lambda \eta = 0$ (cf. Section 9.4.1), and we set $Q([\eta], [\eta']) = \int_K \eta \wedge \eta' \wedge \omega^{n-w}$. Note that $(V_\mathbb{R}, Q)$ depends only on the Kähler class $[\omega]$ and is independent of the complex structure of K. In the most interesting case for the physical applications, $n = w = 3$ with K simply connected, $(V_\mathbb{R}, Q)$ is independent of $[\omega]$ too: Exercise 11.1.3. If we specify a complex structure on K, we may decompose the harmonic w–forms into harmonic forms of definite type (p, q). Setting $H^{p,q}$ to be the vector space of harmonic forms of type (p, q), we get a decomposition

$$H^w(K, \mathbb{C})_{\text{pr}} \equiv H^w(K, \mathbb{R})_{\text{pr}} \otimes \mathbb{C} = \bigoplus_{p+q=w} H^{p,q}, \qquad (11.10)$$

which is easily seen to satisfy all the conditions in definition 11.1. One has $\dim H^{p,q} = h^{p,q} \equiv h^{q,p}$.

11.1.2 Griffiths period domains

Given $(V_\mathbb{R}, Q)$ and a decreasing set of non–negative integers $\{f^p\}_{p=0}^w$, which is self–dual in the sense that $f^p + f^{w-p+1} = f^0 = n$, we are interested in the space D parametrizing *all* flags of the form

$$F^\bullet \equiv F^w \subset F^{w-1} \subset \cdots \subset F^0 \equiv V_\mathbb{C} \quad \text{with } \dim F^p = f^p, \qquad (11.11)$$

and satisfying the two bilinear relations (11.4) and (11.5). The first relation, $(F^p)^\perp = F^{w-p+1}$, says that the flag $F^\bullet$ is determined by its sub–flag

$$F^w \subset F^{w-1} \subset \cdots \subset F^v \subset V_\mathbb{C}, \qquad v = [(w+1)/2]. \qquad (11.12)$$

The space $\check{D}$ that parametrizes the flags of dimensions $\{f^p\}$ satisfying only the first bilinear relation (11.4) is an algebraic submanifold of the product $\prod_{p=v}^w \text{Gr}(f_p, V_\mathbb{C})$. In particular, $\check{D}$ is a compact complex manifold.

We claim that $\check{D}$ is a homogeneous space for the complex group $G_\mathbb{C}$ of the automorphisms of $V_\mathbb{C}$ preserving Q. We show this for $w =$ odd, leaving the even case as Exercise 11.1.3. For $w = 2s - 1$, Q is a symplectic form on $V_\mathbb{C}$, and a flag satisfying (11.4) is specified by the sub–flag

$$F^{2s-1} \subset \cdots \subset F^s \subset V_\mathbb{C}. \qquad (11.13)$$

We have $V_\mathbb{C} = F^s \oplus \overline{F^s}$ and $(F^s)^\perp = F^s$, so F^s is a complex Lagrangian subspace of $(V_\mathbb{C}, Q)$. $Sp(n, \mathbb{C})$ acts transitively on such Lagrangian subspaces, while the subgroup of $Sp(n, \mathbb{C})$ preserving F^s is[3] $GL(n/2, \mathbb{C})$; hence $Sp(n, \mathbb{C})$ acts transitively on the set of sub–flags.

Fixing a reference flag $F_0^\bullet$, $\check{D}$ is identified with its $G_\mathbb{C}$–orbit,

$$\check{D} \equiv G_\mathbb{C}/B, \qquad (11.14)$$

where B is the isotropy group of $F_0^\bullet$. Choosing a basis in $V_\mathbb{C}$ adapted to the filtration $F_0^\bullet$, $G_\mathbb{C}$ becomes a matrix group and B its subgroup of block–triangular matrices. One shows (Exercise 11.1.3) that $\check{D}$ may also be identified with the homogeneous space G_{comp}/V where G_{comp} is the compact real form of $G_\mathbb{C}$ and $V = B \cap G_{\text{comp}}$. This gives an alternative proof of the fact that $\check{D}$ is compact.

The space D that parametrizes the polarized flags is obtained from $\check{D}$ by imposing the second bilinear relation (11.5). Since (11.5) is a positivity condition, D is an open domain in $\check{D}$. We claim (Exercise 11.1.3) that D is homogeneous under the action of the real group G of the automorphisms of $V_\mathbb{R}$ preserving Q. Hence

$$D = G/V, \qquad V = B \cap G. \qquad (11.15)$$

The subgroup V is always compact; indeed, $V = B \cap G_{\text{comp}}$.

Thus the weight w polarized Hodge structures with fixed dimensions $f^p = \sum_{k \geq p} h^{k,w-k}$, $f^0 = n$, are parametrized by the homogeneous spaces D

$$\begin{cases} Sp(n, \mathbb{R})/\prod_{p \leq m} U(h^{p,w-p}) & \text{for } w = 2m+1, \\ SO(s, n-s)/\prod_{p < m} U(h^{p,w-p}) \times SO(h^{m,m}) & \text{for } w = 2m, \end{cases} \qquad (11.16)$$

where

$$s = \sum_{p \text{ even}} h^{p,w-p},$$

which are called the *Griffiths period domains*. $\check{D}$ is called the *compact dual* of D. Since $D \subset \check{D}$ are open, and the $\check{D}$ are complex manifolds, the Griffiths domains have a natural complex structure.

Let K be the *unique* maximal compact subgroup of G containing V. Then G/K is a symmetric space. We have a canonical fibration

$$\varpi : G/V \to G/K \qquad (11.17)$$

with fibers K/V, which are complex submanifolds of G/V.

---

[3] This follows from the computation in Lemma 1.6 by complexification.

11.1 Hodge structures and Griffiths domains

Example. Let us consider the Hodge decomposition for a curve of genus 1 (i.e., a one–dimensional complex torus) T. We have $H^1(T, \mathbb{R}) \simeq \mathbb{R}^2$, while Q is the usual symplectic bilinear form (cf. Section 1.5). Then

$$G = Sp(2, \mathbb{R}) \equiv SL(2, \mathbb{R}), \quad G_{\mathbb{C}} = SL(2, \mathbb{C}), \quad G_{\text{comp}} = SU(2). \quad (11.18)$$

The flag is $F^1 \subset F^0 \equiv \mathbb{C}^2$ with $\dim F^1 = h^{1,0} = 1$, and $\check{D}$ is just the variety which parametrizes the complex lines F^1 in $\mathbb{C}^2$, that is, $\check{D} = P\mathbb{C}^1$, the Riemann sphere on which $SL(2, \mathbb{C})$ acts transitively by complex automorphisms

$$z \mapsto \frac{az + b}{cz + d}, \quad \begin{pmatrix} a & b \\ c & d \end{pmatrix} \in SL(2, \mathbb{C}). \quad (11.19)$$

The isotropy subgroup B_z fixing a point z in the sphere is conjugate to the standard Borel subgroup $\mathbb{C}^\times \ltimes \mathbb{C} \subset SL(2, \mathbb{C})$ of upper triangular matrices with $c = 0$, which is the isotropy group B_∞ of the point at infinity. The conjugate isotropy subgroup of $z = i$ is

$$B_i = \begin{pmatrix} \cos\phi + \beta & \sin\phi - i\beta \\ -\sin\phi - i\beta & \cos\phi + \beta \end{pmatrix}, \quad e^{i\phi} \in \mathbb{C}^\times, \beta \in \mathbb{C}. \quad (11.20)$$

The real group $G = SL(2, \mathbb{R})$ acts transitively on the space D of the Hodge structures. The isotropy group is $V = B_i \cap SL(2, \mathbb{R})$, i.e., in (11.20) we restrict to ϕ real and $\beta = 0$, getting $V = SO(2) \simeq U(1)$. In this case the Griffiths domain D is just the classical upper half–plane $\mathfrak{H} = SL(2, \mathbb{R})/U(1)$, which is an open domain in the Riemann sphere $\check{D} = SL(2, \mathbb{C})/(\mathbb{C}^\times \ltimes \mathbb{C})$, which coincides with the symmetric space $SU(2)/U(1) \equiv SO(3)/SO(2) \equiv S^2$.

11.1.3 Lie algebraic properties of period domains

$D = G/V$ is a Griffiths period domain, $K \subset G$ a maximal compact subgroup, and $\varpi: G/V \to G/K$ the natural projection. The Lie algebras of G, V, K are $\mathfrak{g}$, $\mathfrak{v}$, and $\mathfrak{k}$, respectively. We set $\mathfrak{g} = \mathfrak{v} \oplus \mathfrak{m}$. A point in G/V specifies a weight w Hodge structure $(V_\mathbb{R}, F^\bullet)$ polarized by the $(-1)^w$–symmetric form Q, where $G = GL(V_\mathbb{R}, Q)$ and V is the isotropy subgroup preserving $F^\bullet$. The corresponding complex Lie algebra decomposes as

$$\mathfrak{g}_\mathbb{C} = \text{End}(V_\mathbb{C}, Q) = \bigoplus_k \mathfrak{g}^{-k,k}, \quad (11.21)$$

where $\mathfrak{g}^{-k,k}$ denotes the endomorphisms sending $H^{p,q}$ to $H^{p-k,q+k}$. In particular $\mathfrak{v}_\mathbb{C} = \mathfrak{g}^{0,0}$ and

$$\mathfrak{v} = \mathfrak{g}^{0,0} \cap \mathfrak{g}, \qquad \mathfrak{m} = \mathfrak{g} \cap \bigoplus_{k \neq 0} \mathfrak{g}^{k,-k}. \tag{11.22}$$

One has

$$[\mathfrak{g}^{p,q}, \mathfrak{g}^{k,\ell}] \subset \mathfrak{g}^{p+k,q+\ell} \quad \text{in particular,} \quad [\mathfrak{v}, \mathfrak{m}] \subset \mathfrak{m}. \tag{11.23}$$

Thus $\mathfrak{m}$ is a V-module; by construction it is isomorphic to the tangent space of G/V at the identity. We introduce the splitting $\mathfrak{m}_\mathbb{C} = \mathfrak{m}^+ \oplus \mathfrak{m}^-$ where

$$\mathfrak{m}^+ = \bigoplus_{k>0} \mathfrak{g}^{k,-k}, \qquad \mathfrak{m}^- = \bigoplus_{k<0} \mathfrak{g}^{k,-k}. \tag{11.24}$$

This is a V-invariant decomposition such that $\mathfrak{m}^- = \overline{\mathfrak{m}^+}$. It defines a complex structure on $\mathfrak{m}$ which acts on $\mathfrak{m}^\pm$ as multiplication by $\mp i$. More generally, Eq. (11.23) implies that all subspaces $\mathfrak{g}^{k,-k}$ are V-modules.

Fix a reference point in D. The corresponding Hodge structure defines a Weil operator $C \colon V_\mathbb{C} \to V_\mathbb{C}$ as in Definition 11.1.

Lemma 11.2 *One has $C \in G$.*

Proof Indeed $Q(Cu, Cv) = Q(u,v)$ for all $u, v, \in V_\mathbb{C}$, while $\overline{Cv} = C\overline{v}$. The first equality implies $C \in G_\mathbb{C}$ and the second one that C is real. □

By definition $K \subset G$ is the subgroup preserving the positive–definite Hermitian form $Q(Cv, \overline{u})$. Thus K is the centralizer of C and hence

$$\mathfrak{k}_\mathbb{C} = \bigoplus_{k \text{ even}} \mathfrak{g}^{k,-k}. \tag{11.25}$$

Exercise 11.1.1 Show that for a simply connected compact Kähler three–fold K_3 the polarization Q on $H^3(K_3, \mathbb{R})_{\text{pr.}}$ does not depend on the choice of the Kähler form.

Exercise 11.1.2 Show that $G_\mathbb{C}$ acts transitively on the flags satisfying the first Riemann relation.

Exercise 11.1.3 Show that $G_{\text{comp.}}$ acts transitively on $\check{D}$.

Exercise 11.1.4 Show that the real non–compact group G acts transitively on the flags satisfying both Riemann relations.

Exercise 11.1.5 Show that the isotropy groups of the polarized flags of given dimensions are as in Eq. (11.16).

Exercise 11.1.6 Prove the claim that the fibers K/V in Eq. (11.17) are complex submanifolds.

Exercise 11.1.7 Show that the decomposition $\mathfrak{g} = \bigoplus_k \mathfrak{g}^{k,-k}$ is a polarized Hodge structure with respect to the bilinear form $B(x,y) = \operatorname{tr}(x \cdot y)$.

11.2 Geometry of reductive homogeneous spaces G/V

The Griffiths domains are special instances of *reductive homogeneous spaces*. In this section we focus on the differential geometry of the reductive homogeneous space G/V where $V \subset G$ are Lie groups with Lie algebras $\mathfrak{v} \subset \mathfrak{g}$. *Reductive* means that we have an Ad_V–invariant decomposition of $\mathfrak{g}$, i.e.,

$$\mathfrak{g} = \mathfrak{v} \oplus \mathfrak{m}, \qquad \text{with } [\mathfrak{v}, \mathfrak{m}] \subset \mathfrak{m}. \tag{11.26}$$

For the Griffiths domains this condition follows from Eq. (11.23). G/V is *symmetric* iff, in addition, we have $[\mathfrak{m}, \mathfrak{m}] \subset \mathfrak{v}$ (cf. Sections 3.2 and 5.5)

Canonical connection

The natural projection $\pi : G \to G/V$ defines a principal V–bundle over G/V with a canonical connection, invariant under the left action of G, already described in Chapter 5 for the special case of a symmetric space. It is given by the projection on the subalgebra $\mathfrak{v}$ of the Maurier–Cartan form of G,

$$A = g^{-1} dg \big|_{\mathfrak{v}}. \tag{11.27}$$

We write $\phi = g^{-1} dg|_{\mathfrak{m}}$, so that the full Maurier–Cartan form is $A + \phi$. As in Chapter 5, from the fact that the Maurier–Cartan form is "pure gauge" and hence has vanishing curvature, we get the *structure equations* of G/V,

$$\begin{cases} dA + A \wedge A = -(\phi \wedge \phi)|_{\mathfrak{v}}, \\ d\phi + A \wedge \phi + \phi \wedge A = -(\phi \wedge \phi)|_{\mathfrak{m}}, \end{cases} \tag{11.28}$$

so that the curvature of the canonical connection A is

$$F = -(\phi \wedge \phi)|_{\mathfrak{v}}. \tag{11.29}$$

The principal G–bundle $\mathcal{F}$

On the homogeneous space $D = G/V$ we have an associated G–bundle[4]

$$\mathcal{F} = G \times_V G \simeq D \times G, \qquad (11.30)$$

where the isomorphism is the G–equivariant map $[g, g'] \mapsto ([g], gg')$. The canonical connection $A_\mathcal{F}$ is just obtained by pulling back the Maurier–Cartan form of G and hence is *flat*.

G–invariant metric on D

In the particular case that our homogeneous space is a Griffiths domain D, the identification of the tangent space to D at the origin with the real subspace of $\bigoplus_{k\neq 0} \mathfrak{g}^{k,-k} \subset \bigoplus_{k\neq 0} \bigoplus_p \mathrm{Hom}(H^{p,w-p}, H^{p+k,w-(p+k)})$, together with the positive–definite Hermitian metric h on $H^{p,w-q}$, defines a G–invariant Hermitian metric ds_D^2 on D [172, 173].

11.2.1 Homogeneous bundles on G/V

For the applications to SUGRA we are interested in the *homogeneous vector bundles* over G/V. A bundle $p : \mathcal{X} \to G/V$ is homogeneous if it carries a G–action compatible with the left action on G/V, that is,

$$p(g \cdot x) = g \cdot p(x), \quad \text{for all } g \in G, \ x \in \mathcal{X}. \qquad (11.31)$$

It is elementary to show that the homogeneous bundles $\mathcal{X}$ over G/V are in one–to–one correspondence with the V–modules X (i.e., with the vector spaces X carrying a representation of the Lie group V). Indeed,

$$\mathcal{X} := G \times X \big/ \sim \qquad (11.32)$$

where $\sim$ is the equivalence $(g, x) \sim (gv, v^{-1}x)$. G acts on $\mathcal{X}$ as

$$g' \cdot [g, x] = [g' \cdot g, x] \qquad (11.33)$$

where $[g, x]$ denotes the equivalence class of (g, x). Thus the canonical V–connection (11.27) induces a canonical G–invariant connection ∇ on *all* homogeneous bundles $\mathcal{X}$.

Holomorphic structures on homogeneous bundles

Let us specialize in the case that $G/V = D$ is a Griffiths domain with compact dual $\check{D} = G_\mathbb{C}/B$. By a *holomorphic homogeneous bundle* we mean a bundle with

---

[4] In the math literature [77] this bundle is called the *Higgs* principal bundle.

a compatible $G_\mathbb{C}$-action. By the same argument as before, such holomorphic bundles are determined by a representation of the parabolic subgroup B. Let $\mathfrak{b}$ be the Lie algebra of the subgroup B. One has

$$\mathfrak{b} = \mathfrak{v}_\mathbb{C} \oplus \mathfrak{n}^-, \qquad (11.34)$$

where $\mathfrak{n}^-$ is the (unique) maximal nilpotent ideal of $\mathfrak{b}$ (which is identified with $\mathfrak{m}^+$ in Eq. (11.24)). Let X be a V-module. The representation of its Lie algebra $\mathfrak{v}$ on X may be extended by $\mathbb{C}$-linearity to a representation of $\mathfrak{b}$ by letting $\mathfrak{n}^-$ act trivially. This action may be exponentiated to an action of the group B, since V and B have the same fundamental group. Thus every representation X of V leads to a unique homogeneous holomorphic vector bundle $\mathcal{X}^h$.

On a homogeneous bundle $\mathcal{X}$ we have the canonical connection ∇. If $\mathcal{X}$ has a V-invariant Hermitian metric, there is a unique complex structure on $\mathcal{X}$ such that ∇ is the Chern connection. Of course, it should coincide with the one constructed in the previous paragraph.

Tangent bundles

By construction, the tangent bundle TG/V is the homogeneous bundle on G/V corresponding to the V-module $\mathfrak{m}$ with the action of V induced by the adjoint representation of G, Eq. (11.26).

If G/V is a Griffiths domain, we saw in Eq. (11.24) that $\mathfrak{m}_\mathbb{C}$ decomposes into the direct sum of two V-modules $\mathfrak{m}^\pm$ with $\mathfrak{m}^- = \overline{\mathfrak{m}^+}$, which defines a complex structure on $\mathfrak{m}$ and hence also on the associated homogeneous bundle TG/V. Each of these V-modules defines a homogeneous bundle, and in fact a *holomorphic* homogeneous bundle. By the argument following Eq. (11.24), these homogeneous bundles may further be decomposed into homogeneous bundles associated with the sub–modules $\mathfrak{g}^{k,-k}$.

A Griffiths domain $D = G/V$ is a complex manifold and hence has a *holomorphic tangent bundle* T_D which is obviously homogeneous. We claim that this is precisely the homogeneous bundle associated with the V-module $\mathfrak{m}^-$. This is a consequence of the Lie–theoretic equality $\mathfrak{n}^- = \mathfrak{m}^+$. T_D has a distinguished holomorphic subbundle T_D^{-1} associated with $\mathfrak{g}^{-1,1}$.

Since K is a maximal compact subgroup of G, G/K is symmetric. We may split the tangent space of G/V at the identity into horizontal and vertical subspaces with respect to the projection

$$\varpi : G/V \to G/K. \qquad (11.35)$$

From the arguments of Section 11.1.3,

$$(T_eD)^{\text{vert}} = \left(\bigoplus_{\substack{k \text{ even} \\ k \neq 0}} \mathfrak{g}^{k,-k}\right)_{\mathbb{R}}, \quad (T_eD)^{\text{hor}} = \left(\bigoplus_{k \text{ odd}} \mathfrak{g}^{k,-k}\right)_{\mathbb{R}}, \qquad (11.36)$$

where $(\cdots)_{\mathbb{R}}$ stands for the real subspace with respect to R. The two subspaces are V–modules (under the adjoint action) and hence define two smooth homogeneous subbundles of the tangent bundle TD,

$$TD = (TD)^{\text{vert}} \oplus (TD)^{\text{hor}}, \qquad (11.37)$$

called, respectively, the *vertical* and *horizontal* subbundles. The fibers of $(TD)^{\text{vert}}$ are the tangent spaces to the fiber of ϖ; hence $(TD)^{\text{vert}} \subset TD$ is an integrable G–invariant distribution. The horizontal one, $(TD)^{\text{hor}}$, in general is *not* integrable. Its complexification $(TD)^{\text{hor}} \otimes \mathbb{C}$ contains the distinguished holomorphic subbundle T_D^{-1}, which is sometimes called the *horizontal holomorphic tangent bundle*. In conclusion, as homogeneous bundles

$$T_D = \bigoplus_{k>0}\bigoplus_{p} \text{Hom}(\mathcal{H}^{p,w-p}, \mathcal{H}^{p-k,w-p+k}), \qquad (11.38)$$

$$T_D^{-1} = \bigoplus_{p} \text{Hom}(\mathcal{H}^{p,w-p}, \mathcal{H}^{p-1,w-p+1}). \qquad (11.39)$$

Example 11.3 w is odd, so $D = Sp(2m, \mathbb{R})/\prod_p U(h^{p,w-p})$ while $K = U(m)$. From the "Cayley rotated" perspective, the symmetric space $Sp(2m, \mathbb{R})/U(m)$ parametrizes the complex m–dimensional subspaces $E \subset \mathbb{C}^{2m}$ satisfying $Q(E, E) = 0$ and $iQ(E, \bar{E}) > 0$: to see this, compare with the Hodge structure of a genus m curve. Then E is precisely the i^w–eigenspace of the Weil operator C. Then the projection (11.35) is explicitly (here $H^{p,w-p} \equiv F^p/F^{p+1}$)

$$\varpi(F^w \subset \cdots \subset F^0) = H^{w,0} \oplus H^{w-2,2} \oplus \cdots H^{1,w-1}. \qquad (11.40)$$

11.2.2 Hodge bundles

Let us consider a Hodge structure whose Griffiths domain is the homogeneous space G/V. Fixing a base point $e \in G/V$, we specify the subspaces $\{F^p\}$ and $H^{p,q}$ of $V_{\mathbb{C}}$. By definition, the subspaces F^p and $H^{p,q}$ are V–invariant; hence they define homogeneous bundles over G/V which have natural holomorphic structures by Eq. (11.34).

Definition 11.4 The holomorphic homogeneous bundles $\mathcal{F}^p \to G/V$ and $\mathcal{H}^{p,q} \to G/V$ defined by the V–modules F^p, resp. $H^{p,q}$ ($p, q = 0, \ldots, w$), are called *Hodge bundles*.

Hodge bundles carry a natural Hermitian metric, since the V–modules $H^{p,q}$ carry a natural Hermitian metric, namely $Q(Cu, v)$. The canonical connection is holomorphic and metric with respect to these structures.

11.3 Quick review of Kodaira–Spencer theory

Let us return to the motivating context, the Hodge theory of a compact Kähler manifold K_0. We wish to study the space S of all possible deformations of its complex structure. Then we consider a family $\{K_s\}_{s \in S}$ of complex manifolds, all diffeomorphic to K_0 as smooth manifolds, but having different complex structures for $s \neq 0$. The theory of the deformations of complex structures is due to Kodaira and Spencer. Here we review just the little we need to put our differential geometric discussion into the proper perspective. Detail may be found in the book [217].

Specifying a complex structure on a (even–dimensional) smooth manifold K_{smooth} amounts to specifying, for all open subsets $U \subset K_{\text{smooth}}$, the subring $\mathcal{O}(U) \subset \mathcal{A}(U)$ of the C^∞–functions $U \to \mathbb{C}$ which are *locally holomorphic* in that particular complex structure. In more technical terms, one has to specify the sub-sheaf $\mathcal{O}$ of germs of holomorphic functions inside the sheaf $\mathcal{A}$ of germs of smooth functions. This is done by specifying a *Cauchy–Riemann differential operator*, $\overline{D}_{CR}$, whose kernel consists precisely of the local holomorphic functions; that is,

$$f \in \mathcal{O}(U) \quad \Longleftrightarrow \quad \overline{D}_{CR} f = 0. \tag{11.41}$$

In the original (undeformed) complex structure K_0, the Cauchy–Riemann operator is just the Dolbeault differential, $\overline{\partial} = d\bar{z}^i \, \partial/\partial\bar{z}^i$, which maps smooth functions into smooth $(0, 1)$ forms. The deformed Cauchy–Riemann acts between the same spaces,

$$\overline{D}_{CR} \colon \mathcal{A}(U) \to \Lambda^{0,1}(U). \tag{11.42}$$

Since all complex manifolds K_s are smoothly equivalent, for $s \neq 0$ we may still use the complex coordinates z^i of K_0 as smooth (non–holomorphic !) coordinates, and then we may write $\overline{D}_{CR}$ in the following general form:

$$\overline{D}_{CR} = d\bar{z}^i \left(\frac{\partial}{\partial \bar{z}^i} + \phi_i^j \frac{\partial}{\partial z^j} \right) \equiv \overline{\partial} + \phi. \tag{11.43}$$

$\phi \equiv d\bar{z}^i \phi_i{}^j \partial_j \in \Lambda^{0,1}(\Theta)$ is a $(0, 1)$–form with coefficients in the holomorphic tangent bundle Θ, called the *Kodaira–Spencer* (KS) *vector*. It is the generalization of classical Beltrami differential in one complex dimension.

In order to define a complex structure, we need to have enough independent local holomorphic functions to form a local holomorphic coordinate system. This requires the Cauchy–Riemann operator $\bar{\partial} + \phi$ to satisfy an integrability condition which generalizes the property $\bar{\partial}^2 = 0$ of the undeformed one. Then we need

$$(\bar{\partial} + \phi)^2 \equiv \bar{\partial}\phi + \phi^2 = 0, \qquad (11.44)$$

which is analogous to a zero–curvature condition. The gauge analogy may be pushed further. If ϕ is "pure gauge," that is of the form $\phi = \exp(-\xi)\bar{\partial}\exp(\xi)$ for some $(1, 0)$ vector field[5] ξ, then

$$(\bar{\partial} + \phi)f = e^{-\xi}\bar{\partial}(e^\xi f) = 0, \qquad (11.45)$$

and the deformed complex structure is the same as ("gauge equivalent to") the old one. Thus, the main result of the theory is that the deformations of complex structures are given by the solutions to the integrability condition (11.44) modulo the trivial solutions of the form $e^{-\xi}\bar{\partial}e^\xi$.

To describe the *infinitesimal* deformations we consider the linearized version of Eqs. (11.44), (11.45). We get that the space of (formal) infinitesimal deformations of the complex structure is given by the solutions to the $\bar{\partial}$–closure condition, $\bar{\partial}\phi = 0$, modulo the exact ones, $\phi = \bar{\partial}\xi$. By definition, this is just the cohomology group $H^1(K_0, \Theta)$, where Θ is the (sheaf of germs of holomorphic sections of) holomorphic tangent bundle of K_0. Thus $H^1(K_0, \Theta)$ is the formal tangent space of the deformation space S at its base point $s = 0$. We say "formal" because it is not true, in general, that a solution to the linearized equations may be completed to a solution of the full non–linear Kodaira–Spencer (KS) equation (11.44); in particular, the class of $[\phi, \phi]$ in $H^2(K_0, \Theta)$ is an obstruction to the existence of a solution of the form $\epsilon\phi + O(\epsilon^2)$ for $\phi \in H^1(K_0, \Theta)$. If a solution of this form does not exist, we say that the formal deformation $\phi \in H^1(K_0, \Theta)$ is *obstructed* in higher order.

For the physical applications the following special case is crucial:

Theorem 11.5 (Tian [285]) *Let K_0 be a Calabi–Yau n–fold. Then <u>all</u> infinitesimal deformations in $H^1(K_0, \Theta)$ are <u>unobstructed</u> and the universal deformation space S is smooth.*

Proof By definition the canonical bundle is trivial and we have a *parallel* holomorphic $(n, 0)$ form $\varepsilon_{i_1 i_2 \dots i_n}$ which we may use to rise and lower indices, so that we

---

[5] The exponential of vector fields on a manifold was defined in Section 3.2.

have an isomorphism from $(k, 0)$–vector–valued $(0, q)$–forms and $(n - k, q)$–forms

$$\varepsilon: \Omega^{0,q}(\wedge^k \Theta) \to \Omega^{0,q}(\wedge^{n-k} \Theta^\vee) \equiv \Omega^{n-k,q}, \tag{11.46}$$

which commutes with covariant derivation. The Lie commutator of two vector–valued $(0, 1)$ forms, $\phi^i \partial_i$ and $\psi^i \partial_i$, is the vector–valued $(0, 2)$ form

$$\begin{aligned}[\phi, \psi] &= \phi^i \partial_i \psi^k \partial_k + \psi^i \partial_i \phi^k \partial_k \\ &= D_i(\phi^i \psi^k + \psi^i \phi^k)\partial_k - (D_i \phi^i)\psi^k \partial_k - (D_i \psi^i)\phi^k \partial_k.\end{aligned} \tag{11.47}$$

In Exercise 11.3 you show that

$$\varepsilon([\phi, \psi]) = \bar\partial\, \varepsilon(2\phi^i \psi^k \partial_i \wedge \partial_k) - (D_i \phi^i)\, \varepsilon(\phi^k \partial_k) - (D_i \psi^i)\, \varepsilon(\psi^k \partial_k). \tag{11.48}$$

In particular, suppose that $\partial\varepsilon(\phi) = \partial\varepsilon(\psi) = 0$. Since $\partial\varepsilon(\phi) = 0 \Leftrightarrow D_i\phi^i = 0$, the above equation reduces to $\varepsilon([\phi, \psi]) = \bar\partial\, \varepsilon(2\phi \wedge \psi)$. Let us apply this to the Kodaira–Spencer equation

$$\bar\partial \phi(t) + \frac{1}{2}[\phi(t), \phi(t)] = 0, \tag{11.49}$$

where $\phi(t) = \sum_{a=1}^\infty \phi_a t^a$ is the KS vector corresponding to a one–parameter family of deformations parametrized by a real variable t so that $t = 0$ is the undeformed manifold, i.e., $\phi(0) = 0$. We already know that at the linearized level, i.e., at the first order in t around 0, the solution is $\varepsilon(\phi(t)) = t\xi + O(t^2)$ with ξ a harmonic $(n - 1, 1)$–form. Since K_0 is Kähler, this implies $\partial\xi = 0$. We look for a solution to the full non–linear KS equation such that $\partial\varepsilon(\phi(t)) = 0$. Then, applying ε to the KS equation, we get

$$\bar\partial\varepsilon(\phi(t)) + \partial\varepsilon(\phi(t) \wedge \phi(t)) = 0. \tag{11.50}$$

The $(n - 1, 2)$–form $\partial\varepsilon(\phi(t) \wedge \phi(t))$ is $\bar\partial$–closed and ∂–exact. By the $\partial\bar\partial$–Poincaré lemma (Lemma 9.13), it can be written as $\bar\partial\partial\psi(t)$ for some $(n - 2, 1)$–form $\psi(t)$. So we have to solve the three equations

$$\bar\partial(\varepsilon(\phi(t)) + \partial\psi(t)) = 0, \quad \partial\varepsilon(\phi(t)) = 0, \tag{11.51}$$

$$\partial(\varepsilon(\phi(t) \wedge \phi(t)) + \bar\partial\psi(t)) = 0. \tag{11.52}$$

For *all* $\xi \in \mathcal{H}^{n-1,1}(K_0)$, the first two equations are solved by

$$\varepsilon(\phi(t)) = t\xi - \partial\psi(t) = t\xi - \sum_{k=0}^\infty t^{k+2}\, \psi_{k+2}, \tag{11.53}$$

while the third one is automatically solved up to $O(t^2)$. Suppose, by induction, that the third equation had already been solved up to $O(t^{k+2})$. At order t^{k+2} we have

$$\partial\bar{\partial}\psi_{k+2} = -\sum_{a+b=k+2} \partial\varepsilon(\phi_a \wedge \phi_b), \qquad (11.54)$$

which has a solution by the Poincaré lemma. To complete the argument, one shows that the series has a non–zero convergence radius. □

For a general Kähler space K_0 we may only conclude that the tangent space $T_0 S$ of the complex structure deformation space S at the base point is equal to a *subspace* of $H^1(K_0, \Theta)$.

Weil–Petersson metric on the moduli space

From the Kodaira–Spencer theory we also see that, whenever the universal deformation space exists and is smooth, it comes equipped with a canonical Hermitian metric called the Weil–Petersson metric. Indeed, the KS identification $\iota\colon TS \to H^1(K, \Theta)$ allows us to define a metric

$$\begin{aligned}ds^2(\partial/\partial s^a, \partial/\partial s^b) &\stackrel{\text{def}}{=} \langle \phi_a, \phi_b \rangle_{\text{Hodge}} \\ &= \int_K d^{2n}z\, \sqrt{g}\, g_{i\bar{j}}\, g^{\bar{l}k}\, (\iota(\partial/\partial s^a))^i_{\bar{l}}\, \overline{(\iota(\partial/\partial s^b))^{\bar{j}}_k},\end{aligned} \qquad (11.55)$$

where $\langle \cdot, \cdot \rangle_{\text{Hodge}}$ is the Hodge Hermitian form.

Exercise 11.3.1 Prove the identity (11.48).

11.4 Variations of Hodge structures (VHS)

We consider the following geometric situation: we have a smooth family of compact Kähler manifolds, $\{K_s\}_{s \in S}$, depending holomorphically on complex parameters $s \in S$. In particular the Kähler form $\omega(s)$ depends smoothly[6] on s. More formally, we have complex manifolds K, S, and a holomorphic proper submersion $\varrho\colon K \to S$ such that $\varrho^{-1}(s) \equiv K_s$ are smooth compact complex submanifolds; moreover on K we have a smooth 2–form ω which restricts to the Kähler form $\omega(s)$ on each fiber K_s. For simplicity, and also because this condition is automatically satisfied in the situations relevant for us, we also assume $[\omega(s)]$ to be an integral class.

---

[6] Such a $\omega(s)$ always exists thanks to regularity theorems due to Kodaira and Spencer [218].

Example. To justify the physical interest of the situation, we present the classical example [75], namely the one–parameter family of $\mathbb{Z}_5^4$–symmetric quintic Calabi–Yau 3–fold

$$K = \Big\{ \sum_{j=1}^{5} X_j^5 + (s-5)X_1 X_2 X_3 X_4 X_5 = 0 \Big\} \subset P\mathbb{C}^4 \times \mathbb{C}^\times. \tag{11.56}$$

For fixed $s \in \mathbb{C}^\times$ we get a degree 5 hypersurface $K_s \subset P\mathbb{C}^4$, which is the simplest example of Calabi–Yau 3–fold. The restriction of the standard Kähler–form of $P\mathbb{C}^4$ to each K_s gives the smoothly varying family of Kähler forms. The complex structure of the manifold K_s depends holomorphically on $s \in \mathbb{C}^\times$. The point $s=0$ is excluded since it does not correspond to a smooth hypersurface.

By general differential topology, all manifolds K_s are diffeomorphic. Fixing a base point $o \in S$ and a curve $\gamma(t) \subset S$ connecting the given point $s = \gamma(1)$ to the base one $o = \gamma(0)$, we may construct a smooth family of diffeomorphisms $\phi_t \colon K_0 \to K_t$ by solving the differential equation

$$\frac{d}{dt}\phi_t = \tilde{v}(t), \tag{11.57}$$

where $\tilde{v}(t)$ is any vector field lifting $v(t) \equiv \dot{\gamma}(t)$. The homotopy class of the diffeomorphism ϕ_s depends only on the homotopy class of the path γ.

Since all the manifolds K_s are smoothly equivalent, they have the same topological invariants. In particular, the cohomology groups $H^k(K_s, \mathbb{Z})$ are s–independent. For each s we have the Lefshetz and Hodge decomposition:

$$H^k(K_s, \mathbb{Z}) = H^k(K_s, \mathbb{Z})_{\mathrm{pr}} \oplus L H^{k-2}(K_s, \mathbb{Z}), \tag{11.58}$$

$$H^k(K_s, \mathbb{Z})_{\mathrm{pr}} \otimes \mathbb{C} \equiv H^k(K_s, \mathbb{C})_{\mathrm{pr}} = \bigoplus_{p+q=k} H_s^{p,q}. \tag{11.59}$$

11.4.1 Monodromy representation

Pulling back $H^k(K_s, \mathbb{Z})_{\mathrm{pr}}$ to the base point o by ϕ_s, we may identify $H^k(K_s, \mathbb{Z})_{\mathrm{pr}}$ with a fixed lattice $V_\mathbb{Z} \subset V_\mathbb{R} \equiv V_\mathbb{Z} \otimes \mathbb{R} \subset V_\mathbb{C} \equiv V_\mathbb{R} \otimes \mathbb{C}$. This identification is locally unique, but globally it depends on the homotopy class[7] of the identifying diffeomorphism ϕ_s, that is, on the homotopy class $[\gamma]$ of the defining path $\gamma \subset S$. Suppose γ is a closed loop in S starting and ending at the base point o. γ defines a diffeomorphism $\phi_\gamma \colon K_o \to K_o$, unique up to homotopy, which induces

---

[7] Recall that the cohomology is a homotopy invariant [60].

an automorphism of the lattice $V_\mathbb{Z}$ that leaves invariant the natural bilinear form $Q(\alpha,\beta) \equiv \int L^{n-k} \alpha \wedge \beta$, that is, $\phi^*_{[\gamma]}$ is an element of the arithmetic group $G_\mathbb{Z}$:

$$G_\mathbb{Z} = \begin{cases} Sp(m;\mathbb{Z}) & k \text{ odd}, \\ SO(a,b;\mathbb{Z}) & k \text{ even} \end{cases} \quad (11.60)$$

($m = \dim H^k(K_0, \mathbb{R})_{\text{pr}}$, $a = \sum h^{k-2q,2q}$, $b = \sum h^{k-2q+1,2q-1}$). Hence we have a group representation

$$\phi^*: \pi_1(S, o) \to G_\mathbb{Z}, \quad (11.61)$$

called the *monodromy representation*. The group $\Gamma \equiv \phi^*(\pi_1(S, o)) \subset G_\mathbb{Z}$ is the monodromy group.

11.4.2 Period map. Griffiths infinitesimal relations

The subspaces $\phi^*_s H^{p,q}_s \subset V_\mathbb{C}$ give a Hodge decomposition of the fixed vector space $V_\mathbb{C}$, which varies smoothly with $s \in S$. A crucial property is that the dimensions $h^{p,q}$ of the subspaces $\dim H^{p,q}_s$ are independent of s [77]. Then for each point $s \in S$ we have a flag $F^\bullet_s: F^k_s \subset F^{k-1}_s \subset \cdots \subset F^0_s \equiv V_\mathbb{C}$ of fixed dimensions $f^\bullet$ which satisfies the Riemann bilinear relations, Eqs. (11.4) and (11.5), and hence by Section 11.1.2 corresponds to a point in the appropriate Griffiths domain D. Thus we have a map

$$F^\bullet: S \to D = G/V. \quad (11.62)$$

Strictly speaking this is correct only locally. Globally we have the ambiguity given by the monodromy representation (unless S is simply connected) and only the map

$$F^\bullet: S \to \Gamma\backslash D = \Gamma\backslash G/V \quad (11.63)$$

is well defined.

Definition 11.6 The map $F^\bullet: S \to \Gamma\backslash D$ is called the *period map*. To simplify the notation, we shall denote it simply as p.

The crucial result is:

Theorem 11.7 (Griffiths transversality [173]) *The period map* $p \equiv F^\bullet$ *is holomorphic and horizontal, in fact it satisfies the* <u>stronger</u> *condition*

$$p_*(TS) \subset T_D^{-1}. \quad (11.64)$$

Equation (11.64) is called the *Griffiths infinitesimal period relation*.

11.4 Variations of Hodge structures (VHS)

Proof We write z^i for the local holomorphic coordinates of the undeformed complex manifold K_0. Then locally on K we may find smooth functions $\zeta^i(z^j, \bar{z}^j, s)$, depending holomorphically on s, with $\zeta^i(z^j, \bar{z}^j, 0) = z^i$, which are holomorphic coordinates on K_s. In particular, they satisfy the Kodaira–Spencer equation

$$0 = \bar{\partial}\zeta^i(s) + \phi(s)\,\zeta^i(s) \equiv d\bar{z}^j(\partial_{\bar{z}^j}\zeta^i(s) + \phi(s)_{\bar{z}^j}{}^k \partial_{z^k}\zeta^i(s)), \qquad (11.65)$$

where $\bar{\partial}, \partial$ are defined with respect to the undeformed $s = 0$ complex structure, and $\phi(s)$ is the Kodaira–Spencer vector which depends holomorphically on s and has the form

$$\phi(s) = s^a\,\phi_a + O(s^2), \quad \text{where } \phi_a \in H^1(K_0, \Theta). \qquad (11.66)$$

Then

$$d\zeta^i(s) = \partial\zeta^i(s) + \bar{\partial}\zeta^i(s)\partial\zeta^i(s) - \phi(s)\zeta^i(s)$$
$$= \frac{\partial \zeta^i(s)}{\partial z^j}(dz^j - \phi(s)_{\bar{z}^k}{}^j\,d\bar{z}^k). \qquad (11.67)$$

We see that in complex structure s the subbundle of $(1,0)$–forms, $T_{1,0}^*(s) \subset T_\mathbb{R}^* \otimes \mathbb{C}$, is generated by the forms $\omega(s)^i = dz^j - \phi(s)_{\bar{z}^k}{}^j\,d\bar{z}^k$. Then a harmonic (p,q)-form on K_s has the structure [218]

$$\Phi(s, \bar{s}) = \Phi(s, \bar{s})_{i_1\cdots,i_p \bar{j}_1\cdots\bar{j}_q}\,\omega(s)^{i_1} \wedge \cdots \omega(s)^{i_p} \wedge \overline{\omega}(s)^{\bar{j}_1} \wedge \cdots \wedge \overline{\omega}(s)^{\bar{j}_q}. \qquad (11.68)$$

Let us take the derivatives with respect to s^a and $\bar{s}^a$ at $s = 0$,

$$\partial_{\bar{s}^a}\Phi(s,\bar{s})\big|_{s=0} = (\partial_{\bar{s}^a}\Phi_{i_1\cdots,i_p \bar{j}_1\cdots\bar{j}_q})\,dz^{i_1}\wedge\cdots dz^{i_p}\wedge d\bar{z}^{\bar{j}_1}\wedge\cdots\wedge d\bar{z}^{\bar{j}_q}$$
$$- q\,(\overline{\phi_a})_{i_{p+1}}^{\bar{j}_1}\,\Phi_{i_1\cdots,i_p \bar{j}_1\cdots\bar{j}_q}\,dz^{i_1}\wedge\cdots dz^{i_p}\wedge dz^{i_{p+1}}\wedge d\bar{z}^{\bar{j}_2}\wedge\cdots\wedge d\bar{z}^{\bar{j}_q}, \qquad (11.69)$$

$$\partial_{s^a}\Phi(s,\bar{s})\big|_{s=0} = (\partial_{s^a}\Phi_{i_1\cdots,i_p \bar{j}_1\cdots\bar{j}_q})\,dz^{i_1}\wedge\cdots dz^{i_p}\wedge d\bar{z}^{\bar{j}_1}\wedge\cdots\wedge d\bar{z}^{\bar{j}_q}$$
$$+ (-1)^k q\,(\phi_a)_{\bar{j}_{q+1}}^{i_1}\,\Phi_{i_1\cdots,i_p \bar{j}_1\cdots\bar{j}_q}\,dz^{i_2}\wedge\cdots dz^{i_p}\wedge d\bar{z}^{\bar{j}_1}\wedge\cdots\wedge$$
$$d\bar{z}^{\bar{j}_q}\wedge d\bar{z}^{\bar{j}_{q+1}}, \qquad (11.70)$$

which we may write as

$$\bar{\partial}_{\bar{s}}(\phi_s^* H^{p,q}) \subseteq \phi_s^* H^{p,q} \oplus \phi_s^* H^{p+1,q-1}, \qquad (11.71)$$

$$\partial_s(\phi_s^* H^{p,q}) \subseteq \phi_s^* H^{p,q} \oplus \phi_s^* H^{p-1,q+1}, \qquad (11.72)$$

or, in terms of the filtrations,

$$\partial_s(\phi_s^* F^p) \subseteq \phi_s^* F^{p-1}, \tag{11.73}$$

$$\bar{\partial}_{\bar{s}}(\phi_s^* F^p) \subseteq \phi_s^* F^p. \tag{11.74}$$

This implies the claim of the theorem. Indeed, comparing with Eq. (11.38) and (11.39), we see that Eqs. (11.73)(11.74) are equivalent to

$$p_*(\overline{TS}) \subseteq \overline{TD}, \tag{11.75}$$

$$p_*(TS) \subseteq T_D^{-1} \subset TD. \tag{11.76}$$

$\square$

11.4.3 Gauss–Manin connection. Hodge bundles

On $D = G/V$ we have several natural homogeneous bundles: the trivial complex bundle $\mathcal{V}_{\mathbb{C}}$, with its canonical flat connection d, the holomorphic subbundles $\mathcal{F}^p \subset \mathcal{V}_{\mathbb{C}}$ ($p = 0, \ldots, k$), and their quotient bundles $\mathcal{H}^{p,k-p} \equiv \mathcal{F}^p/\mathcal{F}^{p+1}$. We pull back all these bundles to S via the period map p, getting holomorphic vector bundles which we denote by the same symbols and also call Hodge bundles, since their fibers over $s \in S$ are, respectively, the vector spaces $V_{\mathbb{C}}$, F_s^p, and $H_s^{p,k-p}$.

The bundle $\mathcal{V}_{\mathbb{C}} \to S$ inherits a *flat* connection

$$\nabla : \mathcal{V}_{\mathbb{C}} \to \Lambda^1(\mathcal{V}_{\mathbb{C}}), \tag{11.77}$$

which is called the *Gauss–Manin connection*. Geometrically we can think of it as follows: the integral cohomology groups $H^k(K_s, \mathbb{Z})$ define a lattice $\mathcal{V}_{\mathbb{Z}}$ of integral local section of $\mathcal{V}_{\mathbb{C}}$. We define the connection ∇ by declaring the local sections of $\mathcal{V}_{\mathbb{Z}} \subset \mathcal{V}_{\mathbb{C}}$ to be *parallel*. This produces a flat connection on $\mathcal{V}_{\mathbb{C}}$ whose monodromy is precisely the monodromy representation of Section 11.4.1.

We decompose ∇ into types (1, 0) and (0, 1): $\nabla = \partial_\nabla + \bar{\partial}_\nabla$. We may also restrict ∇ to a subbundle $\mathcal{F}^p$; *a priori* we have $\partial_\nabla(\mathcal{F})^p \subseteq \Lambda^{1,0}(\mathcal{V}_{\mathbb{C}})$ and $\bar{\partial}_\nabla(\mathcal{F})^p \subseteq \Lambda^{0,1}(\mathcal{V}_{\mathbb{C}})$, but Eqs. (11.73)(11.74) give the stronger inclusions

$$\bar{\partial}_\nabla(\mathcal{F})^p \subseteq \Lambda^{0,1}(\mathcal{F}^p), \tag{11.78}$$

$$\partial_\nabla(\mathcal{F})^p \subseteq \Lambda^{1,0}(\mathcal{F}^{p-1}). \tag{11.79}$$

Equation (11.78) says that $\bar{\partial}_\nabla$ is a holomorphic structure on $\mathcal{F}^p$; this, of course, is no surprise since it is the pull–back of a holomorphic bundle with respect to a holomorphic map. Equation (11.79) defines the subbundle connection on $\mathcal{F}^p$, as we are going to discuss.

11.4.4 Curvature properties of Hodge bundles

Again we have our holomorphic bundle $\mathcal{V} \to S$, with its flat Gauss–Manin connection ∇, and its Hodge filtration by holomorphic subbundles

$$0 \subset \mathcal{F}^w \subset \mathcal{F}^{w-1} \subset \cdots \subset \mathcal{F}^0 \equiv \mathcal{V}. \tag{11.80}$$

The Hodge bundles are the quotient bundles $\mathcal{H}^{p,w-q} = \mathcal{F}^p/\mathcal{F}^{p+1}$, $p = 0, \ldots, w$. It is convenient to introduce an *indefinite* Hermitian form on $\mathcal{V}$,

$$h_{\text{ind}}(\cdot, \cdot) \stackrel{\text{def}}{=} i^w Q(\cdot, \bar{\cdot}), \tag{11.81}$$

which is preserved by the Gauss–Manin connection. The positive–definite Hermitian metric is given by $h(\cdot, \cdot) = i^{-w} h_{\text{ind}}(C \cdot, \cdot)$ with C the Weil operator. ∇ is *not* a metric connection with respect to h, since $\nabla C \neq 0$.

We may use h_{ind} to define orthogonal projections. Then, as smooth bundles, $\mathcal{F}^{p-1} = \mathcal{F}^p \oplus \mathcal{H}^{p-1,w-p+1}$. The Griffiths period relations give

$$\nabla \colon \mathcal{H}^{p,w-p} \to \Lambda^{1,0}(\mathcal{H}^{p-1,w-p+1}) \oplus \Lambda^1(\mathcal{H}^{p,w-p}) \oplus \Lambda^{0,1}(\mathcal{H}^{p+1,w-p-1}), \tag{11.82}$$

where $\Lambda^{p,q}(\mathcal{E})$ (resp. $\Lambda^k(\mathcal{E})$) stands for the vector space of C^∞ (germs of) differential forms of type (p,q) (resp. degree k) with coefficients in the sections of the vector bundle $\mathcal{E}$. Then we write

$$\nabla = \sigma + \nabla_H + \sigma^\dagger, \tag{11.83}$$

where

- $\sigma \colon \mathcal{H}^{p,w-p} \to \Lambda^{1,0}(\mathcal{H}^{p-1,w-p+1})$ is a C^∞–section of the bundle

$$\Lambda^{1,0}(\text{Hom}(\mathcal{H}^{p,w-p}, \mathcal{H}^{p-1,w-p+1})), \tag{11.84}$$

 called the *second fundamental form* of $\mathcal{H}^{p,w-p}$;
- $\sigma^\dagger$ is the conjugate of σ with respect to the Hermitian form h;
- $\nabla_H \colon \mathcal{H}^{p,w-p} \to \Lambda^1(\mathcal{H}^{p,w-p})$ is a connection on the Higgs bundle $\mathcal{H}^{p,w-p}$ which, by construction,[8] is metric for the indefinite Hermitian metric h_{ind}. Since

$$h|_{\mathcal{H}^{p,w-p}} = (-1)^{p-w} h_{\text{ind}}|_{\mathcal{H}^{p,w-p}}, \tag{11.85}$$

∇_H is also metric for the positive–definite Hermitian metric. Moreover, from the flatness of Gauss–Manin connection, $\nabla^2 = 0$, it follows that the $(0, 1)$ part of ∇_H, $\nabla_H^{0,1}$, also squares to zero: $(\nabla_H^{0,1})^2 = 0$. Hence ∇_H defines a complex

---

[8] Note that this construction is exactly the Levi–Civita construction of the metric connection.

structure on $\mathcal{H}^{p,w-p}$, and thus ∇_H is the Chern connection, the unique one which is both metric and adapted to the complex structure.

It is convenient to take the direct sum over p and define

$$\nabla: \bigoplus_p \mathcal{H}^{p,w-p} \to \Lambda^1\Big(\bigoplus_p \mathcal{H}^{p,w-p}\Big), \qquad (11.86)$$

and similarly for σ and $\sigma^\dagger$. Going to the holomorphic gauge, we set $\nabla_H^{0,1} = \bar{\partial}$ and $\nabla_H^{1,0} = D$. Then $\nabla^2 = 0$ gives the tt^*–like equations[9]

$$\bar{\partial}\sigma = \sigma \wedge \sigma = D^2 = 0, \qquad (11.87)$$

$$F = \bar{\partial}D + D\bar{\partial} = -\sigma \wedge \sigma^\dagger - \sigma^\dagger \wedge \sigma. \qquad (11.88)$$

Let us consider the curvature F of the first Hodge bundle $\mathcal{H}^{w,0}$. One has $\sigma^\dagger|_{\mathcal{H}^{w,0}} = 0$. Hence

$$F\big|_{\mathcal{H}^{w,0}} = -\sigma^\dagger\big|_{\mathcal{H}^{w-1,1}} \wedge \sigma\big|_{\mathcal{H}^{w,0}}. \qquad (11.89)$$

Recall that an element of $\Lambda^{1,1}(\text{End}\,\mathcal{E})$ is said to be *positive* if for all non–zero C^∞ sections of s of $\mathcal{E}$ the $(1,1)$ form $i\,h(Fs,s)$ is a positive–definite Kähler form. Then Eq. (11.89) shows the following:

Proposition 11.8 *The curvature of the Hodge bundle $\mathcal{H}^{w,0}$ is positive when restricted to T_D^{-1}; that is*

$$i\,h(Fs,s)(\xi,\bar{\xi}) > 0 \qquad (11.90)$$

for all non–zero section s and non–zero $\xi \in T_D^{-1}$.

11.4.5 Infinitesimal variations of Hodge structures

The period map $p\colon S \to D \equiv G/V$ has no algebraic invariants, since G acts transitively on D. However, in the physical applications of the VHS there are such invariants in the form of Yukawa couplings. (Recall from Section 1.2.2 that these couplings are best understood as bundle maps.) The counterpart of Yukawa couplings[10] in Hodge theory are known as *infinitesimal variations of Hodge structures* [173].

Our discussion here will be informal. For a more precise treatment see ref. [173]. Suppose we have a VHS of weight k with period map $p\colon S \to D$. Let $s_0 \in S$ be a

---

[9] Historically the tt^* equations were first motivated by the analogy with the variations of Hodge structures [85].
[10] For Yukawa couplings in heterotic string theory from the VHS viewpoint, see [73].

point and $\{H^{p,q}\}$ the polarized Hodge decomposition at s_0. Then the differential at s_0

$$p_*: T_{s_0}S \to T_p^{-1}(s_0) = \bigoplus_{1\leq p\leq k} \mathrm{Hom}(H^{p,k-p}, H^{p-1,k-p+1}). \tag{11.91}$$

We write δ for this linear map ($T \equiv T_{s_0}S$)

$$\delta: T \to \bigoplus_{1\leq p\leq k} \mathrm{Hom}(H^{p,k-p}, H^{p-1,k-p+1}). \tag{11.92}$$

δ satisfies two properties:

(i) $\delta(\xi_1)\delta(\xi_2) = \delta(\xi_2)\delta(\xi_1) = 0 \quad \forall\, \xi_1, \xi_2 \in T.$ (11.93)

(ii) $Q(\delta(\xi)v, w) + Q(v, \delta(\xi)w) = 0 \quad \forall v \in H^{p,q}, w \in H^{q+1,p-1}.$ (11.94)

The *infinitesimal variations of Hodge structure* are the axiomatization of this situation; we have a polarized Hodge structure $(V_{\mathbb{Z}}, H^{p,k-p}, Q)$, a vector space T, and a linear map δ as in Eq. (11.92) satisfying (i) and (ii).

The basic invariant is the kth iterate of δ:

$$\delta(\xi_1)\delta(\xi_2)\cdots\delta(\xi_k)\colon \otimes^k T \to \mathrm{Hom}(H^{k,0}, H^{0,k}) \equiv \otimes^2(H^{k,0})^\vee. \tag{11.95}$$

It follows from Eq. (11.94) that this map, which we call δ^k, is symmetric in its arguments ξ_ℓ. Then, actually, we have a map

$$\delta^k: \odot^k T \to \odot^2(H^{k,0})^\vee \simeq \odot^2 H^{0,k}. \tag{11.96}$$

This symmetric map is the basic algebraic invariant of an infinitesimal deformation of Hodge structure. It is believed that it fully determines the variation of Hodge structure.

11.4.6 Limit behavior

For the application to supergravity and string theory, in particular to understand the details of the mirror map, it is important to consider the limit behavior of the Hodge variations. The topic is strictly related to the conjectures about the properties of an effective theory which describes the low–energy physics of a *consistent* quantum theory of gravity, as discussed in Section 4.9. An example would be the one–parameter family of quintic 3–folds in Eq. (11.56). The point $s = 0$ was excised from S since it does not correspond to a smooth manifold. One easily checks that the point $s = 0$ is at infinite distance from any regular point $s \in \mathbb{C}^\times$ in the intrinsic Weil–Petersson metric; so here we are discussing the behavior at "infinite complex structure." For simplicity we consider a one–parameter family and focus on a neighborhood of the degeneration point; hence we take S to be a punctured disk Δ^*, with $s = 0$ removed, and consider the limit as $s \to 0$ of the Hodge structure.

Example: Genus 1 curves. Let us consider a simple example, namely the universal family of elliptic curves (i.e., genus 1 Riemann surfaces) in the Weierstrass form

$$y^2 = 4x^3 - \frac{27j}{j-1728}x - \frac{27j}{j-1728}, \qquad (11.97)$$

where $j = j(\tau) \in P\mathbb{C}^1$ is the modular invariant of the curve[11] which, by a change of coordinates $j \mapsto s$, we rewrite as

$$E_s: \qquad y^2 = 4x^3 - 27(s+1)x - 27(s+1), \qquad (11.98)$$

so that the singular curve is at $s = 0$. By definition, the space of holomorphic 1–forms on a genus one curve is one–dimensional; an element of $H^{1,0}(E_s)$ is a multiple of

$$\eta(s) = \frac{dx}{y} = \text{PR}\left\{\frac{dx \wedge dy}{y^2 - 4x^3 + 27(s+1)x + 27(s+1)}\right\}, \qquad (11.99)$$

where PR stands for "Poincaré residue" [171]. $\eta(s)$ is manifestly holomorphic in s. In this example the Hodge filtration reduces to one term

$$H^{1,0}_s \equiv F^1 \subset F^0 \equiv H^1(E_s, \mathbb{C}) \qquad (11.100)$$

and the Griffiths domain $Sp(2, \mathbb{R})/U(1)$ is just the upper half–plane $\mathfrak{H}$, which we see as an open domain in $P\mathbb{C}^1$. Fixing a symplectic basis A, B of $H_1(E_s, \mathbb{Z})$, $Q^\vee(A, B) = 1$, the period map $p \colon \mathbb{C}^* \to P\mathbb{C}^1$, up to $SL(2, \mathbb{Z})$ monodromy, is given by

$$p(s) = \left(\int_B \eta(s) : \int_A \eta(s)\right) \equiv (\tau : 1). \qquad (11.101)$$

The Weil–Petersson metric is just the pull–back of the usual $SL(2, \mathbb{R})$–invariant Poincaré metric on the upper half–plane $\mathfrak{H}$. Since as $\tau \to i\infty$

$$s = 1728\, e^{2\pi i \tau} + O(e^{4\pi i \tau}), \qquad (11.102)$$

---

[11] Compare proposition 4.3 of [202]. Here $j(\tau)$ is the usual modular invariant function $j = q^{-1} + 744 + 196884q + O(q^2)$ where $q = e^{2\pi i \tau}$ [276].

11.4 Variations of Hodge structures (VHS)

we see that $s = 0$ gives precisely the degenerate torus with $\tau = i\infty$, and a loop generating $\pi_1(S) = \mathbb{Z}$ is given by $\tau \mapsto \tau+1$: that is, by the modular transformation $T \in SL(2,\mathbb{Z})$. Thus the monodromy is an integral matrix whose eigenvalues are all equal to 1, but not the identity, since it has a non–trivial Jordan block: T is just *unipotent*. Its minimal polynomial is

$$(T - 1)^2 = 0. \tag{11.103}$$

The logarithm N of the monodromy,

$$N = \log T = \begin{pmatrix} 0 & 1 \\ 0 & 0 \end{pmatrix}, \tag{11.104}$$

is nilpotent of index 1, $N^2 = 0$. Consider the map

$$\hat{p}(s)\colon \Delta^* \to P\mathbb{C}^1, \tag{11.105}$$

$$\hat{p}(s) = \exp\left(\frac{1}{2\pi i} \log s\, N\right) \begin{pmatrix} -\frac{\log 1728}{2\pi i} \\ 1 \end{pmatrix} = \begin{pmatrix} \tau + O(e^{2\pi i\tau}) \\ 1 \end{pmatrix}, \tag{11.106}$$

which differs from the actual period map $p(s) = (\tau : 1)$ only for the exponentially small (as $\tau \to i\infty$) correction $O(e^{2\pi i\tau})$. In physical applications, this exponentially small correction will be interpreted as a non–perturbative term $O(e^{-c/g^2})$, so, in physical language "*$p(s)$ and $\hat{p}(s)$ agree to all orders in perturbation theory.*"

The above example illustrates well the general features of the "infinite deformations" of complex structures. We summarize the basic facts in the following theorem.

Theorem 11.9 *(SL$_2$ orbit theorem) Consider a weight k variation of Hodge structure with parameter space the punctured disk Δ^*, and let T be the image of the generator of $\pi(\Delta^*)$ under the monodromy representation. Then the monodromy T is quasi–unipotent, i.e.,*

$$(T^n - 1)^{m+1} = 0 \quad \text{for some } n, m \in \mathbb{N}, \tag{11.107}$$

with $m \leq k$. By changing the coordinate on Δ^*, $s \mapsto s^n$, we reduce to the case that T is unipotent. We write z for the coordinate on the universal cover $\mathfrak{H}$ of Δ^*, $s = \exp(2\pi i z)$. Then let N be the logarithm of T:

$$N = \log T = (T-1) - \frac{1}{2}(T-1)^2 + \cdots + \frac{(-1)^{m+1}}{m}(T-1)^m. \tag{11.108}$$

There exists a constant element $\psi(0) \in \check{D}$ (the limiting filtration) such that the map

$$\hat{p}(z)\colon \mathfrak{H} \to \check{D}, \qquad z \mapsto \exp(zN)\psi(0), \tag{11.109}$$

has the properties:

(1) *The map $\hat{p}(z)$ is horizontal (i.e., satisfies the Griffiths relations).*
(2) *For $\operatorname{Im} z \gg 0$ $\hat{p}(z) \in D$, and the difference between $\hat{p}(z)$ and the period map $p(z)$ is exponentially small $O(e^{-2\pi \operatorname{Im} z})$.*

We shall not prove this theorem, although at this point we have all the necessary tools. Proofs may be found in [173].

11.5 The case of a Calabi–Yau 3–fold

The instance of Hodge variation most interesting for physics is the weight 3 one which describes the varying Hodge decomposition of $H^3(K_s, \mathbb{C})$ in a universal family of strict Calabi–Yau 3–folds K_s. Recall that a Calabi–Yau 3–fold is a compact three–dimensional Kähler manifold with trivial canonical bundle. The basic result is:

Theorem 11.10 (Yau [326]) *Let K be a compact complex manifold with a Kähler metric of Kähler form ω and trivial canonical bundle. Then there is a <u>unique</u> Ricci–flat Kähler metric whose Kähler class is $[\omega]$.*

Comparing with our discussions in Chapters 3 and 9, we may also state this by saying that the Riemannian metrics on K having holonomy algebra $\subseteq \mathfrak{su}(3)$ are parametrized by the moduli space of complex structures on K together with the space of the positive–definite Kähler classes (which is a convex cone in $\mathbb{R}^{B_2}$). For a proof see [326]. For the application to physics, the following result is also important:

Proposition 11.11 *Let K be a compact Kähler 3–fold with trivial canonical bundle. Then one of the following three possibilities applies:*

(1) *$B_1 = 0$ and all Ricci flat Kähler metrics have holonomy algebra $\mathfrak{su}(3)$.*
(2) *$B_1 = 2$ and all Ricci flat Kähler metrics have holonomy algebra $\mathfrak{su}(2)$.*
(3) *$B_1 = 6$ and all Ricci flat Kähler metrics have holonomy algebra 0.*

Physically, the three possibilities will correspond to low–energy effective 4D theories with $\mathcal{N} = 2, 4$, and 8, respectively.

Proof Let ξ be a harmonic 1–form. From the Weitzenböck formulae, we see that a harmonic 1–form on a Ricci flat compact manifold is parallel, hence invariant under $\mathfrak{hol}$ by Fundamental Principle 3.3. Then the holonomy representation decomposes, and using de Rham's Theorem 3.6 we see that the universal cover $\widetilde{K}$ is a metric product of the form $\mathbb{R}^{B_1} \times X$, with X Ricci–flat Kähler and *indecomposable*. Hence $\mathfrak{hol} = \mathfrak{su}(3 - B_1/2)$ (B_1 is even since K is Kähler). $B_1 = 4$

11.5 The case of a Calabi–Yau 3–fold

is not possible since in that case we would get $\mathfrak{hol} = \mathfrak{su}(1) \equiv 0$, which implies $B_1 = 6$. □

Corollary 11.12 *Let K be a compact strict Calabi–Yau 3–fold (i.e., $\mathfrak{hol} = \mathfrak{su}(3)$). Then $H^3(K, \mathbb{C})_{\text{prim}} = H^3(K, \mathbb{C})$.*

Thus we may forget about the distinction of primitive and non–primitive cohomology, which is a simplification. Moreover, we have already seen that the polarized variation of the Hodge structure of $H^3(K, \mathbb{C})$ is independent of the Kähler class. This means that, in the strict $\mathfrak{su}(3)$ case, the moduli space of Ricci–flat Kähler metrics on K factorizes into a product

$$\text{(moduli of complex structures)} \times \text{(moduli of Kähler structures)}. \tag{11.110}$$

By a *universal deformation space* we mean a deformation space S such that $TS \simeq H^1(K, \Theta)$; that is, depending on the maximal number of parameters $\dim H^1(K, \Theta) \equiv h^{2,1}$. We have already shown, Theorem 11.5, that such a universal space exists and it is smooth.

The isomorphism $H^{2,1}(K) \simeq H^1(K, \Theta)$ has another useful consequence. Let $\Omega_{ijk} dz^i \wedge dz^j \wedge dz^k$ be the holomorphic 3–form. Let $\{\phi_a\}$ ($a = 1, \ldots, h^{2,1}$) be the basis of $H^1(K, \Theta)$ corresponding to the basis $\partial/\partial s^a$ of $T_0 S$. The identification $H^1(K, \Theta) \to H^{2,1}(K)$ takes the form

$$\phi_{a\,ij\bar{k}} = \Omega_{ijl} \phi_{a\,\bar{k}}^{\,l} \qquad a = 1, \ldots, h^{2,1}, \tag{11.111}$$

and the $\phi_{a\,ij\bar{k}}\, dz^i \wedge dz^j \wedge d\bar{z}^k$ form a basis of $H^{2,1}(K)$. We saw in Eq. (11.69) that

$$\partial_{s_a} \Omega\big|_0 = -3\, \Omega_{ijl} \phi_{a\,\bar{k}}^{\,l} dz^i \wedge dz^j \wedge d\bar{z}^k \mod H^{3,0}. \tag{11.112}$$

Let $\Omega(s)$ be any (local) holomorphic section of the Hodge line bundle $\mathcal{L} \equiv \mathcal{F}^3 \to S$; for fixed s, $\Omega(s)$ is a holomorphic $(3, 0)$–form on K_s with some s–dependent normalization. Comparing the two equations (11.111) and (11.112), we deduce that

$$\Omega(s),\ \partial_{s_1}\Omega(s),\ \partial_{s_2}\Omega(s),\ \cdots,\ \partial_{s_{h^{2,1}}}\Omega(s) \text{ span } F_s^2 = H_s^{3,0} \oplus H_s^{2,1}. \tag{11.113}$$

F_s^3 is, of course, spanned by $\Omega(s)$. On the other hand, $F_s^1 = (F_s^3)^\perp$, so the full Hodge filtration $F_s^3 \subset F_s^2 \subset F_s^1 \subset F_s^0 \equiv V_\mathbb{C}$ is completely determined by $\Omega(s)$ and its derivatives. Hence $\Omega(s)$ fully determines the period map

$$p \colon S \to Sp(2 + 2h^{2,1}, \mathbb{R})/U(1) \times U(h^{2,1}). \tag{11.114}$$

Note that for all holomorphic (local) functions $\lambda(s)$, the sections $\Omega(s)$ and $e^{\lambda(s)}\Omega(s)$ give the same period map p. Hence the transformation

$$\Omega(s) \to e^{\lambda(s)} \Omega(s) \tag{11.115}$$

is a kind of "gauge transformation," leaving invariant the Hodge variation and all its physical consequences.

Equation (11.113) is the special version of the Griffiths infinitesimal relations which applies to simply–connected Calabi–Yau 3–folds. These relations may be written in a more convenient way [66], as we are going to show.

11.5.1 Holomorphic contact structures and Legendre manifolds

A *holomorphic contact structure* on a $(2m+1)$–dimensional manifold M is the obvious analogue of a real contact structure: it is a rank 1 holomorphic subbundle $\mathcal{C} \subset T^*M$ such that if κ is a local generator (that is, a nowhere vanishing local holomorphic section) we have

$$\kappa \wedge (d\kappa)^m \neq 0 \quad \text{pointwise.} \tag{11.116}$$

Note that the local 1–forms κ and $e^\lambda \kappa$ represent the same contact structure, for all holomorphic functions λ.

A Legendre submanifold is a holomorphic submanifold $\iota\colon L \to M$, of complex dimension m, such that

$$\kappa|_L \equiv \iota^*\kappa = 0. \tag{11.117}$$

The Darboux theorem [67] states that, given a holomorphic contact structure $\mathcal{C}$ on M, we may find local coordinates (S, p_i, q_i), $i = 1, \ldots, m$, such that

$$\kappa = dS - p_1 \, dq_1 - p_2 \, dq_2 - \cdots - p_m \, dq_m, \tag{11.118}$$

as we know from classical mechanics. Then, locally, *generic* Legendre submanifolds $L \subset M$ are described in terms of a holomorphic function $S(q_i)$, called the Hamilton–Jacobi function in classical mechanics, by the equations

$$S = S(q_i), \quad p_i = \frac{\partial S}{\partial q_i}. \tag{11.119}$$

Note that the function $S(q_i)$ is not globally defined.

11.5.2 The canonical contact structure on $P\mathbb{C}^{2m+1}$

A point $x \in Sp(2m+2, \mathbb{R})/U(1) \times U(m)$ specifies a self–dual flag

$$F^3 \subset F^2 \subset F^1 \subset \mathbb{C}^{2m+2} \tag{11.120}$$

with dimensions $(f^3, f^2) = (1, m+1)$, which is polarized by the symplectic form $Q(\cdot, \cdot)$. In particular, a point x defines a line $F^3 \subset \mathbb{C}^{2m+2}$, that is, a point $f(x) \in P\mathbb{C}^{2m+1}$. Thus we have a map

$$f: Sp(2+2m, \mathbb{R})/U(1) \times U(m) \to P\mathbb{C}^{2m+1}. \tag{11.121}$$

By composing the period map $p: S \to Sp(2+2m, \mathbb{R})/U(1) \times U(m)$ with the map f, we get a map

$$\zeta = f \circ p: S \to P\mathbb{C}^{2m+1}, \tag{11.122}$$

which is a kind of reduced period map. We see a point $z \in P\mathbb{C}^{2m+1}$ as a $(2m+2)$–vector $z = (z_1, \ldots, z_{2m+2})$ defined up to overall non–zero factors, $z \sim e^\lambda z$. The projective space $P\mathbb{C}^{2m+1}$ has a canonical holomorphic contact structure. Let $Q: \mathbb{C}^{2m+2} \otimes \mathbb{C}^{2m+2} \to \mathbb{C}$ be the standard symplectic form on $\mathbb{C}^{2m+2}$. The 1–form

$$\kappa = Q(z, dz) \tag{11.123}$$

defines the canonical holomorphic contact structure on $P\mathbb{C}^{2m+1}$. The structure is clearly well defined; that is, independent of the representative z of the point in $P\mathbb{C}^{2m+1}$. Indeed, let $e^\lambda z$ be another representative

$$Q(e^\lambda z, d(e^\lambda z)) = e^{2\lambda} Q(z, dz), \tag{11.124}$$

since $Q(z, z) = 0$.

Proposition 11.13 ([66]) *The map $\zeta: S \to P\mathbb{C}^{2m+1}$ is a Legendre submanifold L with respect to the canonical contact structure.*

Proof Indeed, $\zeta(s) = \Omega(s)$ and the statement is equivalent to the Griffiths infinitesimal period relations (11.113). □

We stress that the 1–form $\zeta^* \kappa$ has coefficients valued in the line–bundle $\mathcal{L}^2$, where $\mathcal{L} \to S$ is the Hodge line bundle of holomorphic $(3, 0)$–forms.

We may reconstruct the full period map p from the Legendre submanifold by taking the derivatives of $\zeta(s)$, as discussed around Eq. (11.114). We conclude that *the period map is locally specified by a (holomorphic) Hamilton–Jacobi function $S(q_i)$*. Of course, globally we have to pay attention to the monodromy representation. The function $S(q_i)$ changes under the action of the monodromy group Γ.

All other invariants of the variation of Hodge structure are likewise constructed out of κ. From the infinitesimal period relations we get

$$Q(\Omega(s), \partial_{s_a}\Omega(s)) = 0, \tag{11.125}$$

$$Q(\Omega(s), \partial_{s_a}\partial_{s_b}\Omega(s)) = 0, \tag{11.126}$$

$$Q(\Omega(s), \partial_{s_a}\partial_{s_b}\partial_{s_c}\Omega(s)) \neq 0. \tag{11.127}$$

The third line gives a symmetric cubic form

$$\delta^3 \colon \odot^3 TS \to \mathcal{L}^{-2}, \tag{11.128}$$

which is nothing other than the basic invariant (11.96).

11.5.3 The Weil–Petersson metric on S

The above structures gives a simple formula for the Weil–Petersson metric on the moduli space S. The basic fact is:

Theorem 11.14 *Let $K \to S$ be a universal family of strict Calabi–Yau 3–folds. The Weil–Petersson metric on the universal moduli space S is Hodge, its Kähler form being the curvature of the $\mathcal{L} \equiv \mathcal{F}^3 \equiv \mathcal{H}^{3,0}$ Hodge bundle.*

The (holomorphic) Hodge line bundle $\mathcal{L} \to S$ has a natural, positive–definite, Hermitian metric $iQ(\bar{v}, v) > 0$. The curvature of the associated holomorphic connection is

$$-i\partial\bar{\partial} \log iQ(\overline{\Omega(s)}, \Omega(s)), \tag{11.129}$$

which, as we saw in Section 11.4.4, is a *positive* $(1,1)$–form on S, hence defines a Hodge–Kähler metric on the moduli space S. The Kähler form (11.129) is

$$\omega(s) = -i\partial\bar{\partial} \log \int_K \overline{\Omega}(s) \wedge \Omega(s)$$

$$= i\frac{\int_K \overline{\partial\Omega}(s) \wedge \partial\Omega(s)}{\int_K \overline{\Omega}(s) \wedge \Omega(s)} - i\frac{\int_K \overline{\partial\Omega}(s) \wedge \Omega(s) \int_K \overline{\Omega}(s) \wedge \partial\Omega(s)}{(\int_K \overline{\Omega}(s) \wedge \Omega(s))^2}. \tag{11.130}$$

It remains to see that this natural Hodge metric coincides with the Weil–Petersson metric on the moduli space. Indeed, from Eq. (11.112) we have

$$\partial_{s_a}\Omega(s) = f_a(s)\,\Omega(s) - i(\phi_a(s))\Omega(s), \tag{11.131}$$

where $\phi_a(s)$ is the Kodaira–Spencer vector dual to ds^a, $i(\,\cdot\,)$ is the vector–form contraction operation, and $f_a(s)$ are some holomorphic functions. We may write

$\Omega(s) = \rho \, \epsilon(s)$, where $\epsilon(s)$ is a half–volume form, i.e.,

$$i\,\epsilon(s) \wedge \overline{\epsilon}(s) = \frac{1}{3}\omega(s)^3. \tag{11.132}$$

Plugging these expressions into Eq. (11.130), we get for the Kähler metric:

$$G_{a\bar{b}}(s) = \frac{i}{V}\int_K i(\phi_a)\epsilon \wedge \overline{i(\phi_b)\epsilon} = V^{-1} h(\phi_a, \phi_b), \tag{11.133}$$

where h is the canonical Hermitian metric on $\Lambda^{0,1}(\Theta)$. Then, by definition $G_{a\bar{b}}$ is the (properly normalized) Weil–Petersson metric. Since the line bundle $\mathcal{L} \to D$ is homogeneous, we conclude that:

Corollary 11.15 *The Weil–Petersson metric on the moduli space S is the pull-back of a $Sp(2h^{2,1}+2,\mathbb{R})$–invariant metric on $Sp(2h^{2,1}+2,\mathbb{R})/U(1) \times U(h^{2,1})$ by the period map p. In particular, the holomorphic sectional curvatures of S are negative and bounded away from zero.*

11.5.4 Special (projective) coordinates

The geometry of the period map for a strict Calabi–Yau 3–fold simplifies using *special projective* coordinates. In Section 11.5.2 we saw that the period map p defines a map $\zeta: S \to P\mathbb{C}^{2h^{2,1}+1}$. We aim to write ζ explicitly. To do that, we fix a symplectic basis in the lattice $H_3(K,\mathbb{Z})$; that is, we choose 3–cycles A^I, B_J, with[12] $I, J = 0, 1, \ldots, h^{2,1}$, such that

$$A^I \cdot A^J = B_I \cdot B_J = 0, \qquad A^I \cdot B_J = \delta^I{}_J, \tag{11.134}$$

where $\cdot$ stands for the intersection pairing in homology. Then ζ is explicitly given by the periods of $\Omega(s)$,

$$\zeta(s) = \left(\int_{A^0}\Omega(s),\cdots,\int_{A^{h_{2,1}}}\Omega(s), \int_{B_0}\Omega(s),\cdots,\int_{B_{h_{2,1}}}\Omega(s)\right) \in P\mathbb{C}^{2h_{2,1}+1}, \tag{11.135}$$

which specify the position of the line F_s^3 in the vector space $V_\mathbb{C} \simeq \mathbb{C}^{2h^{2,1}+2}$. We write $(X^0, X^1,\ldots,X^{h^{2,1}},P_0,P_1,\ldots P_{h_{2,1}})$ for the homogeneous coordinates of $P\mathbb{C}^{2h^{2,1}+1}$. Then the contact form κ, which is a 1–form with values in the line

---

[12] Capital letters $I, J,\ldots$ take values $0, 1,\ldots, h^{2,1}$, lower-case letters $i, j,\ldots$ values $1, 2,\ldots, h^{2,1}$.

bundle $\mathcal{O}(2) \to P\mathbb{C}^{2h^{2,1}+1}$, takes the form

$$\kappa = \sum_{I=0}^{h^{2,1}} P_I \, dX^I. \tag{11.136}$$

Comparing with Section 11.5.2, we see that a Legendre submanifold in general position for this holomorphic contact structure is locally specified in terms of a Hamilton–Jacobi function $S(X^I)$ which is *a homogeneous function of the homogeneous coordinates X^I of degree* 2, i.e.

$$S(\lambda X^I) = \lambda^2 S(X^I) \quad \text{for all } \lambda \in \mathbb{C}. \tag{11.137}$$

In terms of such an $S(X^I)$, the Legendre manifold is the locus in $P\mathbb{C}^{2h^{2,1}+1}$ given by the homogeneous equations

$$P_I = \frac{\partial S(X^J)}{\partial X^I}, \quad I = 0, 1, \ldots, h^{2,1}. \tag{11.138}$$

Then the special coordinates on S are just the non–homogeneous version of the X^I, i.e., $z^I = X^I/X^0$. There are $h^{2,1} = \dim S$ of them, and they form a good (local!) system of coordinates by the Griffiths infinitesimal relations. Globally, we have to glue together several such local descriptions *via* the monodromy representation.

All quantities have simple expressions in terms of $S(X^I)$.

The Kähler potential of the Weil–Petersson metric. The Kähler potential of the Weil–Petersson metric is

$$\Phi = -\log\left(i X^I \overline{\partial_I S} - i \partial_I S \overline{X}^I\right). \tag{11.139}$$

The projective equivalence $X^I \to e^\lambda X^I$ becomes the Kähler equivalence.

Symmetric 3–form. The symmetric 3–form is simply

$$\delta^3(\xi_1, \xi_2, \xi_3) = \partial_{X^i} \partial_{X^j} \partial_{X^k} S(X^I) \xi_1^i \xi_2^j \xi_3^k, \quad i, j, k = 1, 2, \ldots, h^{2,1}. \tag{11.140}$$

11.5.5 Monodromy and the Hamilton–Jacobi function

We have already mentioned that the Hamilton–Jacobi function $S(X^I)$ will *change* (also in value) under monodromy transformations. A monodromy transformation is an element $\sigma \in Sp(2h^{2,1}+2, \mathbb{Z})$ which acts linearly on the homogeneous coordinates (X^I, P_I) of $P\mathbb{C}^{h^{2,1}+1}$. Comparing with Section 1.4 we see that this is the same action as electromagnetic duality on the field–strengths (F^I, G_I). The role of the Hamilton–Jacobi functions in the two cases is exactly the same, and so to get the variation of $S(X^I)$ under monodromy we may borrow directly Eq. (1.72) from Chapter 1.

11.5.6 "Large" complex structures

Finally, we want to discuss the behavior of the period map, that is, of the homogeneous Hamilton–Jacobi function $S(X^I)$, for "large" complex structures, namely near the boundary of the moduli space S.

For simplicity let us consider first the one–modulus case, $h^{2,1} = 1$. The asymptotic behavior of p is described by Theorem 11.9 in terms of a rational matrix N satisfying $N^4 = 0$. By a $Sp(4, \mathbb{R})$ duality redefinition we may put the matrix N in the form

$$N = \left(\begin{array}{c|c} T & 0 \\ \hline C & -T^t \end{array} \right), \tag{11.141}$$

$$T = \begin{pmatrix} 0 & 0 \\ 1 & 0 \end{pmatrix}, \quad C = \begin{pmatrix} a & b \\ b & c \end{pmatrix}. \tag{11.142}$$

Then, as in the example of Section 11.4.6, *up to exponential small terms* we have

$$\begin{pmatrix} X^I(w) \\ P_I(w) \end{pmatrix}_{\text{asymp}} = \exp(wN) \begin{pmatrix} X^I \\ P_I \end{pmatrix} \tag{11.143}$$

for some constants $X^I, P_I, I = 0, 1$. Explicitly, this gives the period map

$$X^0(w) = X^0,$$

$$X^1(w) = wX^0 + X^1,$$

$$P_0(w) = \left(a - \frac{c}{3}w^3\right)X^0 + \left(bw - \frac{c}{2}w^2\right)X^1 + P_0 + O(e^{-2\pi \operatorname{Im} w}), \tag{11.144}$$

$$P_1(w) = \left(b + \frac{c}{2}w^2\right)X^0 + cwX^1 + P_1 + O(e^{-2\pi \operatorname{Im} w}).$$

We may set $X^1 = 0$ and $b = 0$ by a shift of w and P_1, so that $w = X^1/X^0$ coincides with the special coordinate. This period map follows from the homogeneous Hamilton–Jacobi function

$$S(X^0, X^1) = \frac{1}{X^0}\left(\frac{1}{2}P_0(X^0)^3 + aX^1(X^0)^2 + \frac{c}{3}(X^1)^3\right) + O(e^{-2\pi \operatorname{Im} X^1/X^0}). \tag{11.145}$$

In the case of a multi–dimensional degeneration, S is locally of the form $(\Delta^*)^{h^{2,1}}$ whose universal cover is the domain in $\mathfrak{H}^{h^{2,1}}$ with $\operatorname{Im} w^i \gg 0$. If we set $w^i = tc^i$

and consider the asymptotic behavior of $S(w^i)$ as $t \to \infty$ at fixed c^i, we reduce back to the previous one–modulus analysis. Then S is a cubic polynomial in t. Since this happens for all c^i we conclude that:

Theorem 11.16 *The large complex structure limit of the period map for strict Calabi–Yau 3–folds is described (in some convenient $Sp(2h^{2,1} + 2, \mathbb{R})$ frame!) by a homogeneous Hamilton–Jacobi function $S(X^I)$, $I = 1, 2, \ldots, h^{2,1}$, which, up to exponentially small corrections, is a cubic form in the X^I,*

$$S(X^I) = \frac{C_{IJK} X^I X^J X^K}{3\, X^0} + O(e^{-2\pi \operatorname{Im} X^i / X^0}), \tag{11.146}$$

for some constant symmetric tensor C_{IJK}. In particular, the symmetric cubic form has a constant limit equal to C_{ijk}, $i, j, k = 1, 2, \ldots, h^{2,1}$.

Remark. The construction of the large complex structure asymptotic behavior is based on the nilpotent matrix N. By the Jacobson–Morosov theorem (Proposition 4.11), N defines a $SL(2, \mathbb{R})$–module structure on $V_\mathbb{C} \equiv H^3(K, \mathbb{C})$. This should be compared with the $SL(2, \mathbb{R})$–module structure on the cohomology of K given by the Lefshetz decomposition. In other words, at infinite complex structure, the Hodge decomposition "looks like" the Lefshetz decomposition for a Hodge diamond rotated by $90°$. This should not be a surprise for the cognoscenti: it is mirror symmetry in action.

11.5.7 The physical relevance of VHS

The relevance of Calabi–Yau 3–folds for physics is that they represent supersymmetry–preserving vacua of superstring theory [74]. Suppose we compactify one of the five 10–dimensional superstring theories on a manifold of the form $\mathbb{R}^4 \times K$ with K a strict Calabi–Yau 3–fold. By the arguments of Chapter 6, the resulting 4D effective theory is invariant under a number of supersymmetries equal to the number of Killing spinors in this geometry. For type IIA or IIB this gives a low–energy effective $\mathcal{N} = 2$ supergravity in four dimensions whose physics is controlled by the deformations of the geometry of the Calabi–Yau manifold K. Since the low–energy degrees of freedom correspond to harmonic forms on K, the dynamics of the effective SUGRA is described by variations of Hodge structures. Then it is quite natural that the geometric structures of *all* 4D $\mathcal{N} = 2$ SUGRA's may be described in the language of the variations of Hodge structures. This is the approach we shall use in the next chapter. In Section 12.6 we shall discuss the relations of the geometry of a family of Calabi–Yau 3–folds with $\mathcal{N} = 2$ supergravity in more detail.

11.5 The case of a Calabi–Yau 3–fold

Exercise 11.5.1 Prove that a compact strict Calabi–Yau n–fold with $n > 2$ is always an algebraic manifold.

Exercise 11.5.2 Explain why we added the adjective "generic" in the sentence after Eq. (11.118).

Exercise 11.5.3 Show that the Riemann curvature of the Weil–Petersson metric has a universal expression in terms of the symmetric cubic form.

Exercise 11.5.4 Show that the Weil–Petersson Kähler metric associated to the cubic Hamilton–Jacobi function in Eq. (11.146), with no corrections, has at least $h_{2,1}$ holomorphic isometries.

Exercise 11.5.5 Show, conversely, that whenever the Weil–Petersson metric on the complex moduli space of a would–be CY 3–fold has $h_{2,1}$ *commuting* holomorphic isometries, one may always find a $Sp(2h_{2,1}+2,\mathbb{R})$ frame where S is given purely by a cubic form as in Eq. (11.146).

Exercise 11.5.6 Show that for weight 2 VHS with $h^{2,0} \equiv 1$ the Griffiths infinitesimal period relations are empty.

12
Four–dimensional $\mathcal{N} = 2$ supergravity

In this chapter we discuss *ungauged* 4D $\mathcal{N} = 2$ supergravity. The gauged case is easily obtained by the methods of Chapter 7 where the isometries of Quaternionic–Kähler manifolds were discussed in detail. Standard references for 4D $\mathcal{N} = 2$ SUGRA are [14, 105, 117, 116, 114, 118] and the book [146].

12.1 The four geometric structures of $\mathcal{N} = 2$ supergravity

The bosonic sector of (ungauged) 4D $\mathcal{N} = 2$ supergravity coupled to m vector–multiplets and h hypermultiplets contains, besides the metric vielbein e^a_μ, $m+1$ Abelian gauge vectors A^a_μ (m from the matter multiplets plus one graviphoton), and $2m + 4h$ real scalars. From the general discussion in Chapters 2, 3 we know that the scalar manifold $\mathcal{M}$ is a direct product $\mathcal{M} = \mathcal{M}_V \times \mathcal{M}_H$ endowed with four geometric structures:

H $\mathcal{M}_H$ is a Quaternionic–Kähler manifold of (real) dimension $4h$ and negative Ricci curvature parametrized by the hypermultiplet scalars.

K $\mathcal{M}_V$ is a Kähler m–fold parametrized by the m complex scalars in the vector multiplets. More precisely, $\mathcal{M}_V$ is a Hodge manifold and its Kähler form is the curvature of a Chern connection on a holomorphic line bundle $\mathcal{L} \to \mathcal{M}_V$.

V The coupling between the vectors A^a_μ and the scalars is encoded in a map $\mu \colon \mathcal{M} \to Sp(2m+2, \mathbb{R})/U(m+1)$. In particular, over $\mathcal{M}$ we have a *flat* $Sp(2m+2, \mathbb{R})$–bundle $\mathcal{V}$ whose structure group $Sp(2m+2, \mathbb{R})$ acts on the vectors' field–strengths by electromagnetic dualities. By the same argument as in rigid $\mathcal{N} = 2$ SUSY (cf. Section 2.9), the bundle $\mathcal{V}$ is the pull–back of a bundle over the factor manifold $\mathcal{M}_V$. In fact, μ is independent of the point in $\mathcal{M}_H$ and reduces to a map $\mu \colon \mathcal{M}_V \to Sp(2m+2, \mathbb{R})/U(m+1)$;

Z   We have the central charge geometry as in Sections 2.10.3 and 4.7. For a given point $\phi \in \mathcal{M}_V$ we have a complex line in $\mathcal{L}_\phi \subset \mathcal{V}_\phi$ defined by the central charge $\mathcal{Z}_\phi$. This defines a complex line subbundle $\mathcal{L} \subset \mathcal{V}$.

12.2 K ∩ V ∩ Z, or *projective* special Kähler geometry

The three structures **K**, **V**, and **Z** all refer to the vector–multiplet scalar manifold $\mathcal{M}_V$. This reflects the fact that, in the ungauged theory, there is no direct coupling between the vector fields and the hypermultiplet scalars. The three structures **K**, **V**, and **Z** are not independent: they are related by natural compatibility conditions. The resulting geometry on the complex manifold $\mathcal{M}_V$ is called *local special Kähler geometry* (or *projective* special Kähler geometry [145]). It is the supergravity variant of the *rigid* special geometry for the 4D $\mathcal{N} = 2$ gauge theories discussed in Section 2.9.3. In that case we had just two geometric structures, **K** and **V**, related by two compatibility conditions. In the decoupling limit of gravity the local special geometry should reduce to the rigid one.

Let us specify a bit more the new SUGRA structure, **Z**. The arguments of Section 4.7 imply that this geometric structure is encoded in a map ζ from the relevant scalar manifold $\mathcal{M}_V$ to a suitable polarized *flag manifold*. Let us review the argument. The ordinary gauge field–strengths

$$\mathcal{F}_{\mu\nu}^\pm \equiv (F^I_{\mu\nu}, G_{\mu\nu I}), \qquad I = 0, 1, \ldots, m, \tag{12.1}$$

transform under electromagnetic duality in the defining representation **2m + 2** of $Sp(2m + 2, \mathbb{R})$. As in Section 4.7.1, we may introduce the H–covariant field–strengths $\mathfrak{F}_{\mu\nu}^\pm$, where H is the *local* symmetry acting on the fermions which, neglecting the hypermultiplet fermions, is $H = U(1)_R \times U(m)$. By H–covariance, the vector field–strengths may enter the fermionic SUSY transformations only through the combinations $\mathfrak{F}_{\mu\nu}^\pm$. In particular, the "central charge field–strength" $\mathcal{Z}_{\mu\nu}^\pm$, which is defined by the gravitino SUSY transformations

$$\delta \psi_\mu^A = 2 D_\mu \epsilon^A - \cdots - \frac{1}{4} \sigma^{\rho\sigma} \gamma_\mu \, \mathcal{Z}_{\rho\sigma}^- \, \epsilon^{AB} \epsilon_B + \cdots, \tag{12.2}$$

is the component of $\mathfrak{F}_{\mu\nu}^\pm$ in the trivial representation of $SU(m)$ having $U(1)_R$ charge ∓ 2. Thus the SUSY transformations define a reduction of the structure group of the bundle $\mathcal{V}$ from $G = Sp(2m+2, \mathbb{R})$ to $H = U(1) \times U(m)$; that is, they define a map[1]

$$\zeta \colon \mathcal{M}_V \to Sp(2m+2, \mathbb{R})/U(1) \times U(m). \tag{12.3}$$

---

[1] This is precise if $\mathcal{M}_V$ is replaced by its universal cover; otherwise, we should take into account the monodromy representation of $\pi_1(\mathcal{M}_V)$ as in Chapter 11.

The homogeneous space on the RHS is a holomorphic polarized flag manifold whose geometry was discussed in detail in Chapter 11. Not surprisingly (in view of Section 11.5.7), the target manifold of ζ is the Griffiths period domain parametrizing the Hodge structures in degree 3 of a Calabi–Yau 3–fold with $h^{2,1} \equiv m$.

The two structures **V** and **Z** have a similar form: they are encoded, respectively, in the maps μ and ζ

$$\mathbf{V} \qquad \mu \colon \mathcal{M}_V \to Sp(2m+2,\mathbb{R})/U(m+1), \qquad (12.4)$$

$$\mathbf{Z} \qquad \zeta \colon \mathcal{M}_V \to Sp(2m+2,\mathbb{R})/U(1) \times U(m). \qquad (12.5)$$

For the **Z** structure the $U(1)$ factor in the isotropy group is identified with the $U(1)_R$ local symmetry and the $U(m)$ factor with the holonomy group of the Kähler manifold $\mathcal{M}_V$, while, for both structures, $Sp(2m+2,\mathbb{R})$ is identified with the electromagnetic duality group. The map μ defines the vector–scalar couplings in the Lagrangian, and ζ the SUSY transformations of the fermions. The two structures should be mutually compatible. μ should be determined by ζ, since the couplings may be read from the on-shell SUSY transformations. As in Eqs. (11.17) and (11.35), let

$$\varpi \colon Sp(2m+2,\mathbb{R})/U(1) \times U(m) \to Sp(2m+2,\mathbb{R})/U(m+1), \qquad (12.6)$$

be the natural fibration with fiber $U(m+1)/U(1) \times U(m) \equiv P\mathbb{C}^m$. The first compatibility condition is

$$\mu = \varpi \circ \zeta, \qquad (12.7)$$

that is, ζ and μ are equal in appropriate $U(m+1)$ gauges.

The fact that $Sp(2m+2,\mathbb{R})$ is the electromagnetic duality group entails that (in ungauged SUGRA) all geometric structures are $Sp(2m+2,\mathbb{R})$–*homogeneous*, that is, covariant under the $Sp(2m+2,\mathbb{R})$ action. All relevant bundles, metrics, connections, and curvatures on $\mathcal{M}_V$ are then the pull–back by ζ of the homogeneous bundles, metrics, connections, and curvatures on $Sp(2m+2,\mathbb{R})/U(1) \times U(m)$ which we explicitly constructed in Chapter 11. The map ζ satisfies certain conditions:

- Both $\mathcal{M}_V$ and $D \equiv Sp(2m+2,\mathbb{R})/U(1) \times U(m)$ are complex manifolds. The physical identification of the subgroup $U(m)$ of the isotropy group of the flag manifold D with the holonomy group of the Kähler space $\mathcal{M}_V$, required by gauge–invariance of the fermionic SUSY transformations, implies that $\zeta_* \colon T\mathcal{M}_V \to TD$ should preserve the decomposition of tangent vectors into types $(1,0)$ and $(0,1)$: that is, ζ must be *holomorphic*.
- The requirement that μ contains the same physical information as ζ entails that the image of ζ is horizontal with respect to the fibration ϖ, i.e., $\zeta_*(T\mathcal{M}_V)$ belongs to the horizontal tangent bundle $(TD)^{\text{hor}}$.

Putting the two together, we are led to expect that the map ζ satisfies the slightly stronger – but geometrically deeper – condition that $\zeta_*(T^{1,0}\mathcal{M}_V)$ takes values in the holomorphic horizontal tangent bundle T_D^{-1}, Eq. (11.39),

$$\zeta_*(T^{1,0}\mathcal{M}_V) \subseteq T_D^{-1}\big|_{\zeta(\mathcal{M}_V)}. \tag{12.8}$$

Equation (12.8) coincides with the Griffiths infinitesimal period relations. This is the second compatibility condition.

Finally, the $U(1)$ subgroup of the isotropy group of the polarized flag manifold D canonically defines a homogeneous line bundle $\mathcal{L} \to D$, namely the Hodge line bundle $\mathcal{F}^3$, which is equipped with a unique homogeneous metric and connection (cf. Section 11.2). Since $U(1) = U(1)_R$, it follows from the Hodge structure of $\mathcal{M}_V$ (cf. Section 2.7) that its Kähler form is the curvature of a connection on $\zeta^*\mathcal{L}$. This curvature should be $Sp(2m + 2, \mathbb{R})$–covariant; hence it is the pull-back by ζ of the unique homogeneous one on $\mathcal{L} \to D$, given in Theorem 11.14. Thus the scalars' Kähler metric is determined by the map ζ. This is a check on the claimed compatibility condition between structures **K** and **Z**, Eq. (12.8). Indeed, unitarity requires the metric on $\mathcal{M}_V$ to be positive–definite, which is guaranteed by Eq. (12.8): positivity of the kinetic terms is equivalent to horizontality of ζ. The conclusive evidence for (12.8) is that, decoupling gravity, it gives back rigid special geometry.

12.3 Formulas in projective special coordinates

All the geometric constructions of Section 11.5 apply to $\mathcal{N} = 2$ SUGRA. To connect with the SUGRA literature [105, 116, 117, 118] one should work with the projective special coordinates X^I, which play a preminent role in the superconformal approach. In this section we present the explicit formulae. In the SUGRA literature the (local) Hamilton–Jacobi function $S(X^I)$, with a different normalization, is called the *prepotential* $F(X^I)$

$$S(X^I) = \frac{i}{2} F(X^I). \tag{12.9}$$

As in Section 11.5.4, $F(X^I)$ is a holomorphic homogeneous function of degree 2,

$$F(\lambda X^I) = \lambda^2 F(X^I) \qquad \lambda \in \mathbb{C}. \tag{12.10}$$

$F(X^I)$ is unique only up to the action of $Sp(2m+2, \mathbb{R})$, cf. Section 11.5.4. In particular, it may have singularities which are unphysical, in the sense that they disappear by going to a different duality frame. It is convenient to adopt the standard SUGRA notations. The derivative $\partial_I F$ will be written as F_I and more generally we write

$F_{I_1,I_2,\cdots,I_k}$ for $\partial_{I_1}\partial_{I_2}\cdots\partial_{I_k}F$. The fact that F is homogeneous of degree 2 gives the identities

$$X^I F_I = 2F, \qquad F_{IJ} X^J = F_I, \qquad F_{IJK} X^K = 0. \qquad (12.11)$$

12.3.1 Scalar Kähler metric and all that

Through the construction in Section 11.5, we know that the SUGRA line bundle $\mathcal{L} \to \mathcal{M}_V$ is given by the pull-back of the homogeneous line bundle $\mathcal{O}(1) \to P\mathbb{C}^{2m+1}$. In view of Eq. (12.9), this bundle is given explicitly by

$$(X^I, P_I) = (X^I, \tfrac{i}{2} F_I). \qquad (12.12)$$

The Kähler potential K of the scalar manifold $\mathcal{M}_V$ is given by Eq. (11.129). Explicitly,

$$e^{-K} = -i Q\big((X^I, P_I), (\overline{X}^I, \overline{P}_I)\big) = -\frac{i}{2} \sum_I \left(P_I \overline{X}^I - X^I \overline{P}_I\right)$$
$$= \frac{1}{4}(F_I \overline{X}^I + X^I \overline{F}_I) = \frac{1}{4}(F_{IJ} + \overline{F}_{IJ}) X^I \overline{X}^J, \qquad (12.13)$$

where in the last equality we have used the second identity (12.11). The following notations are standard in $\mathcal{N} = 2$ SUGRA:

$$N_{IJ} = \frac{1}{4}(F_{IJ} + \overline{F}_{IJ}), \quad (NX)_I = N_{IJ} X^J, \quad (N\overline{X})_I = N_{IJ} \overline{X}^J$$
$$XN\overline{X} = N_{IJ} X^I \overline{X}^J, \qquad XNX = N_{IJ} X^I X^J, \qquad \overline{X}N\overline{X} = N_{IJ} \overline{X}^I \overline{X}^J. \qquad (12.14)$$

which, together with Eqs. (12.11) give

$$\partial_I (XN\overline{X}) = (N\overline{X})_I, \qquad \overline{\partial}_I (N\overline{X})_J = N_{IJ}, \qquad (12.15)$$

and their complex conjugates. In particular,

$$G_{I\bar{J}} \equiv \partial_I \overline{\partial}_J K = -\frac{N_{IJ}}{XN\overline{X}} + \frac{(N\overline{X})_I (NX)_J}{(XN\overline{X})^2}. \qquad (12.16)$$

Note that $X^I G_{I\bar{J}} = G_{I\bar{J}} \overline{X}^J = 0$; that is, $G_{I\bar{J}}$ is (as it should be) a metric on the orbit space $\mathcal{M}_V$ of the flow generated by the complex Euler vector $X^I \partial_I$.

N_{IJ} has a transparent meaning: it is the Kähler metric of a *rigid* special geometry with prepotential $F(X^I)$. It is a Hermitian form, but it is not positive–definite; indeed, from Eq. (12.16) we see that positivity of the physical scalar kinetic terms requires N_{IJ} to have signature[2] $(m, 1)$ with X^I "time–like" $XN\overline{X} < 0$. N_{IJ} should

---

[2] The other possibility is signature $(1, m)$ with $XN\overline{X} > 0$, which is the convention used in [118]. The two conventions are related by a sign flip $F(X^I) \leftrightarrow -F(X^I)$.

12.3.2 Vector kinetic terms

The vector kinetic terms are described by the map

$$\mu: \mathcal{M}_V \to Sp(2m+2, \mathbb{R})/U(m+1) \tag{12.17}$$

which associates with a point in $\mathcal{M}_V$ a complex $(m+1)$–dimensional subspace $W \subset \mathbb{C}^{2m+2}$, which satisfies the same Riemann bilinear relations as the Hodge decomposition for a genus $(m+1)$ curve. From Chapter 11 we know that W is an eigenspace of the Weil operator (in the geometric situation it is an eigenspace of the Hodge $*$ operator). From Example 11.3 we see that

$$W = H^{3,0} \oplus H^{1,2}. \tag{12.18}$$

We take advantage of the analogy with the period map for a genus $(m+1)$ curve to simplify the computations. Let

$$(A^I{}_J, B_{IJ}) \in \mathbb{C}^{2m+2}, \quad J = 0, 1, \ldots, m \tag{12.19}$$

be a basis of W. In the curve analogy, $A^I{}_J$ (B_{IJ}) would be the matrix of the A–periods (resp. B–periods). As in the curve case, it is convenient to normalize the basis so that $A^I{}_J = \delta^I{}_J$. Then the bilinear relations say that B_{IJ} is equal to a symmetric matrix with *negative*–definite imaginary part.[3] By the general theory of duality, $B_{IJ} = -2i\mathcal{N}_{IJ}$ where $\mathcal{N}_{IJ}$ is the symmetric matrix appearing in the vector kinetic terms, $\mathcal{N}_{IJ}F^{I+\mu\nu}F^{+J}_{\mu\nu}$ + h.c. Thus, to get $\mathcal{N}_{IJ}$ we have simply to construct a basis of the subspace $H^{3,0} \oplus H^{1,2}$ of the form

$$V_J \equiv (\delta^I{}_J, B_{IJ}) \in \mathbb{C}^{2m+2}, \quad J = 0, 1, \ldots, m. \tag{12.20}$$

The symmetric matrix $\mathcal{N}_{IJ} \equiv iB_{IJ}/2$ is determined by the condition

$$Q(V_I, U) = 0 \quad \text{for all } U \in H^{3,0} \oplus H^{1,2}. \tag{12.21}$$

Indeed, if this condition is satisfied, $V_I \in H^{3,0} \oplus H^{1,2}$ and, since the V_I are linear–independent, they span the subspace.

$H^{3,0}$ is generated by the vector $\Omega = (X^I, S_I)$.[4] Let us construct a spanning set for $H^{1,2}$. From the Griffiths horizontality condition we know that the vectors $U_J =$

---

[3] The reason why the imaginary part of B_{IJ} is *negative*–definite instead of positive as in the curve case, is that the Weil operator i^{p-q} acts on $H^{3,0}$ as $-i$ and not as $+i$ as on $H^{1,0}$.
[4] We write $S_{I_1, I_2, \ldots, I_s} \equiv \partial_{I_1} \partial_{I_2} \ldots \partial_{I_s} S$ for the derivatives of the Hamilton–Jacobi function $S \equiv iS/2$ with respect to the special homogeneous coordinates.

$(\delta^I{}_J, \overline{S}_{IJ})$ span $H^{1,2} \oplus H^{0,3}$; to get an *overcomplete* spanning set $U_J^\perp$ for $H^{1,2}$ we just project out the $H^{0,3}$ component of the U_J using the indefinite Hermitian product

$$U_J^\perp = U_J - \frac{Q(U_J, \Omega)}{Q(\overline{\Omega}, \Omega)} \overline{\Omega} = \left(\delta^I{}_J - \frac{(LX)_J}{\overline{X}LX} \overline{X}^I, \overline{S}_{IJ} - \frac{(LX)_J}{\overline{X}LX} \overline{S}_I \right), \quad (12.22)$$

where $L_{IJ} = S_{IJ} - \overline{S}_{IJ}$. The conditions determing B_{IJ} in Eq. (12.20) become

$$Q(\Omega, V_J) = Q(U_K^\perp, V_J) = 0, \quad (12.23)$$

or, explicitly,

$$X^I B_{IJ} - S_J = 0, \qquad B_{KJ} - \overline{S}_{JK} - \frac{(LX)_K}{\overline{X}LX}\left(\overline{X}^I B_{IJ} - \overline{S}_J\right) = 0, \quad (12.24)$$

whose solution is

$$B_{IJ} = \overline{S}_{IJ} + \frac{(LX)_I (LX)_J}{XLX}. \quad (12.25)$$

Reverting to the SUGRA conventions[5] (12.9), we get

$$\mathcal{N}_{IJ} = \frac{1}{4} \overline{F}_{IJ} - \frac{(NX)_I (NX)_J}{XNX}, \quad (12.26)$$

in agreement with the standard result [118].

12.3.3 The H–covariant field strengths

Next we construct the $H \equiv U(1) \times U(m)$ covariant field–strengths, i.e., the combinations of vectors' field–strengths and scalars which enter into the fermionic SUSY variation. Recall from Section 4.7.1 that half the $Sp(2m + 2, \mathbb{R})$–invariant $U(1) \times U(m)$–covariant linear combinations define the covariant field–strengths, while the other half give bilinears in the Fermi fields (denoted K in Section 4.7.1).

The $U(1)$ covariant field–strength enters in the gravitino variation, Eq. (12.2), and hence defines the central charge geometry. The four possible $U(m)$–scalar $Sp(2m + 2, \mathbb{R})$–invariant combinations are

$$\begin{aligned} P_I F^{I-}_{\mu\nu} - X^I G^-_{I\mu\nu}, & \quad P_I F^{I+}_{\mu\nu} - X^I G^+_{I\mu\nu}, \\ \overline{P}_I F^{I+}_{\mu\nu} - \overline{X}^I G^+_{I\mu\nu}, & \quad \overline{P}_I F^{I-}_{\mu\nu} - \overline{X}^I G^-_{I\mu\nu}. \end{aligned} \quad (12.27)$$

In view of the relations $P_I = iF_I/2$ and

$$G^+_{I\mu\nu} = -2i\, \mathcal{N}_{IJ}\, F^{J+}_{\mu\nu} + \text{Fermi bilinears}, \quad (12.28)$$

---

[5] Except that for us all metrics, including Re $\mathcal{N}_{IJ}$, are positive–definite, while in refs.[116, 118] they are chosen to be negative–definite. In other words, our prepotential has the opposite sign.

the two complex conjugate combinations on the right produce Fermi bilinears and hence the two conjugate combinations on the left are the $U(1)$–covariant field–strengths $\mp 2i\, (\overline{X}NX)\, T^{\pm}_{\mu\nu}$, where

$$T^{+}_{\mu\nu} = \frac{(NX)_I F^{I+}_{\mu\nu}}{\overline{X}NX}. \tag{12.29}$$

To construct the $U(m)$ part of the covariant field–strength, notice that tangent space to the scalar manifold is identified in an $Sp(2m+2, \mathbb{R})$–invariant way with the hyperplane in $\mathbb{C}^{m+1}$ of the vectors v^I such that $(XN)_I v^I = 0$. This is the space in which the gaugini take value, and $U(m)$ is the holonomy acting on this space. Hence the covariant field–strength is a tangent vector which we write as an element of the above hyperplane

$$\mathfrak{F}^{I+}_{\mu\nu} = F^{I+}_{\mu\nu} - \frac{X^I}{\overline{X}NX}(NX)_J\, F^{J+}_{\mu\nu}. \tag{12.30}$$

12.3.4 Pauli couplings

By the general theory of electromagnetic duality developed in Section 4.7.2, the full non–linear Pauli couplings are obtained by replacing in their linearized expression the "bare" field–strength with the H–covariant one. For the two–gaugino–one–field–strength coupling the linearized coupling is the one in rigid $\mathcal{N} = 2$ SUSY, which is just a special case of (rigid) $\mathcal{N} = 1$. Hence the answer may be extracted from the expressions in Chapter 10.

As in the discussion preceding Eq. (12.30), we see the gaugini as Fermi fields Ω^{iI}, $I = 0, 1, \ldots, m$, $i = 1, 2$, subject to the constraint $(XN)_I \Omega^{iI} = 0$; as is customary in SUGRA [118], the upper/lower position of the $SU(2)_R$ index i stands for positive/negative chirality $\gamma_5 \Omega^{iI} = \Omega^{iI}$, $\gamma_5 \Omega^I_i = -\Omega^I_i$. Then the $\Omega^2 F$ coupling reads

$$-\frac{e}{32} \overline{F}_{IJK}\, \overline{\Omega}^{iI} \sigma^{\mu\nu} \Omega^{jJ}\, \epsilon_{ij}\, \mathfrak{F}^{K+}_{\mu\nu} + \text{h.c.}$$

$$= -\frac{e}{32} \overline{F}_{IJK}\, \overline{\Omega}^{iI} \sigma^{\mu\nu} \Omega^{jJ}\, \epsilon_{ij} \left(F^{K+}_{\mu\nu} - \frac{X^K}{\overline{X}NX}(NX)_L F^{L+}_{\mu\nu}\right) + \text{h.c.} \tag{12.31}$$

We see that *the Pauli coupling is proportional to the symmetric cubic form of the period map p*. Indeed, the geometry of the period map fixes the form of this coupling up to the overall numerical factor.

Similar considerations work for the other Pauli couplings, as well as for the 4–Fermi couplings which may be determined (in principle) by the considerations in Chapter 4. We refer to [118] for the complete Lagrangian of $\mathcal{N} = 2$ supergravity coupled to matter multiplets, including the 4–Fermi terms.

12.4 Aspects of projective special Kähler manifolds

Product spaces

Non–Ricci flat Quaternionic–Kähler manifolds are automatically irreducible. We may ask if the analogous statement is true for the projective Kähler manifolds. In fact (simply connected) projective Kähler manifolds may be products if and only if they are symmetric spaces of the form

$$\frac{SU(1,1)}{U(1)} \times \frac{SO(n,2)}{SO(n) \times SO(2)} \qquad n \in \mathbb{N}. \tag{12.32}$$

In particular, we may have the triple product $(SU(1,1)/U(1))^3$, corresponding to $n = 2$ in Eq. (12.32). This result was first shown in ref. [140]. We present a different argument, exploiting the fact that all projective special Kähler manifolds arise from a weight 3 *abstract*[6] variation of Hodge structures with $h^{3,0} = 1$.

Assume $\mathcal{M}_V = \mathcal{M}_1 \times \mathcal{M}_2$ equipped with a direct sum Kähler metric. Since the Kähler form, being Hodge, is the curvature of a line bundle $\mathcal{L}$, this implies that $\mathcal{L} = \mathcal{L}_1 \otimes \mathcal{L}_2$, $\mathcal{L}_a$, being a line bundle over $\mathcal{M}_a$ ($a = 1, 2$), while the section $\Omega(s) \in \Gamma(\mathcal{L})$ is the product $s_1 s_2$ of sections of the two bundles, $s_a \in \Gamma(\mathcal{L}_a)$. Acting on the section s_a with the Kodaira–Spencer vectors associated with $T\mathcal{M}_a$, we construct two vector spaces H_a such that $V_\mathbb{C} = H_1 \otimes H_2$, so that the weight 3 abstract Hodge structure describing the given projective special Kähler manifold $\mathcal{M}_1 \times \mathcal{M}_2$ is the direct product of two Hodge structures of weights w_1, w_2 with $w_1 + w_2 = 3$ and $h^{w_1,0} = h^{w_2,0} = 1$:

$$H_1 = \bigoplus_{p=0}^{w_1} H_1^{p,w_1-p}, \qquad H_2 = \bigoplus_{p=0}^{w_2} H_2^{p,w_2-p},$$

$$V_\mathbb{C} \equiv H_1 \otimes H_2 = \bigoplus_{p=0}^{3} H^{p,3-p}, \tag{12.33}$$

where

$$H^{p,3-p} = \bigoplus_{k_1+k_2=p} H_1^{k_1,w_1-k_1} \otimes H_2^{k_2,w_2-k_2}.$$

Without loss of generality we may take $w_1 = 1$ and $w_2 = 2$. The Griffiths infinitesimal relations are empty for $w = 1$ VHS as well as for the weight 2 VHS with $h^{2,0} = 1$ (Exercise 11.5.7). In other words, in these two cases the period map is a local cover map; going to the universal covers, the period map gets identified with

---

[6] *Abstract* means that the variations of Hodge structure do *not* necessarily arise as period maps for geometric families of CY varieties. Nevertheless, they satisfy all axioms of Griffiths theory.

the identity map. Hence (the universal cover of) $\mathcal{M}_V$ is just the product of the two Griffiths period domains, that is

$$\frac{Sp(2,\mathbb{R})}{U(1)} \times \frac{SO(h^{1,1},2)}{SO(h^{1,1}) \times U(1)}, \qquad (12.34)$$

which is clearly the same space as (12.32) (with $n = h^{1,1}$). The SL_2–orbit theorem then guarantees that all these spaces are produced by a cubic prepotential as in Eq. (11.146).

Symmetric spaces

Next we ask which symmetric spaces are projective special Kähler manifold. The classification of such space was first obtained in [99]; we adopt a different approach.

By the previous result, a symmetric projective special Kähler manifold either is a space in the infinite family (12.32), or it is a negatively curved (i.e., type III) *irreducible* Hermitian symmetric space G/K which should appear in the list of Section 9.6.2. In particular, it must be biholomorphic to an irreducible Siegel domain, $\mathfrak{G}(V, F)$, for some irreducible symmetric convex cone V and some V–Hermitian function F. We also know that the isometry G acts on the vectors by duality transformations, which means that the period map

$$p \colon G/K \to Sp(2m+2,\mathbb{R})/U(1) \times U(m), \qquad m = \dim_{\mathbb{C}} G/K, \qquad (12.35)$$

is induced by a group embedding i,

$$i \colon G \to Sp(2m+2,\mathbb{R}), \qquad m = \dim_{\mathbb{C}} G/K, \qquad (12.36)$$

as in the prototypical example of Eq. (1.96); that is,

$$p = \pi \circ i \qquad (12.37)$$

where π is the canonical projection

$$\pi \colon Sp(2m+2,\mathbb{R}) \to Sp(\mathbf{2m+2},\mathbb{R})/U(1) \times U(m). \qquad (12.38)$$

Then the Lie group G must have a $(2m+2)$–dimensional symplectic representation which is irreducible over the reals, hence either $\mathbb{C}$–irreducible or of the form $W \oplus \overline{W}$ with W $\mathbb{C}$–irreducible and $\overline{W} \not\simeq W$. For instance,

$$E_6/SO(2) \times SO(10) \qquad (12.39)$$

cannot be a special Kähler manifold, since E_6 has no symplectic representation nor representations of dimension **17**. *The would–be period map p in Eq. (12.37)* defines a projective special Kähler geometry if and only if it satisfies the Griffiths infinitesimal period relations; from the Lie algebraic description of the relations (see Section 11.2.1) we know that this condition is equivalent to the statement that the differential of i, $i_*: \mathfrak{g}, \to \mathfrak{sp}(2m+2, \mathbb{R})$ has the property

$$i_*(\mathfrak{g}) \subset \left(\begin{array}{c|c|c|c} * & 0 & 0 & * \\ \hline 0 & * & * & * \\ \hline 0 & * & * & 0 \\ \hline * & * & 0 & * \end{array} \right), \tag{12.40}$$

where the block–matrix is written in a complex Cayley basis adapted to the decomposition $V_\mathbb{C} = (H^{3,0} \oplus H^{1,2}) \oplus (H^{2,1} \oplus H^{0,3})$.

The irreducible symmetric cones V are listed in Table 9.2: by inspection we see that either V is a spherical cone or it is the cone of positive $k \times k$ matrices, $k \geq 3$, having entries in a suitable normed algebra. We extend the rank k to the spherical cones by stating that a spherical cone of dimension ≥ 2 (resp. $= 1$) has $k = 2$ (resp. $k = 1$). We claim that, if the irreducible Siegel domain $\mathfrak{S}(V, F)$ is a projective special Kähler space, it must have $k \leq 3$. More generally, if a product of Siegel domains $\prod_\ell \mathfrak{S}(V_\ell, F_\ell)$ is special Kähler, then $\sum_\ell k_\ell \leq 3$. In addition, if $\sum_\ell k_\ell > 1$ we must have $F \equiv 0$; that is, the space should be a product of domains of the first kind. This follows from comparing the behavior for large $\mathrm{Im}\, Z^a$ of the Hodge fiber metric e^{-K} of $\mathfrak{S}(V, F)$ which has the form[7]

$$e^{-K} = P_k[\mathrm{Im}\, Z^a - F(W, W)^a], \tag{12.41}$$

where P_k is the degree k polynomial

$$P_k(x^a) = \det(x^a H_a), \quad H_a : \begin{cases} \text{a real basis of } k \times k \text{ Hermitian} \\ \text{matrices with entries in } \mathbb{R}, \mathbb{C}, \mathbb{H}, \end{cases} \tag{12.42}$$

and the large complex structure asymptotics of the period map p for a weight 3 variation of Hodge structure, which is a polynomial of degree at most 3 (Section 11.5.6). We remain with the four $k = 3$ Siegel domains of the first kind,

$$\begin{aligned} & Sp(6, \mathbb{R})/U(6), \quad SU(3,3)/[U(3) \times SU(3)], \\ & SO^*(12)/U(6), \quad E_{7(-25)}/[E_6 \times SO(2)], \end{aligned} \tag{12.43}$$

---

[7] Cf. Section 9.6.2. With respect to that section we have rescaled the Kähler metric by setting $K = -\Gamma_V/m$, where m is a suitable integer, so that e^{-K} is an irreducible polynomial instead of a power of a polynomial.

12.4 Aspects of projective special Kähler manifolds

and possibly products of $k = 1, 2$ domains. The $k = 3$ irreducible domains of the first kind are manifestly special Kähler with prepotential (in the standard duality frame)

$$F(X^I) = i\frac{P_3(X^i)}{X^0}, \quad i = 1, \ldots, m, \tag{12.44}$$

with P_3 the cubic polynomial (12.42).

An irreducible $k = 2$ symmetric domain has the form

$$SO(2, n)/[SO(2) \times SO(n)] \quad (n \geq 2), \tag{12.45}$$

which certainly is not special Kähler since $SO(2, n)$ has no symplectic representation of dimension $(2n + 2)$ nor complex representations of dimension $(n + 1)$. A Siegel domain with $k = 1$ has the form

$$U(1, n)/U(n), \tag{12.46}$$

for some n ($n = 1$ being first kind, $n \geq 2$ second kind). The symmetric spaces (12.46) are all special Kähler. Indeed, $U(1, n)$ has a dimension $(n + 1)$ complex representation W, namely the defining one, and this representation embeds $\mathfrak{u}(1, n) \xrightarrow{i_*} \mathfrak{sp}(2n + 2, \mathbb{R})$ as

$$\left(\begin{array}{c|c} u & v \\ \hline -v^\dagger & U \end{array}\right) \xrightarrow{i_*} \left(\begin{array}{c|c|c|c} u & 0 & 0 & v \\ \hline 0 & U^* & -v^t & 0 \\ \hline 0 & v^* & u^* & 0 \\ \hline -v^\dagger & 0 & 0 & U \end{array}\right), \tag{12.47}$$

which defines a special Kähler manifold since the RHS has the required form (12.40). The geometries (12.46), with electromagnetic dualities acting in the decomposable representation $(\mathbf{n + 1}) \oplus \overline{(\mathbf{n + 1})}$, are called *minimal coupling* in the SUGRA literature. Their prepotential is a quadratic form $F(X^I) = a_{IJ}X^I X^J$. By a change of $Sp(2m + 2, \mathbb{R})$ duality frame we may set the imaginary part of a_{IJ} to zero (or any other given value), and then by a real linear redefinition of the X^I we may take $a_{IJ} \equiv N_{IJ} = \text{diag}(-1, 1, 1, \ldots, 1)$, the signature of the quadratic form being fixed by the requirement of positive kinetic terms.

We may also consider products of symmetric cones with $k = 1$ and $k = 2$. This gives back the reducible spaces

$$\frac{Sp(2, \mathbb{R})}{U(1)} \times \frac{SO(n, 2)}{SO(n) \times U(1)}, \tag{12.48}$$

the case $n = 2$ giving the special case of a product of three $k = 1$ domains.

Although Eqs. (12.43),(12.46), and (12.48) give the complete list of symmetric *domains* which are special Kähler, this is not yet the full list of symmetric special Kähler *geometries*. A single symmetric domain G/K may be realized in several *geometrically inequivalent* ways as a projective special Kähler manifold. Indeed, there may be inequivalent group embeddings $i\colon G \to Sp(2\dim G/K + 2, \mathbb{R})$: they would correspond to different Hodge bundle geometries, different duality realizations of the isometry group G, as well as symmetric Kähler metrics differing in their overall normalization. As noted above, each such i is associated either with an irreducible quaternionic G–representation Q of dimension $2\dim_{\mathbb{C}} G/K + 2$ or with a complex representation W of dimension $\dim_{\mathbb{C}} G/K + 1$. There is only one symmetric domain in the above list having both kinds of representations: namely $Sp(2,\mathbb{R})/U(1) \equiv SL(2,\mathbb{R})/U(1)$, the upper half–plane. In this case W is the fundamental **2** representation, while Q is its symmetric cube **4** which is also a quaternionic representation. The corresponding prepotentials of these two distinct geometries read (in a standard duality frame)

$$F_{\mathbf{2}} = X^0 X^1, \qquad F_{\mathbf{4}} = i\frac{(X^1)^3}{X^0}. \qquad (12.49)$$

Homogeneous spaces

The analysis may be extended to projective special Kähler spaces which are homogeneous, i.e., have a transitive group of isometries. There are two infinite families $K(p,q)$ and $H(p+1, q+1)$ with $p, q \in \mathbb{N}$ [84]. All these spaces are rank $k = 3$ Siegel domains of the first kind; the isometry group G is never *unimodular* unless the space is one of the symmetric spaces in Eqs. (12.32) and (12.43). All these spaces (but the minimal coupling ones), being rank 3 first type domains, have cubic prepotentials (in some standard frame). From the VHS standpoint this is obvious: by homogeneity, given any small region U of the domain, we may find an isometry mapping it at infinity, where the large complex structure asymptotics applies.

12.5 Coupling hypermultiplets to $\mathcal{N} = 2$ supergravity

We know that the hypermultiplets in 4D $\mathcal{N} = 2$ supergravity live in a Quaternionic–Kähler manifold $\mathcal{M}_H$ of fixed (for a given dimension) negative Ricci curvature; the argument of Chapters 2, 3, and 4 fix the Lagrangian completely. We review the situation following ref. [31]. [8]

---

[8] In particular, we use their conventions. This implies a rescaling by a factor 2 of the scalar metric, and corresponding factors in the curvatures.

12.5 Coupling hypermultiplets to $\mathcal{N}=2$ supergravity

Let ϕ^i ($i = 1, 2, \ldots, n$) be the scalar fields. Recall that we have

$$\delta\phi^i = \gamma^i_{AZ}\bar{\epsilon}^A\chi^Z, \qquad A = 1,2, \; Z = 1,2,\ldots, 2n, \tag{12.50}$$

where γ^i_{AZ} is the vielbein giving the bundle isomorphism $T\mathcal{M}_H \simeq H_2 \otimes H_{2n}$, where H_2 (resp. H_{2n}) have structure group $USp(2)$ (resp. $USp(2n)$). We write ϵ_{AB} and ϵ_{YZ} for the two parallel invariant antisymmetric tensors of $USp(2)$ and $USp(2n)$. The vielbein and the Quaternionic–Kähler metric g_{ij} are related as in rigid SUSY, Eq. (2.137),

$$\begin{aligned} g_{ij}\,\gamma^i_{AY}\,\gamma^j_{BZ} &= \epsilon_{AB}\,\epsilon_{YZ}, \\ \gamma^i_{AZ}\,\gamma^{jBZ} + \gamma^j_{AZ}\,\gamma^{iBZ} &= g^{ij}\,\delta^B_A, \\ \gamma^i_{AY}\,\gamma^{jAZ} + \gamma^j_{AY}\,\gamma^{IAZ} &= \frac{1}{n}g^{ij}\,\delta^Z_Y, \end{aligned} \tag{12.51}$$

where the indices $A, B, \ldots$ (resp. $X, Y, \ldots$) are raised/lowered using ϵ_{AB} (resp. ϵ_{XY}). The Riemann curvature is the sum of the $\mathfrak{sp}(2)$ and $\mathfrak{sp}(2n)$ ones

$$R_{ijkl}\,\gamma^l_{AY}\gamma^k_{BZ} = \epsilon_{AB}\,R_{ijYZ} + \epsilon_{YZ}\,R_{ijAB}, \tag{12.52}$$

The $\mathfrak{sp}(2)$ curvature (associated with the SUSY automorphism group) is fixed, as in Chapters 2 and 4, to be

$$R_{ijAB} = \gamma_{iAZ}\,\gamma_{jB}{}^Z - \gamma_{jAZ}\,\gamma_{iB}{}^Z \tag{12.53}$$

and then Theorem 4.5 gives

$$R_{ijXY} = \gamma_{iAX}\,\gamma_j{}^A{}_Y - \gamma_{jAX}\,\gamma_i{}^A{}_Y + \gamma_i^{AW}\,\gamma_{jA}{}^Z\,\Omega_{XYZW}, \tag{12.54}$$

with Ω_{XYZW} totally symmetric in its four indices.

The complete Lagrangian of $\mathcal{N}=2$ SUGRA coupled to n hypermultiplets living on the Quaternionic–Kähler space $\mathcal{M}_H$ is [31]

$$\begin{aligned} L = &-\frac{e}{2}R - e\,g_{ij}\hat{D}_\mu\phi^i\hat{D}^\mu\phi^j - \frac{1}{2}\epsilon^{\mu\nu\rho\sigma}\bar{\psi}_{\mu A}\gamma_5\gamma_\nu D_\rho \psi^A_\sigma \\ &-\frac{e}{4}\hat{F}_{\mu\nu}\hat{F}^{\mu\nu} - \frac{e}{2}\bar{\chi}_Z\gamma^\mu D_\mu\chi^Z + \frac{e\sqrt{2}}{16}\bar{\psi}_{\mu A}\gamma_5\psi^A_\nu(\hat{F}^{\mu\nu} + \widetilde{\hat{F}}^{\mu\nu}) \\ &+ e\,\gamma_{iAZ}\,\bar{\chi}^Z\sigma^{\mu\nu}\psi_\mu^A(\partial_\nu\phi^i + \hat{D}_\nu\phi^i) \\ &+ \frac{\sqrt{2}}{4}e\,\bar{\chi}_Z\sigma^{\mu\nu}\chi^Z\hat{F}_{\mu\nu} - \frac{e}{32}(\bar{\chi}_Y\gamma_\mu\gamma_5\chi^Y)(\bar{\chi}_Z\gamma^\mu\gamma_5\chi^Z) \\ &+ \frac{e}{16}(\bar{\chi}_Y\sigma_{\mu\nu}\chi^Y)(\bar{\chi}_Z\sigma^{\mu\nu}\chi^Z) + \frac{e}{16}\Omega_{XYZW}(\bar{\chi}^X_L\gamma_\mu\chi^Y_L)(\bar{\chi}^Z_L\gamma^\mu\chi^W_L), \end{aligned} \tag{12.55}$$

where

$$\hat{F}_{\mu\nu} = F_{\mu\nu} - \frac{\sqrt{2}}{2}\bar{\psi}_{\mu A}\psi_{\nu}{}^{A}, \qquad (12.56)$$

$$\hat{D}_{\mu}\phi^{i} = \partial_{\mu}\phi^{i} - \frac{1}{2}\gamma^{i}_{AZ}\bar{\psi}_{\mu}{}^{A}\chi^{Z}, \qquad (12.57)$$

are the supercovariant field–strength and derivative. The derivative $\mathcal{D}_\mu$ is covariant with respect to the space–time spin connection as well as the $\mathfrak{sp}(2)$ and $\mathfrak{sp}(2n)$ connections according to the respective representations of the Fermi fields ψ_μ^A and χ^Z.

The SUSY transformations which leave the Lagrangian invariant are

$$\begin{aligned}
\delta e_{a\mu} &= \bar{\epsilon}_A \gamma_a \psi_\mu^A, \\
\delta A_\mu &= \sqrt{2}\,\bar{\epsilon}_A \psi_\mu{}^A, \\
\delta \phi^i &= \gamma^i_{AZ}\bar{\epsilon}^A \chi^Z, \\
\delta \chi^Z_L &= 2\partial_\mu \phi^i \gamma_i^{AZ}\gamma^\mu \epsilon_{RA} - \Gamma^Z_{iY}\delta\phi^i \chi_L^Y, \\
\delta \psi_{\mu L}{}^A &= 2\mathcal{D}_\mu \epsilon_L^A + \frac{\sqrt{2}}{2}(\hat{F}_{\mu\nu} - \frac{1}{2}\tilde{\hat{F}}_{\mu\nu}\gamma_5)\gamma^\nu \epsilon_R^A + \text{3-fermions}.
\end{aligned} \qquad (12.58)$$

The most general ungauged $\mathcal{N} = 2$ supergravity coupled to arbitrary vector–multiplets and hypermultiplets may be obtained by combining the previous results. The gaugings may be performed using the methods of Chapters 7 and 8. For brevity we refer to the papers [14, 118].

12.6 Compactifying type II supergravity on a CY manifold

We have seen that the vector–multiplet couplings in a general 4D $\mathcal{N} = 2$ supergravity are described by an abstract (i.e., not necessarily arising from a family of manifolds) weight 3 variation of Hodge structure with $h^{3,0} = 1$. The deep reason why this remarkable fact has to be true is that we may construct a large class of $\mathcal{N} = 2$ supergravities by considering a 10D type II supergravity (either IIA or IIB) on a product manifold of the form

$$\mathbb{R}^4 \times K, \qquad (12.59)$$

where K is a compact Calabi–Yau 3–fold with $B_1 = 0$. Since (Theorem 3.23) on K we have two parallel spinors of opposite chirality, a type II SUGRA on $\mathbb{R}^4 \times K$ has 8 (parallel) Killing spinors; that is, the effective 4D theory on $\mathbb{R}^4$ is an $\mathcal{N} = 2$ SUGRA. The massless 4D fields arise from harmonic forms in K, and their

12.6 Compactifying type II supergravity on a CY manifold

Table 12.1 *Massless bosonic fields in 10D type II superstrings.*

	NS–NS massless bosonic fields	R–R field–strengths[a]
Type IIA	e^A_M, ϕ, B_{MN}	$F^{(2)}, F^{(4)}$
Type IIB	e^A_M, ϕ, B_{MN}	$F^{(1)}, F^{(3)}, F^{(5)} = *F^{(5)}$

[a] $F^{(k)}$ denotes a k–form field–strength. In Type IIB $F^{(5)}$ is self–dual.

couplings are described by the geometry of these forms, that is, by Hodge theory on K. Consistency then requires agreement between the geometry of $\mathcal{N} = 2$ SUGRA and the geometry of Hodge structures.

It is convenient to use the language of string theory, and divide the 10D massless states into the NS–NS and the RR sectors; see Table 12.1. The NS-NS light fields, common to type IIA and IIB, are e^A_M, ϕ and B_{MN}. The 4D light scalar fields arising from the inner components of the metric vielbein e^A_M correspond to the moduli of the CY metrics which, by Theorem 11.10, are given by the deformations of the complex structure and of the Kähler class, making $2h^{2,1} + h^{1,1}$ real fields. The internal component of B_{MN} may be deformed by adding any harmonic 2–form, producing $h^{1,1}$ moduli. The last two NS–NS scalars are ϕ and the dual σ to the 4D 2–form field $B_{\mu\nu}$. The NS–NS sector does not give any 4D massless vector field since K has vanishing first Betti number:

$$\textbf{NS-NS scalars} \begin{vmatrix} e^A{}_M : & h^{1,1} \text{ Kähler} + 2\, h^{2,1} \text{ complex moduli} \\ B_{MN} : & h^{1,1} \text{ moduli} \\ \phi : & 1 \text{ scalar} \\ B_{\mu\nu} : & 1 \text{ scalar}, d\sigma = *dB. \end{vmatrix} \quad (12.60)$$

In the RR sector it is convenient to consider the field–strengths, which are the k–forms $F^{(k)}$ listed in Table 12.1. We expand the closed k–forms $F^{(k)}$ à la Künneth in terms of 4D field–strengths and harmonic ℓ–forms $\omega^{(\ell)}$ on K.

In type IIA we have

$$F^{(2)} \sim dA^0, \qquad F^{(4)} \sim \sum_a \omega_a^{(3)} d\phi^a + \sum_k \omega_k^{(2)} dA^k, \qquad (12.61)$$

which gives $B_3 \equiv 2 + 2h^{2,1}$ real scalars ϕ^a, and $h^{1,1} + 1$ vectors A^0, A^k. Adding the NS–NS sector, in total we have $h^{1,1} + 1$ vectors and $4 + 4h^{2,1} + 2h^{2,1}$ scalars; together with $g_{\mu\nu}$ they are precisely the bosonic degrees of freedom of $\mathcal{N} = 2$ supergravity coupled to $h^{1,1}$ vector multiplets and $h^{2,1} + 1$ hypermultiplets.

In type IIB,

$$F^{(1)} = d\varrho, \qquad F^{(3)} = \sum_k \omega_k^{(2)} d\phi^k + dB^{(2)},$$

$$F^{(5)} = (1+*)\left(\sum_k \omega_k^{(4)} d\varphi^k + \sum_a \omega_a^{(3)} dA^a\right), \qquad (12.62)$$

which gives $h^{1,1} + h^{2,2} + 2 \equiv 2 + 2h^{1,1}$ scalars, $\varrho, \phi^k, \varphi^k$ and the dual $\tilde{\sigma}$ to the 2–form $B^{(2)}$, $d\tilde{\sigma} = *dB^{(2)}$, together with $B_3 = 2 + 2h^{2,1}$ vector fields A^a. Adding the NS–NS, we have $h^{2,1} + 1$ vectors and $4 + 4h^{1,1} + 2h^{2,1}$ scalars, corresponding to $h^{2,1}$ vector multiplets and $h^{1,1}$ hypermultiplets. Note that the type IIA and type IIB spectra are exchanged under $h^{1,1} \leftrightarrow h^{2,1}$.

Let us compute the effective 4D Lagrangian.

Type IIB. We consider first the kinetic terms of the scalars in the vector multiplets, z^a. In type IIB these fields parametrize the deformations of the complex structure of K; then they are related to the deformation of the Calabi–Yau metric $g_{i\bar{j}}$ as

$$\delta g_{\bar{i}\bar{j}} = \delta z^a (\phi_a)_{\bar{i}}{}^k g_{k\bar{j}}, \qquad (12.63)$$

where $(\phi_a)_{\bar{i}}{}^k$ is the Kodaira–Spencer vector. Plugging the deformed metric into the 10D action, and extracting the term quadratic int the derivatives of z^a, one gets for the kinetic term

$$G_{a\bar{b}} \, \partial^\mu z^a \partial_\mu \bar{z}^b = \partial^\mu z^a \, \partial_\mu \bar{z}^b \int_K \sqrt{g}\, d^6 x\, (\phi_a)_{\bar{i}}{}^k (\bar{\phi}_b)_l{}^{\bar{j}} \, g_{k\bar{j}}\, g^{\bar{i}l}, \qquad (12.64)$$

which just says that the vector scalar metric in type IIB is equal to the Weil–Petersson metric on the complex moduli space, which also agrees with the Hodge metric on $H^{2,1}(K)$.

In the same way, using the self–duality of $F^{(5)}$, we see that the matrix appearing in the self–dual part of the vector kinetic terms,

$$\mathcal{N}_{IJ} F^{I+}_{\mu\mu} F^{J+}_{\rho\sigma} g^{\mu\rho} g^{\nu\sigma}, \qquad (12.65)$$

precisely corresponds to the normalized period matrix of the harmonic 3–forms ω_I^- on K which satisfy the duality condition $*\omega_I^- = -i\omega_I^-$,

$$\int_{A^I} \omega_J^- = \delta^I{}_J, \qquad \int_{B_I} \omega_J^- = \mathcal{N}_{IJ}. \qquad (12.66)$$

12.6 Compactifying type II supergravity on a CY manifold

This explains why, physically, the geometry of $\mathcal{N} = 2$ supergravity should coincide with the geometry of the variation of Hodge structures, and hence to the geometry of the polarized flag manifolds and their homogeneous bundles and connections.

It is important to note that the above result is *exact* at the full (nonperturbative) quantum level. Indeed, in principle the above classical result may be corrected at the quantum level by two kinds of quantum effect:

- *Stringy loop corrections* controlled by the string coupling constant e^ϕ, ϕ the dilaton field. From the viewpoint of 4D $\mathcal{N} = 2$ SUGRA, ϕ is a hypermultiplet scalar living on $\mathcal{M}_H$. Since the scalar manifold is a product $\mathcal{M}_V \times \mathcal{M}_H$, the metric on $\mathcal{M}_V$ is independent of the coordinates on $\mathcal{M}_H$; hence, in particular, it is independent of the coupling constant e^ϕ.
- σ-*Model (curvature) corrections* controlled by the overall size of the internal manifold K. The overall size is the class of ω^3 in $H^6(K)$, so depends only on the Kähler moduli, which are hypermultiplets in the type IIB case. The previous argument then applies also to the curvature corrections.

Type IIA. In type IIA the vector–multiplets correspond to deformations of the *complexified* Kähler class, that is, the class of $\omega + iB$ in $H^{1,1}(K)$. At the *classical level* the metric appearing in the 4D effective Lagrangian is again given by the Hodge metric, this time on $H^{1,1}(K)$ instead of $H^{2,1}(K)$. Let ω_A be a basis of harmonic $(1, 1)$ forms on K, representing *fixed* classes $[\omega_A] \in H^2(K_{\text{top}})$. The Kähler form $\omega(t)$ may be written as a function of the Kähler moduli t_A in the form

$$\omega(t) = \sum_A t_A \, \omega_A. \qquad (12.67)$$

We have a t–dependent Lefshetz decomposition of $H^2(K)$ into primitive and non–primitive parts

$$\omega_A = \varpi_A(t) + \lambda_A(t) \, \omega(t), \qquad (12.68)$$

where the coefficient $\lambda_A(t)$ is given by

$$\lambda_A(t) \int_K \omega(t)^3 = \int_K \omega(t)^2 \wedge \omega_A. \qquad (12.69)$$

The Weil formula [203] yields

$$*\varpi_A(t) = -\omega(t) \wedge \varpi_A(t), \qquad *\omega(t) = \frac{1}{2}\omega(t)^2, \qquad (12.70)$$

and hence

$$*\omega_A = -\omega(t) \wedge \omega_A + \frac{3}{2}\lambda_A(t)\,\omega(t)^2. \tag{12.71}$$

Therefore, the normalized Hodge metric is

$$G_{AB}(t) \equiv 6\,\frac{\int_K \omega_A \wedge *\omega_B}{\int_K \omega(t)^3}$$

$$= -\frac{6\int_K \omega_A \wedge \omega_B \wedge \omega(t)}{\int_K \omega(t)^3} + \frac{9\int_K \omega(t)^2 \wedge \omega_A \int_K \omega(t)^2 \wedge \omega_B}{\left(\int_K \omega(t)^3\right)^2}. \tag{12.72}$$

Let us define the topogical intersection numbers

$$K_{ABC} = \int_K \omega_A \wedge \omega_B \wedge \omega_C \tag{12.73}$$

and the Kähler potential

$$\Phi(z^A, \bar{z}^B) = -\log\bigl(K_{ABC}(\mathrm{Im}\,z^A)(\mathrm{Im}\,z^B)(\mathrm{Im}\,z^C)\bigr). \tag{12.74}$$

Under the identification $z^a \equiv it_A$, the corresponding Kähler metric coincides with the Hodge metric (12.72), i.e., with the metric in the effective kinetic terms for the Kähler moduli; we recognize this metric as the projective special Kähler one corresponding to the cubic prepotential

$$F(X^A) = \frac{K_{ABC}X^A X^B X^C}{X^0}. \tag{12.75}$$

The $h_{1,1}$ moduli $B_{i\bar{j}}$ correspond to $\mathrm{Re}\,z^A$; from Eq. (12.74) we see that shifts in these moduli are isometries of the classic Kähler metric.

From the point of view of the VHS, (12.75) corresponds to the "large" complex structure limit (Section 11.5.6). Thus, in the leading classical approximation the Kähler structure behaves as a complex structure near the degeneration point. However, in the type IIA case the prepotential is corrected by quantum "curvature" effects. There are only two kinds of correction compatible with the symmetries: first we may add to F a perturbative term of the form $\lambda(X^0)^2$ which does not spoil the perturbative symmetry under shifts of $\mathrm{Re}\,z^A$. Such a term indeed arises at the four–loop level. Then we may have non–perturbative terms of the form $e^{-n_A \mathrm{Im}\,z^A}$ which are exponentially small in the large volume limit $\mathrm{Im}\,z^A \to +\infty$. These terms exactly reproduce the exponential corrections predicted by the general SL_2–orbit theorem for the asymptotics of the period map. We conclude that, again, the exact expression is given by the period map of some Hodge structure. It is a deep fact that the exact map is the period map for a geometric family of CY spaces, namely the *mirror family* of CYs [188].

Table 12.2 *c–map for symmetric projective special Kähler spaces.*

Special Kähler space	dim$_\mathbb{C}$	Duality representation	Wolf space
$\frac{U(1,n)}{U(1) \times U(n)}$	n	$(n+1) \oplus \overline{(n+1)}$	$\frac{U(2,n+1)}{U(2) \times U(n+1)}$
$\frac{SU(1,1)}{U(1)} \times \frac{SO(n-1,2)}{SO(n-1) \times SO(2)}$	$n \geq 2$	$(2, n+1)$	$\frac{SO(n+1,4)}{SO(n+1) \times SO(4)}$
$\frac{SU(1,1)}{U(1)}$	1	4	$\frac{G_{2(+2)}}{SO(4)}$
$\frac{Sp(6,\mathbb{R})}{U(3)}$	6	14	$\frac{F_{4(+4)}}{USp(6) \times SU(2)}$
$\frac{U(3,3)}{U(3) \times U(3)}$	9	20	$\frac{E_{6(+2)}}{SU(6) \times SU(2)}$
$\frac{SO^*(12)}{U(6)}$	15	32	$\frac{E_{7(-5)}}{SU(6) \times SU(2)}$
$\frac{E_{7(-26)}}{E_6 \times SO(2)}$	27	56	$\frac{E_{8(-24)}}{E_7 \times SU(2)}$

12.7 Hypermultiplets: the *c*–map

In type IIA superstring, compactified to 4D on a strict Calabi–Yau 3–fold K, the complex deformation parameters of K are promoted to light scalars belonging to hypermultiplets. There are $4 + 4h^{2,1}$ hypermultiplet scalars, the first four being the "universal[9] scalars" whose couplings are the same for all K: ϕ, σ, and the two scalars associated in (12.61) with the harmonic (3,0) and (0,3) forms.

As reviewed in Section 12.5, the hypermultiplet scalars parametrize a Quaternionic–Kähler manifold $\mathcal{M}_H$ with a prescribed negative Ricci curvature. In particular, $\mathcal{M}_H$ should be irreducible. We conclude that the deformations of the complex structure of a Calabi–Yau 3–fold K describe – besides the projective special Kähler manifold $\mathcal{M}_V$ of the type IIB vector scalars – also the *quaternonic–Kähler geometry* of the type IIA hypermultiplet scalars (of quaternionic dimension $h^{2,1} + 1$). As we shall see shortly, the statement holds for arbitrary abstract weight 3, $h^{3,0} = 1$ variations of Hodge structures; that is, it holds for all special Kähler geometries whether they arise from a family of CY varieties or not. Therefore, *for each projective special Kähler manifold there is a corresponding Quaternionic–Kähler space* (having the Ricci curvature prescribed by $\mathcal{N} = 2$ supergravity).

---

[9] From the world–sheet superconformal field theory the universal sector corresponds to the identity operator and its spectral flows [91].

The explicit transformation from a projective special Kähler manifold of complex dimension m to the associated Quaternionic–Kähler space of quaternionic dimension $m + 1$ is called the *c–map* [91].

The *c*–map may be constructed as follows. We start from a 4D supergravity coupled to m vector–multiplets and h hypermultiplets whose scalars live on the product space

$$\mathcal{M}_V \times \mathcal{M}_H, \qquad \begin{bmatrix} \mathcal{M}_V \text{ projective special Kähler,} \\ \mathcal{M}_H \text{ quaternionic Kähler,} \end{bmatrix} \qquad (12.76)$$

where $\mathcal{M}_V$ is the given special Kähler geometry. Then we compactify the theory to 3D on a circle, getting a 3D $\mathcal{N} = 4$ supergravity. In 3D we get a new scalar from the metric, $m + 1$ new scalars from the fourth components of the vector fields (including the graviphoton), and $m + 2$ vectors (including a KK vector from the metric). As in Chapter 2 we dualize the vectors into scalars, so that in total we get $2m + 4$ more scalars in 3D than in 4D. These new scalars, as well those of the 4D vector–multiplets, belong to *twisted* hypermultiplets (cf. Section 2.4.1), and the 3D scalar manifold is a product

$$\mathcal{M}_{\text{twisted}} \times \mathcal{M}_H, \qquad (12.77)$$

where the factor spaces are Quaternionic–Kähler manifolds whose tangent spaces realize the two inequivalent $\mathbb{C}\ell^0(4)$–modules. $\mathcal{M}_H$ is the same space as in 4D, while $\mathcal{M}_{\text{twisted}}$ replaces the projective special Kähler manifold $\mathcal{M}_V$ in (12.76). The *c*–map is given by

$$\mathcal{M}_V \xrightarrow{c-\text{map}} \mathcal{M}_{\text{twisted}}. \qquad (12.78)$$

The *c*–map associates an explicit Quaternionic–Kähler metric G_{ij} to a (duality class of) degree 2 homogeneous prepotential $F(X^I)$. The expression of G_{ij} in terms of $F(X^I)$ is quite involved, and will not be given here; it may be found in ref. [138].

According to the general arguments in Chapter 4, the special Kähler manifold $\mathcal{M}_V$ embeds in its *c*–map image $\mathcal{M}_{\text{twisted}}$ as a *totally geodesic submanifold*. Hence, if $\mathcal{M}_{\text{twisted}}$ is a symmetric space, so is $\mathcal{M}_V$. Conversely, if $\mathcal{M}_V$ is symmetric (resp. homogeneous), $\mathcal{M}_{\text{twisted}}$ is symmetric and hence a Wolf space (resp. homogeneous and hence a Alekseeskii space). In Table 12.2 we list the *c*–map action on the symmetric special Kähler manifolds of Section 12.4 following [91]. Note that all Wolf spaces are recovered as *c*–maps of special Kähler symmetric spaces *but* the quaternionic hyperbolic spaces $Sp(2, 2n)/Sp(2) \times Sp(2n)$, which are not in the image of the *c*–map for any special Kähler geometry [91].

The first row in the table corresponds to the *minimal coupling* special geometries (the only ones with a reducible duality representation). In particular, setting $n = 0$

12.7 Hypermultiplets: the c–map

(no vector multiplets) we get the universal sector which parametrizes the Wolf space,

$$\mathcal{M}_{\text{univ.}} = U(2,1)/U(2,1). \tag{12.79}$$

The simplest way to get the c–map in the first row of the table is to realize that, acting on the minimal coupling special geometries, the c–map should produce manifolds which are Quaternionic–Kähler as well as Kähler; we know from Chapter 3 that the *only* spaces with this property are the Wolf spaces $U(2,n)/U(2) \times U(n)$, and this implies the first row of the table. To see that the relevant spaces should be Kähler, consider 4D $\mathcal{N}=2$ ungauged SUGRA with n "minimally coupled" vector multiplets. The bosonic part of the Lagrangian is

$$e^{-1} L_{\text{bosonic}} = -\frac{1}{2}R + g_{i\bar{j}}\, \partial_\mu z^i\, \partial^\mu \bar{z}^j + \left(\mathcal{N}_{IJ} F^{I+}_{\mu\nu} F^{J+}_{\rho\sigma}\, g^{\mu\rho} g^{\nu\sigma} + \text{h.c.}\right) \tag{12.80}$$

where $g_{i\bar{j}}$ is Kähler and $\mathcal{N}_{IJ}$ depends holomorphically on the z^i. Inverting the space–time orientation, we get a bosonic sector which is allowed in 4D $\mathcal{N}=1$ SUGRA; we may take this particular $\mathcal{N}=1$ model and compactify it in 3D, getting a 3D $\mathcal{N}=2$ supergravity whose scalars parametrize a Kähler manifold. However, in the bosonic sector, there is no difference between the computation of the effective 3D Lagrangian for the $\mathcal{N}=2$ and the $\mathcal{N}=1$ models, and we conclude that the scalar space should be both Kähler and Quaternionic–Kähler. On the contrary, for a non–minimal special Kähler geometry $\mathcal{N}_{IJ}$ is *not* holomorphic and the resulting quaternionic–Kähler space is *not* Kähler.

Appendix
G–structures on manifolds

In this appendix we collect some basic definitions and results on G–structures on manifolds. No proofs are provided. Details may be found in [214].

M is a smooth manifold of dimension n and $L(M)$ the bundle of linear frames over M. $L(M)$ is a principal bundle on M with group $GL(n, \mathbb{R})$. G is a (closed) Lie subgroup of $GL(n, \mathbb{R})$.

Definition A.1 A *G–structure* on M is a smooth subbundle P of $L(M)$ with structure group G. A G–structure P on M is said to be *integrable* if every point of M has a neighborhood U with local coordinates $x^1, \ldots, x^n$ such that the local section $(\partial/\partial x^1, \ldots, \partial/\partial x^n)$ of $L(M)$ over U is a local section of P. In this case we say that the local coordinate system $x^1, \ldots, x^n$ is *admissible* for the G–structure P.

If x^i and y^i are two admissible coordinate systems in the charts U and V the Jacobian matrix $\partial y^i / \partial x^j$ is in G at all points in $U \cap V$.

G is a closed subgroup of $GL(n, \mathbb{R})$. The embedding in $GL(n, \mathbb{R})$ defines a representation ρ of G in $\mathbb{R}^n$. Let $\rho(P) \equiv P \times_\rho \mathbb{R}^n$ be the rank n vector bundle (with structure group G) associated to the principal bundle P through ρ. The definition of the G–structure is equivalent to the statement that we have an *isomorphism* of vector bundles

$$\theta : TM \to \rho(P), \qquad (A.1)$$

which reduces the structure group from $GL(n, \mathbb{R})$ to its subgroup G. Conversely, any isomorphism of TM with a rank n vector bundle $\mathcal{V}$ with structure group G defines a G–structure. In Eq. (A.1) θ is a section of the vector bundle $T^*M \otimes \rho(P)$, that is, a 1–form with coefficients in $\rho(P)$. We shall call θ the *G–vielbein* (in the math literature it is also known as the *solder* form). The G–vielbein θ fully specifies the G–structure. This construction may be generalized to any tensor T on $\mathbb{R}^n$. T belongs to a representation ρ_T of $GL(n, \mathbb{R})$ which decomposes into representations of G. In particular, let T be a tensor on $\mathbb{R}^n$ which is invariant under the subgroup $G \subset GL(n, \mathbb{R})$. A G–structure on M naturally defines a tensor field T associated *to* T: at each point $x \in M$ choose a linear frame $u_x \in P_x$; u_x gives a linear isomorphism $\mathbb{R}^n \to T_x M$ which induces an isomorphism of the corresponding tensor algebras. Let T_x be the image of T. The fact that T is G–invariant implies that T_x is independent of the choice of u_x.

Proposition A.2 *Let* T *be a tensor on* $\mathbb{R}^n$ *and* $G \subset GL(n, \mathbb{R}^n)$ *the subgroup leaving invariant it. Then a G–structure P is integrable if and only if each point $x \in M$ has a neighborhood U with local coordinates $x^1, \ldots, x^n$ with respect to which the components of the tensor field T associated to* T *are constant functions.*

Suppose we have a G–structure P_G on M. We may want to further reduce the structure group of TM to a subgroup $H \subset G$. This leads us to consider a subbundle P_H of P_G with structure group H.

Proposition A.3 [215] *The reduction from P_G to P_H is possible if and only if there exists a section $M \to P_G/H$. There is a 1–to–1 correspondence between such sections and reductions of the G–structure P_G to an H–structure.*

Connections and torsion. A connection on a G–structure is a connection on the underlying principal bundle P. The representation ρ induces a connection 1–form $\omega^a{}_b$ with coefficients in the Lie algebra $\mathfrak{g}$ of G on the vector bundle $\rho(P) \simeq TM$. The torsion T^a is defined as

$$\nabla \theta^a \equiv d\theta^a + \omega^a{}_b \wedge \theta^b = T^a. \tag{A.2}$$

Proposition A.4 *If G is integrable P admits a torsion–free connection.*

Intrinsic torsion. The torsion depends on the chosen connection. In view of the above result, one is interested in the part of the torsion, called the *intrinsic torsion*, independent of the connection which is the obstruction to integrability of the G–structure. For the intrinsic torsion see Section 3.6.

Examples

- $G = GL(n, \mathbb{R})^+$, the group of matrices with positive determinant. A $GL(n, \mathbb{R})^+$–structure is an orientation of the manifold, and its exists iff M is orientable. If it exists, a $GL(n, \mathbb{R})^+$–structure is always integrable.
- $G = SL(n, \mathbb{R})$. A $SL(n, \mathbb{R})$–structure defines a volume form on M; it exists iff M is orientable, and it is always integrable.
- $G = GL(n/2, \mathbb{C})$, n even. The invariant tensor is an almost complex structure I, and hence a $GL(n/2, \mathbb{C})$–structure is an *almost complex structure* on M. It is integrable iff I is a complex structure, i.e. M is a complex manifold.

Theorem A.5 (Newlander–Nirenberg) *A $GL(n/2, \mathbb{C})$–structure is integrable if and only if it admits a torsion–free connection.*

- $G = Sp(n, \mathbb{R})$ n even. The invariant tensor is a 2–form Ω everywhere of maximal rank, hence a $Sp(n, \mathbb{R})$–structure is an *almost symplectic structure*. Note that $GL(n/2, \mathbb{R})$ and $Sp(n, \mathbb{R})$ have the same maximal compact subgroup, $U(n/2)$, and hence a manifold M admits an almost symplectic structure if and only if it admits an almost complex structure. An integrable $Sp(n, \mathbb{R})$–structure is called a *symplectic structure*.

Theorem A.6 (Darboux) *(1) An almost symplectic structure is a symplectic structure iff the invariant 2–form Ω is closed. (2) An almost symplectic structure is a symplectic structure iff it admits a torsionless connection.*

- $G = SO(n)$. The invariant tensor is a positive metric tensor and a $SO(n)$–structure is a Riemannian structure. It is integrable only if it is flat.

Theorem A.7 (The fundamental theorem of differential geometry) *On an $SO(n)$–structure there is a unique compatible connection for each given torsion tensor $T^\rho_{\mu\nu}$. In particular, there is a unique torsionless metric connection, the Levi–Civita one.*

- $G =$ (a compact subgroup). Since $SO(n)$ is the maximal compact subgroup of $GL(n, \mathbb{R})$, G is in particular a subgroup of $SO(n)$ and hence leaves invariant a Riemannian metric. Thus, in this case, we get a special Riemannian geometry with the structure group reduced to G. The existence of a torsionless compatible connection (which coincides with the Levi–Civita one by Theorem A.7) gives strong restrictions on G, see chapter 3.
- $G = SO(p, n - p)$. A *pseudo*–Riemannian structure of signature $(p, n - p)$.
- $G = U(n/2)$, n even. Since $U(n/2) \subset GL(n/2, \mathbb{C})$, $U(n/2) \subset SO(n)$, and $U(n/2) \subset Sp(n, \mathbb{R})$, a $U(n/2)$–structure is, in particular, an almost complex structure I, a Riemannian metric g, and an almost symplectic structure Ω. Since the metric is Hermitian with respect to the almost complex structure I, a $U(n/2)$–structure is called an *almost Hermitian structure*. If the underlying almost complex structure is integrable, we say that it is a *Hermitian structure*. A torsionless $U(n/2)$–structure is called a *Kähler structure*. By Theorems A.5 and A.6 in this case I is integrable and Ω closed.
- $G = 1$. A 1–structure is an *absolute parallelism*.

Cartan structure equations. Curvature

Given a G–structure θ^a with connection $\omega^a{}_b$, the torsion T^a and the curvature $R^a{}_b$ are the 2–forms with value, respectively, in the vector bundles $\rho(P)$ and $\text{ad}(P)$,

$$\nabla\theta^a \equiv d\theta^a + \omega^a{}_b \wedge \theta^b = T^a, \tag{A.3}$$

$$d\omega^a{}_b + \omega^a{}_c \wedge \omega^c{}_b = R^a{}_b. \tag{A.4}$$

Equations (A.3) and (A.4) are called the *Cartan structure equations*.

Bianchi identities

Taking the differential of Eqs. (A.3) and (A.4) and using $d^2 = 0$, we get the *Bianchi identities*

$$\nabla T^a \equiv dT^a + \omega^a{}_b \wedge T^b = R^a{}_b \wedge \theta^b, \tag{A.5}$$

$$\nabla R^a{}_b \equiv dR^a{}_b + \omega^a{}_c \wedge R^c{}_b - R^a{}_c \wedge \omega^c{}_b = 0. \tag{A.6}$$

In particular, for a torsion–free G–structure (with connection) the first Bianchi identity (A.5) becomes

$$R^a{}_b \wedge \theta^b = 0, \tag{A.7}$$

or, in components, $R^a{}_{b\,[\mu\nu}\,\theta^b{}_{\rho]} = 0$.

Ricci identity

Let σ be a representation of G. We write $\sigma(\omega)$ for the connection on $\sigma(P)$ induced by ω. Then the covariant differential acting on sections of $\sigma(P)$ is

$$\nabla = d + \sigma(\omega). \tag{A.8}$$

The Ricci identity states

$$(d + \sigma(\omega))^2 = d\sigma(\omega) + \sigma(\omega) \wedge \sigma(\omega) \equiv \sigma(d\omega + \omega^2) \equiv \sigma(R), \tag{A.9}$$

or, in components,

$$\nabla_\mu \nabla_\nu - \nabla_\nu \nabla_\mu = \sigma(R_{\mu\nu}), \tag{A.10}$$

which gives the geometric interpretation of the curvature as the measure of the failure of the covariant derivative to commute.

References

[1] Abbott, L.F., and Deser, S. 1982. Stability of gravity with a cosmological constant. *Nucl. Phys.*, **B195**, 76.
[2] Adams, J.F. 1996. *Lectures on Exceptional Lie Groups*. University of Chicago Press.
[3] Aharony, O., Bergman, O., Jafferis, D.L., and Maldacena, J. 2008. $N = 6$ superconformal Chern–Simons–matter theories, M2-branes and their gravity duals. *JHEP*, **0810**, 091.
[4] Akhiezer, D.N. 1990. Homogeneous complex manifolds. Pages 148–244 of: *Several Complex Variables IV*. (Encyclopaedia of Mathematical Sciences, vol. 10). Springer.
[5] Alekseeskii, D.V. 1975. Classification of quaternionic spaces with transitive solvable group of motion. *Ozv. Akad. Nauk. SSSR Ser. Math.*, **39**, 315–362.
[6] Alekseev, A., Malkin, A., and Meinjrenken, E. 1998. Lie group valued momentum maps. *J. Differential Geom.*, **48**, 445–495.
[7] Alvarez-Gaumé, L. 1983. Supersymmetry and the Atiyah–Singer index theorem. *Commun. Math. Phys.*, **90**, 161–173.
[8] Alvarez-Gaumé, L., and Freedman, D. 1983. Potentials for the supersymmetric non–linear σ–models. *Commun. Math. Phys.*, **91**, 87.
[9] Alvarez-Gaumé, L., and Freedman, D.Z. 1980. Kähler geometry and the renormalization of supersymmetric sigma models. *Phys. Rev.*, **D22**, 846.
[10] Alvarez-Gaumé, L., and Freedman, D.Z. 1981. Geometrical structure and ultraviolet finiteness in the supersymmetric sigma model. *Commun. Math. Phys.*, **80**, 443.
[11] Alvarez-Gaumé, L., and Witten, E. 1983. Gravitational anomalies. *Nucl. Phys.*, **B234**, 269–330.
[12] Alvarez-Gaumé, L., Freedman, D.Z., and Mukhi, S. 1981. The background field method and the ultraviolet structure of the supersymmetric nonlinear sigma model. *Ann. Phys.*, **134**, 85.
[13] Ambrose, W., and Singer, I.M. 1953. A theorem on holonomy. *Trans. Amer. Math. Soc.*, **79**, 428–443.
[14] Andrianopoli, L., Bertolini, M., Ceresole, A., D'Auria, R., Ferrara, S., Fré, P., and Magri, T. 1997. $N = 2$ supergravity and $N = 2$ superYang–Mills theory on general scalar manifolds: symplectic covariance, gaugings and the momentum map. *J. Geom. Phys.*, **23**, 111–189.
[15] Andrianopoli, L., Cordaro, F., Fré, P., and Gualtieri, L. 2001. Nonsemisimple gaugings of $D = 5$ $N = 8$ supergravity and FDA.s. *Class. Quant. Grav.*, **18**, 395–414.
[16] Andrianopoli, L., D'Auria, R., Ferrara, S., and Lledo, M.A. 2002a. Duality and spontaneously broken supergravity in flat backgrounds. *Nucl. Phys.*, **B640**, 63–77.

[17] Andrianopoli, L., D'Auria, R., Ferrara, S., and Lledo, M.A. 2002b. Gauging of flat groups in four-dimensional supergravity. *JHEP*, **0207**, 010.
[18] Arnold, V.I. 1989. *Mathematical Methods of Classical Mechanics*. (Graduate Texts in Mathematics, vol. 60). Springer.
[19] Arnold, V.I., and Givental, A.B. 2001. Symplectic geometry. In: *Dynamical Systems. IV*. (Encyclopaedia of Mathematical Sciences, vol. 4). Springer.
[20] Atiyah, M. 1988a. *Collected Works. Vol. 3. Index Theory*. Oxford Science Publications; Oxford University Press.
[21] Atiyah, M. 1988b. *Collected Works. Vol. 4. Index Theory: 2*. Oxford Science Publications; Oxford University Press.
[22] Atiyah, M.F. 1984. The momentum map in symplectic geometry. Pages 43–51 of: *Durham Symposium on Global Riemannian Geometry*. Ellis Horwood Ltd.
[23] Atiyah, M.F., and Bott, R. 1966. A Lefschetz fixed point formula for elliptic differential operators. *Bull. Amer. Math. Soc.*, **12**, 245.
[24] Atiyah, M.F., and Bott, R. 1982. Yang–Mills equations over Riemann surfaces. *Phil. Trans. R. Soc. Lond.*, **A308**, 523–615.
[25] Atiyah, M.F., and Bott, R. 1984. The momentum map and equivariant cohomology. *Topology*, **23**, 1–28.
[26] Atiyah, M.F., Bott, R., and Shapiro, A. 1964. Clifford modules. *Topology*, **3**, 3–38.
[27] Awada, M., and Townsend, P.K. 1986. Gauged $N = 4$, $d = 6$ Maxwell–Einstein supergravity and "antisymmetric-tensor Chern–Simons" forms. *Phys. Rev.*, **D33**(Mar), 1557–1562.
[28] Baez, J.C. 2002. The octonions. *Bull. Amer. Math. Soc.*, **39**, 145–205.
[29] Bagger, J., and Lambert, N. 2007. Modeling Multiple M2's. *Phys. Rev.*, **D75**, 045020.
[30] Bagger, J., and Lambert, N. 2008. Gauge symmetry and supersymmetry of multiple M2-branes. *Phys. Rev.*, **D77**, 065008.
[31] Bagger, J., and Witten, E. 1983. Matter couplings in $N = 2$ supergravity. *Nucl. Phys.*, **B222**, 1.
[32] Banks, T., and Seiberg, N. 2011. Symmetries and strings in field theory and gravity. *Phys. Rev.*, **D83**, 084019.
[33] Bar, Ch. 1993. Real Killing spinors and holonomy. *Commun. Math. Phys.*, **154**, 509–521.
[34] Baum, H. 2002. *Conformal Killing spinors and special geometric structures in Lorentzian geometry – a survey*. arXiv:math/0202008 [math.DG].
[35] Baum, H. 2008. Conformal Killing spinors and the holonomy problem in Lorentzian geometry – A survey of new results. In: *Symmetries and Overdetermined Systems of Partial Differential Equations*. Springer.
[36] Baum, H. 2011. *Holonomy groups of Lorentzian manifolds – a status report*. Available at the author's webpage: http://www.mathematik.hu-berlin.de/baum/publikationen- fr/Publikationen-ps-dvi/Baum-Holono-my-report-final.pdf. Accessed August 13, 2014.
[37] Baum, H., and Kath, I. 1999. Parallel spinors and holonomy groups on pseudo–Riemannian spin manifolds. *Ann. Global Anal. Geom.*, **17**, 1–17.
[38] Baum, H., Friedrich, T., Grunewald, R., and Kath, I. 1991. *Twistor and Killing Spinors on Riemannian Manifolds*. (Teubner–Texte zur Mathematik, vol. 124). Teubner-Verlag.
[39] Berard-Bergery, L., and Ikenakhen, A. 1993. On the holonomy of Lorentzian manifolds. In: *Differential Geometry: Geometry in Mathematical Physics and Related Topics*. American Mathematical Society.

[40] Berg, M., Haak, M., and Samtleben, H. 2003. Calabi–Yau fourfolds with flux and supersymmetry breaking. *JHEP*, **04**, 3–38.
[41] Berger, M. 1955. Sur les groupes d'holonomie des variétés à connexion affine et des variétés riemanniennes. *Bull. Soc. Math. France.*, **83**, 279–330.
[42] Berger, M. 2000. *A Panoramic View of Riemannian Geometry*. Springer.
[43] Berger, M., Gauduchon, P., and Mazet, E. 1971. *Le spectre d'une varieté riemannienne*. (Lectures Notes in Mathematics, vol. 194). Springer.
[44] Bergman, S. 1950. *The Kernel Function and Conformal Mapping*. (Mathematical Surveys and Monographs, vol. 5). America Mathematical Society.
[45] Bergshoeff, E., Koh, I.G., and Sezgin, E. 1985. Coupling Yang–Mills to $N = 4$ $d = 4$ supergravity. *Phys. Lett.*, **B155**, 71–75.
[46] Bergshoeff, E., Cecotti, S., Samtleben, H., and Sezgin, E. 2010. Superconformal sigma models in three dimensions. *Nucl. Phys.*, **B838**, 266–297.
[47] Bergshoeff, E.A., de Roo, M., and Hohm, O. 2008a. Multiple M2-branes and the embedding tensor. *Class. Quant. Grav.*, **25**, 142001.
[48] Bergshoeff, E.A., Hohm, O., Roest, D., Samtleben, H., and Sezgin, E. 2008b. The superconformal gaugings in three dimensions. *JHEP*, **0809**, 101.
[49] Berndt, R. 2001. *An Introduction to Symplectic Geometry*. (Graduate Studies in Mathematics, vol. 26). American Mathematical Society.
[50] Bershadsky, M., Cecotti, S., Ooguri, H., and Vafa, C. 1994. Kodaira–Spencer theory of gravity and exact results for quantum string amplitudes. *Commun. Math. Phys.*, **165**, 311–428.
[51] Besse, A.L. 1987. *Einstein Manifolds*. Springer.
[52] Birmingham, D., Blau, M., Rakowski, M., and Thompson, G. 1991. Topological field theory. *Phys. Rep.*, **209**, 129–340.
[53] Blau, M., and Thompson, G. 1995. Localization and diagonalization: a review of functional integral techniques for low–dimensional gauge theories and topological field theories. *J. Math. Phys.*, **36**, 2192–2236.
[54] Blencowe, M.P., and Duff, M.J. 1988. Supermembranes and the signature of space-time. *Nucl. Phys.*, **B310**, 387–404.
[55] Borel, A. 1949. Some remarks about Lie groups transitive on spheres and tori. *Bull. Amer. Math. Soc.*, **55**, 580–587.
[56] Borel, A. 1950. Le plan projectif des octaves et les sphéres comme espaces homogènes. *C. R. Acad. Sci. Paris*, **230**, 1378–1380.
[57] Borel, A. 1960. On the curvature tensor of Hermitian symmetric manifolds. *Ann. Math.*, **71**, 508–521.
[58] Born, M., and Infeld, L. 1934. Foundations of the new field theory. *Proc. R. Soc.*, **A144**, 425–451.
[59] Bott, R. 1957. Homogeneous vector bundles. *Ann. Math.*, **66**, 203–248.
[60] Bott, R., and Wu, L.W. 1982. *Differential Forms in Algebraic Topology*. (Graduate Texts in Mathematics, vol. 82). Springer.
[61] Bourbaki, N. 1989. *Elements of Mathematics. Lie Groups and Lie Algebras*. Chapt. 1–3. Springer.
[62] Boyer, C., and Galicki, K. 2008. *Sasakian Geometry*. (Oxford Mathematical Monographs). Oxford University Press.
[63] Boyer, C.P., and Galicki, K. 1999. 3–Sasakian manifolds. *Surveys Diff. Geom.*, **7**, 123.
[64] Boyer, C.P., Galicki, K., and Mann, B.M. 1993. Quaternionic reduction and Einstein manifolds. *Commun. Anal. Geom.*, **1**, 229–279.

[65] Bröker, T., and tom Dieck, T. 1985. *Representations of Compact Lie Groups*. (Graduate Texts in Mathematics, vol. 98). Springer.
[66] Bryant, R., and Griffiths, P. 1983. Some observations on the infinitesimal period relations for regular threefolds with trivial canonical bundle. Pages 77–102 of: Artin, M. and Tate, J. (eds.), *Arithmetic and Geometry. Papers dedicated to I.R. Shafarevich. Vol. II*. Birkhäuser.
[67] Bryant, R.L., Chern, S.S., Gardner, R.B., Goldschmidt, H.L., and Griffiths, P.A. 1991. *Exterior Differential Systems*. (Mathematical Sciences Research Institute Publications, vol. 18). Springer.
[68] Bump, D. 1997. *Automorphic Forms and Representations*. (Studies in Advanced Mathematics, vol. 55). Cambridge University Press.
[69] Bump, D. 2000. *Lie Groups*. (Graduate Texts in Mathematics, vol. 225). Springer.
[70] Cahen, M., and Wallach, N. 1970. Lorentzian symmetric spaces. *Bull. AMS*, **76**, 585–591.
[71] Calabi, E., and Vesentini, E. 1960. On compact, locally symmetric Kähler manifolds. *Ann. Math.*, **71**, 472–507.
[72] Campbell, L.C., and West, P.C. 1984. $N = 2$, $d = 2$ nonchiral supergravity and its spontaneous compactification. *Nucl. Phys.*, **B243**, 112–124.
[73] Candelas, P. 1988. Yukawa couplings between (2,1) forms. *Nucl. Phys.*, **B298**, 458.
[74] Candelas, P., Horowitz, G.T., Strominger, A., and Witten, E. 1985. Vacuum configurations for superstrings. *Nucl. Phys.*, **B 258**, 46–74.
[75] Candelas, P., De La Ossa, X.C., Green, P.S., and Parkes, L. 1991. A pair of Calabi–Yau manifolds as an exactly soluble superconformal theory. *Nucl. Phys.*, **B 359**, 21–74.
[76] Cannas da Silva, A. 2008. *Lectures on Symplectic Geometry*. (Lecture Notes in Mathematics, vol. 1764). Springer.
[77] Carlson, J., Müller-Stach, S., and Peters, C. 2003. *Period Mappings and Period Domains*. (Cambridge Studies in Advanced Mathematics, vol. 85). Cambridge University Press.
[78] Cartan, E. 1935. Sur les domaines bornés homogènes de l'espace de n variables complexes. *Abh. Math. Semin. Univ. Hamb.*, **11**, 116–162.
[79] Castellani, L., Ceresole, A., D'Auria, R., Ferrara, S., Fre, P., and Maina, E. 1985. σ Models, duality transformations and scalar potentials in extended supergravities. *Phys. Lett.*, **B161**, 91.
[80] Castellani, L., Ceresole, A., D'Auria, R., Ferrara, S., Fré, P., and Maina, E. 1986a. The complete $N = 3$ matter coupled supergravity. *Nucl. Phys.*, **B268**, 376–382.
[81] Castellani, L., Ceresole, A., Ferrara, S., D'Auria, R., Fré, P., and Maina, E. 1986b. The complete $N =3$ matter coupled supergravity. *Nucl. Phys.*, **B268**, 317.
[82] Castellani, L., D'Auria, R., and Fré, P. 1991. *Supergravity and Superstrings: A Geometrical Perspective*, vols. 1, 2 & 3. World Scientific.
[83] Cecotti, S. 1988. Homogeneous Kähler manifolds and flat potential $N =1$ supergravities. *Phys. Lett.*, **B215**, 489–490.
[84] Cecotti, S. 1989. Homogeneous Kähler manifolds and T algebras in $N = 2$ supergravity and superstrings. *Commun. Math. Phys.*, **124**, 23–55.
[85] Cecotti, S. 1991. Geometry of $N = 2$ Landau–Ginzburg families. *Nucl. Phys.*, **B 355**, 755–776.
[86] Cecotti, S., and Girardello, L. 1982. Functional measure, topology and dynamical supersymmetry breaking. *Phys. Lett.*, **B110**, 39.
[87] Cecotti, S., and Vafa, C. 1991. Topological antitopological fusion. *Nucl. Phys.*, **B 367**, 359–461.

[88] L. Lusanna (ed.). 1985. John Hopkins workshop on current problems in particle theory 9: new trends in particle theory. Singapore: World Scientific.

[89] Cecotti, S., Girardello, L., and Porrati, M. 1986. Constraints on partial superHiggs. *Nucl. Phys.*, **B268**, 295–316.

[90] Cecotti, S., Ferrara, S., and Girardello, L. 1988. Hidden noncompact symmetries in string theories. *Nucl. Phys.*, **B 308**, 436.

[91] Cecotti, S., Ferrara, S., and Girardello, L. 1989. Geometry of type II superstrings and the moduli of superconformal field theories. *Int. J. Mod. Phys.*, **A4**, 2475.

[92] Chapline, G., and Manton, N.S. 1983. Unification of Yang–Mills theory and supergravity in ten dimensions. *Phys. Lett.*, **B120**, 124–134.

[93] Chern, S.S. 1943. On the Curvatura Integra in a Riemannian manifold. *Ann. Math.*, **46**, 674–684.

[94] Chern, S.S. 1979. *Complex Manifolds without Potential Theory (with an Appendix in the Geometry of Characteristic Classes)*. Springer.

[95] Chow, B., Chu, S.-C., Glickestein, D. Guenther, C., Isenberg, J., Ivey, T., Knopf, D., Lu, P., Luo, F., and Ni, L. 2007. *The Ricci Flow: Techniques and Applications. Part I: Geometric Aspects*. (Mathematical Surveys and Monographs, vol. 135). American Mathematical Society.

[96] Coleman, S., and Mandula, J. 1967. All possible symmetries of the S–matrix. *Phys. Rev.*, **159**, 1251.

[97] Cordaro, F., Fré, P., Gualtieri, L., Termonia, P., and Trigiante, M. 1998. $N=8$ gaugings revisited: an exhaustive classification. *Nucl. Phys.*, **B532**, 245–279.

[98] Cremmer, E., and Julia, B. 1979. The $SO(8)$ supergravity. *Nucl. Phys.*, **B159**, 141–212.

[99] Cremmer, E., and Van Proeyen, A. 1985. Classification of Kähler manifolds in $N=2$ vector multiplet supergravity couplings. *Class. Quant. Grav.*, **2**, 445.

[100] Cremmer, E., Julia, B., Scherk, J., Ferrara, S., Girardello, L., and van Nieuwenhuizen, P. 1979. Spontaneous symmetry breaking and Higgs effect in supergravity without cosmological constant. *Nucl. Phys.*, **B147**, 105.

[101] Cremmer, E., Ferrara, S., Kounas, C., and Nanopoulos, D.V. 1983a. Naturally vanishing cosmological constant in $N=1$ supergravity. *Phys. Lett.*, **B133**, 61.

[102] Cremmer, E., Ferrara, S., Girardello, L., and Van Proeyen, A. 1983b. Yang-Mills theories with local supersymmetry: Lagrangian, transformation laws and super-Higgs effect. *Nucl. Phys.*, **B212**, 413.

[103] de Boer, J., Manshot, J., Papadodimas, K., and Verlinde, E. 2009. The chiral ring of AdS_3/CFT and the atractor mechanism. *JHEP*, **0903**, 030.

[104] de Roo, M. 1985. Matter coupling in $N=4$ supergravity. *Nucl. Phys.*, **B255**, 515–531.

[105] de Roo, M., van Holten, J.W., de Wit, B., and Van Proeyen, A. 1980. Chiral superfields in $N=2$ supergravity. *Nucl. Phys.*, **B173**, 175.

[106] de Wit, B. 1979. Properties of $SO(8)$ extended supergravity. *Nucl. Phys.*, **B158**, 189–212.

[107] de Wit, B. 1982. Conformal invariance in extended supergravity. Pages 267–312 of: *Supergravity 1981*. Cambridge University Press.

[108] de Wit, B. 2002. *Supergravity*. Lectures at the 2001 Le Houches Summer School. arXiv:hep-th/0212245.

[109] de Wit, B., and Freedman, D.Z. 1977. On $SO(8)$ extended supergravity. *Nucl. Phys.*, **B130**, 105–113.

[110] de Wit, B., and Nicolai, H. 1982. $N=8$ supergravity. *Nucl. Phys.*, **B208**, 323–364.

[111] de Wit, B., and Samtleben, H. 2005. Gauged maximal supergravities and hierarchies of nonabelian vector–tensor systems. *Fortsch. Phys.*, **53**, 442–449.

[112] de Wit, B., and Samtleben, H. 2008. The end of the p-form hierarchy. *JHEP*, **0808**, 015.

[113] de Wit, B., and van Nieuwenhizen, P. 1989. Rigidly and locally supersymmetric two–dimensional non–linear σ–models with torsion. *Nucl. Phys.*, **B312**, 58–94.

[114] de Wit, B., and Van Proeyen, A. 1984. Potentials and symmetries of general gauged $N=2$ supergravity–Yang–Mills models. *Nucl. Phys.*, **B245**, 89–117.

[115] de Wit, B., and Van Proeyen, A. 1992. Special geometry, cubic polynomials and homogeneous quaternionic spaces. *Commun. Math. Phys.*, **149**, 307–334.

[116] de Wit, B., van Holten, J.W., and van Proeyen, A. 1981. Structure of $N=2$ supergravity. *Nucl. Phys.*, **B184**, 77.

[117] de Wit, B., Lauwers, P.G., Philippe, R., Su, S.Q., and Van Proeyen, A. 1984. Gauge and matter fields coupled to $N=2$ supergravity. *Phys. Lett.*, **B134**, 37.

[118] de Wit, B., Lauwers, P.G., and Van Proeyen, A. 1985. Lagrangians of $N=2$ supergravity – matter systems. *Nucl. Phys.*, **B255**, 569.

[119] de Wit, B., Tollsten, A.K., and Nicolai, H. 1993. Locally supersymmetric $D=3$ nonlinear sigma models. *Nucl. Phys.*, **B 392**, 3–38.

[120] de Wit, B., Herger, I., and Samtleben, H. 2003a. Gauged locally supersymmetric D = 3 nonlinear sigma models. *Nucl. Phys.*, **B671**, 175–216.

[121] de Wit, B., Samtleben, H., and Trigiante, M. 2003b. On Lagrangians and gaugings of maximal supergravities. *Nucl. Phys.*, **B655**, 93–126.

[122] de Wit, B., Samtleben, H., and Trigiante, M. 2004. Gauging maximal supergravities. *Fortsch.Phys.*, **52**, 489–496.

[123] de Wit, B., Nicolai, H., and Samtleben, H. 2008. Gauged supergravities, tensor hierarchies, and M-theory. *JHEP*, **0802**, 044.

[124] Denef, F. 2008. *Les Houches Lectures on Constructing String Vacua*. e–print arXiv:0803.1194.

[125] Deser, S., and Zumino, B. 1976. Consistent supergravity. *Phys. Lett.*, **62B**, 335–337.

[126] Deser, S., Jackiw, R., and Templeton, S. 1982. Topologically massive gauge theories. *Ann. Phys.*, **140**, 373–411.

[127] Dieudonné, J. 1975. *Eléments d'Analyse*. Tome 5. Gauthier–Villars.

[128] Dirac, P.A.M. 1931. Quantised singularities in the electromagnetic field. *Proc. R. Soc.*, **A133**, 60–73.

[129] Donagi, R., and Witten, E. 1996. Supersymmetric Yang–Mills theory and integrable systems. *Nucl. Phys.*, **B460**, 299–334.

[130] Donagi, R.Y., and Wendland, K (eds.). 2008. *From Hodge Theory to Integrability and TQFT: tt^*–geometry*. American Mathematical Society.

[131] Duistermaat, J.J., and Heckman, G.J. 1982. On the variation in the cohomology in the symplectic form of the reduced phase space. *Invent. Math.*, **69**, 259–268.

[132] Dumitrescu, T.T., and Seiberg, N. 2011. Supercurrents and brane currents in diverse dimensions. *JHEP*, **1107**, 095.

[133] Dumitrescu, T.T., Festuccia, G., and Seiberg, N. 2012. Exploring curved superspace. *JHEP*, **1208**, 141.

[134] Eisenhart, L.P. 1997. *Riemannian Geometry*. Princeton University Press.

[135] Faddeev, L.D., and Reshetikhin, N. Yu. 1986. Integrability of the principal chiral field model in (1+1)-dimension. *Ann. Phys.*, **167**, 227.

[136] Farkas, H.M., and Kra, I. 1992. *Riemann Surfaces*. (Graduate Texts in Mathematics, vol. 71). Springer.

[137] Feger, R., and Kephart, T.W. 2012. *LieART – A Mathematical application for Lie algebras and Representation Theory.* arXiv:1206.6379 [math-ph].
[138] Ferrara, S., and Sabharwal, S. 1990. Quaternionic manifolds for type II superstring vacua of Calabi–Yau spaces. *Nucl. Phys.*, **B 332**, 317–332.
[139] Ferrara, S., and Savoy, C.A. 1982. Representations of extended supersymmetry on one and two–particle states. Pages 47–84 of: *Supergravity 1981.* Cambridge University Press.
[140] Ferrara, S., and Van Proeyen, A. 1989. A theorem on $N = 2$ special Kahler product manifolds. *Class. Quant. Grav.*, **6**, L243.
[141] Figueroa-O'Farrill, J.M., Kohl, C., and Spence, B.J. 1997. Supersymmetry and the cohomology of (hyper)Kahler manifolds. *Nucl. Phys.*, **B503**, 614–626.
[142] Fischbacher, T., Nicolai, H., and Samtleben, H. 2004. Nonsemisimple and complex gaugings of $N = 16$ supergravity. *Commun. Math. Phys.*, **249**, 475–496.
[143] Fomenko, A.T. 1995. *Symplectic Geometry.* 2nd ed. (Advanced Studies in Contemporary Mathematics, vol. 5). Gordon and Breach.
[144] Fré, P. 2013. *Gravity, a Geometrical Course.* Vol 2. Springer.
[145] Freed, D.S. 1999. Special Kähler manifolds. *Commun. Math. Phys.*, **203**, 31–52.
[146] Freedman, D.Z., and van Proeyen, A. 2012. *Supergravity.* Cambridge University Press.
[147] Freedman, D.Z., van Nieuwenhuizen, P., and Ferrara, S. 1976. Progress toward a theory of supergravity. *Phys. Rev.*, **D13**, 3214–3218.
[148] Freund, P.G.O. 1988. *Introduction to Supersymmetry.* Cambridge Monographs on Mathematical Physics. Cambridge University Press.
[149] Fritzsche, K., and Grauert, H. 2002. *From holomorphic functions to complex manifolds.* Graduate Texts in Mathematics, vol. 213. Springer.
[150] Gaillard, M.K., and Zumino, B. 1981. Duality rotations for interacting fields. *Nucl. Phys.*, **B193**, 221–244.
[151] Gaiotto, D., and Witten, E. 2010. Janus configurations, Chern-Simons couplings, and the theta-angle in N=4 Super Yang-Mills Theory. *JHEP*, **1006**, 097.
[152] Gaiotto, D., and Yin, X. 2007. Notes on superconformal Chern-Simons-Matter theories. *JHEP*, **0708**, 056.
[153] Galaev, A. 2006. Isometry groups of Lobachewskian spaces, similarity transformation groups of Euclidean spaces and Lorentzian holonomy groups. *Rend. Circ. Mat. Palermo*, **79**, 87–97.
[154] Galicki, K. and Lawson, H.B. 1988. Quaternionic reduction and quaternionic orbifolds. *Math. Ann.*, **282**, 1–21.
[155] Galicki, K. 1987. A generalization of the momentum mapping construction for quaternionic Kähler manifolds. *Commun. Math. Phys.*, **108**, 117–138.
[156] Garrett, P. *Volume of $SL(n,\mathbb{Z})\backslash SL(n,\mathbb{R})$ and $Sp_n(\mathbb{Z})\backslash Sp_n(\mathbb{R})$.* Available at http://www.users.math.umn.edu/~garrett/m/v/volumes.pdf. Accessed August 13, 2014.
[157] Gates, S.J., Grisaru, M.T., Rocek, M., and Siegel, W. 2001. *Superspace, or One Thousand and One Lessons in Supersymmetry.* Free electronic version of the 1983 book (with corrections). Available as arXiv:hep-th/0108200.
[158] Giani, F., and Pernici, M. 1984. $N = 2$ supergravity in ten dimensions. *Phys. Rev.*, **D30**, 325–333.
[159] Gibbons, G.W., and Hull, C.M. 1982. A Bogomolny bound for general relativity and solitons in $N = 2$ supergravity. *Phys. Lett.*, **B 109**, 190.
[160] Gibbons, G.W., Hull, C.M., and Warner, N.P. 1983. The stability of gauged supergravity. *Nucl. Phys.*, **B218**, 173.

[161] Gindikin, S.G., Pjateckii-Sapiro, I.I., and Vinberg, E.B. 1965. Classification and canonical realization of complex bounded homogeneous domains. Pages 404–437 of: *Transactions of the Moscow Mathematical Society for the Year 1963*. American Mathematical Society.

[162] Giveon, A., Porrati, M., and Rabinovici, E. 1994. Target space duality in string theory. *Phys. Rep.*, **244**, 77–202.

[163] Goldberg, S.I. 1982. *Curvature and Homology*. Dover.

[164] Goldberg, S.I. 1960. Conformal transformations of Kähler manifolds. *Bull. Amer. Math. Soc.*, **66**, 54–58.

[165] Goldfeld, D. 2006. *Automorphic Forms and L–functions for the Group $GL(n, \mathbb{R})$*. (Cambridge Studies in Advanced Mathematics, vol. 99). Cambridge University Press.

[166] Gomis, J., Milanesi, G., and Russo, J.G. 2008. Bagger–Lambert Theory for general Lie algebras. *JHEP*, **0806**, 075.

[167] Gorbatsevich, V.V., Onishchik, A.L., and Vinberg, E.B. 1994. Structure of Lie groups and Lie algebras. In: *Lie Groups and Lie Algebras III*. (Encyclopaedia of Mathematical Sciences, vol. 41). Springer.

[168] Grauert, H., and Remmert, R. 1979. *Theory of Stein Spaces*. Springer.

[169] Green, M.B., Schwarz, J.H., and West, P.C. 1985. Anomaly free chiral theories in six–dimensions. *Nucl. Phys.*, **B254**, 327.

[170] Green, M.B., Schwarz, J.H., and Witten, E. 2012. *Superstring Theory*. (Cambridge Monographs on Mathematical Physics). Cambridge University Press (2 volumes).

[171] Griffiths, P., and Harris, J. 1978. *Principles of Algebraic Geometry*. Wiley Interscience.

[172] Griffiths, P., and Schmid, W. 1969. Locally homogeneous complex manifolds. *Acta Math.*, **123**, 145–166.

[173] Griffiths, P.A. 1984. *Topics in Transcendental Algebraic Geometry*. (Annals of Mathematical Studies.) Princeton University Press.

[174] Gross, D.J., Harvey, J.A., Martinec, E.J., and Rohm, R. 1985. Heterotic string theory. 1. The free heterotic string. *Nucl. Phys.*, **B256**, 253.

[175] Gross, D.J., Harvey, J.A., Martinec, E.J., and Rohm, R. 1986. Heterotic string theory. 2. The interacting heterotic string. *Nucl. Phys.*, **B267**, 75.

[176] Gross, M., Huybrechts, D., and Joyce, D. 2003. *Calabi–Yau Manifolds and Related Geometries*. (Universitext). Springer.

[177] Guillemin, V., and Sternberg, S.. 1990. *Symplectic Techniques in Physics*. Cambridge University Press.

[178] Guillemin, V.W., and Sternberg, S. 1999. *Supersymmetry and Equivariant de Rham Theory*. Springer.

[179] Gunaydin, M., Sierra, G., and Townsend, P.K. 1983. Exceptional supergravity theories and the MAGIC square. *Phys. Lett.*, **B133**, 72.

[180] Gunaydin, M., Romans, L.J., and Warner, N.P. 1986. Compact and noncompact gauged supergravity theories in five dimensions. *Nucl. Phys.*, **B272**, 598.

[181] Gunning, R.C., and Rossi, H. 1965. *Analytic Functions of Several Complex Variables*. Prentice–Hall.

[182] Haag, R., Lopuszanski, J.T., and Sohnius, M. 1975. All possible generators of supersymmetries of the S–matrix. *Nucl. Phys.*, **B88**, 257.

[183] Hawking, S.W. 1983. The boundary conditions for gauged supergravity. *Phys. Lett.*, **B126**, 175–177.

[184] Helgason, S. 2001. *Differential Geometry, Lie Groups, and Symmetric Spaces*. (Graduate Studies in Mathematics, vol. 34). American Mathematical Society.

[185] Hirsch, M.W. 1976. *Differential Topology*. (Graduate Texts in Mathematics, vol. 33). Springer.
[186] Hirzebruch, F. 1995. *New Topological Methods in Algebraic Geometry*. (Classics in Mathematics). Springer.
[187] Hiwaniec, H., and Kowalski, E. 2004. *Analytic Number Theory*. (Colloquium Publications, vol. 53). American Mathematical Society.
[188] Hori, K., Katz, S., Klemm, A., Pandharipande, R., Thomas, R., Vafa, C., Vakil, R., and Zaslov, E. 2003. *Mirror Symmetry*. (Clay Mathematics Monographs, vol. 1). American Mathematical Society.
[189] Hosomichi, K., Lee, K.-M., Lee, S., Lee, S., and Park, J. 2008a. $N = 4$ superconformal Chern–Simons theories with hyper and twisted hyper multiplets. *JHEP*, **0807**, 091.
[190] Hosomichi, K., Lee, K.-M., Lee, S., Lee, S., and Park, J. 2008b. $N = 5, 6$ superconformal Chern–Simons theories and M2-branes on orbifolds. *JHEP*, **0809**, 002.
[191] Howe, P. and West, P.C. 1984. The complete $N = 2$, $d = 10$ supergravity. *Nucl. Phys.*, **B238**, 181–219.
[192] Hull, C.M. 1983. The positivity of gravitational energy and global supersymmetry. *Commun. Math. Phys.*, **90**, 545.
[193] Hull, C.M. 1984a. A new gauging of $N = 8$ supergravity. *Phys. Rev.*, **D30**, 760.
[194] Hull, C.M. 1984b. More gaugings of $N = 8$ supergravity. *Phys. Lett.*, **B148**, 297–300.
[195] Hull, C.M. 1984c. Noncompact gaugings of $N = 8$ supergravity. *Phys. Lett.*, **B142**, 39.
[196] Hull, C.M. 1984d. The spontaneous breaking of supersymmetry in curved spacetime. *Nucl. Phys.*, **B239**, 541.
[197] Hull, C.M. 1985. The minimal couplings and scalar potentials of the gauged $N = 8$ supergravities. *Class. Quant. Grav.*, **2**, 343.
[198] Hull, C.M. 2003. New gauged $N = 8$, $D = 4$ supergravities. *Class. Quant. Grav.*, **20**, 5407–5424.
[199] Hull, C.M., and Townsend, P.K. 1995. Unity of superstring dualities. *Nucl. Phys.*, **B438**, 109–137.
[200] Hull, C.M., and Witten, E. 1985. Supersymmetric sigma models and the heterotic string. *Phys. Lett.*, **160 B**, 398–402.
[201] Husemöller, D. 1994. *Fibre Bundles*. (Graduate Texts in Mathematics, vol. 20). American Mathematical Society.
[202] Husemöller, D. 2004. *Elliptic Curves*. 2nd ed. (Graduate Texts in Mathematics, vol. 111). Springer.
[203] Huybrechts, D. 2005. *Complex Geometry. An Introduction*. (Universitext). Springer.
[204] Intriligator, K., and Seiberg, N. 2013. Aspects of 3d $N = 2$ Chern–Simons–matter theories. *JHEP*, **1307**, 079.
[205] Israel, W., and Nester, J.M. 1981. Positivity of the Bondi gravitational mass. *Phys. Lett.*, **A85**, 259–260.
[206] Jeffrey, L.C., and Kirwan, F.C. 1993. *Localization for non–Abelian gauge actions*. arXiv:alg-geom/9307001 [math.AG].
[207] Ji, L. 2005. Lectures on locally symmetric spaces and arithmetic groups. Pages 87–146 of: *Lie Groups and Automorphic Forms*. American Mathematical Society.
[208] Joyce, D.D. 2000. *Compact Manifolds with Special Holonomy*. (Oxford Mathematical Monographs). Oxford University Press.
[209] Joyce, D.D. 2007. *Riemannian Holonomy Groups and Calibrated Geometry*. (Oxford Graduate Texts in Mathematics, vol. 12). Oxford University Press.

[210] Julia, B. 1981. Group disintegration. In: Hawking, S.W., and Rocek, M. (eds.). *Superspace and Supergravity*. Cambridge University Press.
[211] Kac, V.G. 1977. Lie superalgebras. *Adv. Math.*, **26**, 8.
[212] Kao, H.C., and Lee, K.M. 1992. Self–dual Chern–Simons systems with $\mathcal{N} = 3$ extended supersymmetry. *Phys. Rev.*, **D46**, 4691–4697.
[213] Katz, S. 2006. *Enumerative Geometry and String Theory*. (Student Mathematical Library, vol. 32). American Mathematical Society.
[214] Kobayashi, S. 1995. *Transformation Groups in Differential Geometry*. (Classics in Mathematics). Springer.
[215] Kobayashi, S., and Nomizu, K. 1963. *Foundations of Differential Geometry.* Vol. 1. Wiley.
[216] Kodaira, K. 1954. On Kähler varieties of the restricted type (an intrinsic characterization of algebraic varieties). *Ann. Math.*, **60**, 28–48.
[217] Kodaira, K. 1986. *Complex Manifolds and Deformation of Complex Structures*. (Comprehensive Studies in Mathematics, vol. 283). Springer.
[218] Kodaira, K., and Spencer, D.C. 1958. On deformations of complex analytic structures, I–II. *Ann. Math.*, **67**, 328–466.
[219] Kostant, B. 1955. Holonomy and the Lie algebra of infinitesimal motions of a Riemannian manifold. *Trans. Amer. Math. Soc.*, **80**, 528–542.
[220] Kugo, T., and Townsend, P.K. 1983. Supersymmetry and the division algebras. *Nucl. Phys.*, **B221**, 357–380.
[221] Kumar, V., Morrison, D.R., and Taylor, W. 2010. Global aspects of the space of 6D $N = 1$ supergravities. *JHEP*, **1011**, 118.
[222] Labastida, J.M.F., and Llatas, P.M. 1991. Potentials for topological sigma models. *Phys. Lett.*, **B271**, 101–108.
[223] Lahanas, A., and Nanopoulos, D.V. 1987. The road to no–scale supergravity. *Phys. Rep.*, **145**, 1.
[224] Lang, S. 1995. *Differential and Riemannian Manifolds*. (Graduate Texts in Mathematics, vol. 160). Springer.
[225] Lang, S. 1999. *Fundamentals of Differential Geometry*. (Graduate Texts in Mathematics, vol. 191). Springer.
[226] Langlands, R.P. 1966. Volume of the fundamental domain for some arithmetical subgroup of Chevalley groups. Pages 235–257 of: *Algebraic Groups and Discontinuous Subgroups. Proc. Symp. Pure Math. Vol. IX*. American Mathematical Society.
[227] Leistner, T. 2007a. Towards a classification of Lorentzian holonomy groups. *J. Differential Geom.*, **76**, 423–484.
[228] Leistner, T. 2007b. Towards a classification of Lorentzian holonomy groups. Part II: Semisimple, non-simple weak-Berger algebras. *J. Differential Geom.*, **76**, 423–484.
[229] Libermann, P., and Marle, C.M. 1987. *Symplectic Geometry and Analytical Mechanics*. Reider.
[230] Margulis, G.A. 1991. *Discrete Subgroups of Semisimple Lie Groups*. Springer.
[231] Mariño, M. 2005. *Chern–Simons Theory, Matrix Models, and Topological Strings*. (International Series of Monographs on Physics, vol. 131). Oxford University Press.
[232] McLenaghan, R.G. 1974. On the validity of Huygens' principle for second order partial differential equations with four independent variables. Part I: Derivation of necessary conditions. *Ann. Inst. H. Poincaré*, **A20**, 153–188.
[233] Merkulov, S., and Schwachhöfer. 1999. Classification of irreducible holonomies of torsion–free affine connections. *Ann. Math.*, **150**, 77–150.
[234] Meyers, S.B., and Steenrod, N. 1939. The group of isometries of a Riemannian manifold. *Ann. Math.*, **40**, 400–416.

[235] Milnor, J., and Stasheff, J. 1974. *Characteristic Classes*. (Annals of Mathematical Studies, vol. 76). Princeton University Press.
[236] Mohaupt, T. 2001. Black hole entropy, special geometry and strings. *Fortsch. Phys.*, **49**, 3–161.
[237] Montgomery, D., and Samelson, H. 1943. Transformations groups on spheres. *Ann. Math.*, **44**, 454–470.
[238] Moore, G.W. 2011. *A Minicourse on Generalized Abelian Gauge Theory, Self-Dual Theories, and Differential Cohomology*. Available at the author's webpage: http://www.physics.rutgers.edu/ gmoore/SCGP-Minicourse.pdf. Accessed August 13, 2014.
[239] Morgan, J., and Tian, G. 2007. *Ricci Flow and the Poincaré Conjecture*. (Clay Mathematics Monographs, vol. 3). American Mathematical Society.
[240] Morozov, A. 1994. Integrability and matrix models. *Phys. Usp.*, **37**, 1–55.
[241] Morris, D.W. 2001. *Introduction to Arithmetic Groups*. Available as arXiv:math/0106063.
[242] Morse, M. 1934. *The Calculus of Variations in the Large*. (Colloquium Publications, vol. 18). American Mathematical Society (10th printing 2001).
[243] Narain, K.S. 1986. New heterotic string theories in uncompactified dimension < 10. *Phys. Lett.*, **B169**, 41.
[244] Narain, K.S., Sarmadi, M.H., and Witten, E. 1987. A note on toroidal compactification of heterotic string theory. *Nucl. Phys.*, **B279**, 369.
[245] Nester, J.A. 1981a. A new gravitational energy expression with a simple positivity proof. *Phys. Lett.*, **A83**, 241–242.
[246] Nester, J.A. 1981b. Positivity of the Bondi gravitational mass. *Phys. Lett.*, **A85**, 259–260.
[247] Nester, J.M. 1981c. A new gravitational energy expression with a simple positivity proof. *Phys. Lett.*, **A83**, 241–242.
[248] Nicolai, H., and Samtleben, H. 2001a. Compact and noncompact gauged maximal supergravities in three-dimensions. *JHEP*, **0104**, 022.
[249] Nicolai, H., and Samtleben, H. 2001b. Maximal gauged supergravity in three dimensions. *Phys. Rev. Lett.*, **86**, 1686–1689.
[250] Nicolai, H., and Samtleben, H. 2003. Chern–Simons versus Yang–Mills gaugings in three dimensions. *Nucl. Phys.*, **B668**, 167–178.
[251] Nijenhuis, A. 1953. On the holonomy group of linear connections. *Indag. Math.*, **15**, 233–249.
[252] Nishino, H., and Sezgin, E. 1984. Matter and gauge couplings of $N = 2$ supergravity in six dimensions. *Phys. Lett.*, **B144**, 187–192.
[253] Novikov, S., Munakov, S.V., Pitaevsky, L.P., and Zakharov, V.E. 1984. *Theory of Solitons: The Inverse Scattering Method*. (Contemporary Soviet Mathematics). Plenum.
[254] Olmos, C. 2005. A geometric proof of the Berger holonomy theorem. *Ann. Math.*, **161**, 579–588.
[255] Ooguri, H., and Vafa, C. 2007. On the geometry of the string landscape and the swampland. *Nucl. Phys.*, **B766**, 21–33.
[256] Pernici, M., Pilch, K., and van Nieuwenhuizen, P. 1984. Gauged maximally extended supergravity in seven dimensions. *Phys. Lett.*, **B143**, 103.
[257] Petrov, V.I. 1969. *Einstein Spaces*. Princeton University Press.
[258] Polchinski, J. 1998. *String Theory*. (Cambridge Monographs on Mathematical Physics). Cambridge University Press (2 volumes).

[259] Poletskiĭ, E.A., and Shabat, B.V. 1989. Invariant metrics. In: *Several Complex Variables, III*. (Encyclopaedia of Mathematical Sciences, vol. 9). Springer.
[260] Polyakov, A.M. 1981. Quantum geometry of fermionic strings. *Phys. Lett.*, **B103**, 211–213.
[261] Polyakov, A.M., and Wiegmann, P.B. 1983. Theory of nonabelian Goldstone bosons. *Phys. Lett.*, **B131**, 121–126.
[262] Postnikov, M.M. 1994. *Lectures in Geometry: Lie Groups and Lie Algebras*. Editorial URSS.
[263] Postnikov, M.M. 2001. *Geometry VI. Riemannian Geometry*. (Encyclopaedia of Mathematical Sciences, vol. 91). Springer.
[264] Pyatetskiĭ-Shapiro, I.I. 1960. *Automorphic Functions and the Geometry of Classical Domains*. Gordon and Breach.
[265] Romans, L.J. 1986. Self–duality for interacting fields: covariant field equations for six–dimensional chiral supergravities. *Nucl. Phys.*, **B276**, 71–92.
[266] Salam, A., and Sezgin, E. 1989. *Supergravity in Diverse Dimensions*, vols. 1 & 2. World Scientific.
[267] Salamon, S. 1989. *Riemannian Geometry and Holonomy Groups*. Longman Scientific and Technical.
[268] Schoen, R., and Yau, S.-T. 1979a. Positivity of the total mass of a general space-time. *Phys. Rev. Lett.*, **43**, 1457–1459.
[269] Schoen, R., and Yau, S.-T. 1981. Proof of the positive mass theorem. 2. *Commun. Math. Phys.*, **79**, 231–260.
[270] Schoen, R., and Yau, S.T. 1979b. On the proof of the positive mass conjecture in General Relativity. *Commun. Math. Phys.*, **65**, 45–76.
[271] Schoen, R.M., and Yau, S.-T. 1979c. Proof of the positive action conjecture in quantum gravity. *Phys. Rev. Lett.*, **42**, 547–548.
[272] Seiberg, N., and Taylor, W. 2011. Charge lattices and consistency of 6D supergravity. *JHEP*, **1106**, 001.
[273] Seiberg, N., and Witten, E. 1994a. Electric–magnetic duality, monopole condensation, and confinement in $N = 2$ supersymmetric Yang–Mills theory. *Nucl. Phys.*, **B426**, 19–52.
[274] Seiberg, N., and Witten, E. 1994b. Monopoles, duality and chiral symmetry breaking in $N = 2$ supersymmetric QCD. *Nucl. Phys.*, **B431**, 484–550.
[275] Seiberg, N., and Witten, E. 1996. Comments on string dynamics in six-dimensions. *Nucl. Phys.*, **B471**, 121–134.
[276] Serre, J.P. 1973. *A Course in Arithmetic*. (Graduate Texts in Mathematics, vol. 7). Springer.
[277] Sezgin, E., and Tanii, Y. 1995. Superconformal sigma models in higher than two dimensions. *Nucl. Phys.*, **B443**, 70–84.
[278] Simons, J. 1962. On transitivity of holonomy systems. *Ann. Math.*, **76**, 213–234.
[279] Slansky, R. 1981. Group theory for unified model building. *Phys. Rep.*, **79C**, 1–118.
[280] Sternberg, S. 1983. *Lectures on Differential Geometry*. Chelsea Pub. Co.
[281] Strathdee, J. 1987. Extended Poincaré supersymmetry. *Int. J. Mod. Phys.*, **A2**, 273–300.
[282] Swann, A.F. 1991. Hyperkähler and quaternionic Kähler geometry. *Math. Ann.*, **289**, 421–450.
[283] Tanii, Y. 1984. $N = 8$ supergravity in six dimensions. *Phys. Lett.*, **B147**, 47–51.
[284] Terras, A. 1988. *Harmonic Analysis on Symmetric Spaces and Applications. II*. Springer.

[285] Tian, G. 1987. Smoothness of the universal deformation space of compact Calabi–Yau manifolds and its Petersson–Weil metric. Pages 629–645 of: Yau, S.T. (ed.), *Mathematical Aspects of String Theory*. World Scientific.
[286] Townsend, P.K. 1984. A new anomaly free chiral supergravity from compactification on K_3. *Phys. Lett.*, **B139**, 283–287.
[287] Vafa, C. 2005. *The String Landscape and the Swampland*. e–print arXix:hep-th/0509212.
[288] van Leeuwen, M.A.A., Cohen, A.M., and Lisser, B. 1992. *LiE, A Package for Lie Group Computations*. Computer Algebra Nederland.
[289] Van Nieuwenhuizen, P. 1981. Supergravity. *Phys. Rep.*, **68**, 189–398.
[290] van Proeyen, A. 1999. *Tools for Supersymmetry*. (Lectures at the Spring School in Calimananesti, Romania). arXiv:hep-th/9910033.
[291] van Proyen, A. 1983. Superconformal tensor calculus in $\mathcal{N}=1$ and $\mathcal{N}=2$ supergravity. In: *Supersymmetry and Supergravity 1983*. World Scientific.
[292] Vinberg, E.B. 1965. The theory of convex homogeneous cones. Pages 340–403 of: *Transactions of the Moscow Mathematical Society for the Year 1963*. American Mathematical Society.
[293] Voisin, C. 2007a. *Hodge Theory and Complex Algebraic Geometry I*. (Cambridge Studies in Advanced Mathematics, vol. 76). Cambridge University Press.
[294] Voisin, C. 2007b. *Hodge Theory and Complex Algebraic Geometry II*. (Cambridge Studies in Advanced Mathematics, vol. 77). Cambridge University Press.
[295] Wang, M.Y. 1989. Parallel spinors and parallel forms. *Ann. Global Anal. Geom.*, **7**, 59–68.
[296] Weinberg, S. 1972. *Gravitation and Cosmology: Principles and Applications of the General Theory of Relativity*. Wiley.
[297] Wess, J., and Bagger, J. 1992. *Supersymmetry and Supergravity* (Revised edn.). (Princeton Series in Physics). Princeton University Press.
[298] West, P. 1990. *Introduction to Supersymmetry and Supergravity*. (Revised and extended second edn.). World Scientific.
[299] Witten, E. 1979. Dyons of charge $e\theta/2\pi$. *Phys. Lett.*, **B86**, 283–287.
[300] Witten, E. 1981. A new proof of the positive energy theorem. *Commun. Math. Phys.*, **80**, 381–402.
[301] Witten, E. 1982a. Constraints on supersymmetry breaking. *Nucl. Phys.*, **B202**, 253.
[302] Witten, E. 1982b. Supersymmetry and Morse theory. *J. Differential Geom.*, **17**, 661–692.
[303] Witten, E. 1983a. Current algebra, baryons, and quark confinement. *Nucl. Phys.*, **B223**, 433–444.
[304] Witten, E. 1983b. Global aspects of current algebra. *Nucl. Phys.*, **B223**, 422–432.
[305] Witten, E. 1983c. Some inequalities among hadron masses. *Phys. Rev. Lett.*, **31**, 2351–2354.
[306] Witten, E. 1986. New issues in manifolds of $SU(3)$ holonomy. *Nucl. Phys.*, **B283**, 79.
[307] Witten, E. 1987. Elliptic genera and quantum field theory. *Commun. Math. Phys.*, **109**, 525–536.
[308] Witten, E. 1988a. 2+1 dimensional supergravity as an exactly soluble system. *Nucl. Phys.*, **B 311**, 46–78.
[309] Witten, E. 1988b. Topological quantum field theory. *Commun. Math. Phys.*, **117**, 411–449.
[310] Witten, E. 1988c. Topological σ–models. *Commun. Math. Phys.*, **118**, 411.

[311] Witten, E. 1989. Quantum field theory and the Jones polynomial. *Commun. Math. Phys.*, **121**, 351.
[312] Witten, E. 1992. Two–dimensional gauge theories revisited. *J. Geom. Phys.*, **9**, 303–368.
[313] Witten, E. 1995a. On S–duality in Abelian gauge theory. *Selecta Math.*, **1**, 383.
[314] Witten, E. 1995b. *Some Comments on String Dynamics*. e-preprint arXiv:hep-th/9507121.
[315] Witten, E. 1996. talk given at the Newton Institute for Mathematical Sciences, Cambridge.
[316] Witten, E. 1998. New 'gauge' theories in six dimensions. *JHEP*, **9801**, 001.
[317] Witten, E. 2005. Non–Abelian localization for Chern–Simons theory. *J. Geom. Phys.*, **70**, 183–323.
[318] Witten, E., and Bagger, J. 1982. Quantization of Newton's constant in certain supergravity theories. *Phys. Lett.*, **B 115**, 202.
[319] Wolf, J.A. 1962. Discrete groups, symmetric spaces, and global holonomy. *Amer. J. Math.*, **84**, 527–542.
[320] Wolf, J.A. 1965. Complex homogeneous contact manifolds and quaternionic symmetric spaces. *J. Math. Mech.*, **14**, 1033–1047.
[321] Wu, H. 1964. On the de Rham decomposition theorem. *Ill. J. Math.*, **8**, 291–310.
[322] Wu, H. 1967. Holonomy groups of indefinite metrics. *Pac. J. Math.*, **20**, 351–392.
[323] Xu, Y. 2000. *Theory of Complex Homogenous Bounded Domains*. Science Press/Kluwer Academic Publishers.
[324] Yano, K. 1952. On harmonic and Killing vector fields. *Ann. Math.*, **55**, 38–45.
[325] Yano, K. 2011. *The Theory of Lie Derivatives and its Applications*. Nabu Press.
[326] Yau, S.T. 1978. On the Ricci curvature of a compact Kähler manifold and the complex Monge–Ampere equation. I. *Commun. Pure Appl. Math.*, **31**, 339–411.
[327] Zamolodchikov, A.B. 1986. Irreversibility of the flux of the renormalization group in a 2D field theory. *JETP Lett.*, **43**, 730–732.

Index

3–Sasaki manifolds, 201
(G_1, G_2)–structure, 89
G–structures
   in 3D, 67
R–symmetry, 45
T–tensor, 227
U–dualities, 156
U–duality, 149
$\bar{\partial}$–*Laplacian*, 306
$\mathbb{Q}$–rank, 157
c–map, 386
pp–waves, 126

ABJM model, 266
Alekseevskiĭ manifolds, 275
anti–De Sitter space, 187

back–reaction, 138
Bagger–Lambert model, 265
Bernoulli numbers, 159
beta function
   bosonic σ–model, 5
Betti numbers, 54
black hole, 156
Born–Infeld Lagrangians, 17
Bott's periodicity, 44, 64
bounded domain, 30, 178
BPS bound, 57
BPS object, 192
brane charges, 193
Brinkmann space, 127
bundle
   Hodge, 79
   homogeneous, 25
   Swann, 79
   tautological, 25

Cahen–Wallach space, 126
Calabi–Yau, 305
Calabi–Yau manifolds, 117
canonical bundle, 297

Cartan connection, 161
Cartan involution, 111, 182
Cartan–Killing form, 163
Casimir invariants, 159
Cauchy–Riemann operator, 345
Cayley transformation, 29, 150
central charges, 45
characteristic classes, 103
Chern class, 300
Chern connection, 298
Chern–Simons, 39
Clifford algebra, 63
Clifford modules, 64
compact dual, 337
complex structure, 44
concurrent vector, 237
cone (Riemannian), 200
conformal Killing spinor, 202
conjecture
   Banks–Seiberg, 156
   Vafa–Ooguri, 156
connection
   symmetric, 125
connection form (symmetric spaces), 172
contact structure, 359
convex cone, 315
coset double, 137
Cotton tensor, 203
covolume, 157
Coxeter exponents, 159
critical points, 57

Dedekind function, 324
differential forms of type (p, q), 296
dimensional reduction, 138
Dirac quantization, 37
division algebras, 47
Dolbeault cohomology, 297
dualities
   morphisms, 16
   non–Abelian in 3 D, 39

embedding tensor, 267
equivariant cohomology, 57
exceptional holonomy, 117
exponential map, 109
extreme black holes, 214

fermionic shifts, 217
flag, 335
flag manifolds, 315
flat potential supergravities, 327
form fields, 16
   (anti)self–dual, 18
   chiral, 17
   gauge invariance, 16

Gauss's law, 33
Gauss–Manin connection, 351
geodesics symmetry, 109
Geometric Principle, 5
grading model of E_7, 179
Green's operator, 306
Griffiths period domain, 32, 336, 337
Griffiths period relations, 349
group disintegration, 139

Haar measure, 120
Hamilton–Jacobi function, 19
Hermitian symmetric spaces, 117
hidden symmetry, 136
Hodge bundles, 343
Hodge decomposition, 31, 306, 309
Hodge filtration, 335
Hodge identities, 307
Hodge structure, 335
   infinitesimal variations, 353
holonomy
   of a symmetric space, 114
holonomy group, 103
   restricted, 104
hyperbolic n–space, 155
hyperKähler manifolds, 117

infinitesimal period relations, 349
injectivity radius, 109
isogenous (groups), 157
isometry group, 7
   gauging in σ–models, 7
isotypic, 157

Kähler manifolds, 117
Killing spinor, 185, 189
   in mathematics, 194
Killing vector, 7, 185
   conformal, 195
   multisymplectic, 222
Kodaira–Spencer
   theory, 344
   vector, 346

Lagrangian field theory, 3
Lagrangian subspace, 24
landscape, 155
Langlands volume formula, 159
Laplacian, 306
lattice
   arithmetic, 157
   in Lie groups, 157
   irreducible, 157
Lefshetz decomposition, 311
Lefshetz fixed point formula, 58
Lefshetz's $SU(2)$, 310
left translation, 161
left–invariant
   vector field, 161
Legendre submanifold, 359
Lie algebra isomorphisms, 135
Lie derivative, 6
locally symmetric, 109

M2–branes, 265
magnetic monopoles, 33
mirror family, 385
modular transformations, 27
Monge–Ampere equations, 328
Morita equivalence, 64
Morse theory, 57

Narain lattices, 20
Newton constant
   quantized, 78
normal coordinates, 110
no–scale supergravity, 327

octonions, 48
   projective planes, 135
one parameter subgroup, 161

parallel spinor, 120
parallel transport, 103
Pauli couplings, 147
Peccei Quinn symmetry, 16
Penrose operator, 202
period map, 349
periods
   of a Riemann surface, 27
Poisson summation formula, 38
polarization (of a vector space), 334
polarized flag manifold, 152
primitive cohomology, 311
principal Higgs bundle, 25

Quaternionic–Kähler manifolds, 117
quaternionic structure, 43

rank (of a symmetric space), 177
real masses, 253
real structure, 43
reducible manifold, 106

reductive action, 169
Rho tensor, 203
Ricci form, 304
Ricci solitons, 6
Riemann bilinear relations, 335
Riemannian products, 106

Sasaki manifolds, 201
Sasaki–Einstein manifolds, 201
second fundamental form, 106, 352
Serre duality, 298, 307, 309
Siegel modular group, 28
Siegel's modular form, 33
Siegel's upper half-space, 28
sigma model
   4D current algebra, 3
   gauged, 7
   general, 5
space–time charges, 185
special Kähler condition, 88
Spencer cohomology, 124
spin connection, 120
spin structure, 120
spinors
   Majorana, 44
   Majorana–Weyl, 44
   symplectic–Majorana–Weyl, 44
Stiefel–Whitney class, 120
subgroup
   finite index, 156
submanifold
   totally geodesic, 108, 138
supercurrent, 61
superHiggs effect, 193
   partial, 193
superpotential, 55, 56
supersymmetric background, 189
swampland, 155
Swann bundle, 79
symmetric space
   duality, 175
   in Abelian dualities, 24
symplectic structure
   associated to Abelian dualities, 20
symplectomorphisms, 19

target manifold
   equivalence principle, 8
   general σ-model, 4
tensor supermultiplet, 132, 137
   in 6D (2, 0) SUGRA, 154
   in 6D (4, 0) SUGRA, 155

theorem
   Alekseevskii–Kimelfeld, 176
   Ambrose–Singer, 108
   Bär, 200
   Berger, 115
   Bochner, 7
   Bott–Borel–Weil, 125
   Cartan (Cartan subalgebras), 114
   Cartan (symmetric spaces), 110
   Cartan decomposition, 182
   Chow, 314
   de Rham, 107
   de Rham–Wu, 126
   Dolbeault, 297
   Frobenius, 106, 107
   Gaiotto–Witten, 260
   Griffiths transversality, 349
   Haag–Łopuszanski–Sohnius, 42
   Hodge, 53, 306
   Hopf–Rinow, 107
   Hurwitz, 135
   Jacobson–Morosov, 158
   Kodaira embedding, 78
   Kostant, 233
   Leister, 127
   Meyers, 176, 264
   Myers–Steenrod, 111
   Nijenhuis, 108
   Riemann bilinear relations, 28
   Salamon, 141
   Second Cartan, 110
   SL_2 orbit, 356
   Tian, 345
   uniformization, 157
triality, 135
twisted masses, 253
twistor operator, 202

universal deformation space, 358
upper half-plane, 29

vielbein
   Abelian dualities, 22

Weil operator, 335
Weil–Petersson metric, 347
Weyl tensor, 204
Witten index, 53
Witten spinor, 213
Wolf spaces, 117

For EU product safety concerns, contact us at Calle de José Abascal, 56–1°, 28003 Madrid, Spain or eugpsr@cambridge.org.

www.ingramcontent.com/pod-product-compliance
Lightning Source LLC
LaVergne TN
LVHW081516060526
838200LV00005B/195